# Propulsion Systems for Space Applications

# Propulsion Systems for Space Applications

Dr. Goteti Satyanarayana

**BS Publications**
An imprint of **BSP Books Pvt. Ltd.**
4-4-309/316, Giriraj Lane,
Sultan Bazar, Hyderabad - 500 095.

**Published by:**

**BSP BS Publications**

**An Imprint of BSP Books Pvt. Ltd.**
4-4-309/316, Giriraj Lane,
Sultan Bazar, Hyderabad - 500 095.
Phone: 040 - 23445688
e-mail: info@bspbooks.net
www.bspbooks.net

**ISBN: 978-93-95038-81-2 (Paperback)**

*Dedicated to my wife*

**Maruti Devi**

**My children and their Spouses:**
Neelima & Sreedhar, Sameera and Vinay and Aditya Vamsi and Parul;
my **Grandchildren**: Pranav, Kavya, Anuj, Tejas, Anaya and **little** Advaith

# Foreword

The exploration of space has always captivated humanity's imagination, and the possibility of traveling beyond our planet has been a dream for centuries. As our understanding of physics and engineering has advanced, so has our ability to conceive of new ways of propelling spacecraft through the void of space. Space propulsion is a critical field that has enabled humanity to reach new frontiers and push the boundaries of what we once thought was impossible. From the early days of rocketry to the latest advanced propulsion technologies, the science and engineering behind space propulsion have advanced rapidly, bringing us ever closer to exploring the depths of our solar system and beyond.

The goals for space exploration and utilization have evolved over time, reflecting changes in factors like technology, geopolitical competition, and societal priorities. The initial goal of space exploration (during 1950s-1970s) was driven by geopolitical competition between the United States and the Soviet Union. The goal of this competition was to land humans on the Moon, which the United States achieved in 1969 with the Apollo 11 mission. After the Moon landings, the focus of space exploration during 1970s and 2000s shifted more towards scientific research and exploration. This involved sending probes to study other planets and moons in our solar system, as well as launching telescopes to observe the universe beyond our solar system. In addition, the Space Shuttle program was developed to allow more regular access to the International Space Station, which facilitated collaborative scientific research and space-based technology development. In recent years (2000s-present), there has been a growing interest in commercial space exploration and utilization. Private companies like SpaceX, Blue Origin, and Virgin Galactic have emerged, with the goal of reducing the cost of access to space and enabling new forms of space-based commerce, such as space tourism and space mining. The potential benefits of commercial space also include new opportunities for scientific research and technology development. Another emerging goal (present to future) for space exploration is to protect Earth from potential asteroid impacts. NASA and other countries' space agencies are continuing to develop systems to detect and deflect potentially hazardous asteroids before they can collide with Earth. This goal reflects the growing awareness of the need to protect our planet from natural disasters. The long-term goal is survival of human race by finding safe havens like Mars for habitation. Overall, the future of space exploration will continue to be influenced by a combination of the above factors.

To achieve the above goals, it is important to minimize the cost for launching a space vehicle. The cost of a space vehicle launch is determined by the mission goals, payload weight, launch vehicle size, launch location and frequency, vehicle reliability, and regulatory requirements. But the usual measure used is the cost for launching a unit weight of the payload. The basic metric used is the end-to-end cost for launching a space vehicle to fly a given weight of the payload. These include a combination of powerplant, structure, control systems, logistics, etc. Efforts since the 1950s reduced the cost for launching a unit weight of payload from an astronomical amount to around $15,000 in the early 2000's to an estimated $1000 to $5000 now. Cost reduction over the past several decades is illustrated in Figure 1. This was possible due to establishing launch systems and their technologies along with the emergence of new launch of space systems, including propulsion, are essential to push the boundaries of both robotic and human space exploration and utilization and exploring deeper into the solar system.

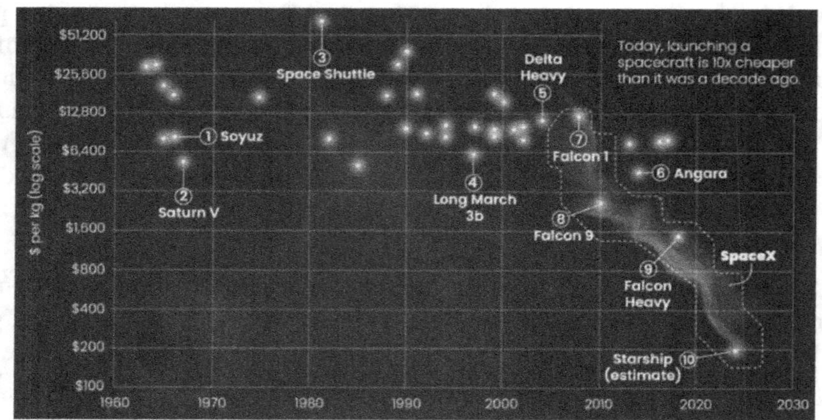


**Figure 1** Cost reduction for launching unit weight of payload.

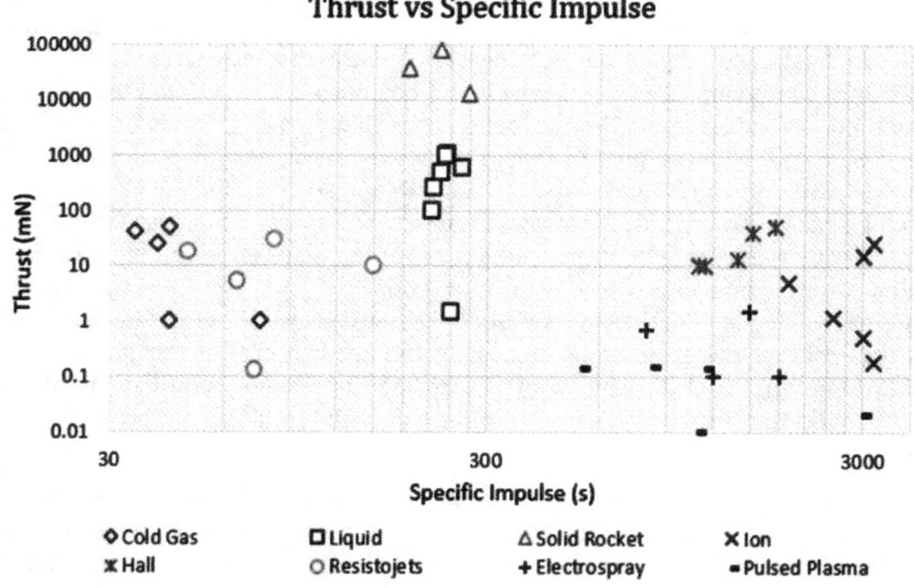


**Figure 2** Comparison between propulsion system thrust and specific impulse.

The combination of propulsion systems for low to high Earth and deep space missions depend on the specific mission requirements and trade-offs between speed, efficiency, cost, and complexity. As spacecraft travel deeper into space, the trade-off between thrust and specific impulse in propulsion becomes increasingly important. Higher thrust allows a spacecraft to accelerate quickly and reach its destination faster, but it also requires more fuel, which increases the spacecraft's mass and decreases its specific impulse. On the other hand, a propulsion system with a higher specific impulse is more efficient and can travel farther with the same amount of fuel, but it may have a lower thrust and take longer to accelerate. A grouping of different propulsion systems may also be used for this reason during different phases of the mission, such as using chemical rockets for launch and advanced electric or nuclear propulsion systems for deep space exploration. Figure 2 compares propulsion systems based on their thrust vs. specific impulse capabilities. Continued advances are needed in space propulsion systems for various reasons. Current space propulsion systems are not very efficient, which limits the amount of payload (including humans) that can be carried into space. Available propulsion systems are also limited in terms of speed. It can take years to travel to distant planets, and even longer to reach other stars. Developing faster propulsion systems could reduce travel times and make human space exploration more feasible. Propulsion systems that rely on renewable energy sources could make space exploration more environmentally friendly. To explore the outer reaches of our solar system or travel to other stars, we will need more powerful propulsion systems than what is available.

Advanced propulsion systems such as electric propulsion, nuclear propulsion, and solar sails are being developed and are selectively being used in space vehicles. Electric propulsion systems, such as ion thrusters and Hall thrusters, are more efficient than traditional chemical rockets and can operate for longer periods of time. These systems are currently being used on some spacecraft and have the potential to significantly reduce the cost of space exploration. Nuclear propulsion systems, such as nuclear thermal propulsion and nuclear electric propulsion, use nuclear reactions to generate thrust. These systems have the potential to provide much higher thrust than chemical rockets and could significantly reduce travel times to distant planets and other stars. Solar sails use the pressure of sunlight to generate thrust, allowing spacecraft to travel without carrying fuel. These systems are currently being tested and could be used for long-duration missions to explore the outer reaches of our solar system and beyond. Spacefaring nations such as the United States, Russia, and China are currently leading several efforts in these areas. NASA is actively researching and developing several advanced propulsion technologies, including nuclear thermal propulsion, solar sails, and electric propulsion systems. Private companies such as SpaceX are also investing in the development of advanced propulsion systems for space exploration. Russia has a long history of developing advanced propulsion technologies for space exploration, including nuclear thermal propulsion and ion thrusters. China has been investing heavily in its space program in recent years, including the development of advanced propulsion technologies. China has successfully tested an ion thruster and is working on the development of nuclear thermal propulsion and solar sails. Japan has been developing electric propulsion systems for space exploration since the 1980s and is a leader in this technology. Japan is also working on the development of nuclear thermal propulsion and is studying the potential for using solar sails for deep space exploration.

There are additional new technologies that could potentially be developed or significantly improved over the next 30 years, which could further advance space exploration and propulsion systems. Propellant less propulsion systems, such as the EmDrive and Mach Effect Thrusters, do not require fuel to generate thrust. These systems are still in the experimental stage and their feasibility is still being debated, but if they can be proven to work, they could revolutionize space propulsion. Fusion propulsion systems, which would generate thrust by fusing atomic nuclei, have the potential to provid e much higher thrust than chemical rockets or even nuclear propulsion. While fusion power for electricity production is still under development, significant progress has been made and it's possible that fusion propulsion could become a reality in the next few decades. Space elevators, which would use a cable or ribbon to lift spacecraft from Earth's surface into space, could revolutionize space travel and drastically reduce the cost of launching payloads into space. While the technology for building a space elevator is still in the early stages of development, significant progress has been made in materials science and nanotechnology that could make it a reality in the coming decades.

This textbook on space propulsion is a welcome addition to the literature on the subject. It provides an in-depth and comprehensive look at the principles, technologies, and applications of space propulsion, covering everything from the basics of rocket science, and links it to orbital mechanics, spacecraft trajectories and maneuvering. It presents cutting-edge propulsion concepts and details on electric propulsion.

The author has done an excellent job of presenting the material in a clear and accessible manner, with a comprehensive description of the systems, their performance parameters, and configurations. This makes it easy for students and researchers to understand and apply the concepts covered in the book. The book also includes numerous examples and case studies to illustrate the real-world applications of space propulsion and to demonstrate the impact it has had on space exploration and research.

One of the great strengths of this textbook is its focus on both established and emerging technologies. It extensively covers traditional rocket propulsion systems, such as solid, liquid, and hybrid engines, as well as a thorough discussion on advanced propulsion technologies like electric propulsion. The author also discusses the challenges and opportunities associated with each technology, providing a well-rounded perspective on the field.

Overall, this textbook is a valuable resource for anyone interested in space propulsion, from students and researchers to professionals in the aerospace industry. It provides a comprehensive overview of the subject, covering both the theory and practical applications of space propulsion, and is sure to be a valuable reference for years to come.

**Dr. Damodar R. Ambur**

Senior Executive for Aerospace Research and Technology (Retired)
NASA Langley Research Center, Hampton, Virginia, U.S.A.

# Preface

The subject space and its exploration are more interesting and matured than it ever was in the twentieth century. It has changed from entirely government activity to private sector business activity, partially supported by funding and extending permission to use its test and launch facilities. Private sector industries have now been actively involved in research and development in this field and some have developed their own space launch vehicles and propulsion systems. Human voyaging of Moon, Mars and Deep Space exploration missions has significant implications on enhancement of present propulsion technology and development of new propulsion systems.

The main objective of this book is to explain fundamental physical principles, and apply rightly to have a deep understanding of rocket vehicle and spacecraft propulsion. This knowledge will help to quantitatively estimate the performance and can realise the possibilities for improvement of propulsion systems. Mathematical complications are kept at minimum level and equations set are derived from fundamentals. More attention is given to explain pertinent physical phenomena. Wherever it is required, data on rocket propulsion systems that are currently in production is presented.

*The presentation in this text is organised into three basic areas*:

1. *Launch Vehicle Propulsion*: It covers the fundamentals of propulsion, the fluid mechanics & thermodynamics of nozzles, propellants used in rocket engines and different types of rocket engines (solid, liquid and hybrid engines) with their characteristics and performance parameters. It discusses the heat transfer aspect of cooling of the thrust chamber. The physical situation that creates the instabilities in the operation of the rocket engine and the methods used to suppress them for smooth operation are discussed briefly.

2. *Orbital Mechanics*: The basic orbital mechanics concepts, spacecraft orbital manoeuvres and interplanetary trajectories are explained briefly. These topics are included to estimate the propellant consumption for various manoeuvres in the space that helps to design the overall requirements of propulsion systems.

3. *Spacecraft Propulsion*: The thrusters (cold gas, mono- and bipropellant) and electric (thermal, electrostatic, electromagnetic) used in spacecraft manoeuvres during the life span of the spacecraft are described. The nuclear propulsion system for deep space applications is briefly explained.

I acknowledge the training and mentoring from my teachers late Dr. H.C. Radha Krishna, Dr. M Ravindran, Dr. S. Kumarswamy and Dr. V. Balabaskaran Department of Mechanical Engineering (Turbomachines) at the Indian Institute of Technology Madras (I I T M), Chennai. I am grateful to Dr. Vasant R Gowariker and Dr. A. E. Muthunayagam in the Indian Space Research Organisation (ISRO) Trivandrum who provided an opportunity to work in the fields of propellant and propulsion engineering. I am indebted to the late Mr. A. T. Mirchandani, MD of M/S Advani-Oerlikon Ltd. continuously encouraged me to conduct my scientific research work at the company. I am grateful to Dr. A. Srikant who provided an opportunity to teach in K L university, Guntur, Andhra Pradesh. Thanks are due to my colleagues at the University, Dr. Anjaneyulu Yerramilli, Dr. B. Nageswara Rao and Dr. K. L. Narayana for their collegiality.

I am thankful to Dr. Damodar R. Ambur, Senior Executive for Aerospace Research and Technology (Retired), NASA Langley Research Centre, Hampton, Virginia, U.S.A., who with his busy schedules, unhesitatingly accepted my request to review and write the Foreword to this book

Finally, I owe a debt of gratitude to my family for their persistent support, timeless inspiration and forbearance for this time-consuming project. No acknowledgement to my wife could possibly state all that for uncountable hours stolen from her, for her tolerance, patience, love and cheerful support, without which this book would not have been completed.

**April 2023**                                                                 **Dr. Goteti Satyanarayana**

# Contents

# CHAPTER 3

## Rocket Principle and Performance Parameters



# CHAPTER 4

## Rocket Nozzles

# CHAPTER 5

## Rocket Propellants

# CHAPTER 6

## Solid Propellant Motors

# CHAPTER 7

## Liquid Rocket Engines

# CHAPTER 8

## Hybrid Rocket Engines



# CHAPTER 9

## Thrust Chamber Cooling



# CHAPTER 10

## Combustion instabilities

# CHAPTER 11

## Spacecraft Orbital Maneuvers



# CHAPTER 12

## Interplanetary Trajectories

# CHAPTER 13

## Chemical Thrusters for SpaceCraft Manoeuvres



# CHAPTER 14

## Electric Propulsion Systems

# Introduction to Space Propulsion

## 1.1  INTRODUCTION

In the early 20$^{th}$ century space flight was a dream with very few countries engaging in space systems development. During the middle of the 20$^{th}$ Century when Yuri Gagarin of the Soviet Union went to outer space and orbited the Earth in 1961, the Space Race heated up within the world. This resulted in the United States developing the Apollo launch vehicle and sending the first astronauts to land on the Moon in July 1969 and successfully bringing them back to earth. Several other nations in the world also got into space technologies and systems development during the early 1960's. The 21$^{st}$ Century saw a rapid increase in the number of launches by government agencies and commercial companies to utilize space for global communications, study the environments of Earth and other planets, explore space to benefit humanity, national defence, and understand the origins of the Universe. Space has become a new frontier which presents humanity with complex challenges and opportunities to research, develop and implement new technologies and methodologies to achieve safe and cost-effective space systems. Advancements in areas such as vehicle configurations, propulsion systems, new materials and structures, sensing and control, communications, radiation protection, space medicine, computing, utilization of in-situ resources in space and on planetary surfaces, and in-space manufacturing are continuing to realize fault-tolerant systems to launch a variety of payloads into low-earth and deep space orbits in the exploration of space for commercial use and manned missions to space, Moon and other Planets. Propulsion systems are required to launch satellites, spacecraft landers, in-space and interplanetary vehicles, and auxiliary controls. Significant opportunities also exist to improve existing propulsion technologies and provide advanced electrical and nuclear propulsion for deep space missions.

Propulsion systems are required to launch satellites, spacecrafts, landers, interplanetary vehicles and to operate auxiliary controls. To achieve a space mission, basic components required are launchers and payloads like satellites or spacecraft.

1. *Launchers*: They are used to move payload from Earth's surface to space in different orbits. Depending on their payload launching capability and performance, they can be classified in different types. The medium expendable launch-vehicle (MLV), like Delta2 of USA, PSLV of India and Ariane 40 to 44L of European Space Agency, are capable transferring payload up to 4000 kg to Low Earth Orbits (LEO) at different latitudes and inclinations. The heavy expendable launch-vehicle (HLV), like Delta 4, Ariane 5 and GSLV are capable of handling

higher payloads of order 10 tons to LEO and 4000 to 6000 kg payload to high Earth orbits (HEO) like Geostationary orbit (GEO). SpaceX's FALCON vehicle is a type of reusable launch vehicle (RLV). NASA's Space Launch System (SLS), the super-heavy lift expendable launch vehicle launched during 2022 has capability to launch 13,000 kg payload to translunar injection with ability to take people to the Moon and Mars.

2. *Satellites*: They are placed in low Earth orbit (LEO)/medium Earth orbit (MEO)/highly inclined orbit and geostationary orbit (GEO). The International Space Station (ISS), is at about 400 km of altitude.

3. *Spacecrafts*: They are space vehicles from Earth orbit to anywhere in the space: interplanetary, planetary orbiter (like Galileo, Magellan and Cassini), planetary landers (like HUYGENS, Rosetta and Mars pathfinder) and outer solar system.

## 1.2  BRIEF HISTORY

Konstantin Tsiolkovsky (1857-1935)    Robert Goddard (1882-1945)

Konstantin Tsiolkovsky (1857-1935), a mathematics teacher from a small city southwest of Moscow, Russia, published in 1903 the article "The Investigation of Space by Means of Reactive Devices". The article explains in detail the relation between fuel (propellant), velocity and total mass of rocket. This is the fundamental equation for rocket propulsion and is known as the Tsiolkovsky equation.

Robert Goddard (1882-1945), a professor of physics at Clark University in Worcester, Massachusetts, USA, was granted patents in 1914 on the design of liquid rockets and its components like combustion chambers and nozzles. In his article, "A Method of Reaching Extreme Altitudes", published in 1919, developed the theory of spaceflight and its design. He also mentioned, though highly ridiculed by the press and people of that time, about the possibility of sending an unmanned rocket to the moon. He launched the very first liquid rocket weighing 5 kg, on the 16th March 1926, propelled by liquid oxygen and petrol.

*Hermann Oberth (1894-1980)   Wernher von Braun (1912-1977)*

Another, the most influential spaceflight pioneer was Hermann Oberth (1894-1989), Hungarian settled in Germany. His doctoral thesis "The Rocket to the Planets" ("Die Rakete zu den Planetenräumen") was rejected at the University of Heidelberg but he published it as a book. It was a best seller book that stimulated interest in many youngsters in rocket science leading to many rocket societies in Germany. The most famous society is *Verein für Raumschifffahrt,* in which Wernher von Braun was a member.

The Treaty of Versailles that followed the end of World War I prohibited Germany from using long-range artillery. As alternative many leaders were interested in the development of rockets to use as a long-range missile to circumvent the prohibition. In 1932, The German army gave a contract to the young Wernher von Braun, Oberth's assistant, to develop a rocket that can be used as a missile. The efforts of von Braun team led to development of the medium range missile, the A-4 rocket, better well known as the V-2 rocket. Advanced versions of the A-4 rocket were ready before the end of World War II. The version A-9/10 would have been the first intercontinental ballistic missile (ICBM) capable of delivering a nuclear warhead from Western France to New York, but could not reach from drawing stage to manufacturing due to Germany's surrender. During the War thousands of V-2 missiles were manufactured and fired by the German Army. After the end of War, teams from the Allied forces the USA, the UK, and the Soviet Union (USSR) raced to the Peenemünde Army Research Center, where the world's first functional large-scale liquid-propellant rocket, the V-2, was developed, to seize key German manufacturing facilities, procure Germany's missile technology, and capture key personnel involved in the missile development. Hundreds of rail-car loads of V-2s and parts were captured and shipped to the United States and USSR. Around 126 of the principal designers, including Wernher von Braun and Walter Dornberger, were in American hands and moved to the USA. After the Nazi defeat, German engineers were moved to the United States, the United Kingdom and the USSR, where they further developed the V-2 rocket for military and civilian purposes. The V-2 rocket also laid the foundation for the liquid fuel missiles and space launchers used later in the USA and USSR. Under the guidance of Von Braun USA continued the research in rocket sciences that led to development of the powerful Saturn-V rocket that brought the first men to the moon in 1969. The Russians under the guidance of Sergei Korolev, the Russian counterpart to Wernher von Braun, developed reliable rocket engines based on V-2 technology.

Now, many countries are involved in space research and developed their own launch vehicles such those of the highly successful Ariane program of the European Space Agency (ESA), the Japanese H-Family, the Chinese Long March and the Indian PSLV and GSLV. The world has witnessed the gradual construction of orbital platforms such as the Russian Mir Space Station and the International Space Station (ISS). Mankind now is living in the satellite communication age enabled by space technology.

Manned missions need high reliability and the best quality. To achieve it requires multiple subsystems with always a backup if one fails. The Saturn V had all of this, but it was used only once, and then discarded. After the success of Apollo 11, the space budget could not be sustained at high levels and to continue the manned exploration of space needed a more cost-effective program. This led to development of the Space Shuttle, the concept to reuse the main vehicle. As of now, the space shuttle main engine (SSME) is the most advanced engine that the world has ever developed. The Space Shuttle was a partially reusable low Earth orbital spacecraft system operated from 1981 to 2011 by the U.S. The disaster of space shuttle Challenger on January 28, 1986, disintegrated 73 seconds after launch due to the low-temperature impairment of an O-ring between segments of the rocket booster casing and another shuttle Columbia on February 1, 2003, disintegrated during re-entry due to damage of the leading edge of the wing caused during launch, killing all the 7 astronauts in each flight, led the U S to review the manned space program and NASA itself.

In January 2004, President George W. Bush called for the retirement of the Space Shuttle on completion of the International Space Station (ISS). It was planned to replace Shuttle with a Crewed Exploration Vehicle, using Apollo style re-entry rather than the complicated system of heat resistant tiles. The last Space Shuttle was launched on July 8, 2011, and landed at the Kennedy Space Center (KSC) on July 21, 2011. From then until the launch of SpaceX's Crew Dragon using Falcon Heavy rocket in 2020, the US launched its astronauts aboard to the ISS with Russian Soyuz spacecraft.

Up to recently, these space accomplishments have been driven and owned by national governments. Now, private industrialists like SpaceX and Boeing, encouraged by national agencies with start-up funds and new technologies, have entered the space business. The space tourism market is now opening with active participation of private industries. With the collaboration of nation and private agencies in the development frontier of space technologies, in the coming a few decades, it is possible to colonization of the Moon, Mars and asteroids by humankind.

## 1.3   SPACE

Space, as we are interested in outer space, begins at about 100 km above sea level where Earth's atmosphere is said to stop. Space is a zone of near vacuum between celestial bodies, a boundless three-dimensional extent in which objects and events have relative position and direction. It contains planets, stars, galaxies, dust, black holes and many other objects. Figure 1.1 shows the variation of atmospheric pressure along altitude. The jet planes fly at about 10 km above the ground level.

Space is very vast. It contains billions of galaxies. Galaxies are concentrations of stars, dust, gas and black matter. All held together by gravity. There are billions of galaxies in endless space. Our Earth belongs to one of the galaxies called the Milky Way galaxy. Our solar system, with the Sun as its center, is a relatively minute section of the vast galactic star system called the Milky Way. This is in turn only one galactic star system among the numerous such systems composing the universe.

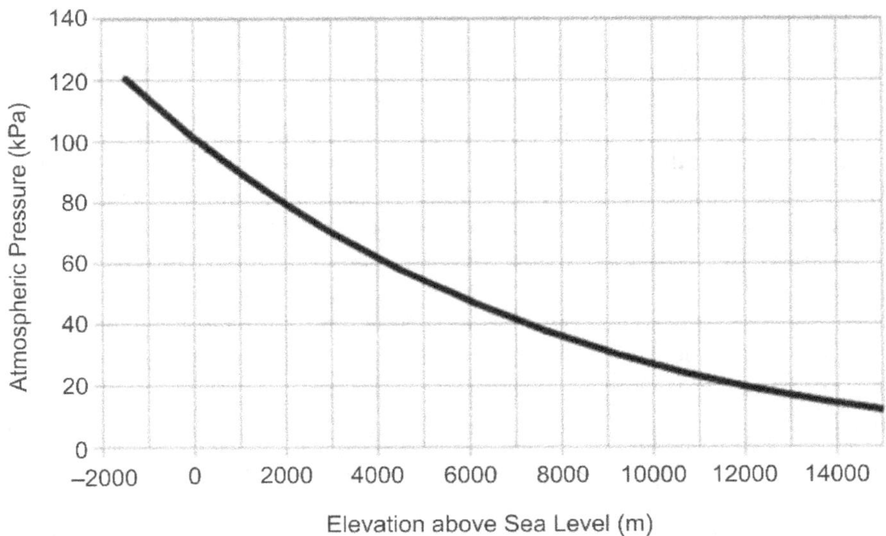

**Figure 1.1** Atmospheric Pressure Vs Elevation from sea level.

A galaxy is a system of stars and can best be visualized as a disc standing on edge. The Milky Way is like a barrel *spiral galaxy,* about two hundred thousand light-years across and thousand light years height. A light year is the distance travelled by light in a year. The speed of light is $3x10^5$ km/s. So, the light year is $3 \times 10^5 \times 365 \times 24 \times 60 \times 60$ is approximately $9.46 \times 10^{12}$ km.

Earth's solar system is located quite far down on the disc. The size of the Milky Way galaxy can easily be visualized: the Sun's light takes 26,080 years to reach the center of the galaxy whereas it takes only eight minutes to reach the Earth.

There are several stars in the Milky Way galaxy and we are particular about one star called the Sun, which we say is the solar system with which we are concerned. Our solar system consists of our star, the Sun, and everything bound to it by gravity the planets – Mercury, Venus, Earth, Mars, Jupiter, Saturn, Uranus and Neptune, dwarf planets such as Pluto, dozens of moons and millions of asteroids, comets and meteoroids. Moons are objects moving around the planets. Earth has got one moon.

All planets move around the sun in elliptical orbits and in the same direction. All nine planets orbit around the sun on nearly the same plane, except dwarf Pluto, but at different distances from the Sun. The dwarf Pluto planet's orbit is highly inclined. The four inner planets Mercury, Venus, Earth, and Mars are relatively small dense bodies known as terrestrial planets. The planets Jupiter,

Saturn, Uranus, and Neptune are called the giant planets and are principally composed of gases with solid ice and rock cores at unknown depths. Humankind's dream to discover a multitude of planets which are comparable to Earth and which could sustain human life depends on exploration of the solar system, development of foolproof means of travel and communication methods of interplanetary and intergalactic space.

## 1.4  PROPULSION

The word Propulsion has come from the Greek word Propellerie, meaning *'pushing forward'*. Space propulsion means pushing forward in space. To move from one place to another, the space vehicle must be accelerated i.e., it gathers incremental velocity (change in velocity $\Delta U$) during the travel. To impart acceleration to the vehicle, a propulsive device is required. One way to give the propulsive force is an internally driven engine, i.e., the propellants are combusted by some means internally within the vehicle and then accelerated to produce a reaction force. This is how conventional rockets work. In gunnery, the energy of the cannonball is given externally to it in the barrel of the gun by the explosion of the gunpowder and expansion of the hot gasses. But, once the ball leaves the barrel, due to air friction or drag, continuously its velocity decreases. On the other hand, the rocket accelerates due to continuous propulsive force. The cannonball is classified as a projectile and the rocket as a vehicle.

Both rockets and jet engines operate on the principle of action and reaction, Newton's third law of mechanics, and achieve the thrust by expanding exhaust gases generated due to burning of fuel and oxygen mixture. The main difference between them is that a jet engine gets the oxygen it needs for combustion from the outside atmosphere whereas a rocket carries its own oxygen source. A jet engine, therefore, can operate only within the earth's atmosphere but a rocket engine can operate anywhere, even in vacuum.

There is a difference between a rocket and a missile. A missile is a powered vehicle designed to carry explosives to a target The guided missile can change its direction by internal or external command at any time during its flight. A ballistic missile is powered and guided during the first part of its flight only, afterwards it proceeds like a projectile or thrown object, without power or guidance. Missiles are rocket engine powered and there is a tendency to call them all rockets. The rockets, we refer, are principally the engines that propel the vehicle into space.

A rocket is basically a thermodynamic system. It converts heat generated through chemical reaction of oxidizer and fuel (propellants) into kinetic energy using a device called nozzle. These propellants can be solid or liquid or gaseous. There are also monopropellant engines which use a catalyzer to decompose and start the reaction. Different propulsion systems generate thrust in different ways, but always through application of Newton's third law of motion. Chemical rocket engines generate a wide range of thrust ranging from a few newtons to millions of newtons. As of present technology levels, these are only powerful enough to use in launch vehicles i.e., to lift the spacecraft from Earth's surface to space i.e., to overcome Earth's gravitational force.

There are two general measures of performance of a rocket engine: (1) the amount of thrust, which determines the amount of propellant ejected or in other words the speed of the rocket that must be used to accomplish a given task, and (2) the fixed weight of the engine including the necessary tankage, power supply, and structure. The propulsion systems that reduce the need of the propellant making the spacecraft lighter or increase the speed of the propellant ejection are considered as advanced propulsion. Long-term, deep-space missions with significant payloads have a critical need for these advanced propulsion systems.

For a space mission, the trajectory analysis provides the required thrust and velocity change $\Delta U$. Other important factors are gravity assists, initial transfer orbit, spacecraft mass and duration of the flight to complete the mission. A manned flight, say to Mars, requires faster transfer time to avoid space travel fatigue and stress. For such missions some key constraints like psychological, safety, and radiation dose are strictly linked to the problem of time. For such flights the selected trajectory is the shortest and the propulsion unit is capable of providing high change in velocity and high thrust. On the other hand, unmanned missions may use a longer duration involving gravity assists and a much less powerful propulsion system.

Electric propulsion (EP) encompasses any propulsion technology in which electrical power is used to generate propellant exhaust velocity. Electric propulsion provides high exhaust velocities that in turn results in reduction of propellant requirement for a given space mission compared to conventional propulsion systems. The reduction in propellant mass can significantly decrease the launch mass of a spacecraft, leading to lower costs from the use of smaller launch vehicles. Electric thrusters use the same basic principle as chemical rockets, i.e., accelerating propellant mass and ejecting it from the vehicle. The ejected mass from electric thrusters, however, is primarily in the form of energetic charged particles. Electric thrusters operate in the power range of hundreds of watts up to tens of kilowatts with very high specific impulse (specific impulse is simply the thrust of the rocket, divided by the mass flow of propellant) of 1000s to 10,000s. The thrust levels are typically of some fraction of a newton. Ion and Hall thrusters generally use heavy inert gases such as xenon as the propellant.

The first extensive application of electric propulsion was by Russia using Hall thrusters for station keeping on communications satellites. Electric propulsion use in spacecraft has grown steadily and advanced electric thrusters have found applications, as an attractive alternative to chemical thrusters, for station-keeping, orbit insertion and communications satellites. The exhaust velocity of chemical rockets is limited by the energy contained in the chemical bonds of the propellant used and of typical values are up to 4 km/s. In electric thrusters, the energy source is power supply, separated from propellant and thus are not subject to the same limitations.

The function of the propulsion system is to produce thrust, which is the force that moves a rocket through air and space. In the space age, spacecrafts are the key to explore space, to conduct experiments in space to understand nature, to place scientific and commercial satellites and to make space travel as a commercial venture. The propulsion system used in rocketry can be categorized into two different types:

**Primary Propulsion:** Used to launch rockets from Earth surface.

**Secondary or Auxiliary Propulsion:** propulsion systems used in spacecraft to propel in space, e.g., satellites and control systems like attitude and guidance units.

Thousands of satellites orbiting the Earth form the backbone of the present communication system. Many countries', like United States, Russia, Europe, China, Japan, India and South

Korea, projections in space programs include to achieve before 2030: manned missions to secure a foothold on the Moon, landing humans on Mars, reaching and intercepting asteroids that might threaten our planet and accomplishing missions to Jupiter and other interplanetary travels. Space launch vehicles, using primary propulsion, must be capable of reaching transfer orbits and detaching the spacecraft, which, driven by secondary propulsion systems with high energy levels, will continue until the destination. An orbit is a path in which a body moves in relation to its source of gravity. A satellite or an interplanetary spacecraft can achieve the designed orbit by launching directly from the Earth's surface or launching from an initial orbit. The major difference is the thrust requirements. An interplanetary trip starting from the initial orbit at low Earth level needs much less power. The thrust required for launching a vehicle from Earth's surface to overcome gravity force will be of the order of Kilonewtons   to meganewtons. The thrust requirement for spacecraft to propel in space requires a few hundreds of newtons only.

*The thrust requirements in reducing order for operation of spacecraft as follows*:

- final orbit acquisition from the initial orbit established by the launch vehicle,
- station-keeping and orbit control,
- attitude control (also called as Reaction controlled system, RCS)

*Orbit changes* include transferring a spacecraft to a desired orbit, plane changes, orbit injection and de-orbit.

*Orbit Maintenance or Station Keeping* requires keeping a spacecraft in the desired mission orbit compensating the disturbing forces like drag, solar wind and gravitational forces.

*Attitude control (RCS)* is required to change the attitude, that is changing the orientation of a spacecraft to the desired direction and keeping it in desired direction by compensating for disturbing torques.

## 1.5   PROPULSION REGIONS AND PROPULSION TECHNOLOGIES

Depending on the sphere of usage and scope of use, the space propulsion can be divided into three different categories as shown in Figure 1.2:

(a)  *Escape propulsion – Earth to orbit*: Thrust is required to overcome the gravitational and drag forces to place the spacecraft in orbit

(b)  *In-space propulsion*: Thrust requirements to achieve orbital movements, maintaining the orbit and travel from an orbit to outer space

(c)  *Deep space Propulsion*: Thrust requirement to travel intergalactic space

The spacecraft launch vehicle should overcome the drag force and Earth gravitational force to escape into space. The Earth monitoring and communication satellites are placed in the in-space starting at 160 km, the start of low earth orbit (LEO), to around 35,000 km, geostationary earth orbit. The medium Earth orbit (MEO) starts around 2000 km and ends below 35000 km. The International Space Station (ISS) is at an orbit of 330 to 410 km. The Lagrangian Point, i.e., where gravitational effects of Earth and Sun balance out is at 1,500,000 km. The Moon is at 384,000 km. Between the Earth and Sun we have planets Venus, Mars and Mercury known as inner planets. In the outer solar system, i.e., after Earth we have Jupiter, Saturn, Uranus etc.

Human exploration beyond Earth to destinations such as the Moon, Mars or Near-Earth Objects, are daunting unless more efficient space propulsion technologies are developed and fielded. There is no single propulsion technology for space use that fits all space missions and mission types, and varies widely according to their intended applications. Selection of technology for in-space propulsion depends on thrust level requirement, specific impulse ($I_{sp}$), power, specific mass (or specific power), volume, system mass, system complexity, previous experience in other spacecraft systems, durability and cost.

| Distance (km) | Space | Space Regions | |
|---|---|---|---|
| | | Observation | |
| 0 | Escape | **Earth** | |
| 100 | Propulsion | End of atmosphere | |
| 160 | | Beginning of Low Earth Orbit (LEO) | *Harbors all Earth* |
| 406 | | International Space Station (ISS) | *Monitoring and* |
| 2,000 | | End of LEO - Beginning of Medium Earth Orbit(MEO) | *Communication* |
| 20,000 | | GPS Satellites | *Systems* |
| 35786 | | Geostationary orbit- Communications Satellites | |
| 384,000 | | **Moon** | |
| 1,500,000 | | Lagrangian Point 1 | |
| 38,200,000 | | Venus | |
| 55,700,000 | In-Space | Mars | *Inner Solar System* |
| 77,300,000 | | Mercury | |
| 149,600,000 | | **SUN** | |
| 588,390,000 | | Jupiter | |
| 1,200,000,000 | | Saturn | |
| 2,580,000,000 | | Uranus | *Outer Solar System* |
| 4,280,000,000 | | Pluto | |
| 4,300,000,000 | | Neptune | |
| 4,500,000,000 | | **Outer Solar System** | |
| 12,560,000,000 | Deep Space | Denser and Hotter Electrically Charged Particles | |
| 21,240,000,000 | | End of Heliosphere | |

**Figure 1.2** Space regions.

The launch vehicle propulsion system performs the function of primary propulsion to lift the payload from Earth surface following the designed trajectory, reaction control, placing the payload in the initial designed orbit, station keeping and orbital maneuvering. From there the space propulsion begins, the main engines used in space provide the primary propulsive force for orbit transfer, planetary trajectories and extra planetary landing and ascent. The auxiliary propulsion systems provide propulsive force to the reaction control and orbital maneuvering systems for orbit maintenance, position control, station keeping, and spacecraft attitude control.

Space exploration requires safe travel in space, arriving quickly to mission targeted places, and transporting maximum mass with low cost. Based on the physics of the propulsion system and how it derives thrust, space propulsion technologies can be classified into three basic types:

(1) Chemical Propulsion, (2) Non-Chemical or Electric Space Propulsion (3) Advanced Propulsion. Figure 1.3 shows the basic propulsion types and their sub classification.

The launch vehicle (rocket) propulsion is generally based on chemical propulsion and the technology is highly matured. Thrust-to-weight ratios greater than unity are required to launch from the surface of the Earth, and chemical propulsion is currently the only flight-qualified propulsion technology capable of producing the magnitude of thrust necessary to overcome Earth's gravity. There are significant technology advancements of in-space propulsion. Once in space, higher specific impulse propulsion systems can be used to reduce total mission propellant mass requirements. Electric propulsion is commonly used for station keeping on commercial communications satellites and for prime propulsion on some scientific missions, robotic deep-space exploration to space stations and human missions to Mars.

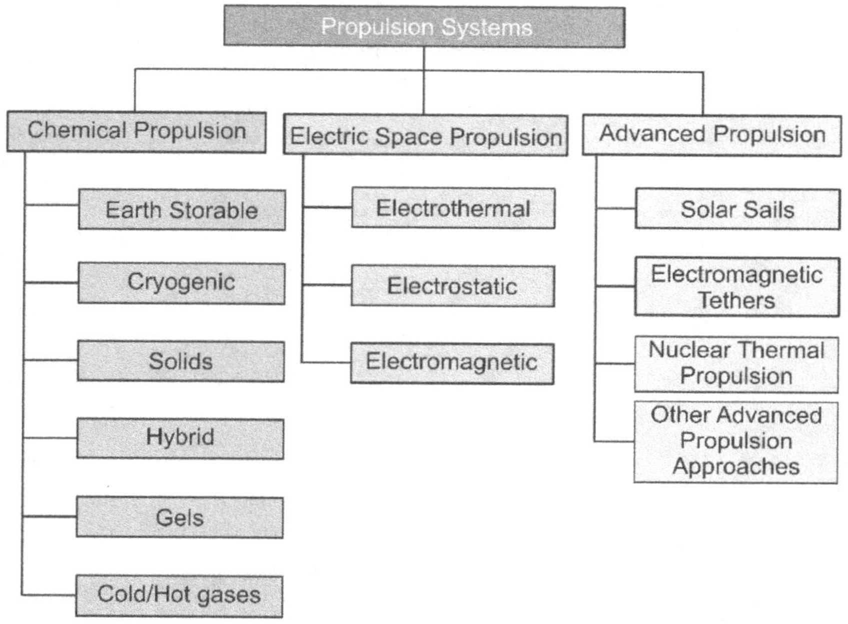


**Figure 1.3** Space Propulsion Technologies.

## 1.5.1 Chemical Propulsion

It includes systems that operate through chemical reactions that produce hot gasses and expand or use a fluid dynamic expansion as in a cold gas to provide thrust.

1. *Earth Storable*: Earth storable propellants remain stable over a range of Earth terrestrial pressures and temperatures and can be stored in a closed vessel for long periods of time.

   *Examples*: Kerosene, Hydrazine, Monomethyl hydrazine, Nitrogen tetroxide, Mixed oxides of nitrogen, Hydroxylammonium nitrate (HAN)-based propellants.

2. *Cryogenic*: Propellants that are liquefied gases at low temperatures. These systems provide high performance, but can present storage and handling challenges to prevent vaporization losses.

   *Examples*: Liquid oxygen (LOX), Liquid hydrogen ($LH_2$)

3. *Solid*: Solid propellants are pre-mixed oxidizers and fuels that are then cast into a particular shape. When ignited the surface area burns at a predetermined and tailored burn rate to generate the thrust and duration required for the mission.

   *Examples*: Polybutadiene Acrylic Acid Acrylonitrile (PBAN), Hydroxyl Terminated Polybutadiene (HTPB).

4. *Hybrid*: Hybrid rockets utilize a solid fuel and liquid oxidizer. They are also less complex and cheaper than liquid rockets.

   *Examples*: Acrylonitrile butadiene styrene thermoplastic, Paraffin-based fuels

5. *Gels*: Gelled and metallized fuels are a class of thixotropic (shear-thinning) fuels that improve the performance of rocket and air-breathing systems.

   *Examples*: Gelled oxygen ($O_2$)/hydrogen ($H_2$), Gelled MMH/IRFNA propellants, Nano Gelled propellants

6. *Cold Gas/Warm Gas*: Gas propulsion systems are typically used for small rocket engines or when small total impulse is required.

## 1.5.2 Electric Space Propulsion

Electric energy is converted to interact with and accelerate a reaction mass (e.g., Propellant) to generate thrust.

1. *Electrothermal*: The propellant is heated and expanded through a nozzle.

   *Example*: Resisto jet, Arc jet

2. Electrostatic: This area covers electric propulsion systems that use electrostatic fields to ionize and accelerate a propellant.

   *Examples*: Ion engines, Hall thrusters, Electrospray propulsion

3. *Electromagnetic*: The electromagnetic fields interact with a reaction propellant to generate thrust.

   *Example*: Pulsed inductive thruster, Magneto plasma dynamic (MPD) thruster

## 1.5.3  Advanced Propulsion

This area includes propellant-less and emerging technologies and physics concepts.

1. *Solar Sails*: Sail propulsion uses lightweight structures with a large surface area to produce thrust by reflecting solar photons or atmospheric molecules (drag), thereby transferring much of their momentum to the sail.

2. *Electromagnetic Tethers*: They are long, lightweight cables that produce thrust through the Lorentz force by carrying electrical current and interacting with a planetary magnetosphere.

3. *Nuclear Thermal Propulsion*: The engine uses a fission reactor (solid, liquid or gas) to heat a large mass flow of propellant to extremely high temperatures for high specific impulse at high thrust.

4. *Other Advanced Propulsion Approaches*: These advanced propulsion technologies include Beamed energy, Fusion Propulsion and Advanced Fission technologies and are in a research stage that could result in breakthroughs for future space missions. Other technologies like High Energy-density Materials and Antimatter propulsion are physics concepts that are actively under consideration for advancement of propulsion technologies.

Beamed energy propulsion uses laser or microwave energy from a ground- or space-based energy source and beams it to an orbital vehicle, which uses it to heat a propellant or reflected photon momentum exchange. Advanced fission and Fusion propulsion systems produce energy required by using the thermal/ kinetic energy resulting from the reactions to accelerate a propellant.

The aim of the book is to provide the reader with knowledge to better understand the fundamental concepts of propulsion technologies for design, analysis and programming of rocket and spacecraft propulsion systems. The book covers the syllabus of undergraduate and postgraduate engineering students of most of the universities. practicing professionals would find the book helpful for solving specific problems related to rocket propulsion systems. The approach and explanation with worked examples provide a comprehensive capability to  the reader to conduct an initial design and sizing of propulsion devices. Chapter 2 deals with the basic laws and relations for the objects to propel into different orbits or escape from the planet. Chapter 3 provides the understanding of rocket propulsion and derivation of the fundamental rocket equation. It explains the basic importance of staging in the design of launch vehicles. The compressible flow expansion of high temperature and pressure gases through rocket nozzle is explained from fundamentals and derived various relations useful for the estimation of performance in chapter 4.  In chapter 5 chemical propellants used in rocket engines, their properties, performance and thermodynamic procedures for estimation of energy release are discussed. Chapters 6, 7 and 8 provide in-depth discussion of solid, liquid, and hybrid rocket propulsion systems. The merits and limitations of the different systems are included in these chapters. The solid rocket motor (SRM), covering all aspects of design and development are included in chapter 6. Chapter 10 provides an overview of the complex topic of combustion instability that occurs due to intensive energy release rates in the thrust chamber and the dynamics of interactions of the different subsystems of a rocket. The mitigation techniques and use of devices to control the instabilities are presented. Chapter 11 and 12 deals with various orbit maneuvering techniques and interplanetary trajectories. These chapters provide the methods to estimate propellant requirements of spacecraft propulsion. Chapter 13 discusses chemical propulsion systems used in spacecraft and their merits and demerits. The final chapter 14 of the book discusses the different electric propulsion systems used in spacecraft.

# 2 Basic Orbital Mechanics

## 2.1 INTRODUCTION

This chapter lays the foundation to understand orbital motion of the spacecraft. The motion of two bodies exclusively due to their own mutual gravitational attraction forces is described. The Keplerian equation of motion is derived that shows the orbital path of one of the masses relative to the other is a conic section. The properties of different conical orbit paths are developed from the laws of conservation of angular momentum and energy. The frame of references to describe the motion of the spacecraft are discussed. The chapter concludes with the explanation of the essential orbital elements.

## 2.2 KEPLER'S LAWS

Many Scientists have been watching the motion of planets around the Sun and postulated their theories. Copernicus, Galileo, Kepler and Newton (with the contribution of many others), provided a scientific model capable of describing why and how our planets move. Copernicus (1467–1543) put forward his model that the planets were moving in epicycles about the Sun. Similarly, as per his model, the Moon moves about the Earth in a circular orbit. This theory opened the way for more proper theories and accurate experimental observations of celestial bodies.

Tycho Brahe (1546–1601), a Danish astronomer, meticulously recorded the motion of various celestial bodies at his astronomical observatory over a period of 13 years. Johannes Kepler (1571–1630) was fascinated by the Copernican theory, tried extensively to uncover additional geometric properties inherent in this heliocentric universe to develop the general empirical laws of planetary motion but his efforts were unproductive. He became an assistant to Tycho Brahe, in his observatory at Prague. On his deathbed, Brahe released his tabulated measurements to his young assistant, Johannes Kepler. His analysis and mathematical brilliance helped to unlock the hidden secrets of Brahe's measurements. His attempt to fit the observations of Brahe regarding Mars motion by using a little-known geometric shape of his time, an ellipse that had principally been studied by Greek mathematicians, resulted to describe the orbit of the planet with remarkable precision.

**Figure 2.1** Johannes Kepler (1571-1630).

Kepler's most important work, *Astronomia Nova de Motibus Stellae Martis,* was published in 1609 and contained the first valid approximations to the kinematic relations of the solar system; Kepler's first two laws arise from this work. After ten years, in 1619 he was able to quantify his third law uniting the field of celestial mechanics so far in an unheard-of fashion. The concepts associated with planetary motions developed by Kepler describe the positions and motions of objects in our solar system. He formulated three laws that are known as Kepler's laws: The first law defines the shape for the planetary orbits, second law pinpoints the changing velocity at which the planets travel about the Sun and the third law provides a relationship between the periods of the planets and their mean distance from the Sun.

1. Planets describe an elliptical path around their centers of gravitation.

2. A line joining a planet and the Sun sweeps out equal areas during equal intervals of time.

3. The square of the orbital period of a planet about a center of attraction is directly proportional to the cube of the semi-major axis of its orbit.

The first law precisely describes that the shape of the planet's orbit is in the form of an ellipse. An ellipse, as shown in Figure 2.2, is a closed plane curve generated in such a way that for any point on the curve the sum of its distances from two fixed points is constant. The distance A+B = constant. The two fixed points are called foci.

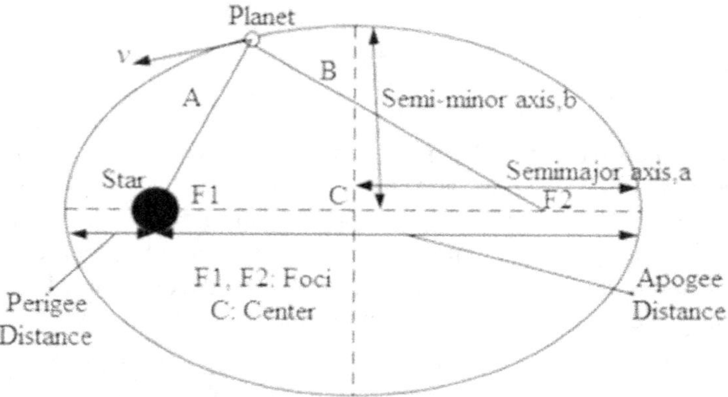


**Figure 2.2** Elliptical Orbit.

In Figure 2.2 both the semi-major and semi minor axes are shown. The shape of the ellipse depends on the distance between the foci and the major axis. This ratio is called the eccentricity and its value lies between 0 to 1. If the foci distance is zero, the eccentricity is zero, the shape becomes a circle. Also, as shown, the nearest distance of a star, say Sun, to the orbit is called perigee and the farthest distance is apogee.

The second law addresses the velocity of the planet in orbit. For a given unit time the planet in the orbit sweeps equal areas. As shown in the figure 2.3, areas generated by arc A and B are equal. This means that Earth moves slowest along arc B and fasted along arc A. Similarly, if a satellite orbits the Earth in an elliptical orbit, it has fastest velocity at perigee and slowest velocity at apogee.

The third law relates the time period of the planet with the size of the orbit. It is expressed as

$$\tau^2 = k\, a^3$$

where $\tau$ is the period of revolution, $a$ is the semimajor axis of the orbit, and $k$ is a constant of proportionality. The third law simply states that for objects further away from the Sun, they will take longer to complete one orbit. These laws are fundamental to of satellite or spacecraft orbit

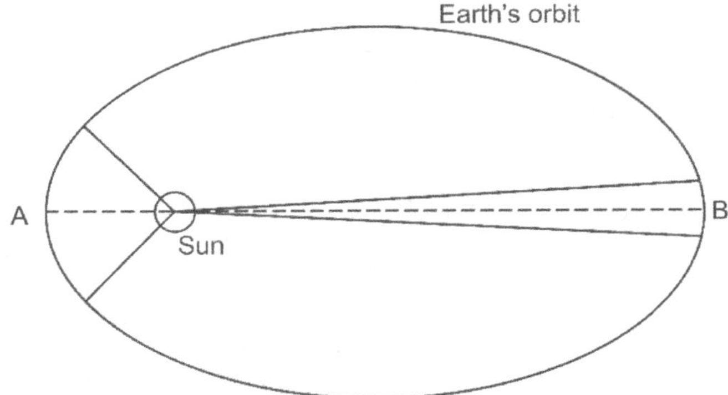


**Figure 2.3** Kepler's 2nd Law.

around any central mass. Obviously, these statements apply to a two-dimensional picture of planetary motion, which is all that is needed for describing orbits. A three-dimensional picture of motion would include the path of the sun through space.


## 2.3   NEWTON'S LAWS OF MOTION

Newton published in 1687 the three laws of motion in *The Mathematical Principles of Natural Philosophy,* or commonly known as the *Principia* and are considered the axioms of mechanics. These three give the important concepts in mechanics: inertia, momentum and reaction respectively.

1. Every particle persists in a state of rest or of uniform motion in a straight line unless acted on by a force.

2. If F is the external force acting on a particle of mass $m$, which, as a consequence, is moving with velocity $v$, then

$$F = \frac{d}{dt}(mv)$$

This can be restated as the rate of change of momentum is proportional to the force and is in the same direction.

For a given mass $m$

$$F = m\frac{dv}{dt} = m\alpha$$

where $\alpha$ is the acceleration. The acceleration imparted to a body is directly proportional to the force acting on the body and inversely proportional to its mass. With mass in kg and acceleration in m/s$^2$, the unit of force is Newton, represented by N (kg-m/s$^2$).

3. To every action there is an equal and opposite reaction. In other words, if particle 1 acts on particle 2 with a force $F_{12}$ in a direction along the line adjoining the particles, while particle 2 acts on particle 1 with a force $F_{21}$, then $F_{12} = F_{21}$.

These laws are not valid when the velocities of the bodies approach the velocity of light such as the quasars in deep space for which relativistic effects become important. In the case of bodies having extremely small dimensions of the order of atomic nuclei, quantum mechanics begins to play a role and the above laws would be inapplicable

## 2.4   UNIVERSAL GRAVITATIONAL LAW

Isaac Newton (1643-1727) explained why Kepler's laws worked, by showing they depend on gravitation. He realized that the force that makes apples fall to the ground is the same force that makes the planets "fall" around the sun. Gravitation is the mutual attraction of all masses in the universe. He was able to explain the Kepler Laws through his famous Law of Universal Gravitation, one of the greatest achievements of humankind.

**Figure 2.4** Isaac Newton (1643-1727).

The law states that every object attracts every other object in the universe with a force which is directly proportional to the product of their masses and inversely proportional to the square of the distance between their centers. Referring to Figure 2.5, the law is expressed as:

$$F = -G\frac{Mm}{r^2}$$

.....(2.1)

where F is the force on mass m due to mass M and $r$ is the vector from M to m. The force F is in Newtons(N), mass m is in kg and distance r is in m, G is universal gravitational constant $G = 6.67408 \times 10^{-11}$ N m²/ kg². The negative sign in the equation is indicative of the fact that the force is attractive. The gravitational field is a weak force and it pursues a very long distance. Since its effect decreases in proportion to distance squared, it never disappears completely. Because of this, it applies, to some extent, regardless of the sizes of the masses or their distance apart.

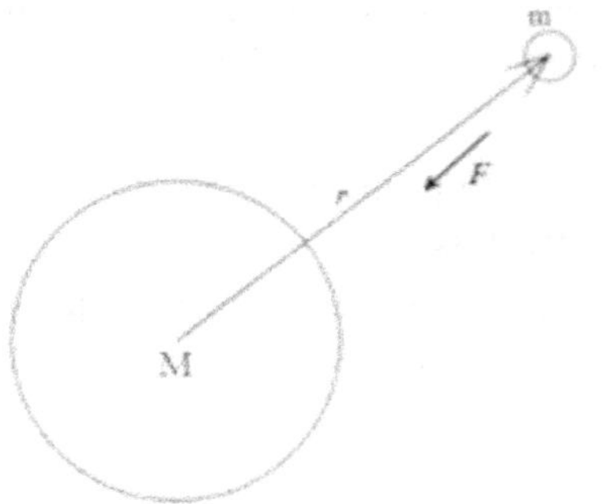


**Figure 2.5** Gravitational force between two masses.

## 2.5   GRAVITATIONAL FIELD

Consider a mass, m is at distance of R from the center of Earth and at a distance of h from the surface of the Earth as shown in Figure 2.6. $R_e$ is the radius of Earth and $M_e$ is the mass of Earth.

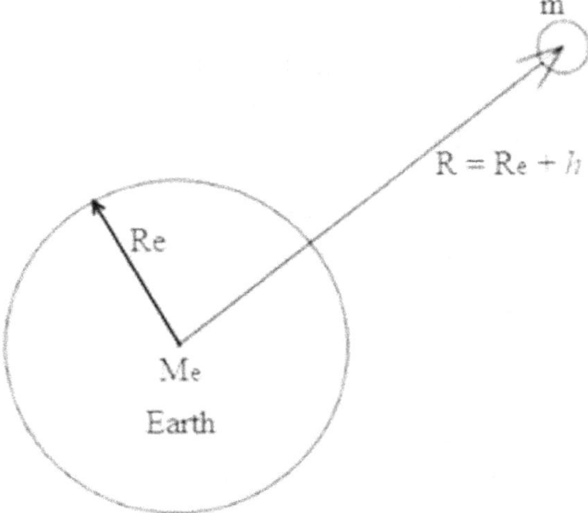


**Figure 2.6** Earth's gravitational force.

As per the Universal law of gravity, the force with which earth attracts body of mass m is

$$F = -G \frac{M_e m}{(R_e + h)^2} \ N$$

$$= -G \frac{M_e m}{R_e^{\,2}} \left[ \frac{1}{\left(1 + \frac{h}{R_e}\right)^2} \right]$$

$$= -G \frac{M_e m}{R_e^{\,2}} \left(1 + \frac{h}{R_e}\right)^{-2}$$

Using Binomial theorem

$$F = -G \frac{M_e m}{R_e^{\,2}} \left(1 - \frac{2h}{R_e} + \cdots . \right)$$

As h<<$R_e$

$$F = -G \frac{M_e m}{R_e^2} \qquad\qquad\qquad .....(2.2)$$

Generalizing the equation

$$F = -G \frac{Mm}{R} \qquad\qquad\qquad .....(2.3)$$

where M mass of a Planet and R is its radius.

Substituting the values of G = $6.671 \times 10^{-11}$ N m$^2$/kg$^2$, mass of the Earth $Me$ = $5.97 \times 10^{24}$ kg and radius of Earth R$_e$ = (12,756,000/2) m =6,378,000 m in Eq. (2.3)

$$F = -6.671 \times 10^{-11} \frac{5.97 \times 10^{24} \times m}{(6378000)^2}$$

$$F = -9.81m \ \ Newton$$

Earth has exerting gravitational force on a mass m on its surface. Comparing with $F= mg$, the unit of g comes has

$$\frac{N}{kg} = \frac{kgm}{s^2} \frac{1}{kg} = \frac{m}{s^2}$$

which is unit of acceleration, and generally g is called acceleration due to gravity on Earth. Table 2.1 gives the surface gravity of different planets.

**Table 2.1** Surface Gravity of Planets.

| Planet Name | Mass (kg) | Radius (km) | Surface gravity (m/s$^2$) | Equivalent of g |
| --- | --- | --- | --- | --- |
| Earth | $5.97 \times 10^{24}$ | 6378 | 9.81 | 1.000g |
| Moon | $7.347 \times 10^{22}$ | 1737 | 1.62 | 0.165g |
| Mercury | $3.285 \times 10^{23}$ | 2440 | 3.70 | 0.380g |
| Mars | $6.417 \times 10^{23}$ | 3389 | 3.71 | 0.380g |
| Jupiter | $1.899 \times 10^{27}$ | 69911 | 24.79 | 2.530g |
| Saturn | $5.684 \times 10^{26}$ | 58232 | 10.44 | 1.065g |
| Uranus | $8.680 \times 10^{25}$ | 25360 | 8.69 | 0.886g |
| Neptune | $1.024 \times 10^{26}$ | 24622 | 11.15 | 1.140g |

**Example 2.1:** Find the value of gravitational force on the Moon. Take relevant data from Table 2.1

The mean radius of Moon R = 1737000 m = 1737 km

Mass of Moon M = $7.3477 \times 10^{22}$ kg

Substituting in Eq. (2.3),

$$F = -6.671 \times 10^{-11} \frac{7.3477 \times 10^{22} \times m}{(1737000)^2} = -1.625 \, m \ Newton$$

The acceleration due to gravity on Moon = 1.625 m/s$^2$ = 1.625/9.81 = 0.165 g, i.e., the gravitational force on the Moon is 0.165 times of Earth.

## 2.6 MOMENTUM

Let us look into a few definitions of quantities that are useful to understand the principles of propulsion.

Velocity, $v$ is the rate of change of distance and its unit is m/s. It is a vector quantity. Momentum, $p$ is product of mass of the object and its velocity

$$p = mv \ kg \frac{m}{s} \qquad \qquad \ldots..(2.4)$$

To push or move or give acceleration to a body, a change of momentum is required and is known as impulse, $I$. If at time $t$, the momentum of the body is p(t) and after a small interval of time $\Delta t$ its momentum is $p(t+\Delta t)$.

The change of momentum= Impulse $I = p(t+\Delta t) - p(t)$ kg m/s

As per Newton's second law, the rate of change of momentum is force.

$$F = \frac{dp}{dt}$$

For a small time period $\Delta t$

$$F\Delta t = dp = I \quad \text{N s} = (\text{kg m/s}^2) \text{ s} = \text{kg m/s}$$

Momentum is a more fundamental quantity than force. The reason being, considering that an iron ball and a feather are moving with the same velocity, the iron ball is more harmful.

## 2.7   ORBIT TYPES

An orbit is a path in which a body moves in relation to its source of gravity. There are four types of orbits, named after the conic sections, i.e., the four basic curves derived by intersecting a cone with a plane as depicted in Figure 2.7.

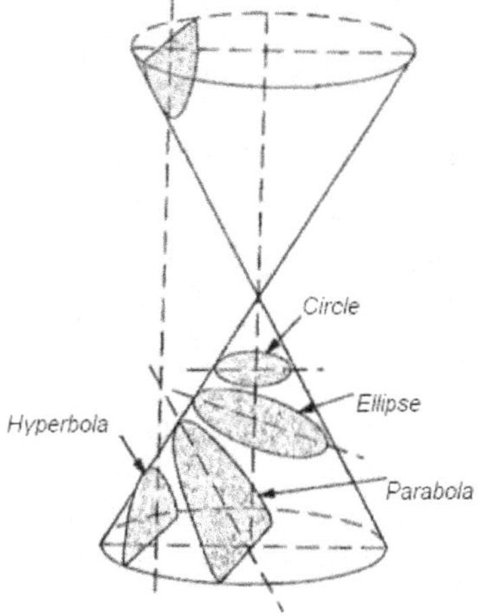


**Figure 2.7** Conic sections.

1. Circular orbit: A plane intersecting a cone, perpendicular to the axis of revolution generates a circle. A spacecraft traveling around the planet at a constant speed and always is equidistant from the planet's center of gravity follows a circular orbit.

2. Elliptical orbit: A plane that intersects one cone at an angle to the axis generates an ellipse. A spacecraft traveling a closed path having major and minor axes follows elliptical orbit.

3. Parabolic orbit: A plane intersecting the cone parallel to the generating line generates parabola. A spacecraft traveling at such a high velocity that it no longer follows a closed path but escapes into space follows a parabolic orbit

4. Hyperbolic orbit: A plane is parallel to the axis of revolution intersecting the cone generates a hyperbola. A spacecraft traveling in essentially a parabolic orbit, but which has the ability to change its position with respect to the sun, follows a hyperbolic orbit.

There are many different satellite orbits, either circular or elliptical, that can be used. The orbit is selected depending upon the application. Some of the important parameters of the orbit are height above the surface Earth, orbit velocity and angle of inclination. Orbits are classified as (Ref. Figure (2.8):

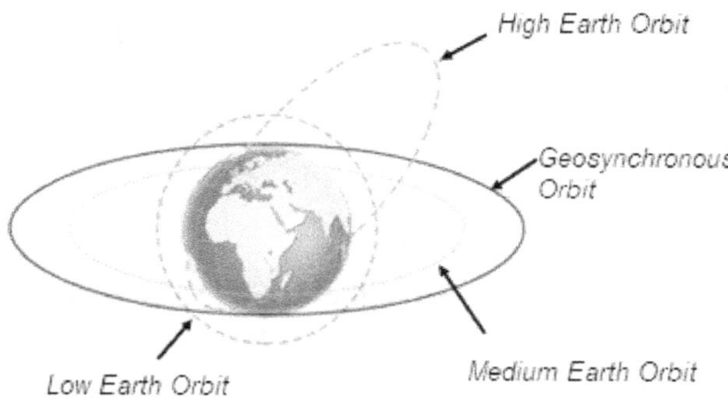


**Figure 2.8** Types of Orbits.

1. Low Earth Orbits (LEO) having the altitude from Earth surface in between 160 to 2000 km

2. Medium Earth Orbits (MEO) having altitude between 2000 to less than 35800 km

3. Geosynchronous orbits (GSO) having the same angular rotation of Earth— the speed of orbit synchronizes with Earth rotation. They are at an altitude of 35800 km.

4. High Earth Orbits (HEO) having altitude above 35800 km.

The geostationary satellite orbit can be any plane of the Earth. One particular form of geosynchronous orbit is known as a geostationary orbit. Its plane of rotation coincides with the equatorial plane. This means that it revolves at the same angular velocity as the Earth and in the same direction and therefore remains in the same position relative to the Earth.

A polar orbit is one in which a satellite passes above or nearly above both poles of the Earth being orbited on each revolution. It has an inclination of 90 degrees to the equator. A satellite in a polar orbit will pass over the equator at a different longitude on each of its orbits. If the launch is near the equator, low inclination, and in the direction of rotation of Earth, the rotational speed of the surface contributes a significant part of the final speed required for orbit. A polar orbit will not be able to take advantage of the Earth's rotation, and thus the launch vehicle must provide all of the energy for attaining orbital speed. Depending on the application of the satellite, the orbit parameters are selected. Generally, communication and television streaming satellites use a geostationary orbit. The systems used for satellite phones and satellite navigation like Global Positioning (GPS) occupy a relatively low Earth orbit. There are also many other types of satellite from weather satellites to research satellites that use LEO or MEO.

To put the satellite or a body in a required orbit, efficient methods are used to ensure that the amount of energy or propellant required is kept to a minimum. The size and cost of the launch vehicle depend on the amount of propellant used. Generally, to control the cost of satellite launching, it is placed into a low Earth orbit with an altitude of around 300 km. Once in the correct position in this orbit rocket engines are fired to put the satellite into an elliptical orbit with the perigee at the low Earth orbit and the apogee at the geostationary orbit. When the satellite reaches the final altitude the rocket engine is again fired to retain it in the geostationary orbit with the correct velocity. If powerful launch vehicles like Ariane are used the satellite is launched directly into the elliptical transfer orbit. Again, when the satellite is at the required altitude the rocket engines are fired to transfer it into the required orbit with the correct velocity.

## 2.8   TWO BODY MOTION

A body or spacecraft motion in space is influenced by numerous attractive forces of all celestial bodies as per gravitational law. A rigorous analysis of this network would be impossible. However, it is observed that the motion of the spacecraft in the solar system is dominated by one central body at a time. This leads to two-body motion in the space. The two-body problem is to determine the motion of the two bodies that interact only with each other. The examples are: a satellite orbiting a planet, a planet orbiting a star, two stars orbiting each other (binary star).

*The assumptions made in this theory are*:

1.  The motion of a spacecraft is governed by attraction to a single central body.
2.  The mass of the spacecraft, $m$ is negligible compared to that of central body mass $M$, $m<<M$
3.  Bodies are spherical and homogeneous with the masses concentrated at the centers.
4.  No forces act on the bodies except gravitational forces and centrifugal forces acting along the line of centers of the two bodies.

The Table 2.2 shows the significant relative accelerations acting on a body in LEO. The gravitational force of Earth is more significant than any other planet forces. The major perturbations at low Earth orbit are the drag due to the residual atmosphere and the Sun. Multi-body analysis is required only when the two-body solution is suspect (e.g., a Mercury orbiter) or when high accuracy is required. It is complex and requires a large computing capacity. Nearly all our applications of the orbital equations will be for the analysis of man-made spacecraft, all of which have a mass that is insignificant compared with the Sun and planets.

**Table 2.2** Acceleration on Lower Earth Orbit.

| Body | Acceleration, g |
| --- | --- |
| Earth | 0.98 |
| Sun | $6 \times 10^{-4}$ |
| Moon | $3 \times 10^{-6}$ |
| Jupiter | $3 \times 10^{-8}$ |
| Venus | $2 \times 10^{-8}$ |

With the given coordinate system as a reference frame, we can describe the kinematic (the motion) of the object, i.e., it's the position, the velocity and the acceleration of the object as a function of time. Newton's laws are valid in a stationary reference frame or a frame moving with constant velocity. They are called inertial frames of reference. In the accelerating frame of reference these laws are not valid. For describing the planet and satellites motion, the inertial frame of reference is used. The Earth is considered as an inertial frame.

## 2.9 ORBITAL VELOCITY AND TIME PERIOD

If an object, say a satellite, is moving in a circular orbit around the Earth, as shown in the Figure 2.9, the pull of gravity attracts it and causes it to fall down freely while the forward orbital velocity component shifts away from falling and the body gets into circular orbit. Let us demystify the origin of forces that appeared to act on the object in a rotating frame. Let a satellite of mass m is orbiting the Earth of mass $M_e$ in circular orbit with a tangential velocity $v$. The forces acting on it are shown. For steady state motion to be maintained, the gravitational force F is to be balanced by centrifugal force. The applied force is fictitious or pseudo force and is called centrifugal force. So, to describe the motion of the body in a rotating frame a pseudo force is to be considered. In an accelerating frame of reference in addition to centrifugal force another pseudo force called Coriolis force to describe the motion correctly is to be considered. However, as planets or satellites move in a rotating frame with a constant velocity, we are confined to an inertial frame of reference. The centripetal force $F_c$ is

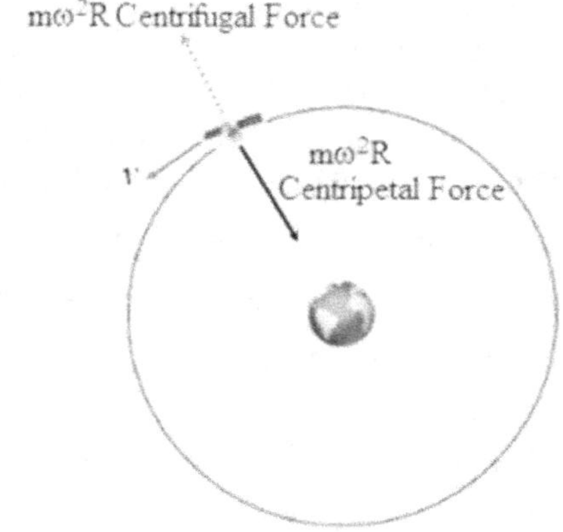


**Figure 2.9** Forces on the Earth Orbiting Body.

$$F_c = \frac{mv^2}{R} = m\omega^2 R \qquad \qquad .....(2.5)$$

where $\omega$ angular velocity of the satellite.

From the universal law of gravitational force, the force F that the body experiences due to the mass of the Earth is towards the center of the Earth.

$$F = \frac{GM_e m}{R^2} \qquad \qquad .....(2.6)$$

Equating F and $F_c$

$$\frac{GM_e m}{R^2} = \frac{m\,v^2}{R}$$

$$v = \sqrt{\frac{GM_e}{R}} = \sqrt{\frac{\mu}{R}}\,\frac{m}{s} \qquad \qquad ....(2.7)$$

where $\mu$ is a product of the mass of the planet and the universal gravitational constant. The orbital velocity decreases as the radius of the orbit increases. The time period $\tau$ of the orbit is the ratio of the orbit circumference and orbital velocity.

The period of circular orbit is

$$\tau = \frac{2\pi R}{v} = 2\pi \sqrt{\frac{R^3}{\mu}}\ \ sec \qquad \qquad .....(2.8)$$

## 2.10  ORBITAL ENERGY

The total energy of a body of mass, $m$ at an altitude, $h$ orbiting a planet consists of both kinetic and potential energy. The kinetic energy is due to rotation of the mass around the central body with a velocity $v$. The kinetic energy of the satellite of mass m orbiting the Earth is:

$$E_k = \frac{1}{2}mv^2 \qquad \qquad .....(2.9)$$

Substituting the value of $v$ from Eq. (2.7)

$$E_k = \frac{m\mu}{2R} \qquad \qquad .....(2.10)$$

The specific energy is energy for unit mass. So, the specific kinetic energy is

$$\varepsilon_k = \frac{\mu}{2R} \qquad \qquad .....(2.11)$$

In addition to this, for the object to reach the height $h$ from Earth surface, a certain amount of energy is to be given. i.e., to put an object in orbit around the Earth, the object has to be lifted to the level of the orbit from the Earth surface and impart a velocity to keep it in the orbit. The potential energy is defined as the *negative* of the work done by the force associate with the potential energy.

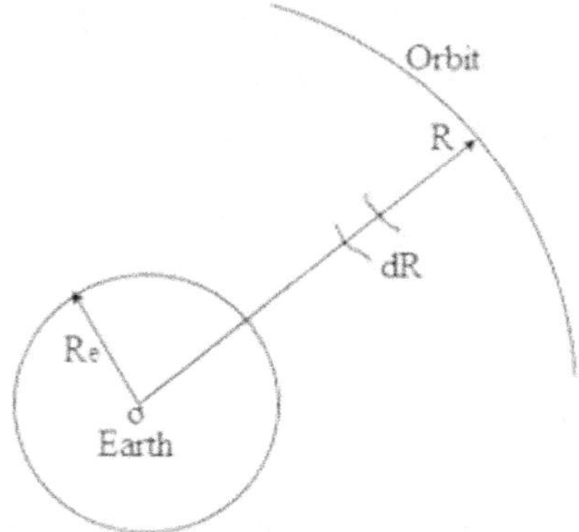


**Figure 2.10** Derivation of Orbit Energy.

As shown in Figure 2.10 let $R$ is the radius of the orbit and $R_e$ the radius of the Earth.

Radius of orbit from Earth center $R = R_e + h$

Let the mass m is raised to a small distance $dR$ along the radius; the work done is:

$$dW = F.dR = -G\frac{M_e m}{R^2}dR$$

Therefore, the work done to raise the mass m from $R_e$ to R is:

$$W = -GM_e m \int_{R_e}^{R} \frac{1}{R^2}\, dr = -GM_e m \left(\frac{1}{R} - \frac{1}{R_e}\right)$$

Potential energy is the difference between the energy of an object in a given position and its energy in a reference position. When work is done against a conservative field, such as gravity, then the energy of that work is converted to potential energy. Setting the reference position as infinitely far away, the gravitational potential is always negative since you must do positive work to lift an object to infinity. The potential energy can be expressed as:

$$E_p = -\frac{GMm}{R} = -\frac{\mu m}{R} \qquad \qquad .....(2.12)$$

We can make two important observations. As $R \to \infty$, $E_p \to 0$, the potential energy is zero when two bodies are infinitely far apart. Second, note that $E_p$ becomes increasingly more negative as the masses get closer. As the two masses are separated, positive work must be done against the force of gravity, and hence, $E_p$ increases (i.e., it becomes less negative).

The specific potential energy $\varepsilon_p$ i.e., energy per unit mass is

$$\varepsilon_p = -\frac{\mu}{R} \qquad \qquad ......(2.13)$$

The total energy per unit mass or specific energy of the orbiting object $\epsilon$ is

$$\varepsilon = \varepsilon_k + \varepsilon_p \qquad \qquad .....(2.14)$$

Substituting the values of the specific kinetic and potential energy from Eq. (2.11) & (2.13)

$$\varepsilon = \frac{\mu}{2R} - \frac{\mu}{R} = -\frac{\mu}{2R}$$    .....(2.15)

In general, for conic section with semi major axis a,

$$\varepsilon = -\mu/2a$$    .....(2.16)

The total energy of any orbit depends on the semimajor axis of the orbit only. For a circular orbit, $a = R$ and specific energy is negative. Also, for an elliptical orbit, $a$ is positive and specific energy is negative. Thus, for all closed orbits specific energy is negative. The specific energy is zero for parabolic orbits as $a = \infty$. The semi-major axis for hyperbolic, $a$ is negative and so, the specific energy is positive. Thus, a parabolic orbit is a boundary condition between hyperbolas and ellipses.

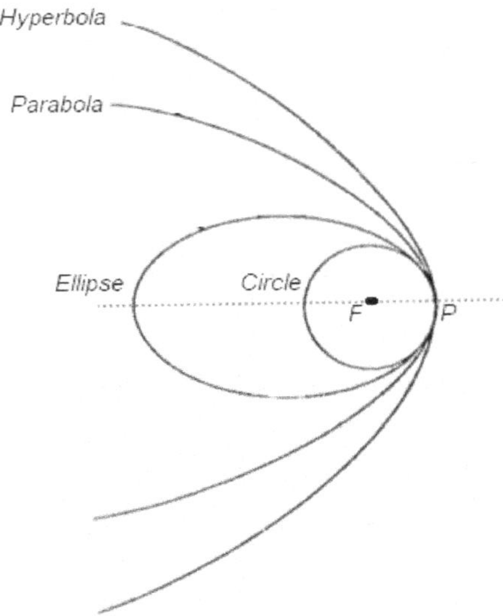


**Figure 2.11**  Orbits Having a Common Focus F and Periapsis.

Figure 2.11 shows a range of trajectories, from a circle through hyperbolas, all having a common focus and periapsis. The parabola is the demarcation between the closed, negative energy orbits (ellipses) and open, positive energy orbits (hyperbolas). At a given radius, velocity and specific energy increase in the following order: circular, elliptical, parabolic, hyperbolic; total spacecraft energy increases in the same order. To change the orbit of the spacecraft from circular to elliptical, energy has to be added. Similarly, to change the orbit from elliptical to circular energy has to be removed. Both adding and removing energy requires a force on the vehicle that consumes from propulsion system. Table 2.3 summarizes the points discussed on conic sections regarding specific energy.

**Table 2.3** Specific Energies of various Orbits.

| Parameter | Circle | Ellipse | Parabola | Hyperbola |
|---|---|---|---|---|
| Eccentricity, e | Zero | < 1 | 1 | > 1 |
| Semimajor axis, a | = Radius | Positive | ∞ | Negative |
| Specific energy, ε | Negative | Negative | Zero | Positive |

To place a spacecraft of mass m in a circular orbit from Earth's surface to a radius of R, the total energy required is

$$E = GM_e m \left(\frac{1}{R_e} - \frac{1}{R}\right) + \frac{1}{2}mv^2 \qquad \dots\dots(2.17)$$

If the total energy given by above equation is provided in the form of kinetic energy, the total velocity $v_t$ required for orbiting the body is obtained by:

$$\frac{1}{2}mv_t^2 = GM_e m \left(\frac{1}{R_e} - \frac{1}{R}\right) + \frac{1}{2}m\frac{GM_e}{R}$$

$$v_t^2 = 2GM_e \left(\frac{1}{R_e} - \frac{1}{2R}\right)$$

Substituting $R = R_e + h$ and simplifying

$$v_t = \sqrt{\frac{GM_e}{R_e}\left(\frac{R_e + 2h}{R_e + h}\right)} \qquad \dots\dots(2.18)$$

As seen the total velocity required to place a body in an orbit increases with height of the orbit from the surface of Earth.

**Example 2.2**

Find the total velocity of a satellite orbiting the Earth in a circular orbit at an altitude of 400 km from Earth's surface.

**Given:** Altitude h = 400 km

Radius from Earth's center R = Altitude + Radius of Earth = 400 + 6378 = 6778 km

Total velocity from Eq. (2.18)

$$v_t = \sqrt{\frac{GM_e}{R_e}\left(\frac{R_e + 2h}{R_e + h}\right)}$$

$$v_t = \sqrt{\frac{6.67 \times 10^{-11} \times 5.97 \times 10^{24}}{6378000}\left(\frac{6378 + 800}{6378 + 400}\right)} = 8131 \; m|s = 8.131 \; km/s$$

Table 2.4 gives the range of launch vehicle velocity required for ideal and approximate actual conditions for different missions.

**Table 2.4** Vehicle Mission Velocities for Typical Interplanetary Missions.

| Mission | Ideal Velocity (km/s) | Approx. Velocity (km/s) |
|---|---|---|
| Satellite orbit around Earth (no return) | 7.9-10 | 9.1-12.5 |
| Escape from Earth (no return) | 11.2 | 12.9 |
| Escape from moon | 2.3 | 2.6 |
| Earth to Moon (soft landing on Moon and no return) | 13.1 | 15.2 |
| Earth to Moon (landing on Moon ad return to Earth) | 15.9 | 17.7 |
| Earth to Mars (soft landing) | 17.5 | 20.0 |
| Earth to Mars (landing on Mars and return to Earth) | 22.9 | 27.0 |

## 2.11  ESCAPE VELOCITY

It is the minimum velocity required for a free, non-propelled object to escape from the gravitational influence of a planet. As the height of orbit increases, the gravitational force reduces. To escape from the Earth this force should vanish. The force is zero when R is infinity. For a body of mass m, the work done to escape against gravity force from Earth's surface is

$$W = \int_{R_e}^{\infty} -\frac{GM_e m}{R_e^2}\, dR = \frac{GM_e m}{R_e} \tag{2.19}$$

If $v_e$ is the escape velocity,

$$\frac{1}{2}mv_e^2 = \frac{GM_e m}{R_e}$$

$$v_e = \sqrt{\frac{2GM_e}{R_e}} = \sqrt{\frac{2\mu}{R_e}} \qquad \dots\dots (2.20)$$

The value of escape velocity $v_e$, is therefore $\sqrt{2}$ times the orbital velocity, $v$ at the surface of Earth,

Eq. (2.7). For Earth escape velocity, substituting the values G, $M_e$ and $R_e$ in Eq. (2.20):

$$v_e = \sqrt{\frac{2 \times 6.67 \times 10^{-11} \times 5.97 \times 10^{24}}{6378000}} = 11.2 \ km/s.$$

Therefore, the escape velocity from Earth's surface $v_e = 11.2$ km/s. It is approximately 33 times the speed of sound (the speed of sound is 0.343 km/s). Consider the energy required to launch a mass of 500 kgs out of Earth's gravitational pull:

$$K.E = \frac{1}{2}mv_e^2 = 0.5 \times 500(kg) \times 11200 \left(\frac{m}{s}\right) \times 11200 = 3.1 \times 10^{10} \ Joules$$

Suppose the mass is launched instantaneously, say in one sec time,

$$The\ power\ required\ =\ 3.1\ \times\ 10^{10}\ /1\ =\ 3.1\ \times\ 10^{10}\ (J/s)\ Watts$$

This is a very huge power requirement, higher than a super thermal plant power generation capacity.

## 2.12  KEPLERIAN MOTION – ORBIT EQUATION

As the motion of the planet lies in a plan, it will be easier to work in polar coordinates. To express the velocity and acceleration in polar coordinates, let $\hat{r}$ and $\hat{\theta}$ be the unit vectors in the direction of increasing r and $\theta$ respectively as shown in Figure 2.12.

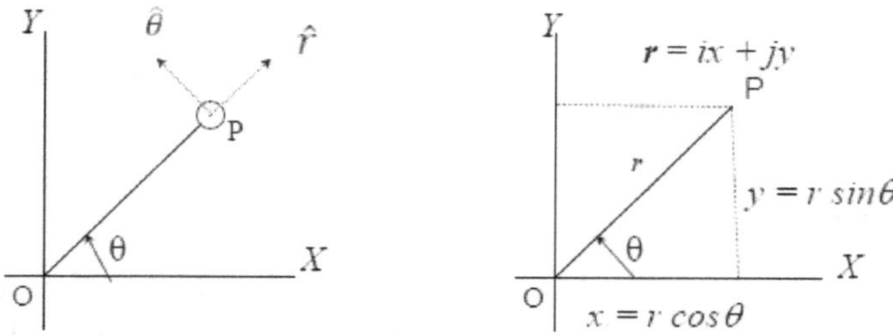


**Figure 2.12** Velocity vectors in Polar Coordinates.

In Cartesian form, these vectors are

$$\hat{r} = i cos\theta + j sin\theta$$

$$\hat{\theta} = -i sin\theta + j cos\theta$$

These vectors are orthogonal at any point and depend on angle $\theta$. Moving in the radial direction doesn't change the position vector, but moving in the angular direction the change is given

$$\frac{d\hat{r}}{d\theta} = -i sin\theta + j cos\theta = \hat{\theta} \qquad .....(2.21)$$

$$\frac{d\hat{r}}{d\theta} = -i sin\theta + j cos\theta = \hat{\theta} \qquad .....(2.22)$$

This means that if the particle moves in a way such that $\theta$ changes with time, then the position vectors themselves will also change with time. The position of a particle is given by:

$$x = r\hat{r}$$

Here, both r and the position vector $\hat{r}$ can change with time. Differentiating with respect to time the above equation gives velocity and second differentiation yields acceleration. The velocity is given by:

$$\frac{dx}{dt} = \frac{dr}{dt}\hat{r} + r\frac{d\hat{r}}{d\theta}\frac{d\theta}{dt} \qquad ......(2.23)$$

Using Eq. (2.21)

$$\dot{x} = \dot{r}\hat{r} + r\dot{\theta}\hat{\theta} \qquad ........(2.24)$$

The second term in the above equation is due to the position vector change with time and is proportional to the angular velocity, $\dot{\theta}$.

Differentiating the above equation results in acceleration in polar coordinates.

$$\frac{d^2x}{dt^2} = \frac{d^2r}{dt^2}\hat{r} + \dot{r}\frac{d\hat{r}}{d\theta}\frac{d\theta}{dt} + \frac{dr}{dt}\dot{\theta}\hat{\theta} + r\frac{d^2\theta}{dt^2}\hat{\theta} + r\dot{\theta}\frac{d\hat{\theta}}{d\theta}\frac{d\theta}{dt}$$

Using Eq. (2.21) & (2.22)

$$\ddot{x} = \ddot{r}\hat{r} + \dot{r}\dot{\theta}\hat{\theta} + \dot{r}\dot{\theta}\hat{\theta} + r\ddot{\theta}\hat{\theta} - r\dot{\theta}^2\hat{r} \qquad \ldots\ldots(2.25)$$

$$\ddot{x} = \left(\ddot{r} - r\dot{\theta}^2\right)\hat{r} + \left(r\ddot{\theta} + 2\dot{r}\dot{\theta}\right)\hat{\theta} \qquad \ldots\ldots(2.26)$$

In the above equation,

$$Radial\ component\ of\ acceleration = \left(\ddot{r} - r\dot{\theta}^2\right)$$

$$Tangential\ component\ of\ acceleration = \left(r\ddot{\theta} + 2\dot{r}\dot{\theta}\right)$$

In equation (2.26), the term $\ddot{r}\hat{r}$ is radial acceleration in radial direction due to change in radial velocity. The term $r\ddot{\theta}\hat{\theta}$ linear acceleration in the tangential direction due to change in the magnitude of the angular velocity. The term $-r\dot{\theta}^2\hat{r}$ is the centripetal acceleration acting towards the center. The term $2\dot{r}\dot{\theta}\hat{\theta}$ is Coriolis acceleration and is present when both r and $\theta$ change with time.

If a particle mass, m moves in the plane under an attractive force $F$ of the planet of mass M acting radially to its center

$$F = -\frac{GMm}{r^2}\hat{r} \qquad \ldots\ldots(2.27)$$

Using Newton's second law $force = mass \times acceleration$,

$$F = m\ddot{x}$$

With Eq. (2.26) and (2.27)

$$-\frac{GMm}{r^2}\hat{r} = m\left(\ddot{r} - r\dot{\theta}^2\right)\hat{r} + m\left(r\ddot{\theta} + 2\dot{r}\dot{\theta}\right)\hat{\theta} \qquad \ldots\ldots(2.28)$$

Equating the components of the vectors on each side of the equality

$$\left(\ddot{r} - r\dot{\theta}^2\right) = -\frac{GM}{r^2}$$

$$\left(r\ddot{\theta} + 2\dot{r}\dot{\theta}\right) = 0$$

The term $\left(r\ddot{\theta} + 2\dot{r}\dot{\theta}\right)$ can be rewritten as

$$\left(r\ddot{\theta} + 2\dot{r}\dot{\theta}\right) = \frac{1}{r}\frac{d}{dt}\left(r^2\dot{\theta}\right) = 0 \qquad \ldots\ldots(2.29)$$

The Eq. (2.28) gives an important factor. The term $r^2\dot{\theta}$ is the expression for the angular momentum $(r.r\,\dot{\theta}, i.e., the\ moment\ of\ momentum)$ of a particle of unit mass at a distance $r$

from the origin with angular velocity. This means that angular momentum is conserved. The specific angular momentum, refer Figure 2.13, is represented by h

$$h = r^2\dot{\theta} = \omega r^2 \qquad .....(2.30)$$

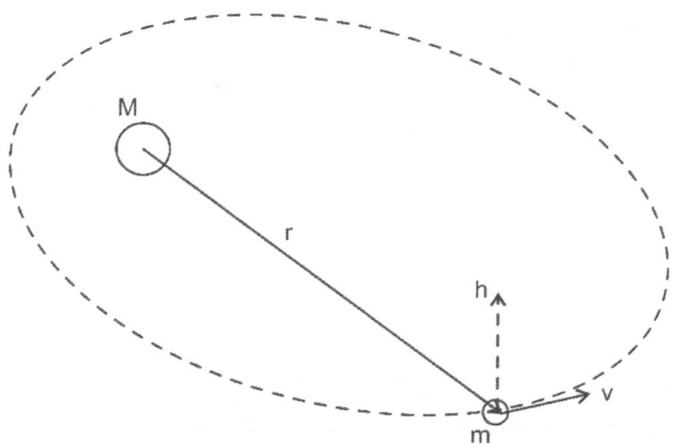

**Figure 2.13** Specific Angular Momentum, h of Mass of m.

where $\omega$ is the angular velocity. For a mass m, the angular momentum, H is

$$H = mr^2\dot{\theta} \qquad .....(2.31)$$

Now, the equation of motion can be written as

$$\left(\ddot{r} - r\dot{\theta}^2\right) = -\frac{GM}{r^2} \qquad ......(2.32)$$

$$\dot{\theta} = \frac{h}{r^2} \qquad .....(2.33)$$

To solve the above equations in a simple way, let u = 1/r. Using chain rule

$$\dot{r} = \frac{d}{dt}\left(\frac{1}{u}\right) = \frac{d}{du}\left(\frac{1}{u}\right)\frac{du}{dt} = -\frac{1}{u^2}\frac{du}{d\theta}\frac{d\theta}{dt} = -r^2\dot{\theta}\frac{du}{d\theta}$$

Using Eq. (2.30)

$$\dot{r} = -h\frac{du}{d\theta} \qquad ......(2.34)$$

Differentiating above equation again using chain rule

$$\ddot{r} = -h\frac{d}{dt}\left(\frac{du}{d\theta}\right) = -h\frac{d^2u}{d\theta^2}\frac{d\theta}{dt} = -h\frac{h}{r^2}\frac{d^2u}{d\theta^2} = -h^2u^2\frac{d^2u}{d\theta^2} \qquad .....(2.35)$$

Substituting the Eq. (2.33) & (2.35) in Eq. (2.32) and replacing 1/r with u

$$-h^2u^2\frac{d^2u}{d\theta^2} - h^2u^3 = -GMu^2$$

$$\frac{d^2u}{d\theta^2} + u = \frac{GM}{h^2} = \frac{\mu}{h^2} \qquad .....(2.36)$$

This is the equation for a harmonic oscillator, with its center displaced by $\mu/h^2$.

The most general solution is

$$u = A\cos(\theta - \theta_0) + \frac{\mu}{h^2} \qquad .....(2.37)$$

with $A$ and $\theta_0$ integration constants. This is the equation of the orbit of mass, m with a central body of mass, M at the focus. At the point where the orbit is closest to the origin (the periapsis), u is largest and its value is $A + \frac{\mu}{h^2}$. This occurs when the polar coordinate orient such that the periapsis occurs at $\theta = 0$, i.e., $\theta_0 = 0$. Replacing u with 1/r and multiplying with $h^2 / \mu$,

$$\frac{h^2}{\mu r} = \frac{Ah^2}{\mu}\cos\theta + 1$$

The integration constant $Ah^2/\mu$ is represented by $e$, is called the eccentricity and it determines the shape of the orbit. It is the vector constant of integration called the eccentricity vector, which lies in the plane of the orbit and $\theta$ is the angle between **r** and **e.** Rearranging the above equation

$$r = \frac{h^2/\mu}{1+e\cos\theta} \qquad .....(2.38)$$

Equation (2.38) is the orbit equation, and it defines the path of the body of mass, m around a planet of mass, M relative to the planet. This is the equation for a conic section in polar coordinates, and the shape of it depends on the value of $e$. The radius $r$ is the distance of a point on the curve from a focus. Figure 2.14 shows the e value of different conic orbits.

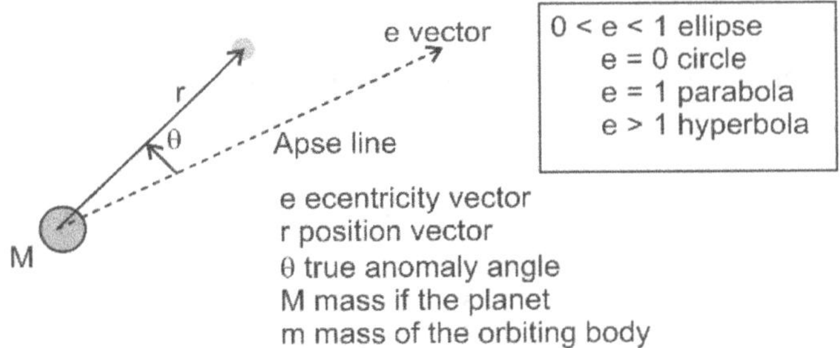


**Figure 2.14** Eccentricity, *e* Value for Conic Sections.

## 2.12.1  Circular Orbits

In section 2.8, the orbital velocity and time period of circular orbit are derived and the Eq. (2.7) and (2.8) give the values for an orbit radius R. The same equations can be derived using orbit equation Eq. (2.38). For circular orbits the eccentricity e = 0. The orbital equation for $e = 0$ is:

$$r = \frac{h^2}{\mu}$$

As $\mu$ and $h$ are constant, $r =$ constant, which means the orbit of mass m around the planet of mass M is a circle. The radial velocity is zero. So,

$$h = rv$$

$$v = \sqrt{\frac{\mu}{r}}$$

The time required for one orbit is known as the time period, T:

$$T = \frac{Circumference}{Speed} = \frac{2\pi r}{\sqrt{\mu/r}} = \frac{2\pi}{\sqrt{\mu}} r^{\frac{3}{2}}$$

The specific energy of a circular orbit is given by Eq. (2.15)

$$\varepsilon = -\frac{\mu}{2r}$$

Obviously, the energy of a circular orbit is negative. As the radius goes up, the energy becomes less negative (i.e., it increases). In other words, the larger the orbit is, the greater is its energy. To launch a satellite from the surface of the earth into a circular orbit requires increasing its specific energy ε. This energy comes from the rocket motors of the launch vehicle. Since the energy of a body of mass m is E= mε, a propulsion system that can place a large mass in a low earth orbit (LEO) can place a smaller mass in a higher earth orbit. Communication satellites and global weather satellites are generally placed in geostationary equatorial orbit (GEO). The speed of the satellite in its circular GEO of radius $r$ is:

$$v_{geo} = \sqrt{\frac{\mu_{earth}}{r_{geo}}}$$

In the circular path the speed of the satellite is related with the absolute angular velocity ω

$$v_{geo} = \omega r_{geo} = \sqrt{\frac{\mu_{earth}}{r_{geo}}}$$

$$r_{geo} = \left(\frac{\mu_{earth}}{\omega^2}\right)^{\frac{1}{3}}$$

Gravitational Parameter $\mu_{earth}$ = 398,600 km³/s²

Angular velocity of Earth ω = 72.9219 × 10⁻⁶ rad/s

$$r_{geo} = \left(\frac{398600}{72.9219 \, x \, 10^{-6}}\right)^{\frac{1}{3}} = 42,164 \; km$$

The distance of the satellite from Earth surface, i.e., height is:

$$height = r_{geo} - Radius \; of \; Earth = 42164 - 6378 = 35,786 \; km.$$

The speed of the satellite is

$$v_{geo} = \sqrt{\frac{\mu_{earth}}{r_{geo}}} = \sqrt{\frac{398600}{42164}} = 3.075 \; km/s$$

**Example 2.3**

Find the orbital velocity and time period of a satellite orbiting the Earth in a circular orbit above at an altitude of 400 km from Earth's surface.

The radius of the satellite from center of Earth $R = R_e + h = 6378 + 400 = 6778$ km

The mass of Earth $M_e = 5.97 \times 10^{24}$ kg

Velocity of the satellite, Eq. (1.10)

$$v = \sqrt{\frac{GM_e}{R}} = \sqrt{\frac{6.67 \times 10^{-11} \times 5.97 \times 10^{24}}{6778000}} = 7665 \; \frac{m}{s} = 7.665 \; km/s$$

The time period of the satellite

$$\tau = \frac{2\pi R}{v} = 2 \times \pi \times \frac{6778000}{7665} = 5553 \; sec = 1.543 \; hrs$$

**Example 2.4**

Find the radius and orbital velocity of the satellite from Earth center having the same time period as Earth rotation, 24 hours (solar day).

Time period, $\tau = 24 \times 60 \times 60$ sec

$G = 6.67 \times 10^{-11}$ N m/kg$^2$; $M_e = 5.97 \times 10^{24}$ kg, $R_e = 6378$ km

Substituting the above values in Eq. (1.11) and solving for radius R

$$R = \sqrt[3]{\left(\frac{\tau}{2\pi}\right)^2 GM_e}$$

$$= \sqrt[3]{\left(\frac{24 \times 60 \times 60}{2\pi}\right)^2 6.67 \times 10^{-11} \times 5.97 \times 10^{24}}$$

$$R = 42241187 \; m = 42241 \; km$$

$$R = R_e + h$$

The height of the satellite from Earth surface $h = 42241 - 6378 = 35863 \; km$

The orbit velocity of the satellite is

$$v = \frac{2\pi R}{\tau} = \frac{2 \times \pi \times 42241000}{24 \times 60 \times 60} = 3070 \frac{m}{s} m/s$$

### 2.12.2 Elliptical Orbits (0 < e < 1)

The ellipse is the general form of a closed orbit and is shown in Figure 2.15. The major axis is called the apse line. Elliptical orbits are by far the most common orbits. All planets and most spacecraft move in elliptical orbits. The periapsis of an orbit is the point of closest approach to the central body or the point of minimum radius. The apoapsis is the point of maximum radius. The apoapsis, periapsis, and center of mass of the central body are joined by the apse line.

Periapsis and apoapsis are general terms for orbits about any central body; there are also body-specific terms:

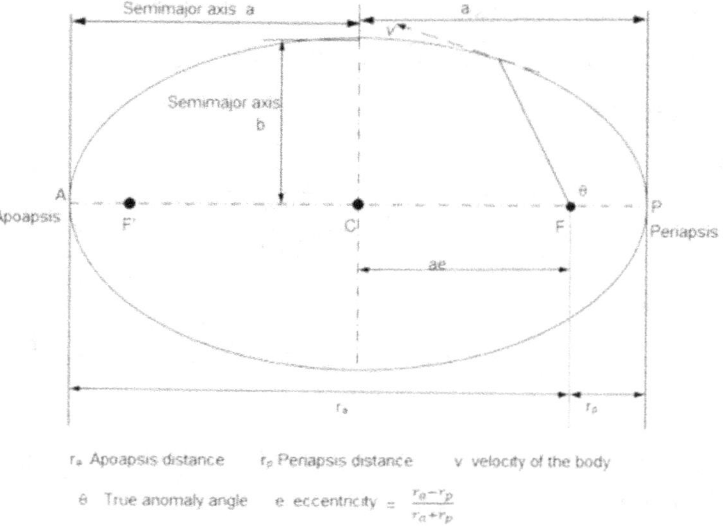


**Figure 2.15** Elliptical Orbit.

General: Periapsis, Apoapsis

Sun: Perihelion, Aphelion

Earth: Perigee, Apogee

Moon: Perilune, Apolune

Elliptical orbit equation is given by Eq. (2.38)

$$r = \frac{h^2/\mu}{1 + e\cos\theta}$$

The position vector radius is minimum when the true anomaly $\theta = 0$, and is equal to periapsis distance $r_p$.

$$r_p = \frac{h^2}{\mu}\frac{1}{1+e} \qquad\qquad \ldots.(2.39)$$

The maximum value of radius $r$ occurs when $\theta = 180$, and its apoapsis distance $r_a$.

$$r_a = \frac{h^2}{\mu}\frac{1}{1-e} \qquad\qquad (2.40)$$

The semimajor axis a is

$$a = \frac{r_a + r_p}{2}$$

$$a = \frac{h^2}{\mu}\frac{1}{1-e^2} \qquad \ldots(2.41)$$

Using above two equations,

$$r_a = a(1 + e) \qquad \ldots(2.42)$$

Using Eq. (2.38) and (2.41), the radius of position vector r

$$r = \frac{a(1-e^2)}{1+ecos\theta} = \frac{r_p(1+e)}{1+ecos\theta} \qquad \ldots(2.43)$$

$$r_a = r_p \frac{(1+e)}{(1-e)} \qquad \ldots(2.44)$$

Solving for $e$, the value expressed in terms of periapsis and apoapsis is:

$$e = \frac{r_a - r_p}{r_a + r_p} = \frac{Distance\ betwee\ the foci}{Length\ of\ the\ major\ axis} \qquad \ldots\ldots(2.45)$$

If $v_t$ is tangential velocity of orbit, the magnitude of specific angular velocity is

$$h = rv_t$$

Figure 2.16 illustrates the flight path angle of the body orbiting, $\gamma$ the angle between the local tangent of the orbit and velocity vector. It also shows the value of $\gamma$ as a function of position on the orbit path. The specific angular momentum, $h$ now can be expressed as of function of flight path angle:

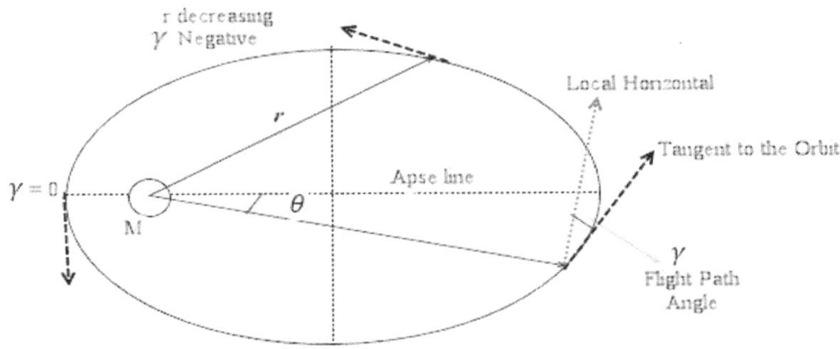


**Figure 2.16** Flight Path Angle.

$$h = rv\ cos\gamma \qquad \ldots(2.46)$$

where

$h$ = magnitude of the specific angular momentum, km²/s

$r$ = magnitude of the radius vector (the distance from the spacecraft to the center of mass of the central body M), km

$v$ = magnitude of the velocity vector, km/s

$\gamma$ = flight path angle, deg

The specific angular momentum is conserved (i.e., constant):

$$rv \cos\gamma = constant$$

Since flight path angle is zero at periapsis and apoapsis

$$r_p v_p = r_a v_a \qquad \qquad .....(2.47)$$

where $v_p$ and $v_a$ are velocities at periapsis and apoapsis.

The specific energy of the orbit is $\varepsilon = specific\ kinetic\ energy + specific\ potential\ enrgy$

Using Eq. (2.14)

$$-\frac{\mu}{2a} = \frac{v^2}{2} - \frac{\mu}{r}$$

Rearranging the above equation, the velocity $v$ of the elliptical orbit at radius r is

$$v = \sqrt{\left(\frac{2\mu}{r} - \frac{\mu}{a}\right)} \qquad \qquad .....(2.48)$$

From Figure 2.16, the flight path angle, $\gamma$ is

$$tan\gamma = \frac{v_r}{v_t}$$

Using Eq. (2.39)

$$v_t = \frac{h}{r} = \frac{\mu}{h}(1 + e\ cos\theta) \qquad \qquad .....(2.49)$$

Since $v_r = \dot{r}$, differentiating the orbit Eq. (2.38) and using the relation $\dot{\theta} = h/r^2$ from Eq. (2.33)

$$\dot{r} = \frac{d}{dt}\left[\frac{h^2}{\mu}\frac{1}{1+ecos\theta}\right] = \frac{h^2}{\mu}\left[-\frac{e(-\theta sin\theta)}{(1+ecos\theta)^2}\dot{\theta}\right] = \frac{h^2}{\mu}\frac{esin\theta}{(1+ecos\theta)^2}\frac{h}{r^2} \qquad \qquad .....(2.50)$$

$$v_r = \frac{\mu}{h}esin\theta \qquad \qquad .....(2.51)$$

Therefore, using Eq. (2.49) and (2.51)

$$tan\gamma = \frac{v_r}{v_t} = \frac{esin\theta}{1+ecos\theta} \qquad \qquad .....(2.52)$$

As shown in Figure 2.16, the flight path angle, $\gamma$ is positive when the spacecraft is moving away from periapsis and is negative when the spacecraft is moving towards periapsis.

The orbit Eq. (2.38) gives the position of the rotating body (spacecraft) mass m in the orbit around mass M as a function of true anomaly. For many practical applications like ground station passes, maneuvers and other mission activities the position of the body as a function time is required. It is convenient to express the true anomaly in terms of an auxiliary angle E called the eccentric anomaly, as shown in Figure 2.17. The relation can be established by drawing an auxiliary circle having a radius equal to the semimajor axis of the ellipse. The point S is on the ellipse with true anomaly $\theta$. Through point S we pass a perpendicular to the apse line, intersecting the auxiliary circle at point Q and the apse line at point V. The angle between the apse line and the radius drawn from the center of the circle to Q on its circumference is the eccentric anomaly E. Some of the important relations are presented here without derivation.

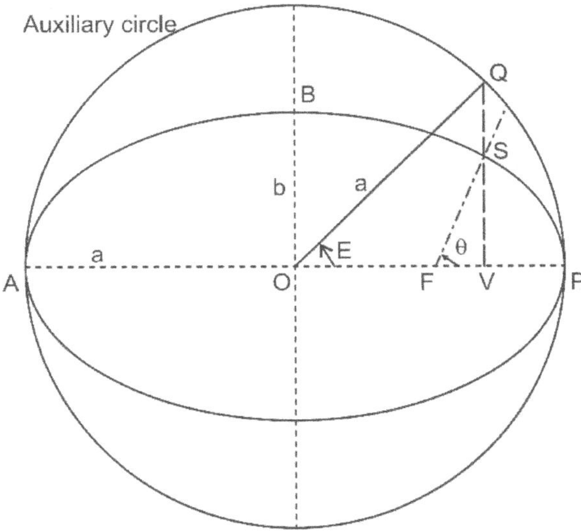


**Figure 2.17**  Eccentric Anomaly Angle E.

The relation between E and $\theta$ can be expressed as

$$\cos\theta = \frac{e - cosE}{ecosE - 1} \qquad\qquad .....(2.53)$$

$$E = \left( \sqrt{\frac{1-e}{1+e}}\ tan\frac{\theta}{2} \right) \qquad\qquad .....(2.54)$$

The angular velocity of the position vector of an elliptical orbit is not constant, but since $2\pi$ radians are swept out per period T, the ratio $2\pi/T$ is the average angular velocity, which is given the symbol $n$ and called the mean motion,

$$n = 2\pi/T \qquad\qquad ......(2.55)$$

The time taken by a spacecraft to move from periapsis to a given true anomaly (time since periapsis) is computed using the famous Kepler equation

$$t = \frac{(E - esinE)}{n} \qquad\qquad ....(2.56)$$

where $t$ = time since periapsis, $E$ = eccentric anomaly, rad, $e$ = eccentricity of orbit and $n$ = mean motion.

### 2.12.3    Parabolic Orbit (e = 1)

If the eccentricity equals 1, then the orbit equation Eq. (2.38) becomes

$$r = \frac{h^2}{\mu}\frac{1}{1+\cos\theta} \qquad .....(2.57)$$

As the true anomaly $\theta$ approaches 180 degrees, the denominator approaches zero, so that r tends toward infinity. A parabola can be considered an ellipse with an infinite semimajor axis. The arms become parallel as $r$ approaches infinity and $a = \infty$. The specific energy of the orbit $\varepsilon$ is zero as per Eq. (2.16).

$$\epsilon = 0 = \frac{v^2}{2} - \frac{\mu}{r}$$

The speed of the mass m anywhere in the orbit is

$$v = \sqrt{\frac{2\mu}{r}} \qquad .....(2.58)$$

If a body of mass m is launched on a parabolic trajectory, it will coast to infinity. Parabolic paths are called escape trajectories. If $v_c$ is the velocity in the circular orbit

$$v_{esc} = \sqrt{\frac{2\mu}{r}} = \sqrt{2}v_c \qquad .....(2.59)$$

The above equation means, to escape from a circular orbit a velocity boost of 41.4% is required. In two body theory, it is assumed that only two masses m and M are in the universe and the effect of other masses on them are neglected. A spacecraft launched from the earth with a velocity $v_{esc}$ will not coast to infinity. i.e., will not leave the solar system. This is because during its travel it will ultimately yield to the gravitational influence of the sun. It will merely end up in the same orbit as the Earth.

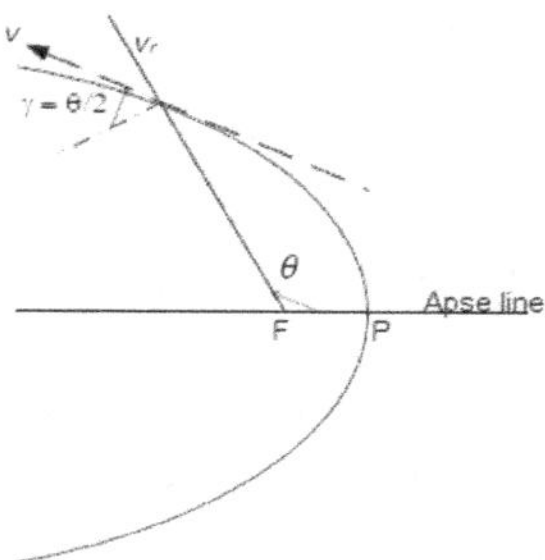


**Figure 2.18** Parabolic Trajectory.

The flight path angle $\gamma$ for parabolic orbit will be obtained by substituting e = 1 in Eq. (2.52).

$$\tan\gamma = \frac{\sin\theta}{1+\cos\theta} \qquad .....(2.60)$$

$$= \frac{2sin\frac{\theta}{2}cos\frac{\theta}{2}}{1+2\frac{\theta}{2}-1}$$

$$tan\gamma = tan\frac{\theta}{2}$$

It follows that

$$\gamma = \frac{\theta}{2} \qquad\qquad .....(2.61)$$

i.e., on parabolic trajectories the flight path angle is always one-half of the true anomaly as shown Figure 2.18.

### 2.12.4    Hyperbolic Orbit (e > 1)

Hyperbolic orbits are used for Earth departure on planetary flights and for planetary arrival. All hyperbolas are escape trajectories as at any point on its trajectory the velocity is higher than it would be on a parabolic orbit. The orbit formula of a hyperbolic orbit trajectory from Eq. (2.38) is

$$r = \frac{h^2}{\mu}\frac{1}{1+ecos\theta}$$

As shown in Figure 2.19, it consists of two symmetric curves. The orbiting body occupies one of them and is known as physical hyperbola. The other one is a mathematical image and it is empty. The point P on the apse line of the physical hyperbola is periapsis. The apoapsis A lies on the apse line of image hyperbola. The center C of hyperbola is halfway between periapsis and apoapsis. The two asymptotes of the hyperbola intersect at C making an angle β to apse line. The eccentricity, e is CF/a, greater than one. Important parameters of the hyperbola are marked in the Figure 2.19.

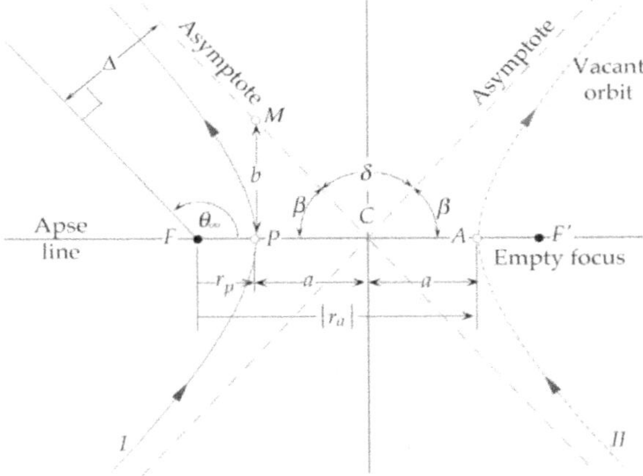


**Figure 2.19** Hyperbola Trajectory.

The radius of the orbit, $r$ reaches infinity when $1 + ecos\theta = 0$. i.e.,

$$cos\theta = -1/e$$

This value of true anomaly is represented by

$$\theta_\infty = \left(\frac{1}{e}\right) \qquad \qquad .....(2.62)$$

The physical trajectory lies in the range of $-\theta\infty < \theta < \theta\infty$,

The asymptote angles

$$\beta = 180 - \theta_\infty$$

$$cos\beta = cos\ (180 - \theta_m) = -cos\theta_m$$

$$\beta = (1/e) \qquad \qquad .....(2.63)$$

i.e.,

The distance of periapsis from focus, $r_p$ is obtained from the orbit equation.by substituting $\theta$=0.

$$r_p = \frac{h^2}{\mu}\frac{1}{1+e} \qquad \qquad .....(2.64)$$

The apoapsis distance, $r_a$ is obtained by taking $\theta = 180$ deg.

$$r_a = \frac{h^2}{\mu}\frac{1}{1-e} \qquad \qquad .....(2.65)$$

As $e > 1$, $r_a$ is negative and it means that apoapsis lies to the right of the focus.

The major axis $2a = -r_a - r_p$, substituting the values and simplifying

$$a = \frac{h^2}{\mu}\frac{1}{e^2-1} \qquad \qquad .....(2.66)$$

Other important relations that can be derived easily are

$$r = a\frac{e^2-1}{1+ecos\theta} \qquad \qquad .....(2.67)$$

$$r_p = a(e - 1) \qquad \qquad ......(2.68)$$

$$r_a = -a(e + 1) \qquad \qquad .....(2.69)$$

The other important parameter is the aiming radius, $\Delta$ the distance between the asymptote and parallel line through the focus as shown in Figure 2.19.

$$\Delta = aesin\beta = aecos\theta_\infty = ae\sqrt{1 - \theta_\infty}$$

$$\Delta = ae\sqrt{1 - \frac{1}{e^2}} = a\sqrt{e^2 - 1} \qquad \qquad .....(2.70)$$

The specific energy for hyperbolic trajectory is

$$\epsilon = \frac{\mu}{2a} = \frac{v^2}{2} - \frac{\mu}{r}$$

The specific energy of a hyperbolic orbit is clearly positive and independent of the eccentricity. If $v_\infty$ is denote the speed at which a body on a hyperbolic path arrives at infinity, i.e., $r = \infty$

$$v_\infty = \sqrt{\frac{\mu}{a}} \qquad\qquad\qquad .....(2.71)$$

$v_\infty$ is called the hyperbolic excess speed. In terms $v_\infty$ of the specific energy relation is

$$\frac{v^2}{2} - \frac{\mu}{r} = \frac{v_\infty^2}{2}$$

The escape velocity $v_{esc}$ is

$$v_{esc} = \sqrt{\frac{2\mu}{r}}$$

Therefore,

$$v^2 = v_{esc}^2 + v_\infty^2 \qquad\qquad\qquad ......(2.72)$$

The hyperbolic excess velocity, $v_\infty$ represents the excess kinetic energy over that is required to simply escape from the center of attraction. The square of $v_\infty$ is denoted C₃, and is known as the characteristic energy,

$$C_3 = v_\infty^2 \qquad\qquad\qquad .....(2.73)$$

C₃ is a measure of the energy required for an interplanetary mission. The time since periapsis can be determined in a manner analogous to that for elliptical orbits with the aid of the hyperbolic eccentric anomaly $F$.

$$t = (e \sinh F - F)/n \qquad\qquad\qquad .....(2.74)$$

In terms of true anomaly $\theta$ and eccentricity $e$,

$$\sinh F = \frac{\sin\theta \, \sqrt{e^2-1}}{1+e\cos\theta}$$

## Example 2.5

As an orbit-raising maneuver of a spacecraft, a liquid rocket engine is fired at an altitude 336 km at perigee. The altitude at apogee has been raised to 74,715 km. Find the velocities of spacecraft at perigee and apogee

The radius Earth $R_e = 6378$ km

The perigee $r_p = altitude + radius\ of\ Earth = 336+6378 = 6714$ km $= 6714000$ m

The apogee $r_a = 74715 + 6378 = 81093$ km $= 81093000$ m

Major axis of elliptical orbit

$$r_p + r_a = 2a = 87807\ km = 87807000\ m$$

Gravitational parameter of Earth

$$\mu = G\ Me = 6.67 \times 10^{-11} \times 5.97 \times 1024 = 3.982 \times 10^{14}$$

The velocity of the spacecraft at perigee from Eq. (2.48)

$$v_p = \sqrt{\frac{2\mu r_a}{r_p(r_a+r_p)}}$$

$$v_p = \sqrt{\frac{2\times3.982\times10^{14}\times81093000}{6714000*87807000}} = 10467 \; m/s$$

The velocity of spacecraft at apogee is

$$v_a = \sqrt{\frac{2GM_e r_p}{r_a(r_a+r_p)}}$$

$$v_a = \sqrt{\frac{2\times3.982\times10^{14}\times6714000}{81093000\times87807000}} = 867 \; m/s$$

Eccentricity of the elliptical path from Eq. (2.42)

$$r_a = a(1+e)$$

$$e = \frac{r_a}{a} - 1 = \frac{81093}{\frac{87807}{2}} - 1 = 0.847$$

## 2.13  LAUNCH OF A SPACE VEHICLE

The launch of a satellite or space vehicle consists of a period of powered flight during which the vehicle is lifted above the Earth's atmosphere and accelerated to orbital velocity by a launch vehicle. Powered flight concludes at burnout of the rocket's last stage at which time the vehicle begins its free flight. During free flight the space vehicle is assumed to be subjected only to the gravitational pull of the Earth. A vehicle's position and velocity can be described by the

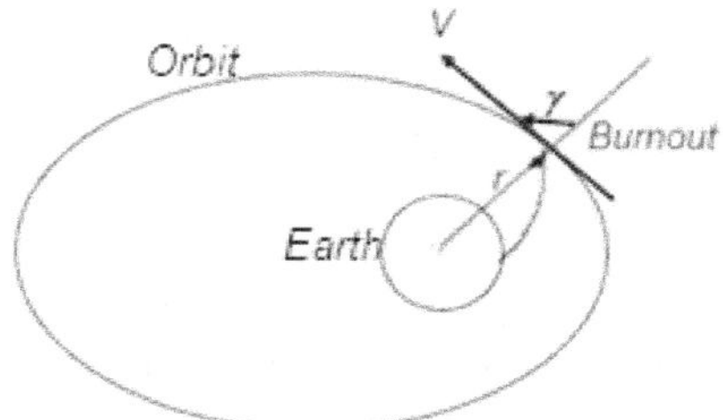


**Figure 2.20** Burnout Position.

variables $r$, $\gamma$ and $v$, where $r$ is the vehicle's distance from the center of the Earth, $v$ is its velocity, and $\gamma$ is the angle between the position and the velocity vectors called zenith angle as shown in Figure 2.20. As general practice, the spacecraft launches are usually terminated at either perigee or apogee, i.e., $\gamma = 90$. This condition results in the minimum use of propellant.

Let $r_1, \gamma_1$ and $v_1$ are the launch values. The specific angular momentum, h, is constant and is given by Eq. (2.46). At perigee $\gamma$ is zero.

$$h = v_p r_p = r_1 v_1 sin\gamma_1$$

$$v_p = \frac{r_1 v_1 sin\gamma_1}{r_p}$$

The specific energy

$$\varepsilon = \frac{v^2}{2} - \frac{\mu}{r} = constant$$

Applying for perigee and at burnout point

$$\frac{v_p^2}{2} - \frac{v_1^2}{2} = \mu\left(\frac{1}{r_p} - \frac{1}{r_1}\right) \qquad\qquad .....(2.75)$$

$$\left(\frac{r_1 v_1 sin\gamma_1}{r_p}\right)^2 - v_1^2 = 2\mu\left(\frac{1}{r_p} - \frac{1}{r_1}\right)$$

By simplifying and rearranging terms, we get the quadratic equation in

$$\left(\frac{r_p}{r_1}\right)^2 (1 - K) + \left(\frac{r_p}{r_1}\right) K - \gamma_1 = 0$$

where

$$K = \frac{2\mu}{r_1 v_1^2}$$

K is a non-dimensional parameter of the orbit.

The solution of this equation gives the values of $r_p$ and $r_a$.

$$\frac{r_p}{r_1} = \frac{-K \pm \sqrt{K^2 - 4(1-K)(-\gamma_1)}}{2(1-K)}$$

Like any quadratic, the above equation yields two answers. The smaller of the two answers corresponds to $r_p$ , the perigee radius. The other root corresponds to the apogee radius, $r_a$ .

**Example 2.6**

A spacecraft is launched into Earth orbit where its launch vehicle burns out at an altitude of 300 km and launch velocity is 8200 m/s with a zenith angle of 85 degrees. Find the altitude of the spacecraft at perigee and apogee.

**Given**

Radius of the spacecraft at burnout

$$r1 = altitude + radius\ of\ the\ Earth = 300 + 6378 = 6678\ km = 6678000\ m$$

Launch velocity at burnout $v_1 = 8200\ m/s$

Zenith angle $\gamma_1 = 85$ degrees

The constant K

$$K = \frac{2\mu}{r_1 v_1^2}$$

$$K = \frac{2 \times 3.982 \times 10^{14}}{6678000 \times 8200^2} = 1.7736$$

$$\frac{r_p}{r_1} = \frac{-K \pm \sqrt{K^2 - 4(1-K)(-\gamma_1)}}{2(1-K)}$$

$$\frac{r_p}{r_1} = \frac{-1.7736 \pm \sqrt{1.7736^2 - 4 \times -0.7736 \times (-85)}}{-2 \times 0.7736}$$

$$\frac{r_p}{r_1} = 0.9699 \; and \; 1.323$$

$$r_p = 0.9699 \times 6678000 = 6477012 \; m = 6477 \text{ km}$$

Altitude at perigee = 6477 – 6378 = 99 km

$$r_a = 1.3228 \times 6678 = 8828 \; km$$

Altitude at apogee = 8828 – 6378 = 2450 km.

The major axis 2a = 8828 + 99 = 8927 km

$$r_a = a(1+e)$$

$$e = \frac{r_a}{a} - 1 = \frac{8828}{\dfrac{8927}{2}} - 1 = 0.977$$

## 2.14  REFERENCE FRAMES

The earth's axis of rotation is tilted away by an angle approximately 23.4°. The earth's equatorial plane and the ecliptic orbit intersect along a line, which is known as the vernal equinox line. The Equinox is defined as the point where the equatorial plane intersects the ecliptic plane. There are two equinoxes in a year, one in the spring and one in the fall. The vernal equinox vector is the vector from the center of mass of the Earth to the center of mass of the Sun on the equinox day as shown in Figure 2.21. For many practical purposes, the vernal equinox line may be considered fixed in space.

A frame of reference consists of a coordinate system and the set of physical reference points that uniquely fix the system. The coordinate system used to describe earth orbits in three dimensions is defined in terms of earth's equatorial plane, the ecliptic plane, and the earth's axis of rotation. Each system is defined by the selection of the origin, selection of axes, and the determination of what is fixed. A body-fixed coordinate system measures all motion relative to that body with the assumption that the body is stationary. An inertially fixed coordinate system is one which is referenced to stellar positions. The vernal equinox vector is the primary reference in such systems.

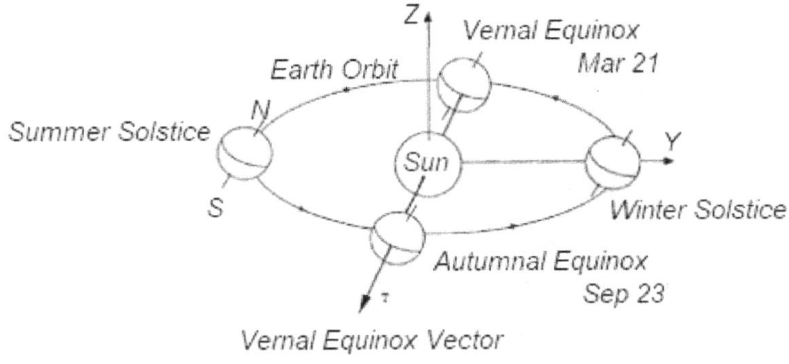


**Figure 2.21**  Vernal Equinox Vector.

## 2.14.1    Geocentric-Inertial Coordinate System

It is the generally used coordinate system for orbital calculations and details are shown in Figure 2.22.

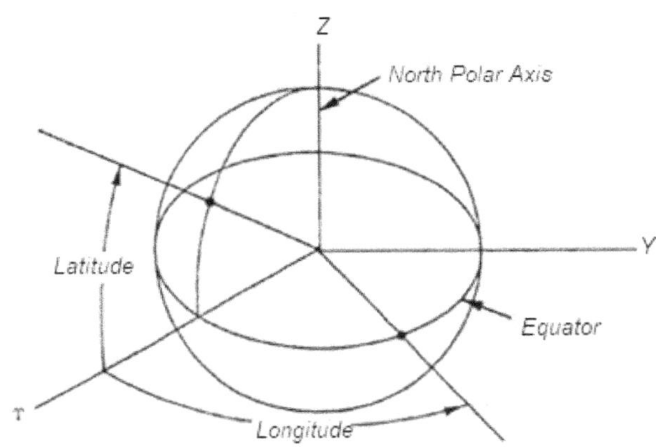


**Figure 2.22**  Geocentric-inertial Coordinate System.

| | |
|---|---|
| Origin | Center of mass of the central body, the Earth. |
| Reference plane | Equatorial plane |
| X axis | Vernal equinox vector |
| Z axis | Spin axis of the Earth, positive towards north. |
| Y axis | Perpendicular to x axis and z axis. |

Positions are measured by latitude and longitude; longitudes are measured from the vernal equinox vector and centered in the Earth. North latitudes are measured in the positive $Z$ direction from the equatorial plane, and south latitudes are measured in a negative $Z$ direction.

### 2.14.2 Heliocentric-Inertial System

It used for interplanetary mission design.

| | |
|---|---|
| Origin | Center of the Sun |
| Reference Plane | The ecliptic plane, the plane that contains the center of mass of the sun and the orbit of the Earth |
| X axis | Vernal equinox vector |
| Z axis | Perpendicular to ecliptic, positive towards north. |
| Y axis | Perpendicular to x axis and z axis. |

Latitude and longitude are measured as in the geocentric system.

## 2.15 ORBITAL ELEMENTS

A standard way of specifying an orbit is to use orbital elements, which refer the orbit to a frame of reference that is fixed relative to the stars. The six orbital elements of a spacecraft's orbit around the Earth as shown in Figure 2.23 represents different features. The orbital plane intersects the equatorial plane in a line called the line of nodes. The ascending node is the point on the equator at which the spacecraft moves from the southern into the northern hemisphere. The descending node is at the other end of the node line at which the spacecraft moves from the northern to southern hemisphere.

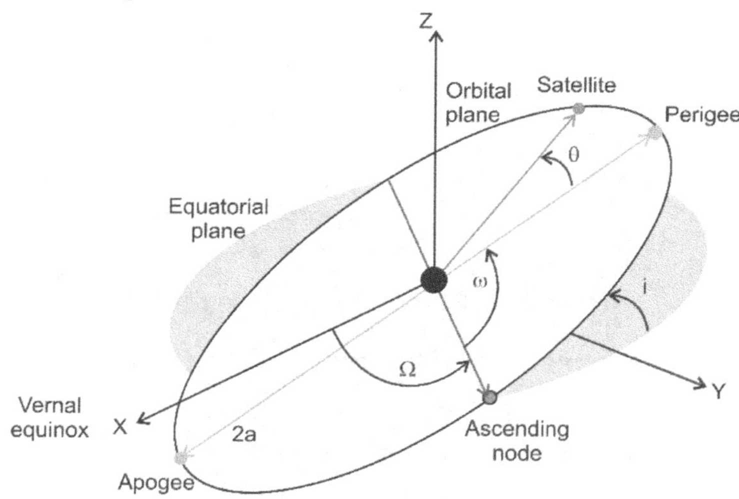


**Figure 2.23** Orbital Elements.

Referring to the Figure 2.23 the important orbital elements are:

Eccentricity of the orbit, e

Semimajor axis of the orbit, a

Inclination, i: the angle between the orbit plane and the reference plane or the angle between the normal to the two planes.

*Argument of periapsis*, $\omega$: the angle from the ascending node to the periapsis, measured in the orbital plane in the direction of spacecraft motion.

*Longitude of the ascending node* $\Omega$: the angle between the vernal equinox vector and the ascending node measured in the reference plane in a counterclockwise direction as viewed from the northern hemisphere.

True anomaly, $\theta$: this element locates the spacecraft position on the orbit. The time $t_p$ of last passage through the periapsis or the mean anomaly $M$ may be used instead of $\theta$.

The orbital parameters $a$ and $e$ define the shape of the orbit; $i$ and $\omega$ define the plane of the orbit.

If the inclination is 90 deg the orbit is polar orbit. For orbits about the Earth or planets, the elements are located with respect to the geocentric system. For interplanetary orbits the elements are given with respect to the heliocentric system. The coordinate system, the orbital elements, and the orbit itself are fixed in inertial space and do not rotate with the central body.

## Nomenclature

$a$   semi-major axis

$e$   eccentricity of a conic section

$E$   energy (J), Eccentric anomaly angle (rad, deg)

$E_k$ kinetic energy (J)

$E_p$ potential energy (J)

$F$   force (N, kg m/s$^2$)

$F_c$ centrifugal force (N)

$g$   acceleration due to gravity (m/s$^2$)

$G$   universal gravitational constant ($6.67408 \times 10^{-11}$ N m$^2$/ kg$^2$)

$h$   height (m), convective heat transfer coefficient

$h$   specific angular momentum (km$^2$/s)

$I$   impulse (Ns or kg m/s)

$K$   constant

$M, m$   mass (kg)

$M_e$ Mass of the Earth ($5.97 \times 10^{24}$ $kg$)

$n$   mean motion (rad/s)

$P$   pressure (Pa), momentum (kg m/s)

$R$   Radius (m)

$R_e$ Radius of the Earth (6,378,000 m)

$T$  time period of an orbit (s)

$v$  velocity (m/s)

$v_a$  velocity at apoapsis (m/s)

$v_e$  escape velocity; exhaust velocity (m/s)

$v_p$  velocity at periapsis (m/s)

$v_t$  total velocity; tangential velocity (m/s)

$v_\infty$ hyperbolic excess velocity (m/s)

$\tau$  time (s)

$\alpha$  acceleration (m/s$^2$), growth rate of the wave

$\omega$  angular velocity (rad/s)

$\mu$  gravitational parameter of the planet (km$^3$/s$^2$)

$\varepsilon$  specific energy (J/kg)

$\varepsilon_k$  specific kinetic energy (J/kg)

$\varepsilon_p$  specific potential energy (J/kg)

$\theta$  polar coordinate; angular rotation (rad)

## PROBLEMS

2.1  Calculate the altitude and speed of a geostationary Earth satellite.

2.2  A spacecraft is placed in a circular orbit at an altitude $h$ from Earth's surface. Examine and plot the variation of spacecraft speed and time period as a function of altitude $h$.

2.3  A spacecraft is in a circular orbit of Mars at an altitude of 300 km. Calculate its speed and its period.

2.4  A spacecraft moving in an Earth's orbit. At two points of the orbit, with reference to geocentric coordinate system, the altitude and true anomaly are 1500 km, 120° and 800 km, 50°. Calculate the eccentricity, semimajor axis, the period and altitude of perigee.

2.5  A particular point on the surface of planet Mars is to be viewed continuously from a spacecraft orbiting Mars. Determine the height of the orbit of this spacecraft above the surface of Mars. You can assume the period of rotation of Mars about its axis to be 24 hours and 40 minutes.

2.6  A satellite is placed in an elliptical orbit around the Earth. The altitude of the satellite at perigee 200 km and at apogee 800 km. Calculate the eccentricity, orbital velocity at perigee and apogee and the period of the orbit.

2.7  A satellite is in polar orbit around the Earth. The perigee is at 300 km of the north pole and arrives at this point once every 90 minutes. Find the eccentricity of the orbit.

2.8  At a given point on the spacecraft's geocentric trajectory, the radius is 15,000 km and the flight path angle is 50°. Find the type of the orbit and its different parameters.

# Rocket Principle and Performance Parameters

## 3.1   INTRODUCTION

The propulsion Engineer, regularly, involved in preliminary sizing of the entire vehicle to achieve the objectives of a well-defined space mission. The propulsion system makes up the bulk of the size and weight of the rocket vehicle. This chapter illustrates the basic working principle of rocket propulsion. The famous Tsiolkovsky's rocket equation is derived and important performance parameters are explained. The advantage of multistage rocket vehicles is illustrated with an example.

## 3.2   BASIC PRINCIPLE OF ROCKET

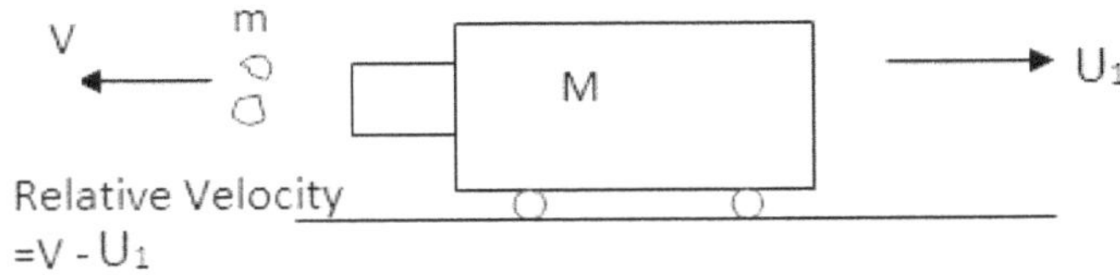


**Figure 3.1** Motion of the Box with Two Stones Ejected simultaneously (Case 1).

To understand how a rocket achieves continuous increment of velocity, i.e., to achieve acceleration, consider a box with stones and mechanism to eject stones as shown in Figure (3.1). Let the total mass of the box with stones be $M$, the mass of each stone is $m$ and it is moving on a frictionless surface. The effect of ejecting stones from the box is assessed in two cases: (1) two stones are simultaneously ejected and (2) one after the other stone is ejected.

***Case* 1:** The box is initially at rest on the frictionless surface. After ejecting the two stones simultaneously with a velocity $V$, the box moves with a velocity $U_1$. The value of the box velocity $U_1$ will be expressed in terms of masses $M$ and $m$. The momentum of the system is conserved. i.e.,

$$Initial\ momentum\ =\ final\ momentum$$

Initially the box is at rest. Therefore,

$$Initial\ momentum = 0$$

The relative velocity of the stones with respect to the box is

$$velocity\ of\ the\ stones\ V - velocity\ of\ the\ box\ U_1 = (V - U_1)$$

The final momentum is the sum of momentum of the box and the stones.

$$Final\ momentum = momentum\ of\ box + momentum\ of\ stones$$

$$= (M - 2m)U_1 - 2m(V - U_1) = 0$$

$$MU_1 - 2mU_1 - 2mV + 2mU_1 = 0$$

$$U_1 = \frac{2m}{M}V \qquad \qquad \qquad .....(3.1)$$

Rewriting above equation

$$U_1 = \left(\frac{m}{M} + \frac{m}{M}\right)V \qquad \qquad .....(3.2)$$

The relative velocity of the stones using

$$V - U_1 = V\left(1 - \frac{2m}{M}\right) \qquad \qquad .....(3.3)$$

**Case 2:** The stones are thrown one after another with a velocity $V$ each as shown in Figure 3.2. The velocity box after the first stone of mass $m$ is ejected can be calculated using Eq. (3.1)

$$U_1 = \frac{m}{M}V$$

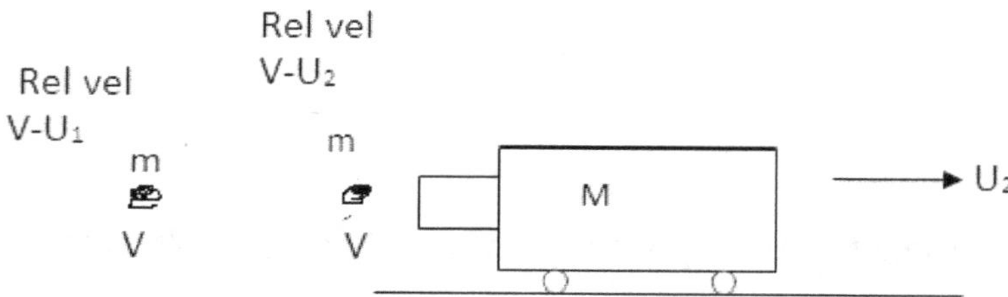


**Figure 3.2** Motion of the Box with the Stones Ejected one after Another (Case 2).

Relative velocity of first stone using Eq. (3.3)

$$V - U_1 = V\left(1 - \frac{m}{M}\right)$$

Let $U_2$ be the velocity of the box after the second stone of mass $m$ is ejected. Balancing the momentum.

$$(M - 2m)U_2 - mV\left(1 - \frac{m}{M}\right) - m(V - U_2) = 0$$

$$(M - m)U_2 = mV \left(\frac{M - m}{M}\right) + mV$$

Rearranging

$$U_2 = V \left(\frac{m}{M} + \frac{m}{M-m}\right) \qquad\qquad .....(3.4)$$

In the RHS of the above equation, the value of the second term is higher than the second term of Eq. (3.2):

$$\frac{m}{M - m} > \frac{m}{M}$$

Therefore,

$$U_2 > U_1$$

i.e., the velocity achieved by the box by ejecting the mass $m$ one after another is greater than the velocity ejecting the two stones at a time. Hence, if a mass $m$ is ejecting continuously the velocity achieved is

$$U = V \left(\frac{m}{M} + \frac{m}{M - m} + \frac{m}{M - 2m} + \frac{m}{M - 3m} + \cdots\cdots\right)$$

This is the basic principle of a rocket. A rocket is a device, continually ejecting a mass $m$ (known as propellant) to achieve high velocity and thereby acceleration to propel forward.

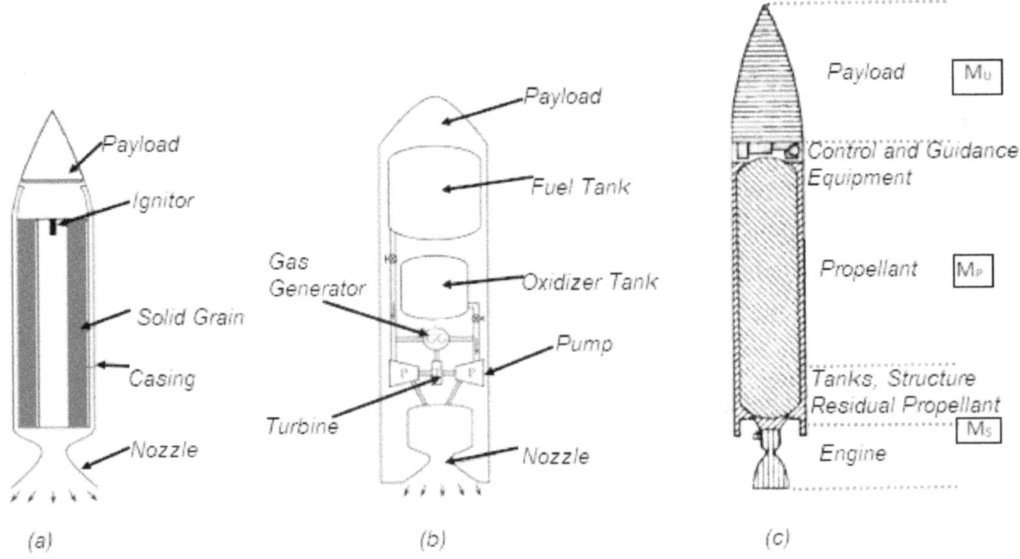


**Figure 3.3** (a) Solid Rocket (b) Liquid Rocket (c) Equivalent Masses Definition.

## 3.3   MASS RATIOS OF ROCKET

The launch vehicle consists of either solid or liquid propulsion unit, payload and other supporting systems like control and guidance unit. The general layout of the vehicle is shown in Figure 3.3. The solid propulsion unit mainly consists of the propellant grain, ignition unit, nozzle and supporting structural elements like casing and assembling units. The liquid propulsion unit mainly consists of the fuel, oxidizer, nozzle, the supporting structures like propellant tanks, piping and turbopump (pumps and turbine).  The total mass of a rocket vehicle is considered to be divided into three effective masses. The figure shows the general layout of the rocket vehicle and definition of masses.

$M_u$ is the useful mass, i.e., the payload mass to put in orbit or land on a planet.

$M_p$ is the propellant mass consisting of both oxidizer and fuel.

$M_s$ is the structural mass, i.e., mass of propellant, engine, guidance and control system and other structural mass.

The initial mass of the system is $M_i$

$$M_i = M_u + M_p + M_s \qquad \qquad .....(3.5)$$

If all the mass of propellant is consumed during firing, the resultant left out mass is called burn out mass or the final mass $M_f$.

$$M_f = M_u + M_s \qquad \qquad .....(3.6)$$

The mass ratio $R$ is defined as:

$$R = \frac{M_i}{M_f} \qquad \qquad .....(3.7)$$

For convenience the payload mass , structural mass , and propellant mass are expressed as ratio of initial mass $M_i$.

$$\alpha = \frac{M_u}{M_i} \qquad \qquad .....(3.8)$$

$$\beta = \frac{M_s}{M_i} \qquad \qquad .....(3.9)$$

$$\gamma = \frac{M_p}{M_i} \qquad \qquad .....(3.10)$$

The mass ratio R can be written as

$$R = \frac{1}{\alpha + \beta} \qquad \qquad .....(3.11)$$

For a chemical rocket the structural coefficient is a measure of the vehicle designer's skill in designing very light tanks and support structure.

## 3.4   ROCKET EQUATION

One of the most fundamental results of rocketry is the equation that relates the impulse provided by a rocket (more accurately, the change in velocity or $\Delta U$) with the amount of propellant spent in a launch. This is the so-called rocket equation or Tsiolkovsky's equation. Consider a rocket vehicle of mass $m$, expelling combustion products at the rate of $\dot{m}$, with constant exhaust velocity $V_e$. The mass of the vehicle is decreasing at the rate of $\dot{m}$. Due to the thrust $F$, developed by the exhaust, the rocket is accelerating. A schematic diagram of position of the rocket at time $t$ and $t + \delta t$ is shown in Figure (3.4). At time $t$, the total mass of the rocket is $m$ and the velocity is $U$. At time $t + \delta t$ its velocity is increased to $U + \delta U$ and the mass is reduced to $m - \delta m$.

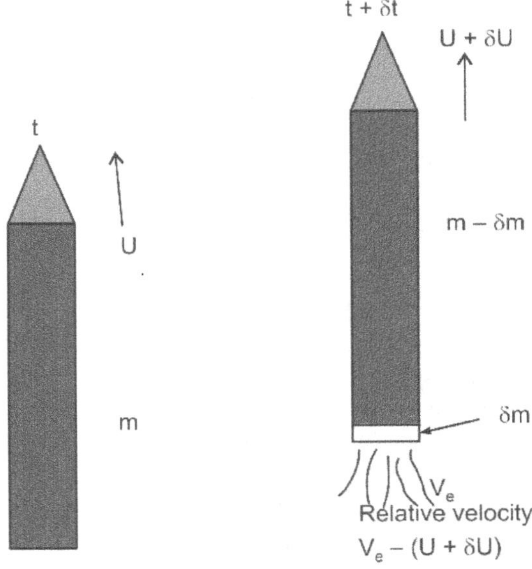


**Figure 3.4** Rocket Mass and Velocity at time $t$ and $t + \delta t$.

Applying the conservation of momentum

$$Initial\ momentum = mU$$

The relative velocity of the jet with respect to the rocket

$$= velocity\ of\ the\ jet - velocity\ of\ the\ rocket = V_e - (U + \delta U)$$

$$Final\ momentum = (m - \delta m)(U + \delta U) - \delta m(V_e - U - \delta U)$$

$$mU = mU + m\delta U - U\delta m - \delta m\delta U - V_e\delta m + U\delta m + \delta m\delta U$$

$$m\delta U = V_e\delta m$$

The $\delta m$ is associated with a negative sign because the mass is reduced. i.e.,

The mass at time t, $m(t) = m$

At time t + δt, $m(t + \delta t) = m - \delta m$

The change in mass $\delta m = Final\ mass - initial\ mass = (m - \delta m) - m = -\delta m$

Therefore,

$$m\delta U = -V_e\,\delta m$$

In limiting case where quantities are very small

$$mdU = -V_e dm \qquad \text{.....(3.12)}$$
$$dU = -V_e \frac{dm}{m}$$

Integrating both sides with limits between initial and final condition: velocity changes from initial value of $V_i$ to final value of $V_f$, and mass from initial value $M_i$ to final value of $M_f$.

$$\int_{V_i}^{V_f} dU = -V_e \int_{M_i}^{M_f} \frac{dm}{m}$$

$$V_f - V_i = V_e ln \left( \frac{M_i}{M_f} \right)$$

Increment in velocity, $\Delta U = V_f - V_i$ is

$$\Delta U = V_e ln \left( \frac{M_i}{M_f} \right) \qquad \text{.....(3.13)}$$

$$\frac{M_i}{M_f} = e^{\frac{\Delta U}{V_e}} \qquad \text{.....(3.14)}$$

This is the famous rocket equation or Tsiolkovsky's equation, after the Russian scientist who derived it for the first time. The equation shows that the final speed depends upon only two parameters, the mass ratio and exhaust velocity. Figure 3.5 shows the variation of rocket velocity, $\Delta U$ as function of mass ratio for a given exhaust velocity, $V_e$. The figure shows the strong dependence of vehicle velocity on exhaust velocity $V_{e.}$. The velocity increases with the time as the propellant is burned, i.e., the final mass, $M_f$ decreases with time.

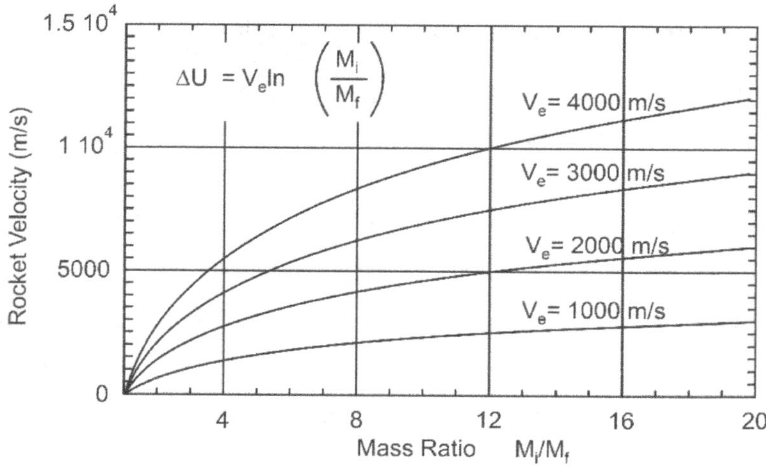


**Figure 3.5** Velocity as Function of Mass Ratio.

Rearranging Eq. (3.14)

$$R = \frac{M_i}{M_f} = e^{\frac{\Delta U}{V_e}}$$

$$\frac{1}{\alpha + \beta} = e^{\frac{\Delta U}{V_e}}$$

$$\alpha + \beta = e^{-\frac{\Delta U}{V_e}}$$

$$\alpha = e^{-\frac{\Delta U}{V_e}} - \beta \qquad\qquad .....(3.15)$$

To have a higher payload fraction, the structural mass fraction should be low and should have high jet velocity for a given velocity increment. Space missions require multiple velocity increments $\Delta U$. Consider a mission from Earth orbit to the surface of the moon. The mission requires at least six velocity increments, $\Delta U$ as illustrated below:

1. Exit Earth orbit and reach transit speed

2. Enter lunar orbit

3. Land on the moon

4. Reach lunar orbit for return journey

5. Exit lunar orbit and reach transit speed

6. Enter Earth orbit.

Additional $\Delta U$ will be required for trajectory correction, vehicle orientation etc. The total $\Delta U$ of the mission is sum of the individual velocity increments.

**Example 3.1**

A rocket engine ejects mass at a rate of 310 kg/s with an exhaust velocity of 3,000 m/s. The initial mass is 215,000 kgs. What is the change in velocity if the spacecraft burns its engine for 540 s.

**Given**

Initial Mass of the vehicle $M_i = 215{,}000$ kg

Mass flow rate $\dot{m} = 310$ kg/s

Exhaust velocity $V_e = 3000$ m/s

Burn time $t = 540$ s

The mass propellant burnt, $M_P = \dot{m}t = 310 \times 540 = 167{,}400 \ kg$

Final mass is, $M_f = M_i - M_P = 47{,}600 \ kg$

Ideal Thrust developed by the engine $F = \dot{m} V_e = 310 \times 3000 = 930000 \ N = 930 \ kN$

From Eq. (3.13), velocity change, $\Delta U$ is $\Delta U = V_e ln\left(\frac{M_i}{M_f}\right)$

$$\Delta U = 3000 \ lnln \ ln\left(\frac{215000}{47600}\right) = 4523 \frac{m}{s} = 4.523 \frac{km}{s}$$

**Example 3.2**

The 2$^{nd}$ stage of a space vehicle has a dry mass of 15400 kg and the effective exhaust gas velocity of its main engine is 4300 m/s. How much propellant must be carried if the propulsion system is to produce a total velocity increment of 2900 m/s? Calculate the duration of burn time when the engine expels mass at a rate of 16 kg/s.

**Given**

Dry mass $M_f = 15400$ kg

Exhaust velocity $V_e = 4300$ m/s

Velocity increment required $\Delta U = 2900$ m/s

Mass flow rate $\dot{m} = 16$ kg/s

The initial mass $M_i$ of the vehicle can be calculated from Eq. (3.14)

$$M_i = M_f e^{\frac{\Delta U}{V_e}} = 15400 \times exp\left(\frac{2900}{4300}\right) = 30228\ kg$$

Mass of the propellant $M_P$ is $M_p = M_i - M_f = 30228 - 15400 = 14828\ kg$

Burn time $t$ is

$$t = \frac{M_P}{\dot{m}} = \frac{15728}{16} = 983\ sec$$

## 3.5   DISTANCE TRAVELED

In the absence of gravity, the distance traveled along the trajectory is obtained by integrating the Eq. (3.13) over time. For constant thrust, the time of burn of propellant is given by

$$t = \frac{M_i - M}{\dot{m}} = \frac{M_i}{\dot{m}}\left(1 - \frac{M}{M_i}\right)$$

where  $M$ is the mass remaining  after time t, and $\dot{m}$ is rate of mass ejecting or burning rate. The distance traveled is:

$$s = \int_0^t U(t)\ dt = \int_0^t V_e\ ln\left(\frac{M_i}{M_i - \dot{m}t}\right) dt$$

By integrating and simplifying with substituting for $t$

$$S = V_e \frac{M_i}{\dot{m}}\left[1 - \frac{M}{M_i}\left(ln\frac{M_i}{M} + 1\right)\right] \qquad \text{.....(3.16)}$$

The above expression for the distance traveled with zero initial velocity. If the spacecraft already has some initial velocity $V_i$, say from the previous stage, a distance traveled due to initial is $V_i$ is to be added to above expression.

$$S = V_e \frac{M_i}{\dot{m}}\left[1 - \frac{M}{M_i}\left(ln\frac{M_i}{M} + 1\right)\right] + V_i \frac{M_i}{\dot{m}}\left(1 - \frac{M}{M_i}\right) \qquad \text{......(3.17)}$$

The effect of the Earth gravity is to be considered for the initial segment of the launch vehicle. During the vertical motion, the thrust, velocity and gravitational field are all aligned and the motion is one dimensional.

From Eq. (3.12) $mdU = -V_e dm$, i.e., $m\dfrac{dU}{dt} = -V_e\dfrac{dm}{dt}$

$$F = -V_e \dot{m} \qquad \qquad ....(3.18)$$

The negative sign indicates the direction of thrust. It shows that the thrust is a product of exhaust velocity and mass flow rate of propellant. The acceleration of the rocket with thrust and against gravity force of Earth is

$$m\frac{dU}{dt} = F - mg$$

$$\frac{dU}{dt} = \frac{F}{m} - \frac{m}{m}g$$

Using Eq. (3.18) and substituting $\dot{m} = \dfrac{dm}{dt}$

$$\frac{dU}{dt} = \frac{V_e}{m}\frac{dm}{dt} - g$$

$$dU = V_e\frac{dm}{m} - gdt$$

Integrating and substituting for t

$$\int_0^U dU = V_e \int_{M_i}^M \frac{dm}{m} - g\int_0^t dt$$

$$\Delta U = V_e \ln\frac{M_i}{M} - g\frac{M_i}{\dot{m}}\left(1 - \frac{M}{M_i}\right) \qquad \qquad .....(3.19)$$

The first term of the expression is the ideal velocity, i.e., without gravity, and the 2nd term is gravity loss. The gravity loss depends upon the rate of burning. A very short acceleration with high thrust and high mass flow rate leads to less gravity loss. If the mass flow rate is high, less propellant has to be carried to high altitudes or accelerated to high velocity, before it burns. That means, exhausting most of the propellant early in the launch is beneficial. Of course, there are a few limitations to high mass flow rates, which will be considered in later chapters. In addition to gravity, aerodynamic force (drag) must be considered and the velocity increment $\Delta U$ is accordingly adjusted to compensate for velocity reduction. To achieve orbit around a planet requires high horizontal velocity. Most of the flight path of the launch vehicle is inclined to the gravitational field to gain horizontal velocity. In general, the computation of the flight is complex and requires numerical solutions. These topics are out of scope of this book.

When the vehicle is in orbit the only force acting on the rocket is the thrust developed by the engine. Gravity, if it is present, is assumed to act in a direction orthogonal to the direction of the thrust, and therefore to have no effect on the vehicle acceleration. This situation pertains for orbital maneuvers, but not in general during space vehicle launching.

## 3.6 ROCKET PERFORMANCE PARAMETERS

(a) *Impulse*: The momentum associated with ejected mass of propellant provides velocity to the rocket. The impulse is force acting for a short period of time, $dt$. Using Eq. (3.18)

$$I = F dt = \dot{m} V_e dt = m V_e \quad \text{Ns}$$

where $\dot{m}$ is the mass flow rate (kg/s) and $m$ is mass (kg) of the propellant.

The units of impulse I are

$$I = m V_e \left( kg \frac{m}{s} = \frac{kg\,m}{s^2} s = Ns \right)$$

(b) *Specific Impulse*: The specific impulse $I_{sp}$ is defined as the impulse given to a rocket by unit *weight* of propellant.

$$I_{sp} = \frac{I}{mg} = \frac{V_e m}{mg} = \frac{V_e}{g} \quad sec \qquad \qquad ......(3.20)$$

where $g$ is acceleration due to gravity m/s$^2$. The value of g is nearly 10, and so multiply the specific impulse by 10 to get exhaust velocity. The specific impulse can be considered as the momentum imparted to the rocket per kilogram of propellant expelled. It is a measure of fuel efficiency of the rocket and is proportional to exhaust velocity. The reason to use specific impulse instead of exhaust velocity is more historical. The other reason may be that the unit if specific impulse is sec and it is same both in SI and imperial. If the specific impulse is defined as the impulse given to a rocket by unit *mass* of propellant.

$$I_{sp} = \frac{I}{\dot{m} dt} = V_e \frac{\dot{m} dt}{\dot{m} dt} = V_e \frac{m}{s} \qquad \qquad .....(3.21)$$

So, the specific impulse defined on mass basis, it is equal to exhaust velocity of the rocket.

(c) Thrust: The force or thrust developed by a rocket is the rate of change of impulse.

$$F = \frac{dI}{dt} = \frac{d}{dt}(m V_e) = \frac{dm}{dt} V_e = \dot{m} V_e \quad N \qquad \qquad .....(3.22)$$

(d) *Impulse to Mass Ratio*: The ratio of the total impulse, $I$, divided by the initial mass of the rocket $M_i$ is known as the impulse to mass ratio.

(e) *Thrust to Mass Ratio*: It represents the initial acceleration of the rocket and is defined as the ratio of the thrust to the initial mass.

(f) *Propulsive Efficiency*: The propulsive efficiency determines how much of the kinetic energy of the exhaust jet is useful for propelling the vehicle.

$$\eta_p = \frac{vehicle\ power}{vehicle\ power + residual\ kinetic\ jet\ power}$$

Vehicle power = force × velocity of vehicle = $\dot{m}\ V_e\ U$

Residual kinetic energy in the jet = $\frac{1}{2} \dot{m}\ (V_e - U)^2$

$$\eta_p = \frac{2V_e U}{2V_e U + (V_e - U)^2}$$

$$\eta_p = \frac{2\frac{U}{V_e}}{1+\left(\frac{U}{V_e}\right)^2} \qquad\qquad \dots\dots(3.23)$$

The propulsive efficiency is maximum when the vehicle velocity is exactly equal to exhaust velocity. Then the residual kinetic energy and the absolute velocity of the jet are the same and the exhaust gasses are still in space.

### Example 3.3

The booster of a rocket vehicle burns for 20 s and develops a thrust of 10,200 kN. The propellant mass of the stage is 65,000 kg. Assuming constant consumption rate of the propellant and constant thrust, determine impulse developed, specific impulse and exhaust velocity. If the initial mass of the stage is 465,000 kg, find the initial acceleration generated. Find the acceleration of the vehicle at time 10 sec.

### Given

Thrust $F = 10200 \times 10^3 N$

Propellant burning time $t = 20\ s$

Propellant mass $M_p = 65000\ kg$

Initial Mass $M_i = 465{,}000$ kg

Impulse developed $I$:

$$I = Thrust\ (F) \times duration\ (t) = 10200 \times 10^3 \times 20 = 204000 \times 10^3\ Ns$$

Specific Impulse $I_{sp}$:

$$I_{sp} = \frac{I}{M_p} = \frac{204000 \times 10^3}{65000} = 3138\ \frac{Ns}{kg}$$

Exhaust velocity $v_e$

$$v_e = I_{sp} = 3138\ \frac{m}{s}$$

$$Acceleration\ at\ the\ time\ of\ take-off = \frac{Thrust - mass\ of\ rocket \times g}{mass\ of\ rocket}$$

$$= \frac{10200 \times 10^3 - 465000 \times 9.81}{465000}\ 12.13\ \frac{m}{s^2}$$

The take-off acceleration is 12.13 m/s², i.e., 1.24 g.

Mass flow rate of propellant

$$\dot{m} = \frac{Mass\ of\ propellant}{Burn\ time} = \frac{65000}{20} = 3250\frac{kg}{s}$$

The propellant consumed in a duration of $10\ s\ =\ 3250 \times 10 = 32500\ kg$

Therefore,

Mass of the vehicle after 10s $= 465000 - 32500 = 432{,}500\ kg$

The vehicle acceleration after 10 s

$$= \frac{10200 \times 10^3 - 432500 \times 9.81}{432500} = 13.77\,\frac{m}{s^2}$$

## 3.7  MULTISTAGE ROCKETS

Different mass ratios, i.e., the payload mass $\alpha$, structural mass $\beta$ and propellant mass $\gamma$, associated with rocket vehicles, are defined by Eq. (3.8) to (3.10). The payload fraction is expressed as in Eq. (3.15) is

$$\alpha = e^{\left(-\frac{\Delta U}{V_e}\right)} - \beta$$

Using Eq. (3.11), the Eq. (3.13) can be rewritten as

$$\Delta U = V_e\, ln\left(\frac{M_i}{M_f}\right) = I_{sp}g\, ln\left(\frac{1}{\alpha+\beta}\right) \quad\quad\quad .....(3.24)$$

It is desirable to have a high payload fraction for a space mission. As noted from the above equations, the payload fraction, $\alpha$ is higher when the exhaust velocity, $V_e$ and structural fraction, $\beta$ are low. For the space missions with high velocity increment demand high specific impulse or jet velocity and low payload and structural fractions are required. The selection of the structural fraction depends upon the strength requirement of the vehicle and the size of propellant tanks and associated piping. For high thrust requirements, an increase in propellant mass makes the rocket unwieldy and heavy and leads to smaller values of thrust to mass ratio. In such cases the total mass of the rocket engine will be dictated by the propellant mass. The variation of structural fraction with heavy engines is negligible and is generally limited to about less than 3% of initial mass, $M_i$.

For a given payload, when large energy change, i.e., high incremental velocity $\Delta U$ is required, the initial mass of a single stage rocket is very high and the cost associated is enormous. Instead of a single large rocket, a multistage rocket, a series of individual vehicles or stages each with its own structure and engines are used to enhance the velocity increment for the entire vehicle. As the propellant is consumed in each stage, portions of the structure and tankage are discarded and thus the energy is not wasted in accelerating the unnecessary masses. The stages are so connected that each operates in turn, accelerating the remaining stages and payloads before being detached from them. Figure 3.6 shows the general arrangement of 3 stages in series of a rocket launch vehicle. The last or top stage, which is usually the smallest, carries the payload. The 1$^{st}$ stage, generally referred to as the booster stage, needs to impart high thrust as it has the largest mass at launch time. The total mass of the 1$^{st}$ stage is the mass of its stage and sum of the masses of all upper stages including payload. The ejection timing of a used stage and ignition timing of the next stage are critical for its successful operation. To provide extra initial thrust for the mission requirement and to overcome the initial gravity losses and atmospheric drag, two or more strap-on booster rockets are attached to vertical stage rocket configuration. All the engines ignite at launch and the strap-on motors fall away when their propellant is consumed.

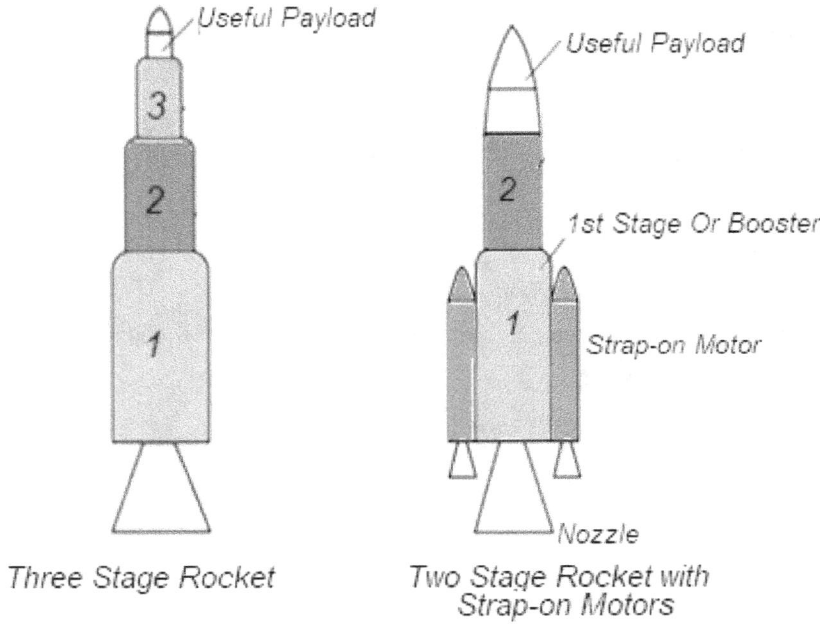



**Figure 3.6**  General Arrangement of Multistage Rocket.

For Multistage vehicle with N stages, the payload of a lower stage *[Mpl]j* is the mass of all higher stages *[Mi]j+1*

$$\left[M_{pl}\right]_j = [M_i]_{j+1} \quad \text{for } j < N$$

From Eq. (3.15) is

$$\alpha = e^{\left(-\frac{\Delta U}{V_e}\right)} - \beta$$

The payload ratio of stage is

$$\alpha_j = exp\left(-\frac{\Delta U}{V_e}\right)_j - \beta_j$$

The initial mass of stage j is given by

$$(M_i)_j = \left[\frac{1}{exp\left(-\frac{\Delta U}{V_e}\right)_j - \beta_j}\right](M_u)_j$$

The total increment of velocity is

$$\Delta U = \sum_{j=1}^{N} (V_e)_j \, ln(R_j)$$

The following example clearly explains the advantages of multistage launch vehicles.

**Example 3.4**

Consider a liquid rocket vehicle with liquid hydrogen (LH$_2$) as fuel and liquid oxygen (LOX) as oxidizer to place payload in an orbit. The specifications are as follows:

Payload, $M_U = 900$ kg

The exhaust velocity, $V_e = 4115$ m/s

Required velocity increment $\Delta U = 14,300$ m/s

The total structural mass, $M_S$ can be taken as 3% of initial mass of the stage, $M_i$.

*Case 1: Single stage Rocket Vehicle*

Consider a single stage for the launch vehicle.

The mass ratio using Eq. (3.14)

$$R = e^{\frac{\Delta U}{V_e}} = exp\left(\frac{14300}{4115}\right) = 32.3$$

The final mass or burn out mass

$$M_f = M_u + M_s = 900 + 0.03 \times M_i$$

Therefore, the initial mass of the rocket $M_i$ is

$$R = \frac{M_i}{M_f}$$

$$32.3 = \frac{M_i}{900 + 0.03 \times Mi}$$

$$M_i = 937,742 \, kg$$

So, to place the 900 kg payload in orbit the vehicle initial mass, $M_i$ required is 937742 kg.

The propellant mass

$$M_p = initial \ mass \ M_i - useful \ payload \ mass \ M_u - structural \ mass \ M_s$$

$$M_p = 9377 \, 42 - 900 - 0.03 \times 937742 = 908,710 \, kg$$

The structural mass

$$M_s = 0.03 * 937742 = 28132 \, kg$$

The ratio of propellant in the vehicle mass = 96.9%

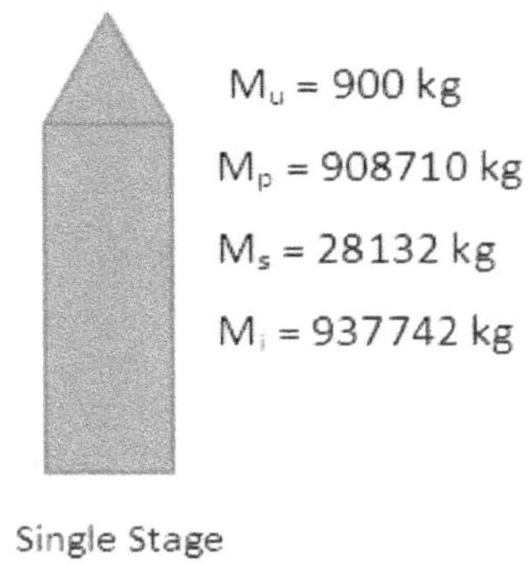

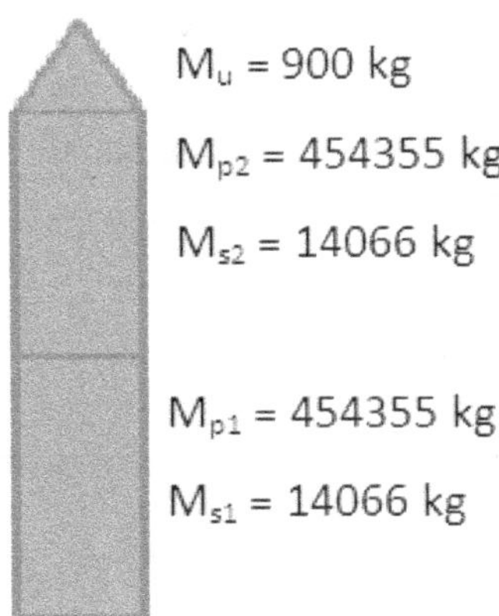


**Figure 3.7** Single Stage Rocket for the Example 3.3.

The mass distribution for a single stage rocket configuration is shown in Figure 3.7.

*Case 2: Two Stage rocket vehicle*

Consider two different situations.

I. The propellant of the single stage is equally divided in two stages. Consider improvement in incremental velocity $\Delta U$.



**Figure 3.8** Two Stage Rocket for Example 3.3.

The mass distribution in two stages is shown in Figure 3.8.

Mass ratio, $R_1$ of first stage is:

$$R_1 = \frac{M_i}{M_u + M_p + M_s}$$

$$R_1 = \frac{937742}{900 + 454355 + 28132} = 1.94$$

Velocity Increment of stage 1

$$\Delta U_1 = V_e ln R_1$$

$$\Delta U_1 = 4115 \times ln 1.94 = 2726 \ \frac{m}{s}$$

The initial mass for the second stage is

$$(M_i)_2 = M_u \ (useful \ payload) + \left(M_p\right)_2 (propellant \ mass \ of \ second \ stge)$$
$$+ (M_s)_2 (structural \ mass \ excluding \ first \ stage \ structural \ mass)$$

Then, the mass Ratio, $R_2$ of second stage is

$$R_2 = \frac{900+454355+14066}{900+14066} = 31.36$$

Velocity increment of stage 2

$$\Delta U_2 = 4115 \times ln 31.36 = 14178 \frac{m}{s} h$$

*The* total increment in velocity

$$\Delta U = \Delta U_1 + \Delta U_2 = 2726 + 14178 = 16904 \frac{m}{s}$$

So, the increment in velocity is improved in a two-stage rocket, from 14300 m/s to 16904 m/s, around 18% increase.

**II.** In this case the required increment velocity is equally divided in the two stages. i.e., each stage develops $\Delta U = 7150$ m/s.

The payload ratio

$$\alpha_j = exp \left(-\frac{\Delta U}{V_e}\right)_j - \beta_j$$

As jet velocity and structural fraction $\beta$ is same in the two stages, $\alpha$ is also constant.

$$\alpha = exp \left(-\frac{7150}{4115}\right) - 0.03 = 0.146.$$

The mass ratio each stage is:

$$R = \frac{1}{\alpha + \beta} = \frac{1}{(0.146 + 0.03)} = 5.68$$

For 1st stage R

$$R = \frac{initial \ mass \ (M_i)_1}{(initial \ mass \ of \ 2nd \ stage \ (M_i)_2 + Structural \ mass \ 0.03 \times (M_i)_1)}$$

For 1st stage

$$5.68 = \frac{(M_i)_1}{(M_i)_2 + 0.03 * (M_i)_1}$$

For 2$^{nd}$ stage

$$R = \frac{initial\ mass\ (M_i)_2}{\left(useful\ payload + Structural\ mass\ of\ the\ 2nd\ stage\ 0.03 \times (M_i)\right)}$$

$$5.68 = \frac{(M_i)_2}{900 + 0.03(M_i)_2}$$

Solving the above two equation/s

$$(M_i)_1 = 42189\ kg$$

$$(M_i)_2 = 6162\ kg$$

So, there is a large reduction in initial mass in a two-stage rocket vehicle compared to a single stage. The initial mass in a two-stage vehicle is 4.5% of single stage. We can see that the lower stage must have a bigger share of the propellant than the upper stage, for the mass ratio to be the same. The example is presented to clearly bring the advantage of the multistage rocket. However, the optimization of stages is complex and involves complex mathematics. The disadvantage of multistage rockets is that each stage will have a separate engine and associate accessories and controls. The cost may be higher. Figure 3.9 shows the Polar Satellite Launch Vehicle (PSLV) of Indian Space Research Organization (ISRO) for C51 Mission to launch main satellite Amazonia 1 along 18 other satellites in Sun Synchronous Polar Orbit.

**Nomenclature**

| | |
|---|---|
| $F$ | force, thrust (N) |
| $I$ | impulse (Ns) |
| $I_{sp}$ | specific impulse (sec, Ns/kg) |
| $M, m$ | mass (kg) |
| $M_f$ | final mass (kg) |
| $M_i$ | initial mass (kg) |
| $M_p$ | propellant mass (kg) |
| $M_s$ | structural mass (kg) |
| $M_u$ | useful mass (kg) |

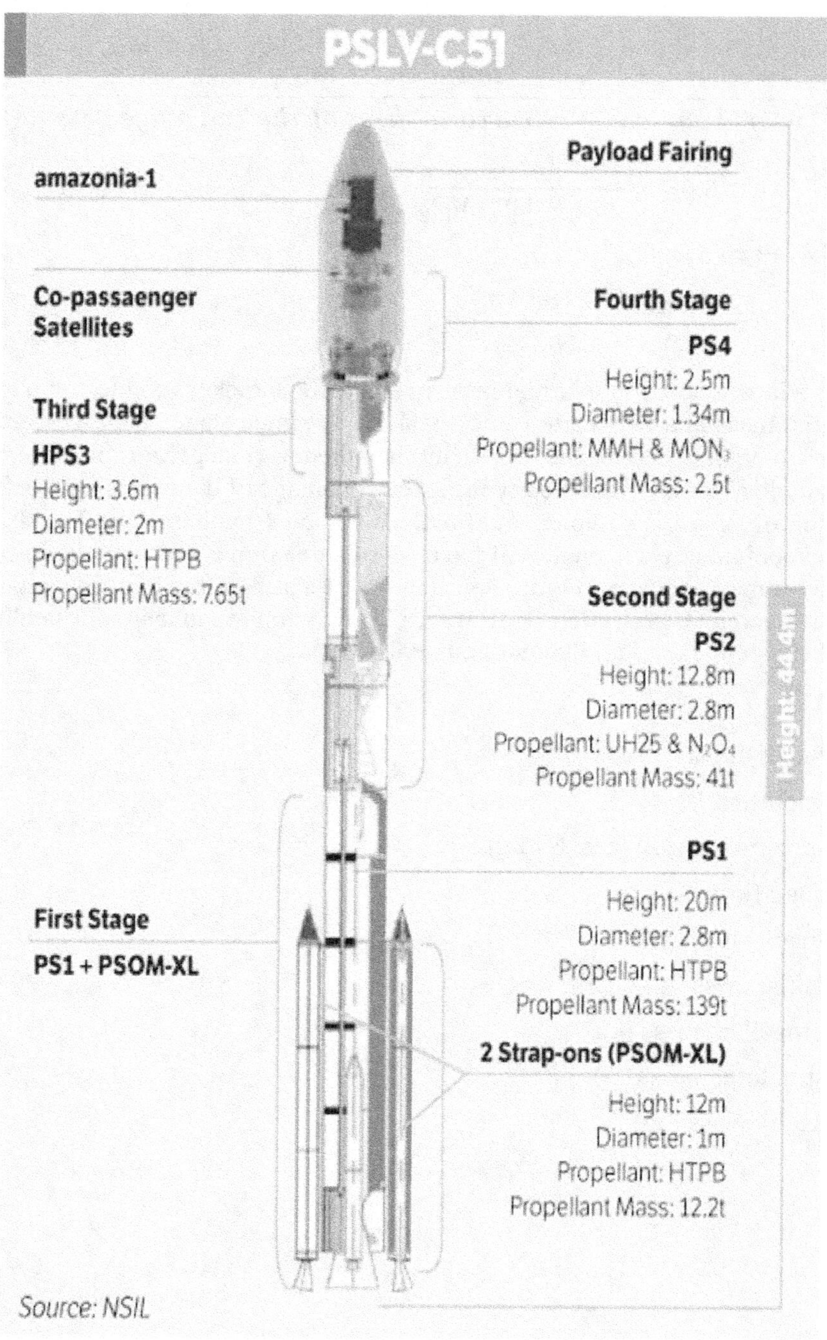



**Figure 3.9** ISRO Launch Vehicle-PSLV. (Credit NSIL).

$\dot{m}$      mass flow rate (kg/s)

$R$      mass ratio

$S$      distance travelled (m)

$t$      time (s)

$U$      velocity (m/s)

$V$      velocity (m/s)

$V_e$      exhaust velocity of the jet (m/s)

$\alpha$      payload mass ratio

$\beta$      structural mass ratio

$\gamma$      propellant mass ratio

$\Delta U$      velocity increment (m/s)

$\eta_p$      propulsive efficiency

## PROBLEMS

3.1      A certain stage of a rocket vehicle develops a thrust of 9.12 kN. The mass flow rate is 4.05 kg/s and the velocity of the vehicle is 390 m/s. Calculate the exhaust velocity of the jet, the kinetic jet energy per unit flow of propellant, the propulsive efficiency, the specific impulse and the specific propellant consumption.

3.2      A rocket carrying a payload of mass 1000 kg requires velocity increment of 900 m/s. The exhaust velocity of the rocket is 3040 m/s. The structural mass fraction of the vehicle is 0.22. Find the propellant mass required and overall mass.

3.3      During the boost phase of the Polar Satellite Launch Vehicle, the initial (booster) stage solid motor rocket (SRM) develops a thrust of 4846 kN. The booster stage operates together with six solid propellant strap-on motors each developing a thrust of 703.5 kN. Find the overall exhaust velocity of the vehicle during the combined thrust operation.

3.4      The first stage of a rocket vehicle burns for 50 s. The rate of burning of propellant is constant. The mass of the propellant is 9000 kg and the specific impulse of the stage is 225 s. The initial mass of the rocket vehicle is 16 MT (metric tons). Calculate the initial acceleration of the vehicle. Also find the vehicle acceleration at time 20 s of travel and at burnout time.

3.5      A strap-on motor has an inert mass of 1,920 kg. With the loading of propellant its mass is 14,120 kg. The igniter mass is 120 kg of which 98 kg is igniter propellant. After a static test of the motor, it is found that the unburned residual propellant is 425 kg. The hardware of the igniter is fully burnt during the testing. Calculate the mass ratio and the mass of propellant used to generate thrust.

3.6      A two-stage rocket vehicle is to be designed to provide an ideal velocity increment of 3.1 km/s to place a payload of 250 kg. The specific impulse of the first stage is 2650 Ns/kg and propellant mass fraction is 0.92. The specific impulse of the second stage is 3140 Ns/kg and propellant mass fraction is 0.84. Calculate the propellant and inert masses of each stage and the initial mass of the vehicle (gross lift off weight). Assume the velocity increment is equally distributed between the two stages.

3.7　A four-stage launch vehicle is used to place a satellite of mass 1,200 kg in orbit. Six strap-on motors are attached to the first stage. The propellant mass, the mass of structure including inert, and specific impulse of each stage are given in the following Table:

| Stage | Mass of Propellant (kg) | Mass of Structure and inert (kg) | Specific Impulse (sec) |
|---|---|---|---|
| Strop-on Motor | 12,200 | 2074 | 262 |
| First stage | 138,200 | 22,000 | 232 |
| Second stage | 42,000 | 8000 | 293 |
| Third stage | 7,600 | 1300 | 295 |
| Fourth stage | 2,500 | 1800 | 305 |

Assume the strap-motors and first stage fire together. The thrust developed by each strap-om motor is 704 kN and the first stag is 4846 kN. Calculate payload fraction, initial acceleration of the vehicle, velocity increment of each stage and total velocity increment of the vehicle.

3.8　Sounding rockets are used to make measurements of Earth's atmosphere. The electronic equipment used in the package can withstand a maximum acceleration of 9 g. One of such sounding rockets, launched vertically, develops a thrust of 7 kN and total impulse of 80 kN s. If the mass of the empty rocket is 35 kg, what is the payload that the sounding rocket can carry so that its acceleration does not exceed 9 g?

# Rocket Nozzles

## 4.1  INTRODUCTION

The thrust developed by the rocket engine is primarily dependent upon the momentum given to the jet of the combustion gasses. It is directly proportional to the exhaust velocity of the jet $V_e$. In chemical rockets, the thermal energy developed due to combustion reactions in the thrust chamber of the rocket is converted into kinetic energy in a device called a nozzle. During their passage through the nozzle, the exhaust gasses are continuously accelerated from low subsonic velocities to high supersonic velocities at the nozzle exit. It is a one of the major components of the rocket engine having a significant influence on the overall engine performance and representing a large fraction of the engine structure. In this chapter, applying the concepts of thermodynamics and fluid dynamics, a relation is derived to relate the exhaust velocity, $V_e$ to geometrical parameters of the nozzle and measurable properties like pressure and temperature of the exhaust gasses. Also, the effect of the nozzle shape on the performance of the engine is presented. The major parameters of nozzle are the shape of the nozzle contour and the nozzle area expansion ratio. A careful shaping of the nozzle contour can lead to a high gain in its performance. The theory applies to any propulsion system that uses the expansion of a gas as a propulsive mechanism (like, chemical rockets, nuclear rockets, solar heated and electrical rocket systems) for ejecting matter at high velocity.

## 4.2  HIGH PRESSURE GAS EXPANSION

In a rocket engine the propellant mixture (oxidizer and fuel) reacts in a thrust chamber and produces high temperature and pressure gasses. Let the chamber contain combustion gasses at a high-pressure $p_c$ and temperature $T_c$. The gas is exhausted through a vent to outside at pressure $p_e$ and temperature $T_e$ as shown in Figure 4.1. Consider a control volume around the vent as shown in figure. The following are main assumptions made to study the expansion of high-pressure gas.

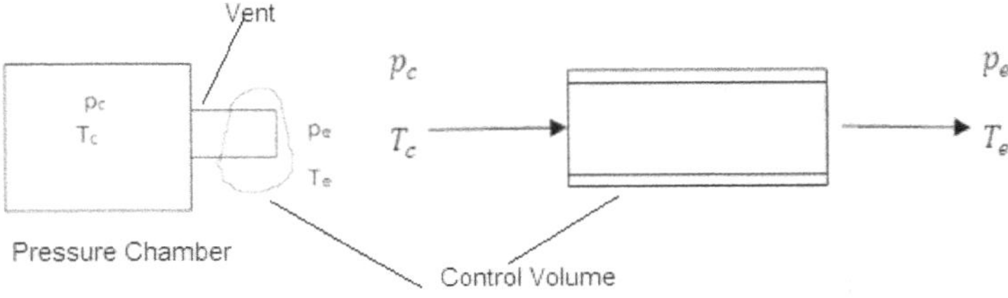


**Figure 4.1** Flow of High Pressure Gasses Through a Vent.

a. There is no heat flow, Q across the vent. (Q=0)

b. The flow is frictionless

c. The vent is rigid, i.e., no work, W is done on the vent. (W=0)

d. The flow is steady.

e. The gas is an ideal gas.

f. The flow is one dimensional

g. No chemical reaction occurs in the vent/duct.

The above assumptions make the flow as adiabatic and isentropic. The steady flow means mass entering is equal to mass leaving the vent.

From the first law of thermodynamics

$$Q - W = E \qquad \qquad .....(4.1)$$

Because Q and W are zero, energy E is zero. This means energy $E$ at inlet of the vent equal to energy at the exit of the vent.

$$(enthalpy + KE + PE)_{inlet} = (enthalpy + KE + PE)_{outlet}$$

Considering unit mass flow rate

$$\left( h_c + \frac{V_c^2}{2} + g z_c \right) = \left( h_e + \frac{V_e^2}{2} + g z_e \right)$$

The units of specific enthalpy, h are J/kg and kinetic energy

$$\frac{m^2}{s^2} = \frac{m^2}{s^2} \frac{kg}{kg} = \frac{N\,m}{kg} = J/kg$$

As the area of the chamber is high compared to the vent, the velocity at the inlet of the vent is negligible and can be taken as zero. The change in potential energy can also be neglected due to the size of the chamber.

$$\frac{V_e^2}{2} = (h_c - h_e)$$

$$V_e = \sqrt{2(h_c - h_e)} \qquad \qquad .....(4.2)$$

The assumption that the gas is an ideal gas means the enthalpy of the gas is a function of temperature, $h = f(T)$. For a perfect gas the variation of h with temperature is linear. i.e., the slope of the line dh/dT is constant and it is called the specific heat of gas at constant pressure, $c_p$.

$$\frac{dh}{dT} = c_p \qquad \qquad .....(4.3)$$

$$dh = c_p dT$$

Also, the gas obeys the gas law,

$$pV = mRT \qquad \qquad .....(4.4)$$

where $p$ is pressure (N/m$^2$), $V$ is the volume (m$^3$), $m$ is the mass (kg) and $T$ is temperature in K and $R$ is the specific gas constant. The units of $R$ are:

$$\frac{N}{m^2}\, m^3\, \frac{1}{kg}\frac{1}{K} = \frac{N\,m}{kg\,K} = \frac{J}{kg\,K}$$

The value of the specific gas constant $R$ is different for different gasses.

The above equation can be written as

$$pV = nR_uT \qquad\qquad \dots\dots (4.5)$$

where $n$ is the number of moles of the gas and $R_u$ is the universal gas constant.

$$R_u = 8.314\ \frac{J}{mole\,K}$$

One mole of gas is equivalent to the molecular weight of the gas. Suppose a chamber contains 1000 gm (1 kg) of oxygen (molecular weight 32 gm), the number of moles oxygen in the chamber are 1000/32=31.25.

The specific gas constant $R$ is

$$R = \frac{R_u}{M} \qquad\qquad \dots\dots (4.6)$$

where $M$ is the molecular mass of the gas.

The specific heat at constant pressure and at constant volume ($c_p$ and $c_v$) for a perfect gas are constant and are related by

$$c_p - c_v = R \qquad\qquad \dots\dots (4.7)$$

The ratio of specific heats $\gamma$ is

$$\gamma = \frac{c_p}{c_v} \qquad\qquad \dots\dots(4.8)$$

Eq. (4.7) can be written as

$$c_p\left(1 - \frac{c_v}{c_p}\right) = c_p\left(1 - \frac{1}{\gamma}\right) = R$$

$$c_p = \frac{R\gamma}{\gamma - 1}$$

Using Eq. (4.6)

$$c_p = \frac{R_u\gamma}{M(\gamma-1)} \qquad\qquad \dots\dots(4.9)$$

As the flow is isentropic, it obeys isentropic relations

$$pv^\gamma = constant \qquad\qquad \dots\dots(4.10)$$

$$\left(\frac{p_e}{p_c}\right)^{\frac{\gamma-1}{\gamma}} = \frac{T_e}{T_c} \qquad\qquad \dots\dots(4.11)$$

As enthalpy for ideal gas is a function of temperature, Eq. (4.2) can be written as

$$V_e = \sqrt{2(T_c - T_e)} \qquad \qquad .....(4.12)$$

$$V_e = \sqrt{2c_p T_c \left(1 - \frac{T_e}{T_c}\right)} \qquad \qquad .....(4.13)$$

$$V_e = \sqrt{\frac{2\gamma R_u T_c}{M(\gamma-1)}\left[1 - \left(\frac{p_e}{p_c}\right)^{\left(\frac{\gamma-1}{\gamma}\right)}\right]} \qquad \qquad .....(4.14)$$

This gives the relation for exhaust velocity in terms of gas properties. The term

$$\frac{2\gamma R_u T_c}{M(\gamma - 1)}$$

is primarily a function of the combustion process, which is influenced by the propellants, and the mixture ratio of propellants. The term $1 - \left(\frac{p_e}{p_c}\right)^{\left(\frac{\gamma-1}{\gamma}\right)}$ is a function of the expansion of the exhaust gases through the vent, which is influenced by back pressure and area ratio (to be discussed later). The following conclusions can be arrived observing above equation:

1. The exhaust velocity, $V_e$, is proportional to chamber temperature. Higher the temperature of the chamber higher the exhaust velocity i.e., relating to rocket engine, higher the thrust chamber combustion temperature, $T_c$ higher the exhaust velocity. The limitation is that the material should withstand the temperatures. Normally the combustion temperatures are in the range of 3000 to 4000 K.

2. The ratio $p_e/p_c$ should be small to get higher exhaust velocity. High chamber pressure, and low exit pressure, $p_e$ results higher exhaust velocity, $V_e$.

3. Lower molecular weight of chamber gas, (i.e., combustion gasses of the thrust chamber of the rocket engine), gives higher exhaust velocity, $V_e$. The hydrogen as fuel combines with oxygen gives water of molecular weight 18, that is smaller than $CO_2$ molecular weight 44.

The other parameter is ratio of specific heats, $\gamma$ and it appears in the equation as

$$\frac{2\gamma}{\gamma - 1} = \frac{2}{1 - \frac{1}{\gamma}}$$

i.e., $\gamma$ should be small so that the fraction becomes large. Lower molecular weight gasses have higher value of $\gamma$, like monoatomic ($\gamma = 1.63$) and diatomic ($\gamma = 1.4$) gasses. Higher molecular weight gasses like $CO_2$, have lower value $\gamma$ ($=1.35$). However, the influence of $\gamma$ value is negligible compare to molecular weight $M$.

Summarizing above points, to achieve high exhaust velocity, $V_e$ of rocket engine, the combustion must take place at high temperature and pressure, the released combustion gasses should have low molecular weight and the gasses are to be expanded to low exhaust pressure. The effect of specific heat ratio is not appreciable on exhaust velocity.

The maximum exhaust velocity is obtained when the pressure $\frac{p_e}{p_c}$ is zero, i.e., when $p_e=0$. The value of maximum velocity is

$$V_{emax} = \sqrt{\frac{2\gamma R_u T_c}{M(\gamma-1)}} \qquad\qquad .....(4.15)$$

$$V_e/V_{emax} = \sqrt{\left[1-\left(\frac{p_e}{p_c}\right)^{\left(\frac{\gamma-1}{\gamma}\right)}\right]} \qquad\qquad .....(4.16)$$

The ratio of exhaust velocity to maximum exhaust velocity is given by

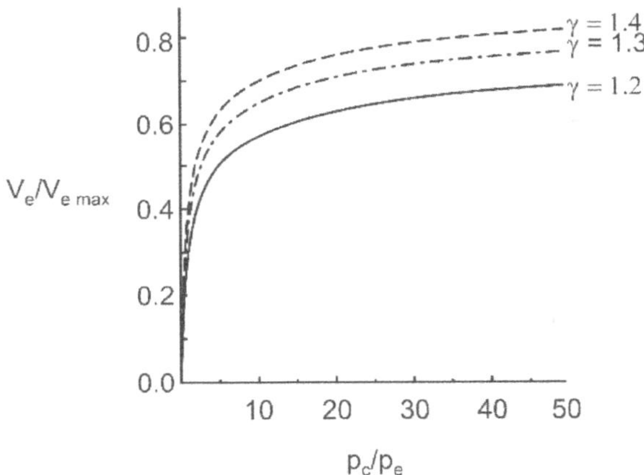


**Figure 4.2** Variation of $V_e/V_{emax}$ ratio with pressure ratio.

Figure 4.2 shows the variation of exhaust velocity ratio with three values of specific heat ratio, $\gamma$. It may be observed that the velocity ratio increases initially at a faster rate up to 25 and thereafter the increase is not appreciable. To keep the structural mass ratio within the practical limits, generally, the pressure ratio of 25 to 50 is preferred. However, for deep space missions higher pressure ratios are preferred to achieve higher performance.

## 4.3   SPEED OF SOUND & MACH NUMBER

Sound can be said to be waves of density and pressure variations in space and in time. In a compressible flow, small pressure changes can affect appreciable density changes, and vice versa. Sound propagates at some finite speed through the compressible medium. In incompressible flow it is generally assumed that sound propagates at infinite speed. The Mach number $M$ is defined as the ratio of flow velocity $V$ to that of the speed of sound, an associated with the flow.

$$M = \frac{V}{a} \qquad\qquad .....(4.17)$$

It is a measure of the density changes for compressible fluids. To relate the speed of sound with gas properties, the following assumptions are made:

a. The disturbance is *weak* enough that only small changes are produced.

b. The flow properties are considered to be *continuous*.

c. The disturbance is fast enough that the process is *adiabatic*; no heat transfer is allowed to take place.

Consider the sound pulse, $dV$ caused by a piston moving in still air slowly in a duct of constant area as shown in Figure 4.3. As in figure (a), for an observer in an inertial frame, the sound wave travels forward in the undisturbed space. As in figure (b), if the observer is on the wave front, the flow is as shown and the problem becomes that of steady state flow.

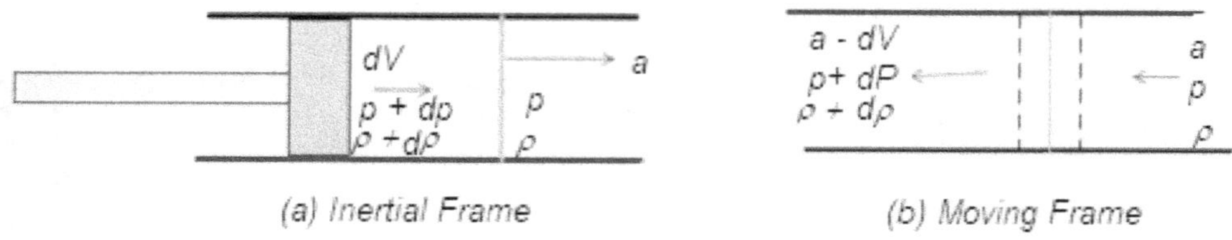


(a) Inertial Frame

(b) Moving Frame

**Figure 4.3** Propagation of Sound Wave in a Duct.

The continuity equation for unit mass in steady state flow gives:

$$(\rho + d\rho)(a - dV) = \rho a$$

$$a d\rho - \rho dV = 0 \qquad\qquad .....(4.18)$$

Applying momentum equation for unit mass

$$[(\rho + d\rho)(a - dV)](a - dV) + (P + dP) = (\rho a)a + P$$

$$(a\rho - \rho dV + a d\rho - d\rho dV)(a - dV) + (P + dP) = \rho a^2 + P$$

$$(\rho a^2 - a\rho dV + a^2 d\rho - a\rho dV + \rho dV dV - a d\rho dV) + (P + dP) = \rho a^2 + P$$

Simplifying and neglecting higher order differential products

$$a^2 d\rho - 2a\rho dV + dP = 0$$

Substituting $\rho dV = a d\rho$ from Eq. (4.18)

$$a^2 d\rho - 2a^2 d\rho + dP = 0$$

$$-a^2 d\rho + dP = 0$$

$$\frac{dP}{d\rho} = a^2 \qquad\qquad .....(4.19)$$

The adiabatic flow relation is

$$\frac{p}{\rho^\gamma} = c \text{ or } p = c\rho^\gamma$$

Differentiating w.r.t $\rho$

$$\frac{dp}{d\rho} = c\gamma\rho^{\gamma-1} = \frac{p}{\rho^{\gamma}}\gamma\rho^{\gamma-1} = \frac{\gamma p}{\rho}$$

Using gas law for unit mass $pV = RT$

$$a^2 = \gamma\frac{\rho RT}{\rho} = \gamma RT$$

$$a = \sqrt{\gamma RT} \qquad\qquad\qquad .....(4.20)$$

Based on the value of Mach number, fluid flows can often be roughly divided into following regimes:

(a) Sonic flow, Mach number $M = 1$

(b) Subsonic flow, Mach number $M < 1$

(c) Super-sonic flow, Mach number $M > 1$

(d) Transonic flow, $0.8 < M < 1.25$

(e) Hypersonic flow, usually said to be $M > 5$, but mostly defined by the high degree of real gas effects, such as dissociation of gas molecules.

## 4.4   NOZZLE SHAPE

Consider an arbitrary shape of nozzle as shown in Figure 4.4. At section x the area of the nozzle is A, velocity V and density $\rho$. At section x + dx the values are A - dA, V + DV and $\rho + d\rho$. The mass flow rate is $\dot{m}$. For steady flow condition, $\dot{m}$ is constant

$$d(\rho AV) = 0$$

$$\rho A dV + \rho V dA + AV d\rho = 0$$

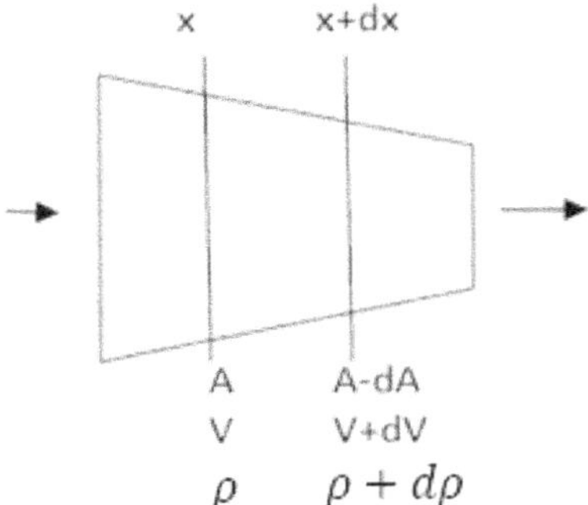


**Figure 4.4** Gas Properties at Two sections in an Arbitrary Shape Nozzle.

Dividing by $\rho AV$ and using Eq.(4.19)

$$\frac{dV}{V} + \frac{dA}{A} + \frac{d\rho}{\rho} = 0$$

$$\frac{dV}{V} + \frac{dA}{A} + \frac{1}{\rho}\frac{dP}{a^2} = 0 \qquad \qquad .....(4.21)$$

Applying momentum equation

$$mass * change\ of\ of\ velocity = force$$

$$\rho AV(V + dV - V) = -dP\ A$$

$$\rho V dV + dP = 0$$

$$\frac{dP}{\rho} = -V dV \qquad \qquad .....(4.22)$$

Substituting above relation in Eq. (4.21)

$$\frac{dV}{V} + \frac{dA}{A} - \frac{V dV}{a^2} = 0$$

$$\frac{dV}{V}\left[1 - \frac{V^2}{a^2}\right] + \frac{dA}{A} = 0$$

$$\frac{dV}{V} = -\frac{dA}{A}\frac{1}{(1-M^2)} \qquad \qquad .....(4.23)$$

With the variation of the value of Mach number M, the following cases are considered:

a.   $M < 1$, the fraction $\frac{1}{(1-M^2)}$ is positive and $dA$ is negative, i.e., the area is decreasing along the length of the nozzle. The velocity increment is positive and velocity increases as shown in Figure 4.5 (a).

b.   In subsonic flow ($M < 1$), when area decreases, the flow accelerates. It is called a converging nozzle.

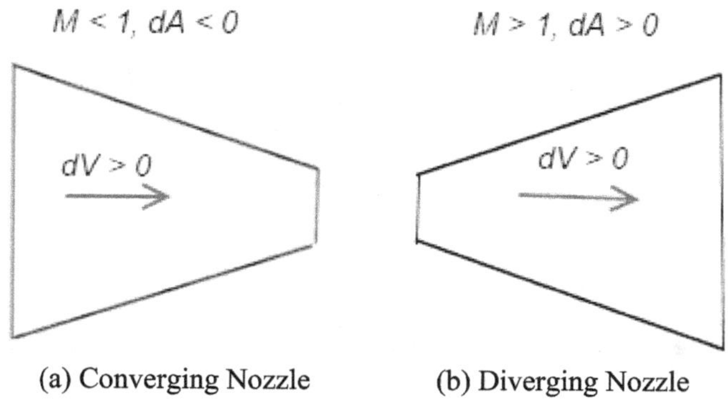


**Figure 4.5** Variation of Area of a Nozzle for change in Velocity.

c.   M > 1, the fraction $\dfrac{1}{(1-M^2)}$ is negative and $dA$ is positive, i.e., the area is increasing:

d.   $dA > 0$, then $dV$ is positive i.e., the velocity increases. This is shown in Figure 4.5 (b).

e.   In supersonic flow $(M > 1)$, when area increases, the flow accelerates. This portion is called a diverging nozzle.

f.   M = 1, to have a meaningful equation, $dA$ should be zero ie., there is no change in area.

This portion is called the throat area.

So, to have accelerating flow i.e., to increase the exhaust velocity of combustion gasses of rocket engine, the shape nozzle should be first convergent area, then constant area, followed by divergent area. This is known as convergent divergent nozzle or de Laval, named after the Swedish Engineer who devised it for the first time. The flow velocity increases from subsonic flow to super-sonic flow. The velocity at constant area, known as throat, reaches sonic $(M=1)$ velocity. The shape of the rocket nozzle is shown in Figure 4.6.

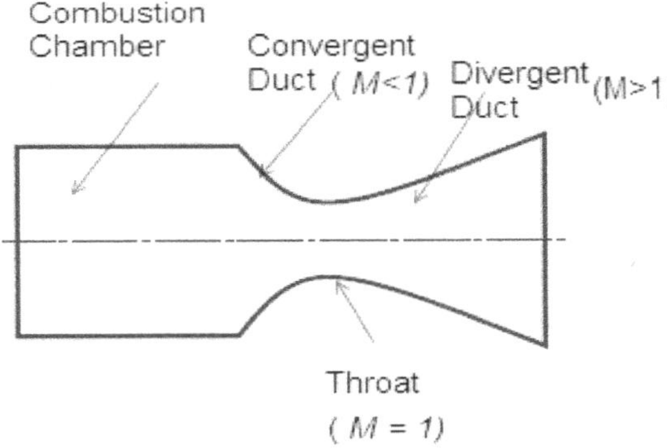


**Figure 4.6** Convergent Divergent Rocket Nozzle.

When the velocity at throat reaches the sonic velocity $(M = 1)$, along the nozzle the variation of the gas properties is:

(i)   Velocity of flow, $V$ increases

(ii)   Pressure, $p$ decreases

(iii)   Density, $\rho$ decreases

(iv)   The density drops rapidly at the throat.

(v)   Temperature, $T$ decreases

(vi)   Speed of sound, $a$ decreases

(vii) Mach number, $M$ increases

If the velocity at the throat is not reached sonic velocity at the throat, the velocity along the nozzle after the throat decreases and the flow remains subsonic. It acts like a diffuser. In this case along the divergent portion of the nozzle the flow properties are:

(i)   Velocity of flow, $V$ decrease

(ii)  Pressure, $p$ increases

(iii) Density, $\rho$ increases

(iv)  Temperature, $T$ increases

(v)   Speed of sound, $a$ increases

(vi)  Mach number, $M$ decreases

## 4.5   CHOKING OF NOZZLE

For a given thrust chamber pressure ($p_c$) and temperature ($T_c$), the variation of mass flow rate ($\dot{m}$) through the convergent divergent nozzle with the change in exhaust pressure ($p_e$) is shown in Figure 4.7. Any pressure disturbances (pressure change) in the downstream will be communicated

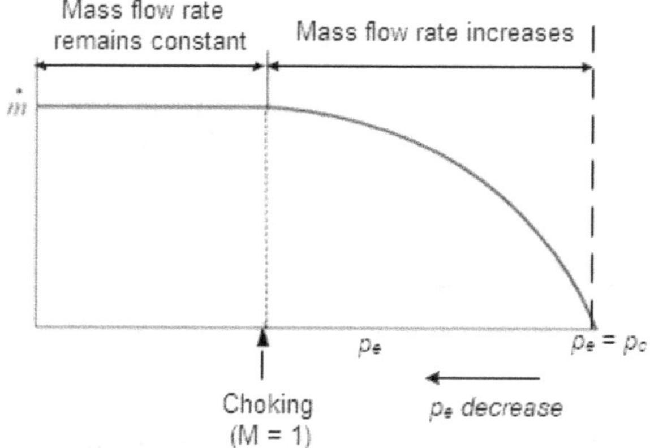


**Figure 4.7** Variation of Mass Flow Rate with Exhaust Pressure.

to upstream with the velocity of sound in the medium. As the exhaust pressure is progressively decreased, the available pressure drop ($\Delta p = p_c - p_e$) increases and the mass flow rate increases.

This holds true until the flow rate is increased to the point that the fluid reaches the local sonic velocity ($M=1$) at the throat of the nozzle. After the velocity at the throat reaches sonic (M=1), any further decrease in exhaust velocity ($p_e$) does not increase the mass flow rate as shown in Figure 4.7. As the throat reaches the sonic speed of sound (i.e., M=1), pressure changes can no longer be communicated upstream as the speed of which these pressure changes are propagated is limited by the speed of sound. In a nozzle this has the effect of isolating the upstream side from the downstream side at the throat. Because of this effect any reduction in downstream pressure will have no effect on the flow rate, as the increased pressure differential is not 'felt' upstream of the throat. When the flow through the nozzle reaches a maximum limit, the nozzle is said to be choked and places an upper limit of the maximum possible mass flow rates through the nozzle.

Since rocket nozzles are configured for achieving maximum exhaust velocities, the flow through it is always designed to be choked.

## 4.6   NOZZLE AREA RATIO

The nozzle area ratio, $\varepsilon$ is defined as

$$\varepsilon = \frac{Area\ of\ nozzle\ exit}{Area\ of\ throat} = \frac{A_e}{A_t}$$

Figure 4.8 shows the combustion gas parameters and areas of the nozzle inlet, throat and exit. The throat conditions of the nozzle can be related with chamber conditions whose values are known after combustion of propellant. Energy balance at chamber and throat

$$h_c + \frac{V_c^2}{2} = h_t + \frac{V_t^2}{2}$$

Where $h$ is enthalpy and $V$ is the velocity of gasses. As the chamber area is comparatively higher than throat area, the velocity, $V_c$ of the gasses is neglected. The gas is considered as ideal gas, the enthalpy is a function of temperature only. Then, the above equation is simplified as:

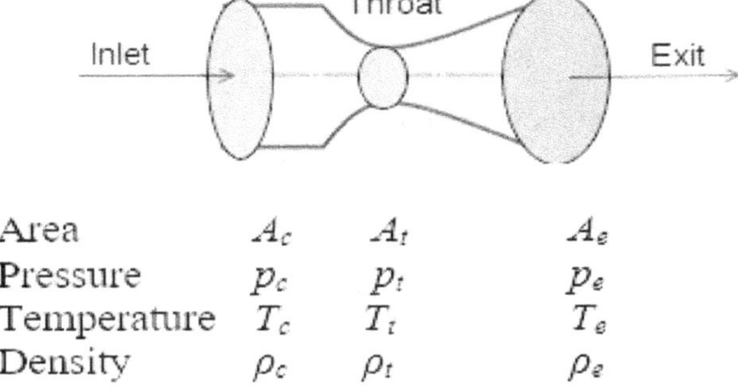


**Figure 4.8** Combustion Gas Parameters at Inlet, Throat and Exit of a Nozzle.

$$V_t^2 = 2c_p(T_c - T_t)$$

Using Eq. (4.9), expressing the specific heat in terms $\gamma\ and\ R$

$$V_t^2 = 2\frac{\gamma}{\gamma - 1}R(T_c - T_t)$$

$$(T_c - T_t) = \frac{\gamma - 1}{2}\frac{V_t^2}{\gamma R}$$

$$T_c = T_t\left(1 + \frac{\gamma - 1}{2}\frac{V_t^2}{\gamma R T_t}\right)$$

To achieve supersonic flow at exit, the Mach number $M = \frac{V_t}{a} = 1$ at throat and the sonic speed is $a^2 = \gamma R T_t$. Then,

$$T_c = T_t \left(1 + \frac{\gamma - 1}{2} M^2\right)$$

$$T_c = T_t \left(1 + \frac{\gamma - 1}{2}\right) \qquad \qquad .....(4.24)$$

$$\frac{T_t}{T_c} = \left(\frac{2}{\gamma + 1}\right) \qquad \qquad ......(4.25)$$

This equation relates the throat temperature to combustion chamber temperature. Assuming the flow is adiabatic and using adiabatic relations of Eq. (4.10) & (4.11), the pressure and density at the throat can be related with combustion chamber pressure and density.

$$\frac{p_t}{p_c} = \left(\frac{T_t}{T_c}\right)^{\frac{\gamma}{\gamma-1}} = \left(\frac{2}{\gamma+1}\right)^{\frac{\gamma}{\gamma-1}} \qquad \qquad (4.26)$$

$$\frac{\rho_t}{\rho_c} = \left(\frac{T_t}{T_c}\right)^{\frac{1}{\gamma-1}} = \left(\frac{2}{\gamma+1}\right)^{\frac{1}{\gamma-1}} \qquad \qquad (4.27)$$

The throat pressure $p_t$, for which an isentropic mass flow rate attains its maximum, is called the critical pressure. Typical values of the above critical pressure ratio range between 0.53 and 0.57 of the nozzle inlet pressure. The mass flow rate is:

$$\dot{m} = \rho_t A_t V_t = \rho_e A_e V_e$$

Now, the nozzle area ratio, $\varepsilon$ can be expressed as:

$$\epsilon = \frac{A_e}{A_t} = \left(\frac{V_t}{V_e}\right)\left(\frac{\rho_t}{\rho_e}\right) \qquad \qquad .....(4.28)$$

$$\frac{\rho_t}{\rho_e} = \left(\frac{\rho_t}{\rho_c}\right)\left(\frac{\rho_c}{\rho_e}\right) \qquad \qquad ......(4.29)$$

From ideal gas law $\frac{p}{\rho} = RT$

$$\frac{p_c}{\rho_c} = RT_c$$

$$\frac{p_e}{\rho_e} = RT_e$$

$$\frac{\rho_c}{\rho_e} = \frac{p_c}{RT_c} \frac{RT_e}{p_e}$$

Eq. (4.29) can be rewritten using the above relation and Eq. (4.27)

$$\frac{\rho_t}{\rho_e} = \left(\left(\frac{2}{\gamma+1}\right)^{\frac{1}{\gamma-1}}\right)\left(\frac{p_c}{RT_c}\frac{RT_e}{p_e}\right) = \left(\frac{2}{\gamma+1}\right)^{\frac{1}{\gamma-1}} \frac{p_c}{p_e}\frac{T_e}{T_c} \qquad .....(4.30)$$

Using adiabatic relation

$$\frac{p_c}{p_e}\frac{T_e}{T_c} = \frac{p_c}{p_e}\left(\frac{p_c}{p_e}\right)^{-\frac{\gamma-1}{\gamma}} = \left(\frac{p_c}{p_e}\right)^{\frac{1}{\gamma}}$$

Then,

$$\frac{\rho_t}{\rho_e} = \left(\frac{2}{\gamma+1}\right)^{\frac{1}{\gamma-1}}\left(\frac{p_c}{p_e}\right)^{\frac{1}{\gamma}} \qquad \dots\dots(4.31)$$

The sonic velocity, at throat is

$$V_t = \sqrt{\gamma R T_t} = \sqrt{\gamma R}\sqrt{T_t}$$

The exit velocity, $V_e$ of the nozzle is given by Eq. (4.14)

$$V_e = \sqrt{\frac{2\gamma R T_c}{(\gamma-1)}\left[1-\left(\frac{p_e}{p_c}\right)^{\left(\frac{\gamma-1}{\gamma}\right)}\right]}$$

Substituting the value of $T_c$ in terms of throat temperature $T_t$ using Eq. (4.25)

$$V_e = \sqrt{\frac{2\gamma R}{(\gamma-1)}\left(\frac{\gamma+1}{2}T_t\right)\left[1-\left(\frac{p_e}{p_c}\right)^{\left(\frac{\gamma-1}{\gamma}\right)}\right]}$$

So, the ratio of velocities at throat and exit of nozzle is

$$\frac{V_t}{V_e} = \frac{\sqrt{\gamma R T_t}}{\sqrt{\frac{2\gamma R}{(\gamma-1)}\frac{\gamma+1}{2}T_t\left[1-\left(\frac{p_e}{p_c}\right)^{\left(\frac{\gamma-1}{\gamma}\right)}\right]}} = \frac{1}{\sqrt{\frac{\gamma+1}{\gamma-1}\left[1-\left(\frac{p_e}{p_c}\right)^{\left(\frac{\gamma-1}{\gamma}\right)}\right]}} \qquad \dots\dots(4.32)$$

Substituting Eq. (4.31) and (4.32) in Eq. (4.29) to get area ratio in terms throat and exit pressures:

$$\epsilon = \frac{\left(\frac{2}{\gamma+1}\right)^{\frac{1}{\gamma-1}}\left(\frac{p_c}{p_e}\right)^{\frac{1}{\gamma}}}{\sqrt{\frac{\gamma+1}{\gamma-1}\left[1-\left(\frac{p_e}{p_c}\right)^{\left(\frac{\gamma-1}{\gamma}\right)}\right]}} \qquad \dots\dots(4.33)$$

This relation gives the area ratio of exit of nozzle to throat as function of pressure ratio and ratio of specific heats. Higher the chamber pressure higher is the area ratio. For a supersonic nozzle the ratio between any downstream area $(A_x)$ and throat can be found from the above equation by replacing $p_e$ with $p_x$, the pressure at the point $x$. Similarly, Eq. (4.32 can be used to find the velocity at any downstream point $x$.

$$\frac{A_x}{A_t} = \frac{\left(\frac{2}{\gamma+1}\right)^{\frac{1}{\gamma-1}}\left(\frac{p_c}{p_x}\right)^{\frac{1}{\gamma}}}{\sqrt{\frac{\gamma+1}{\gamma-1}\left[1-\left(\frac{p_x}{p_c}\right)^{\left(\frac{\gamma-1}{\gamma}\right)}\right]}} \qquad \dots(4.34)$$

Figure 4.9 shows the variation of area ratio and velocity ratio in a supersonic section of the nozzle in the divergent portion. It can be observed that the exhaust velocity gain is marginal above area ratios of 350. Additionally, higher area ratios create design difficulties and the heavy inert nozzle mass. Figure 4.10 shows the variations of pressure, temperature, velocity, and Mach number along with the length of nozzle. The pressure and temperature decrease while velocity and Mach number increase along the nozzle.

## 4.7   MASS FLOW RATE AND CHARACTERISTIC VELOCITY

The mass flow rate from the nozzle is given by

$$\dot{m} = \rho_t A_t V_t$$

The velocity at throat is

$$V_t = \sqrt{\gamma R T_t}$$

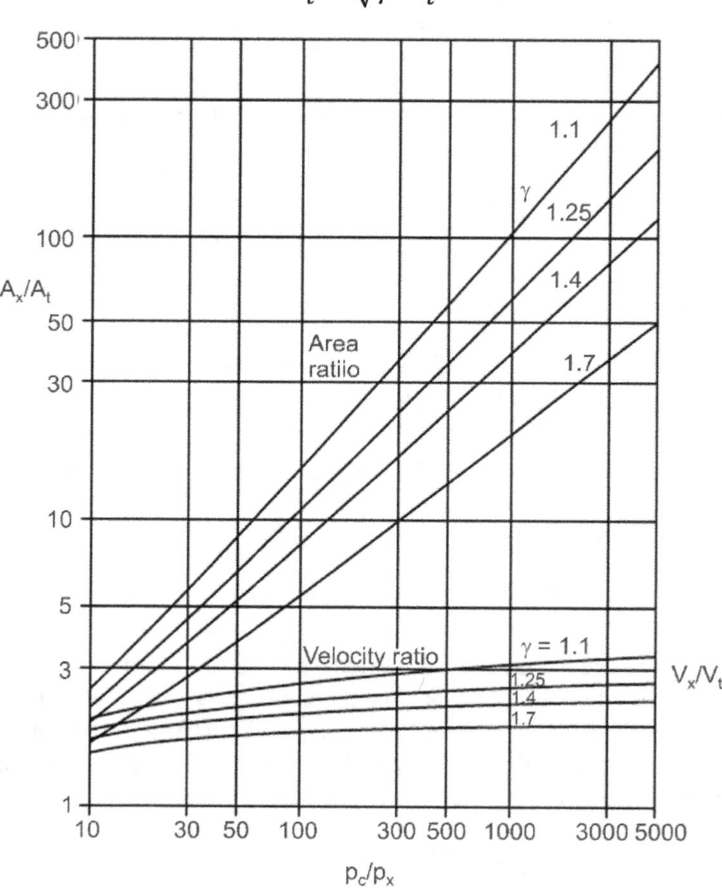


**Figure 4.9** Area and Velocity Ratios as a Function of Pressure Ratio for the Diverging Section of a Supersonic Nozzle (Source: George P Sutton).

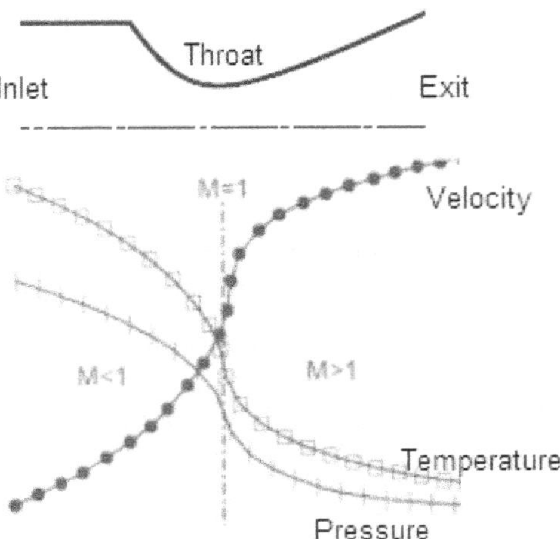


**Figure 4.10** Variations in Pressure, Temperature, Velocity and Mach number along with the length of a nozzle.

Using above relation and Eq. (4.27), the mass flow rate in the nozzle can be written as:

$$\dot{m} = \rho_t A_t V_t = A_t \rho_c \left(\frac{2}{\gamma+1}\right)^{\frac{1}{\gamma-1}} \sqrt{\gamma R T_t}$$

Using gas law $p_c = \rho_c R T_c$

$$\dot{m} = A_t \frac{p_c}{RT_c} \left(\frac{2}{\gamma+1}\right)^{\frac{1}{\gamma-1}} \sqrt{\gamma R T_t}$$

Rearranging the terms and simplifying

$$\dot{m} = A_t p_c \sqrt{\frac{\gamma}{RT_c}} \sqrt{\frac{T_t}{T_c}} \left(\frac{2}{\gamma+1}\right)^{\frac{1}{\gamma-1}}$$

$$= A_t p_c \sqrt{\frac{\gamma}{RT_c}} \left(\frac{2}{\gamma+1}\right)^{\frac{1}{\gamma-1}} \left(\frac{2}{\gamma+1}\right)^{\frac{1}{2}}$$

$$\dot{m} = A_t p_c \sqrt{\frac{\gamma}{RT_c}} \left(\frac{2}{\gamma+1}\right)^{\frac{\gamma+1}{2(\gamma-1)}} \qquad \dots\dots(4.35)$$

$$\dot{m} = \frac{\Gamma}{\sqrt{RT_c}} A_t p_c$$

Where $\Gamma$ is

$$\Gamma = \sqrt{\gamma} \left(\frac{2}{\gamma+1}\right)^{\frac{\gamma+1}{2(\gamma-1)}}$$

$$\dot{m} = \frac{A_t p_c}{c^*} \qquad\qquad .....(4.36)$$

$$c^* = \frac{1}{\Gamma}\sqrt{RT_c} = \frac{1}{\Gamma}\sqrt{\frac{R_u T_c}{M}} \qquad\qquad .....(4.37)$$

where $c^* = \frac{\sqrt{RT_c}}{\Gamma}$ $m/s$ is called characteristic velocity. The units are

$$\sqrt{\frac{J\,K}{kg\,K}} = \sqrt{\frac{Nm}{kg}} = \sqrt{\frac{kg\,m}{s^2}\frac{m}{kg}} = m/s$$

$\Gamma$ is a function of specific heat ratio and is a constant. As can be observed from Eq. (4.37) the $c^*$ value is higher as the combustion temperature, $T_c$, increases and the molecular mass of the combustion gasses reduces. $c^*$ is weakly dependent on the specific heat ratio $\gamma$, with a smaller value of $\gamma$ providing somewhat higher $c^*$ values. The characteristic velocity, $c^*$ is dependent on combustion chamber parameters and so, is a measure of the combustion performance of a rocket engine independent of nozzle performance, and is used to compare different propellants and propulsion systems. It is also a measure of mass flow generation in combustion reaction. The value of $c^*$ range from about 1333 m/s for monopropellant hydrazine up to 2360 m/s for cryogenic propellants oxygen/hydrogen.

## 4.8  THRUST GENERATED BY ROCKET ENGINE

The thrust developed by the rocket engine is derived by the rate of change of momentum of combustion gasses and is given by

$$F = \dot{m}V_e$$

Substituting the values of mass flow rate and exhaust velocity from Eq. (4.35) & (4.14)

$$F = A_t p_c \sqrt{\frac{\gamma}{RT_c}\left(\frac{2}{\gamma+1}\right)^{\frac{\gamma+1}{2(\gamma-1)}}} \sqrt{\frac{2\gamma RT_c}{(\gamma-1)}\left[1-\left(\frac{p_e}{p_c}\right)^{\left(\frac{\gamma-1}{\gamma}\right)}\right]}$$

Simplifying

$$F = A_t p_c \sqrt{\frac{2\gamma^2}{\gamma-1}\left(\frac{2}{\gamma+1}\right)^{\left(\frac{\gamma+1}{\gamma-1}\right)}\left[1-\left(\frac{p_e}{p_c}\right)^{\frac{\gamma-1}{\gamma}}\right]} \qquad\qquad .....(4.38)$$

Consider the Figure 4.11 showing the combustion chamber and nozzle and the forces acting on it. In addition to momentum thrust, another component of thrust called pressure thrust, comes from the force exerted by external pressure differences on the system. The component of the force acting in the direction of momentum thrust is due to the difference of the pressure of the flow leaving the nozzle, $p_e$ through the exit area, $A_e$ of the nozzle and the ambient pressure, $p_a$. Combining the two thrust components

$$F = \text{Momentum thrust} + \text{Pressure thrust}$$

$$F = \dot{m}V_e + (p_e - p_a)A_e \qquad \qquad ....(4.39)$$

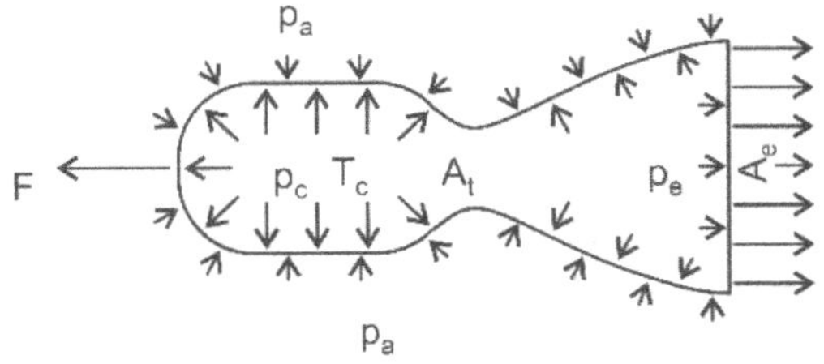

**Figure 4.11** Forces acting on Thrust Chamber and Nozzle.

The first term on the right-hand side of the above equation is the momentum thrust, while the second term is the pressure thrust. The momentum thrust contributes the major overall force, while the pressure thrust is a minor term, but is non-negligible in most instances. For a rocket ascending through the atmosphere, the ambient pressure will decrease with altitude and in general all three regimes, positive, zero, negative pressure thrust, will be realized within a given flight. It can be observed that the maximum thrust for a given nozzle operation is obtained in vacuum where the ambient pressure, $p_a$ is zero. Between sea level and the vacuum of space the thrust varies with altitude. Substituting the value of momentum thrust from Eq. (4.38) in Eq. (4.39)

$$F = A_t p_c \sqrt{\frac{2\gamma^2}{\gamma-1}\left(\frac{2}{\gamma+1}\right)^{\left(\frac{\gamma+1}{\gamma-1}\right)}\left[1-\left(\frac{p_e}{p_c}\right)^{\frac{\gamma-1}{\gamma}}\right]} + (p_e - p_a)A_e \qquad .....(4.40)$$

Rearranging the terms

$$F = A_t p_c \left[\sqrt{\frac{2\gamma^2}{\gamma-1}\left(\frac{2}{\gamma+1}\right)^{\left(\frac{\gamma+1}{\gamma-1}\right)}\left[1-\left(\frac{p_e}{p_c}\right)^{\frac{\gamma-1}{\gamma}}\right]} + \left(\frac{p_e}{p_c} - \frac{p_a}{p_c}\right)\epsilon\right] \qquad ....(4.41)$$

$$F = C_F A_t p_c \qquad \qquad .....(4.42)$$

where $C_F$ is called thrust coefficient and it is the thrust divided by chamber pressure and throat area.

$$C_F = \frac{F}{A_t p_c} \qquad \qquad .....(4.43)$$

The thrust coefficient is a function of gas constant, nozzle area ratio and pressure ratio across the nozzle but independent of chamber temperature. The optimum value of thrust coefficient is obtained when exit pressure of the nozzle, $p_e$ is equal to ambient pressure, $p_a$. The value of optimum thrust coefficient is

$$C_F^o = \sqrt{\frac{2\gamma^2}{\gamma-1}\left(\frac{2}{\gamma+1}\right)^{\left(\frac{\gamma+1}{\gamma-1}\right)}\left[1-\left(\frac{p_e}{p_c}\right)^{\frac{\gamma-1}{\gamma}}\right]} \qquad .....(4.44)$$

The thrust coefficient can be thought of as representing the amplification of thrust due to the gas expanding in the supersonic nozzle as compared to the thrust that would be exerted if the chamber pressure acted over the throat area only, i.e., suppose the nozzle is terminated at throat the thrust developed is

$$F = p_c A_t$$

By adding the divergence part of nozzle, the thrust developed is

$$F = C_F p_c A_t$$

i.e., thrust increased by a factor $C_F$. Figure 4.12 (a) shows the variation of optimum thrust coefficient, $C_F^o$ as a function of pressure ratio, nozzle area ratio and specific heat ratio. The thrust coefficient increases with the increase in pressure ratio. The effect of specific heat ratio on $C_F$ is not appreciable for low pressure ratios. Figure 4.12 (b) shows the variation of the thrust coefficient, $C_F$ as function of area ratio, $\epsilon$ for different pressure ratios $p_c/p_a$. These types of curves are useful in examining various nozzle problems for they permit the evaluation of under and overexpanded nozzle operation and flow separation, to be explained in later sections. As the expansion ratio increases the thrust coefficient approaches an asymptotic maximum as shown in the graph. Generally, a nozzle pressure ratio from 5 to 100 is used for rocket engines. It can be observed for a given pressure ratio ($p_c/p_a$) a humped shape curve is present. The value of the optimum value of thrust coefficient is at the maximum (top) point of this curve. The points to the left of the maximum represented under-expanded nozzles and points to the right of the maximum representing over-expanded nozzles. The ultimate achievable value for $C_F$, for a particular specific heat ratio $\gamma = 1.2$, is 2.246 with finite expansion area ratio.

## 4.9   CHARACTERISTIC VELOCITY AND SPECIFIC IMPULSE

The equations for mass flow rate, Eq. (4.36), and thrust, Eq. (4.42) are

$$\dot{m} = \frac{A_t\, p_c}{c^*}$$

$$F = C_F A_t p_c$$

Combining these two

$$\frac{F}{\dot{m}} = C_F c^*$$

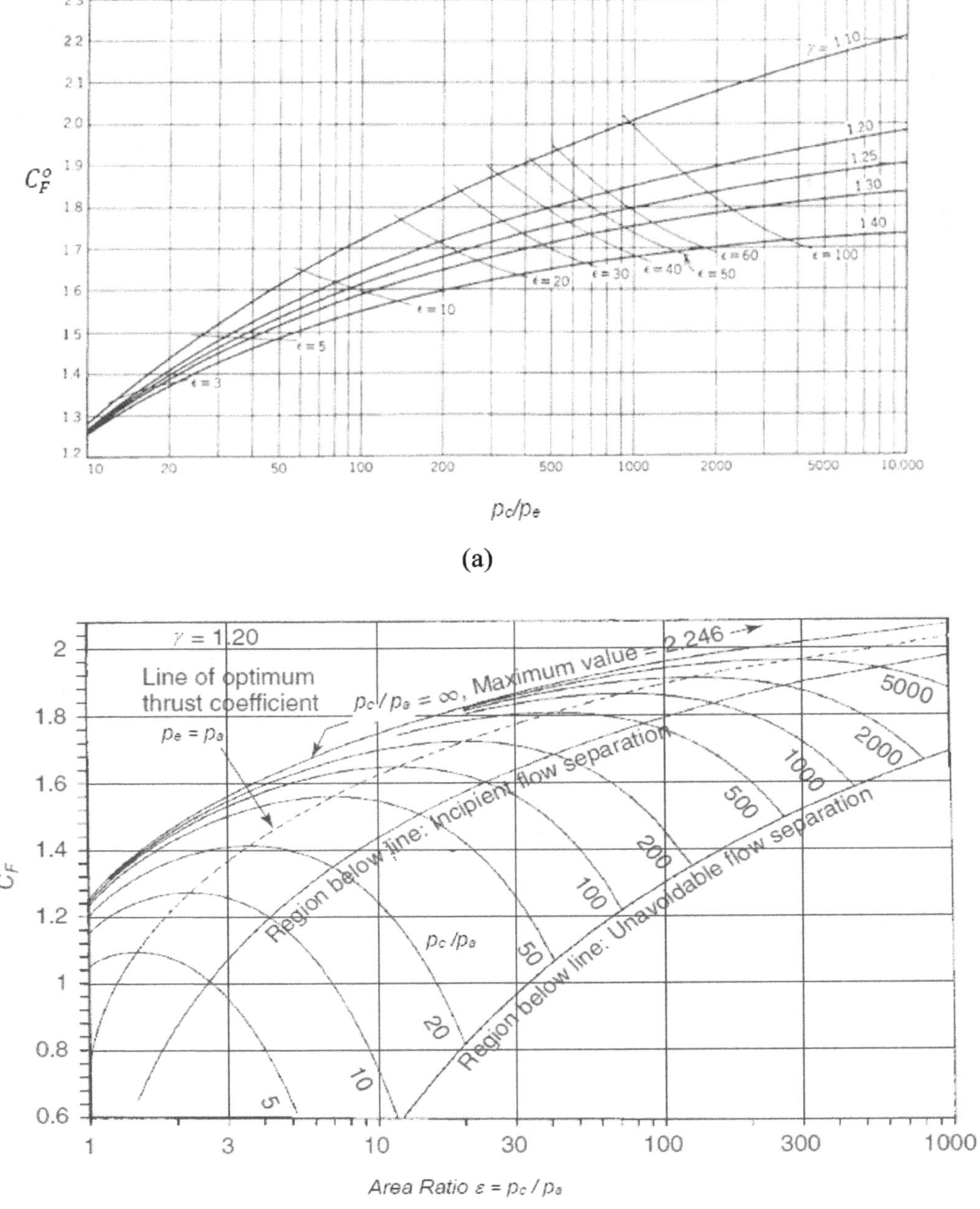


**Figure 4.12**  (a) Optimum Thrust Coefficient as a Function of Pressure Ratio, Nozzle Area Ratio for Different Specific Heat Ratio (b) Thrust Coefficient versus Nozzle Area Ratio for γ =1.2 (Source: George P Sutton, NASA Sp-8039).).

The LHS is thrust per unit mass flow rate which is equal to specific impulse $I_{sp}$.

$$I_{sp} = C_F c^*$$

$$\qquad\qquad\qquad\qquad\qquad .....(4.45)$$

The specific impulse is the product of thrust coefficient, $C_F$ and characteristic velocity, $c^*$. The $c^*$ represents the thrust chamber performance i.e., how efficiently it is producing the high-pressure gasses. The $C_F$ represents the nozzle performance i.e., how efficiently it is converting high pressure gasses into high velocity gasses. So, $c^*$ represents the figure of merit of thrust chamber and $C_F$ represents the figure of merit of nozzle. $I_{sp}$ is the combination of thrust chamber and nozzle performance. The thrust Eq. (4.39) is

$$F = \dot{m}V_e + (p_e - p_a)A_e$$

Rearranging the terms

$$I_{sp} = \frac{F}{\dot{m}} = V_e + (p_e - p_a)\frac{A_e}{\dot{m}}$$

This is at sea level for $p_a = 1$ atm or 0.1 MPa. The $I_{sp}$ at vacuum is when $p_a=0$.

$$I_{sp} = V_e + p_e A_e / \dot{m}$$

Therefore, comparing above two equations

$$(I_{sp})_{vacuum} > (I_{sp})_{sea\ level}$$

Generally, $I_{sp}$ is specified either at sea level or at vacuum. Also, higher chamber pressure $p_c$ gives higher expansion ratio, higher value of $I_{sp}$. As standard, $I_{sp}$ is mentioned at chamber pressure, $p_c$ at 70 atm or 7 MPa. In developing the relations for mass generation and thrust developed, the gas is considered ideal and the flow is one dimensional. In practice it is not, it is three dimensional and real flows. Experimental results are used to define the following efficiencies.

Efficiency of mass generation

$$= \frac{experimentally\ determined\ value\ of\ c^*}{calculated\ ideal\ value\ of\ c^*}$$

Efficiency of thrust generation

$$= \frac{experimentally\ determined\ value\ of\ C_F}{Calculated\ Ideal\ value\ of\ C_F}$$

Typically, these values are of the order 0.95 to 0.98.

## 4.10  EFFECT OF AMBIENT PRESSURE

The expansion of combustion gasses in the convergent divergent nozzle, considered above discussion, is an ideal and isentropic expansion process, i.e., within the nozzle the gasses expand from the constant chamber pressure, $p_c$ to the ambient pressure, $p_a$. In this case, the exit pressure of the nozzle, $p_e$ is equal to ambient pressure, $p_a$ (i.e., $p_e = p_a$). The nozzle is designed for a fixed chamber pressure, $p_c$ and an ambient pressure, $p_a$ i.e., for a particular altitude from the surface of Earth. However, the ambient pressure, $p_a$ varies with the altitude and for the given area ratio (ref. Eq. (4.33)) the performance of the nozzle varies. To get the optimum performance over a wide range of altitudes, the nozzle area ratio, $\varepsilon$ is to be varied. Due to high temperatures involved in the nozzle, to vary the area ratio using a mechanical device is not easy. Thus, considering the performance requirements over the complete trajectory of the rocket vehicle, an appropriate nozzle area ratio is selected in the design.

Consider a fixed geometry convergent divergent nozzle, discharging gas from a chamber with constant pressure, $p_c$ and temperature, $T_c$. The flow control parameter is ambient pressure, $p_a$. By reducing the ambient pressure, $p_a$ from no flow conditions ($p_c = p_a$), various flow regimes are possible as shown in Figure 4.13. The different possible flow conditions in the nozzle with reducing the ambient pressure, $p_a$ may be stated as follows: (ref. Figure 4.13)

1. Curve IE shows pressure variation in ideal expansion, i.e., the exit nozzle pressure is equal to ambient pressure ($p_e = p_a$). The flow is supersonic and adheres to nozzle walls, i.e., no separation of the flow occurs in the nozzle. The thrust coefficient, $C_F$ is optimum given by Eq. (4.44) and the specific impulse, $I_{sp}$ is maximum.

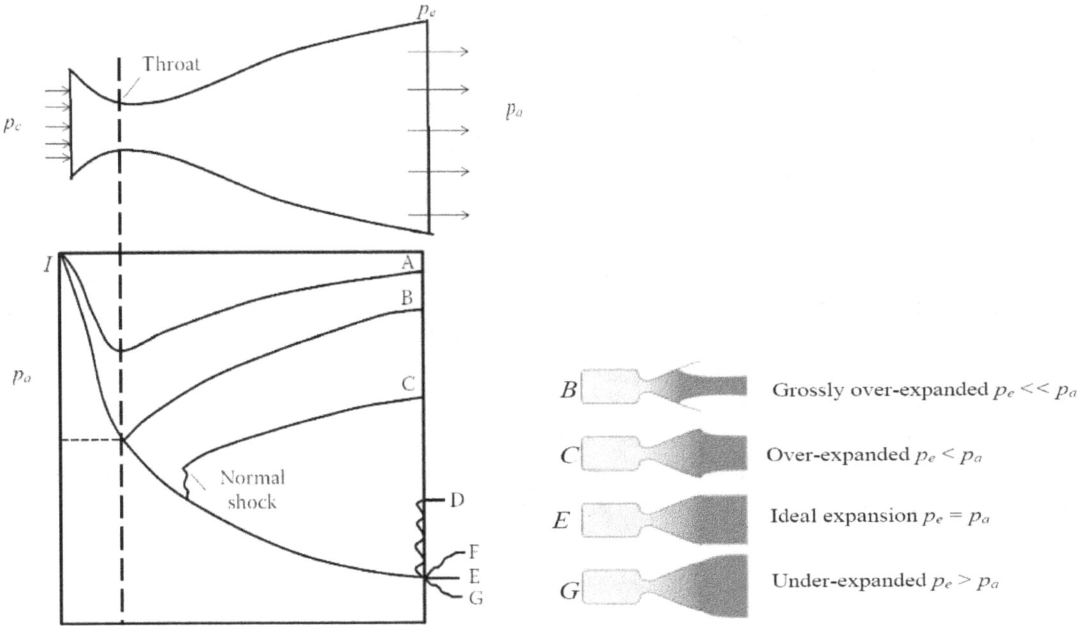


**Figure 4.13** Flow variation in convergent divergent nozzle.

2. If the ambient pressure, $p_a$ is just below the inlet pressure $p_c$, as shown in curve IA, the pressure ratio is below critical pressure ratio given by Eq. (4.26). In such cases, subsonic flow prevails throughout the nozzle. The nozzle experiences this condition normally during start and stop transients.

3. If the ambient pressure, $p_a$ is further reduced, as shown in curve IB, for a certain value of $p_a$, the throat will attain the sonic flow and the nozzle is choked, i.e., any further decrease of $p_a$ cannot increase the flow rate. The flow is fully supersonic in the divergent portion of the

nozzle. If the ambient pressure is further reduced, say to the point C as shown, the flow cannot attain isentropic condition. The pressure decreases from the throat until a normal shock is formed at point S in the divergent portion of the nozzle as shown in the curve IC. Normal shock waves are normal to the flow direction and across the shock wave the pressure, temperature and density of the gas increases instantaneously. The flow decelerates after normal shock in the divergent portion of the nozzle with a flow separation. At the nozzle exit, flow remains supersonic in the center portion but is surrounded by an annular-shaped section of subsonic flow. With the further decrease in pressure, the normal shock moves downward. This type of flow (C, D and F in Figure) with shock wave formed within the divergent portion of the nozzle is known as 'over-expanded' because the supersonic flow in the diverging part of the nozzle has lowered pressure so much that a recompression is required to match the discharge pressure. This is a normal situation for nozzles designed for higher altitudes and working at low altitude. The thrust coefficient, $C_F$ and specific impulse, $I_{sp}$ are lower than the optimum values.

4. When the ambient pressure, $p_a$ is less than the nozzle exit pressure $p_e$, the flow is supersonic and full. However, the expansion of the gas inside the nozzle is incomplete and the values of $C_F$ and $I_{sp}$ will be less than at optimum expansion. Expansion waves appear at the exit, to expand the exhaust to the lower back ambient pressure. This type of flow is named 'under-expanded' because exit pressure, $p_e$, is not low-enough, and additional expansion takes place after exhaust (i.e., outside the nozzle).

### 4.10.1 Under-expansion Flow in the Nozzle

As discussed earlier, the nozzle is designed for a particular altitude i.e., for a particular exit pressure, $p_e$ at the nozzle end. At higher altitude, though the ambient pressure, $p_a$ is less than the nozzle exit pressure $p_e$, there will not be any change in the nozzle flow conditions as it is fully expanded and isentropic flow. The exit pressure $p_e$ remains the same as that of the design value for isentropic flow and the supersonic flow expands further from $p_e$ to $p_a$, through a fan of Prandtl–Meyer-type expansion downstream of the nozzle exit in the free jet wave as shown in Figure 4.14. These waves further process the gasses at pressure $p_a$ and reduce their pressure to a value less than $p_a$ in the region downstream of $COC_1$. To adjust to the ambient pressure, these expansion

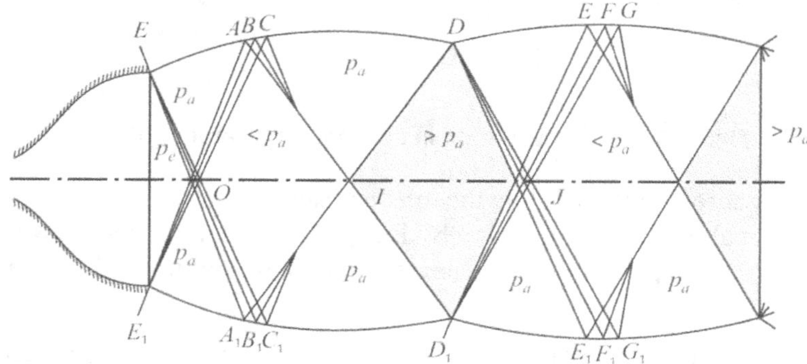


**Figure 4.14** Under-expansion flow pattern downstream of the nozzle exit.

waves get reflected from the free jet boundary downstream of the nozzle and reflected back as compression waves. These compression waves converge and coalesce to form oblique shock waves as shown. These oblique shock waves intersect at the point I as shown, and downstream of this point the pressure increases to values greater than $p_a$. The jet also contracts due to increase of pressure. The oblique shock waves intersect at the jet boundaries at D and $D_1$ and are reflected as a family of expansion waves and the jet boundaries diverge. Downstream of the intersecting point, J of expansion waves the pressure values again falls below ambient pressure $p_a$. This flow pattern, expansion waves followed by oblique shock waves, is repeated downstream of the nozzle till its effect reduces. When the flow gets across a compression shock wave, the abrupt compression causes a pressure increase. These regions of high pressure have high temperatures and therefore are more luminous that are easily observed in the exhaust flow. As the shape of these luminous zones in the exhaust jet are in the form of diamond, they are known as shock diamonds.

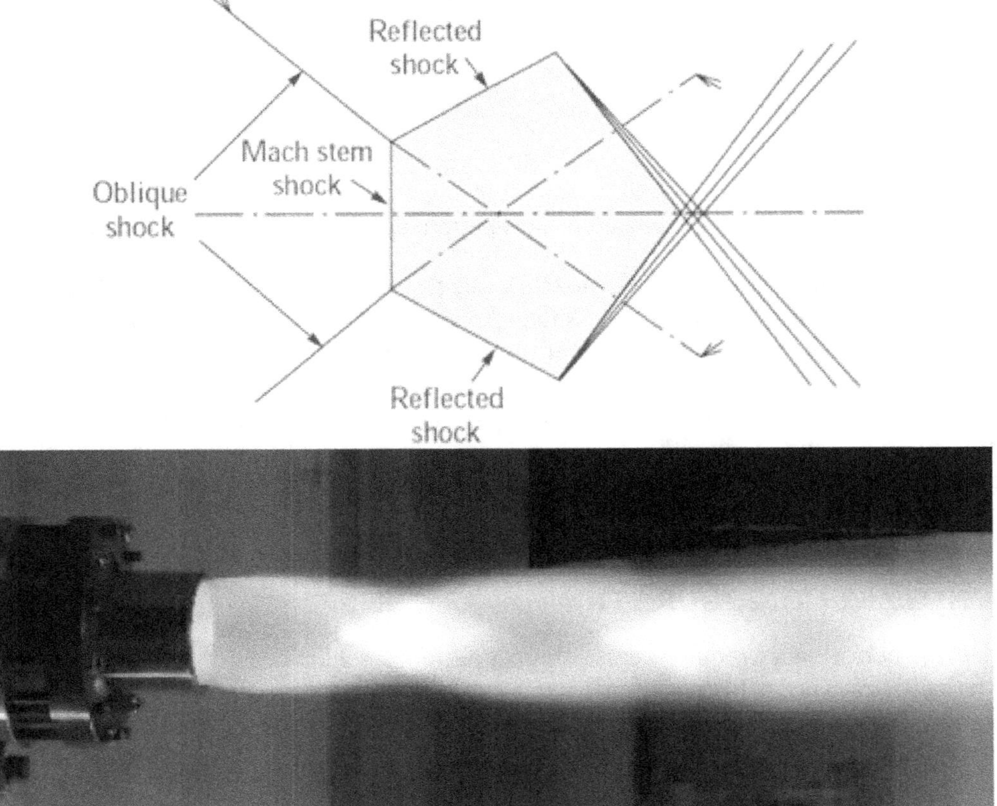


**Figure 4.15** Mach Waves and Shape of the Shock Diamond.

While oblique shock waves interact, a strong shock known as Mach stem shock and a reflected shock wave are generated. Figure 4.15 shows a typical Mach stem shock formed by oblique shocks and the shape of the diamond pattern. The shock diamonds are also known as Mach diamonds.

### 4.10.2 Over-expansion Flow in the Nozzle

While the nozzle is operated in lower altitudes than the designed altitude, overexpansion occurs within the nozzle, i.e., the gas is expanded in the nozzle such that the exit pressure, $p_e$ is less than the ambient pressure $p_a$. As the nozzle is choked condition, the increase in back pressure will not

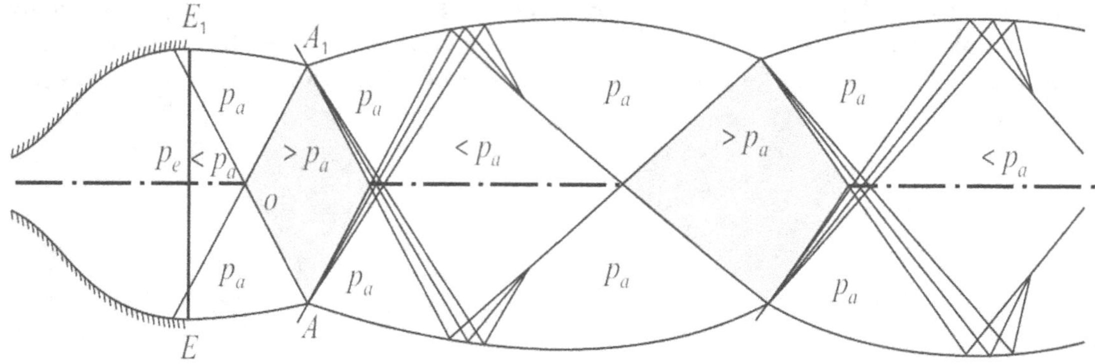


**Figure 4.16** Over-expansion: Flow pattern downstream of the nozzle exit.

change upstream flow conditions. the nozzle exit pressure remains the same as that of the design value for isentropic flow. Since $p_e$ is less than $p_a$, an oblique shock is formed inside the nozzle to compress the gasses to the higher value of the ambient pressure $p_a$ as shown in Figure 4.16. The plume converges as the pressure increases $p_a$. The sequence of compression and expansion is identical to that above-described for an under-expanded nozzle, except that it begins with the creation of an oblique shock rather than expansion waves. This behavior causes the flow spindle inward initially rather than to billow outward. Table 4.1 summarizes the different possible flow conditions in convergent divergent nozzles.

## 4.11 NOZZLE CONFIGURATION

To meet the performance requirements at higher altitudes, rocket nozzles are designed with higher area ratios. However, at lower altitude, over-expanded flow conditions prevail leading to flow separation. During transients, like start and stop, the flow separation may not be axially symmetric and may generate large side loads that may cause damage to nozzle exit cone structure and thrust.

**Table 4.1** Summary of Flow types in Convergent Divergent Nozzle

| Type of Flow | Characteristics | Observation |
|---|---|---|
| Under expanded nozzle. | Exit pressure is higher than atmospheric pressure ($p_e > p_a$). Expansion of the gas inside the nozzle is incomplete. Nozzle will flow full but will have expansion waves at exit of the nozzle. Values of $C_F$ and $I_{sp}$ are less than at optimum expansion. | *Expansion fan* |
| Optimum expansion | Exit pressure is equal to atmospheric pressure ($p_e = p_a$). Expansion of the gas is completed within the nozzle. Nozzle will flow full. No compression or expansion waves. Values of $C_F$ and $I_{sp}$ values are optimum. | *Pressure matched* |
| Over expanded nozzle | Exit pressure is lower than the atmospheric pressure ($p_e < p_a$). Separation of flow takes place inside the nozzle. The diameter of the supersonic jet is smaller than the nozzle exit diameter. Oblique Shock waves at exit of the nozzle. Values of $C_F$ and $I_{sp}$ will be less than at optimum expansion. | *Oblique Shock* |

vector control gimbal actuators. The design considerations of optimum nozzle shape for a given expansion ratio are:

(a) uniform, parallel, axial gas flow at the nozzle exit,

(b) minimum separation and turbulence losses within the nozzle,

(c) shortest possible nozzle length for minimum space envelope, weight, wall friction losses, and cooling requirements,

(d) ease of manufacturing.

The flow process in the nozzle can be divided in three parts: the convergent subsonic section, the throat section, and the divergent supersonic section. The mass flow rate through the nozzle depends on the area of the throat section and conditions of the thrust chamber. The shape of the converging portion depends upon the size of the thrust chamber. In the converging nozzle section, as the flow is subsonic and the pressure gradients are high; hence, the flow adheres to the walls.

The sonic velocity attained at the throat by the gasses is fixed by the combustion chamber conditions. This velocity can be further increased by the expansion occurring in the divergent supersonic portion of the nozzle. This additional velocity, leading to thrust increment, is dependent only on the configuration of the diverging nozzle walls and the exit area. Because of the very high velocities involved, the design of the contour of the divergent section of the nozzle is critical to avoid the separation, turbulence losses and heavy axial loads. The rocket nozzles used in propulsion engines can be grouped into four main types:

(i)   Cone (conical, linear)

(ii)  Bell (contoured, shaped, converging-diverging)

(iii) Annular (spike, aerospike, plug, expansion-deflection)

(iv)  Multi nozzle grid

The typical shapes of some of the nozzles are shown in Figure 4.17. The design of nozzle wall configuration to produce the desired uniform exit flow can be accomplished by the well-known method of characteristics using high speed computers. Figure 4.18 shows the minimum length of nozzles required to obtain uniform exit flow for different values of area ratios ($\varepsilon$) for typical types of nozzles. It is evident from the figure, the ideal nozzles that give maximum thrust performance are long, consequently, heavy.

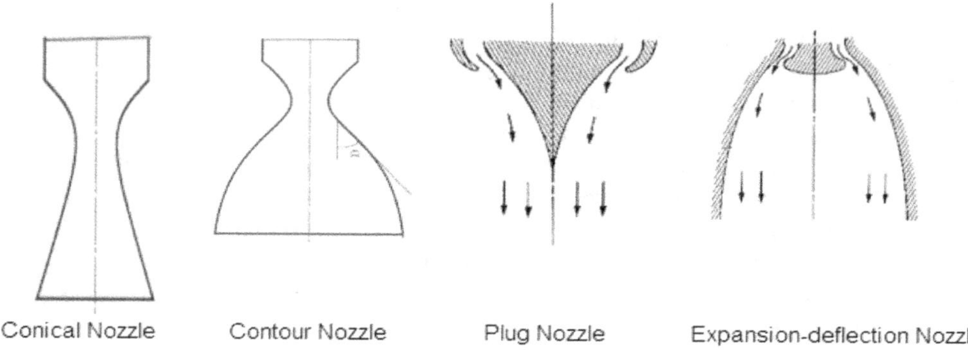


**Figure 4.17** Typical Nozzle Shapes.

### 4.11.1   Conical Nozzle

The use of conical nozzles was used almost exclusively in early rocket engines. The principal appeal of the conical nozzle is that it is easy to manufacture and it has the flexibility of converting an existing design to lower or higher area ratio without much redesign. A typical conical nozzle configuration is depicted in Figure 4.19. $\alpha$ is the half divergence angle of the cone and the performance of the nozzle depends on its value. The flow can be considered as one dimensional along the axis and the thrust is maximum in the direction if the entire flow is parallel to the axis for an optimum performance nozzle as shown in Figure 4.20.  Across the exit the flow direction is almost uniform. To reduce the bulkiness of the nozzle, if the divergence half angle, $\alpha$ is increased keeping the area ratio, $\varepsilon$ constant, the exit flow will have a significant radial component of velocity and highly subjected to flow separation reducing the thrust performance.

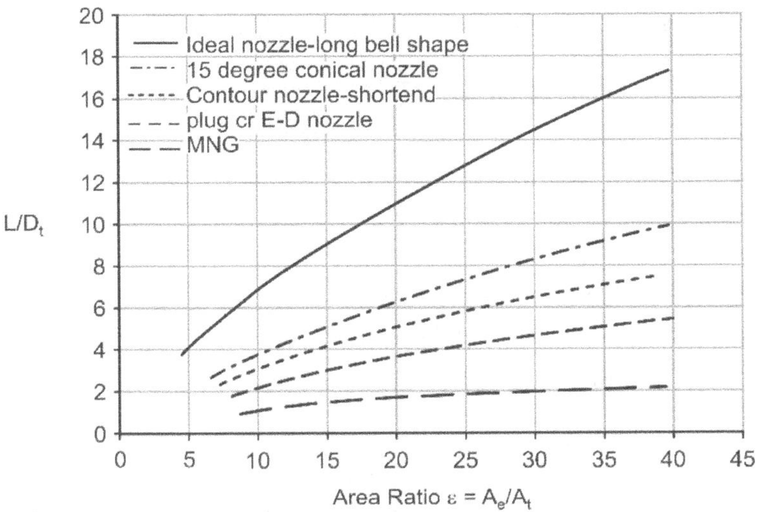


**Figure 4.18**  Length to throat diameter ratio for optimum performance of various nozzle types (Source: Rao GVR).

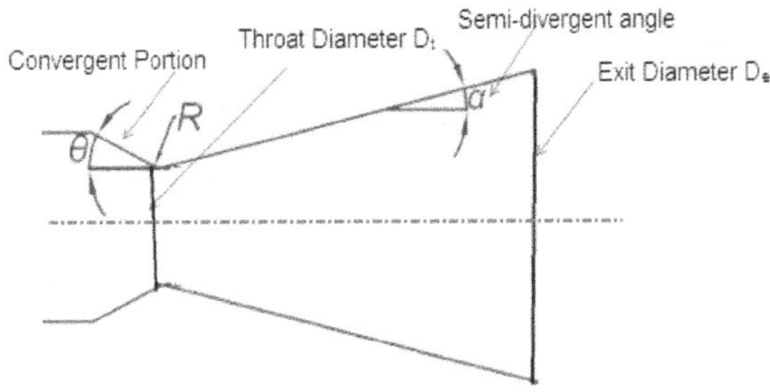


**Figure 4.19**  Typical shape of the conical nozzle.

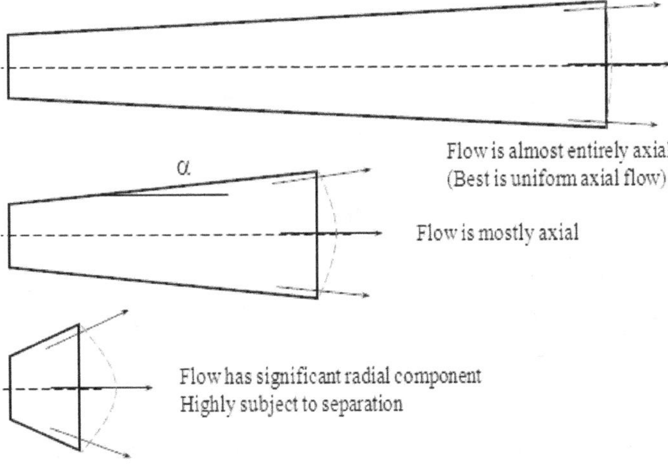


**Figure 4.20**  Flow pattern at the exit of the conical nozzle at various angle $\alpha$.

Assume that across the section the flow averaged at an inclined angle $\frac{\alpha}{2}$. Now, the thrust is

$$F = \dot{m}\cos\frac{\alpha}{2}\, V_e \cos\frac{\alpha}{2} = \dot{m}\, V_e \frac{\alpha}{2} \qquad \dots..(4.46)$$

Using the trigonometric relation $\cos 2\theta = 2\theta - 1$

$$F = \dot{m}V_e \left(\frac{1+\cos\alpha}{2}\right) \qquad \dots..(4.47)$$

$$F = \dot{m}V_e \lambda \qquad \dots..(4.48)$$

$\lambda$ is called is divergence loss factor. The thrust not available due to divergence is $(1-\lambda)$ of thrust. The correction factor applies only on ideal thrust term $(\dot{m}V_e)$ not on pressure term $(p_e\text{-}p_a)\, A_e$. The length of the cone L for exit radius $R_e$ and throat $R_t$ is given by

$$L = \frac{R_e - R_t}{tan\alpha}$$

The value of the correction factor, $\lambda$ decreases as $\alpha$ increases as shown in Figure 4.21.

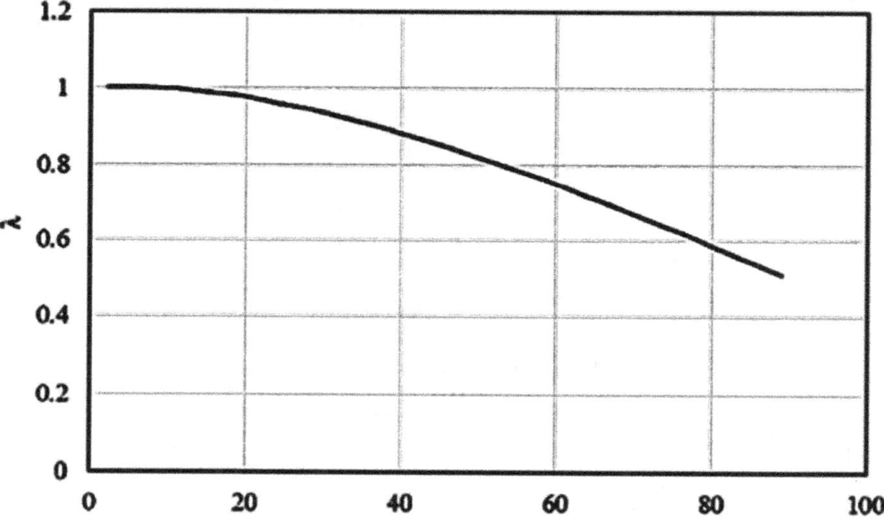

**Figure 4.21** Nozzle angle correction factor.

A small divergence angle causes most of the momentum to be axial and thus gives high specific impulse, but the long nozzle has a penalty in vehicle mass and design complexity. A large divergence angle gives short, lightweight design but the performance is low due to flow separation and turbulence. Usually, a half cone angle of 15° is used for the divergent section of the nozzle. The decrease in thrust coefficient of a 15° conical nozzle compared to one-dimensional is only about 1.7% and varies slightly with altitude conditions. Thus, a 15° conical nozzle is often used as a reference in comparing lengths and thrust performance of new types of nozzles. In the case of small rocket engines fitted with small area ratio nozzles, simple fabrication methods are preferred, and it is a common practice to use conical nozzles.

### 4.11.2    Bell Nozzle

The loss in thrust of a conical nozzle due to flow divergence becomes large with increasing cone angles associated with reduction in nozzle length. By contouring the nozzle wall, as shown in Figure 4.22, the flow can be turned closer to axial direction, and the loss in thrust due to flow divergence can be reduced to some extent. This is the most common nozzle shape used today.

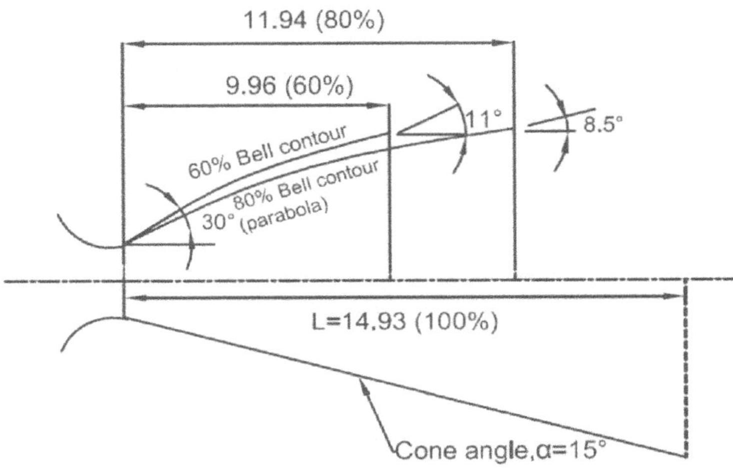


**Figure 4.22** Comparison of a 15° conical nozzle (reference nozzle) with 80% and 60% bell nozzles, all at an expansion ratio ε of 25(Source: George Sutton).

Because of its shape it is called a bell nozzle or Rao nozzle after its major design contributor. As shown in Figure 4.22 it has a high angle expansion section 20 to 50 degrees right behind the nozzle throat, followed by a gradual reversal of nozzle contour slope so that at the exit the divergence angle is small, usually less than 10 degrees. It is possible to have large divergence angles immediately behind the throat (20°–50°) because of the high relative pressure, the large pressure gradient, and the rapid expansion of the working fluid that do not permit separation in this portion unless there are discontinuities in the nozzle contour. As a practice, an equivalent 15-degree half-angle conical nozzle is commonly used as a standard to specify bell nozzles. Figure 4.22 shows a comparison of 80% and 60% of bell nozzles having the same area ratio with standard 15° conical nozzles. An 80% bell nozzle configuration that has the same expansion ratio ($\varepsilon$) as a 15° half angle conical nozzle is 20% smaller in length.

Several methods have been suggested for contouring the nozzle walls. However, the most widely accepted and convenient way of designing a near optimum thrust bell nozzle contour uses the parabolic approximation procedures suggested by G.V.R. Rao as shown in Figure 4.23. The nozzle contour immediately upstream of the throat T is a circular arc with a radius of 1.5 $R_t$. The divergent section nozzle contour is made up of a circular entrance section with a radius of 0.382 $R_t$ from the throat T to the point N and parabola from there to the exit E. The angles $\theta_N$ and $\theta_e$ are shown as function of area ratio $\varepsilon$ for different percentage of bell nozzles is shown in Figure 4.23.

### 4.11.3 Advanced Nozzle

The conical and bell nozzles described above suffer from losses due to their non-adaptability with altitudes. To enhance the nozzle performance over an entire range of flight are being developed to intend to reduce the nozzle length and weight. These new nozzles are not yet flight qualified for commercial use. Nozzles of type extensible, dual bell shaped, expansion-deflection and plug will be briefly described below.

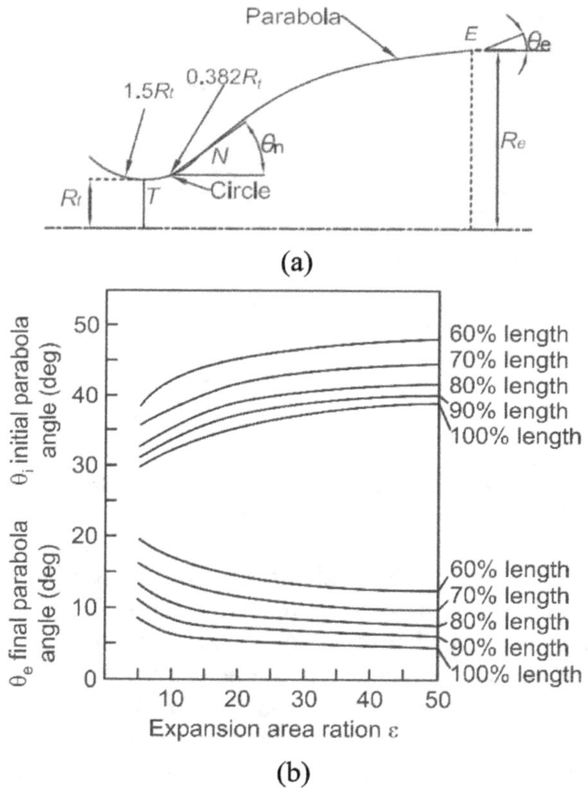


**Figure 4.23** (a) Parabolic approximation bell nozzle design configuration (b) angles $\theta_N$ and $\theta_e$ as function of area ratio $\varepsilon$ (Source: Rao GVR).

### 4.11.3.1 Extendible Nozzle

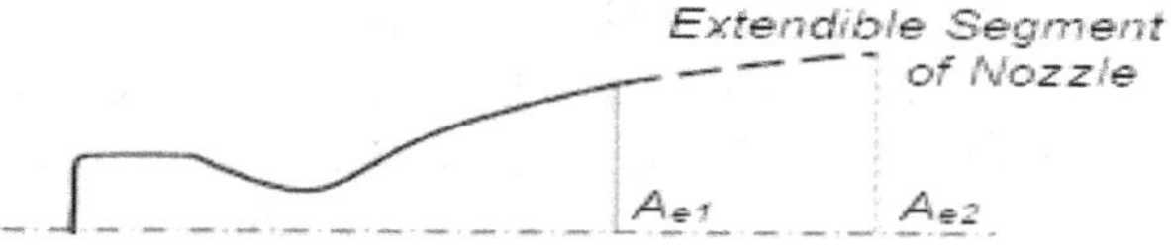


**Figure 4.24** Schematic of an extendible nozzle.

To adapt to the variation of altitude pressure, the nozzle area ratio must change continuously. As it is quite difficult to change the exit area of the nozzle, the extendible nozzle consisting of two or three segments has been designed and developed. Figure 4.24 shows a two-step extendible bell nozzle with exit areas $A_{e1}$ and $A_{e2}$ ($A_{e2} > A_{e1}$). Initially, the extendible nozzle with a smaller exit area $A_{e1}$ is used for a certain range of low altitude. Subsequently, a second segment is actuated using a special mechanism such that the nozzle exit area is increased to $A_{e2}$ at higher altitude for better performance. The problems associated with the actuation mechanism are the seal between nozzle segments and its added weight.

### 4.11.3.2 Dual Bell-Shaped Nozzle

In a dual bell nozzle, the two segments of the nozzle are formed with a hump between them as shown in Figure 4.25. In low altitude trajectory, the initial bell segment with area $A_{e1}$ is operated for gasses expansion. The flow separation occurs at the hump point and the flow in second segment does not adhere to the walls. At higher altitude, due to lower ambient pressure, gas expands further for a particular chamber pressure even beyond the hump point by getting attached up to the larger exit area $A_{e2}$. The performance improves as the nozzle operates at higher area ratio at higher altitudes.

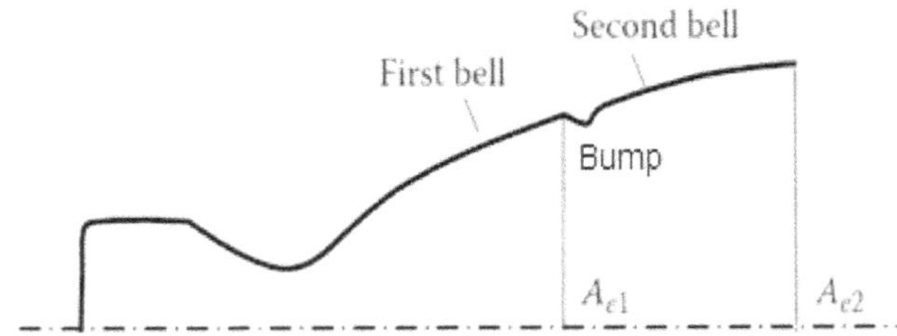


**Figure 4.25** Schematic of a dual bell–shaped nozzle .

### 4.11.3.3 Plug or Aerospike Nozzle

In the plug or aerospike nozzle (ref. Figure 4.17), flow moves inward toward its axis guided by the centrally placed plug and the outer gas boundary is interfaced directly with ambient air. The flow around the plug is adjusted with the changes in ambient pressure due to changes in altitude. The boundary is parallel to the axis at the designed pressure. The boundary moves inward at higher ambient pressure raising the exit pressure. At lower ambient pressure, the boundary moves outward and the gasses expand. The full parabolic plug nozzle, to achieve the optimum performance, is long and heavy. Hence, the plug is truncated to such an extent it provides least performance loss.

### 4.11.3.4 Expansion-deflection Nozzles

As shown in Figure 4.17 an aerodynamic plug is placed at the center of the nozzle. The expansion of the gasses around the plug gets adjusted with ambient pressure due to change in altitude. The flow will be parallel to the axis at the design pressure. The gas flow from the combustion chamber is turned around the curved contour of the plug and moves outward away from its central axis along the curved and diverging surface of the bell nozzle. The purpose of the plug is to force the flow to remain attached to, or to stick to, the nozzle walls. An aerodynamic interface is formed between the inner gas layer along the curved diverging section and the ambient air. When ambient pressure changes with altitude, hot gasses fill the larger portion of the diverging section of the nozzle.

### 4.11.3.5 Multi Nozzle Grid (MNG)

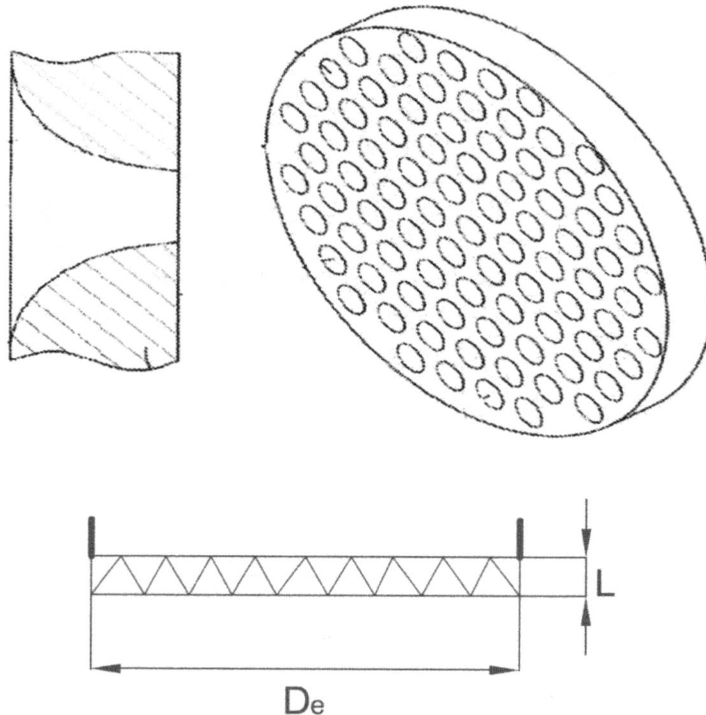


**Figure 4.26** Schematic of Multi Nozzle Grid.

The MGN technology packages an array of small "nozzlettes" that individually exhibit high performance expansion ratios and collectively provide large delivered thrust. An MGN nozzle provides the benefit of reducing conventional nozzle lengths (up to 10:1), reducing system weight while potentially improving overall delivered performance. MNG is the only nozzle configuration where the nozzle length (i.e., plate thickness) to throat diameter can be less than one, yet can provide extremely high area ratio.

## 4.12 LOSSES IN NOZZLES

Compared to an ideal nozzle, the real nozzle has energy losses and energy that is unavailable for conversion into kinetic energy of the exhaust gas. The main losses are:

(a) The divergence of the flow in the nozzle exit sections causes a loss.

(b) Wall friction can reduce the effective exhaust velocity by 0.5 to 1.5%

(c)  Presence of solid particles or liquid droplets in the gas can cause losses up to 5%

(d)  Unsteady combustion and oscillating flow can account for a small loss

(e)  Chemical reactions in nozzle flow change gas properties and gas temperatures, giving typically, a 0.5% loss

(f)  Losses during transient pressure operation, for example during start, stop, or pulsing

(g)  Erosion of the throat. This will reduce the chamber pressure and thrust and cause a reduction in specific impulse

(h)  Non-uniform gas composition can reduce performance (due to incomplete mixing, turbulence, or incomplete combustion regions)

## Example 4.1

Design a nozzle for ideal conditions to meet the following specifications:

| | |
|---|---|
| Thrust at sea level | 678 kN |
| Thrust chamber pressure | 58 bar |
| Chamber temperature | 3350 K |
| Ratio of specific heats | 1.22 |
| Molecular weight | 23 gm/mole |

What is the percentage variation of coefficient of thrust between sea level and 25 km altitude for the same nozzle for the same chamber pressure and temperature.

**Given**:

$$F = 678 \text{ kN (Sea Level)}$$

$$p_c = 58 \text{ bar}$$

$$T_c = 3350 \text{ K}$$

$$\gamma = 1.22$$

$$M = 23 \text{ gm/mole}$$

The throat conditions can be calculated taking Mach number $M = 1$

Temperature at throat, Eq. (4.25)

$$T_t = T_c \frac{2}{\gamma + 1} = 3350 \frac{2}{1.22 + 1} = 3018 \, K$$

Velocity at throat

$$V_t = \sqrt{\gamma R T_c} = \sqrt{1.22 \times \frac{8.31}{\frac{23}{1000}} \times 3018} = 1158 \, \frac{m}{s}$$

$$\left(\frac{kg\,m^2}{s^2K}\frac{1}{kg}\,K\right)$$

The density of gases at chamber, using ideal gas equation

$$\rho_c = \frac{p_c}{RT_c} = \frac{5.8 \times 10^6}{\left(\frac{8.31}{23}\right) \times 1000 \times 3350} = 4.792\,\frac{kg}{m^3}$$

The density of gases at the throat, Eq. (4.27)

$$\rho_t = \rho_c \left(\frac{2}{\gamma+1}\right)^{\frac{1}{\gamma-1}} = 4.792 \times \left(\frac{2}{2.22}\right)^{\frac{1}{0.22}} = 2.982\,\frac{kg}{m^3}$$

The exhaust velocity at the exit of the nozzle, Eq. (4.14)

$$V_e = \sqrt{\frac{2\gamma R_u T_c}{M(\gamma-1)}\left[1 - \left(\frac{p_e}{p_c}\right)^{\left(\frac{\gamma-1}{\gamma}\right)}\right]}$$

$$V_e = \sqrt{\frac{2 \times 1.22 \times |8.314 \times 3350}{\left(\frac{23}{1000}\right) \times 0.22}\left[1 - \left(\frac{0.1}{5.8}\right)^{\frac{0.22}{1.22}}\right]} = 2554\,\frac{m}{s}$$

Density of gases at the exit (using isentropic relations)

$$\rho_e = \rho_c \left(\frac{p_e}{p_c}\right)^{\frac{1}{\gamma}} = 2.982 \times \left(\frac{0.1}{5.8}\right)^{\frac{1}{1.22}} = 0.1068\,kg/m^3$$

Temperature of gases at exit

$$T_e = T_c \left(\frac{p_a}{p_c}\right)^{\frac{\gamma-1}{\gamma}} = 3550 \times \left(\frac{0.1}{5.8}\right)^{\frac{0.22}{1.22}} = 1707\,K$$

Mass flow rate in the nozzle

$$\dot{m} = \frac{F}{V_e} = \frac{678 \times 1000}{2554} = 265.5\,\frac{kg}{s}$$

Area of throat

$$A_t = \frac{\dot{m}}{\rho_t V_t} = \frac{265.5}{2.982 \times 1158} = 0.0769\,m^2$$

Diameter of throat

$$D_t = \sqrt{\frac{4A_t}{\pi}} = \sqrt{\frac{4 \times 0.0769}{\pi}} = 0.313\,m = 313\,mm$$

Area of the nozzle at exit

$$A_e = \frac{\dot{m}}{\rho_e V_e} = \frac{265.5}{0.1069 \times 2554} = 0.9725 \; m^2$$

Diameter of the nozzle at exit

$$D_e = \sqrt{\frac{4A_e}{\pi}} = \sqrt{\frac{4 \times 0.9725}{\pi}} = 1.113 \; m = 1113 \; mm$$

Area ratio of nozzle

$$\epsilon = \frac{A_e}{A_t} = 12.65$$

Characteristic velocity of the nozzle

$$c^* = \frac{A_t p_c}{\dot{m}} = \frac{0.0769 \times 5.8 \times 10^6}{265.5} = 1680 \; \frac{m}{s}$$

Thrust coefficient of the nozzle

$$C_F = \frac{F}{A_t p_c} = \frac{678 \times 1000}{0.0769 \times 5.8 \times 10^6} = 1.52$$

The atmospheric pressure at 25 km altitude $p_a = 0.002549$ MPa. The thrust coefficient at this altitude is, (ref Eq. (4.41) & Eq. (4.42)) calculated by substituting values in the equation

$$C_F = \left[ \sqrt{\frac{2\gamma^2}{\gamma - 1}\left(\frac{2}{\gamma + 1}\right)^{\left(\frac{\gamma + 1}{\gamma - 1}\right)} \left[1 - \left(\frac{p_e}{p_c}\right)^{\frac{\gamma - 1}{\gamma}}\right]} + \left(\frac{p_e}{p_c} - \frac{p_a}{p_c}\right)\epsilon \right]$$

$$C_F = 1.74$$

Percentage increase in $C_F$

$$= \frac{1.74 - 1.52}{1.52} \times 100 = 14.5\%$$

Thrust at 25 km level

$$F = C_F A_t p_c = 1.74 \times 0.0769 \times 5.8 \times 10^6 = 776 \; kN$$

**Nomenclature**

$A_e$  exit area of a nozzle (m²)

$A_t$  throat area of a nozzle (m²)

$a$  sound speed (m/s)

$c^*$  characteristic velocity (m/s)

$C_F$ thrust coefficient

$c_p$  specific heat at constant pressure (J/kg K)

$c_v$  specific heat at constant volume (J/kg K)

$D_e$ nozzle exit diameter (m)

$D_t$ nozzle throat diameter (m)

$h$   specific enthalpy (J/kg)

$M$  molecular mass (kg/mole)

$m$  mass (kg)

$M$ Mach number

$n$   number of moles

$p_c$  combustion pressure (N/m$^2$)

$p_e$  exit pressure (N/m$^2$)

$p_t$  pressure at throat (N/m$^2$)

$R$   specific gas constant (J/kg K)

$R_u$ universal gas constant (8.314 J/mole K)

$T_c$  combustion temperature (K)

$T_e$  exit temperature (K)

$T_t$  temperature at throat (K)

$V$   velocity (m/s), volume (m$^3$)

$V_e$ exhaust velocity (m/s)

$\alpha$   semi-divergent angle of a conical nozzle

$\gamma$   ratio of specific heats

$\varepsilon$   nozzle area ratio

$\lambda$   divergence loss factor of a conical nozzle

$\rho$   density (kg/m$^3$)


## PROBLEMS

4.1   The chamber pressure of a rocket operating at sea level is 22 bar at a temperature of 2300 K. The propellant mass flow rate is 1.2 kg/s. Calculate the thrust, specific impulse and exhaust velocity for optimum expansion.

4.2  Design an ideal nozzle to deliver 4500 N thrust at 20 km altitude. The thrust chamber pressure is 3.02 MPa at a temperature of 3100 K. The specific heat ratio of combustion gas is 1.25 and specific gas constant is 345 J/kg K. How much thrust will it develop at sea level for the same mass flow rate of propellant?

4.3  The characteristic velocity of an ideal rocket is 1250 m/s and the thrust coefficient is 1.50. The mass flow rate is 75.4 kg/s. The nozzle throat area is 0.023 m². Calculate the chamber pressure, exhaust velocity, thrust and specific impulse.

4.4  A solid rocket motor operating at combustion chamber pressure 7 .0 MPa generates gases at a temperature of 3000 K. The gases are expanded in the nozzle to the ambient pressure of 0.1 MPa. Calculate the following:

(a) Characteristic velocity

(b) Thrust coefficient of the rocket

(c) Ideal thrust coefficient.

(d) Ideal Specific Impulse

Take a molecular mass of the gas 22 gm/mole and the specific heat ratio of 1.27.

4.5  A nozzle operates at sea level having a mass flow rate of 4.02 kg/s at a chamber pressure of 2.5 MPa. The chamber temperature is 2750 K. Calculate exhaust velocity and temperature. Find the thrust coefficient of the nozzle. Take the specific heat ratio as 1.30 and molecular mass of the gas as 21 gm/mole. Assume the optimum expansion of the nozzle at sea level.

4.6  The F-1 engine of the Saturn launch vehicle operates at a chamber pressure of 70 bar. The throat diameter of the nozzle is 915 mm. The area ratio is 16. The combustion temperature is 3400 K. The specific impulse recorded during testing is 265 sec. Take the specific heat ratio is 1.23 and molecular mass 22.0 gm/mole. Calculate the following values:

(a)  momentum thrust

(b)  pressure thrust

(c)  total thrust

(d)  exhaust velocity

(e)  characteristic velocity

(f)  propellant mass flow rate

(g)  theoretical specific impulse

(h)  specific impulse efficiency

(i)  thrust coefficient

4.7 A nozzle having throat and exit diameters 131 and 986 mm operates in vacuum. This engine utilizes LOX/LH2 propellants which produce characteristic velocity of 2340 m/s at specific heat ratio of 1.2. The chamber pressure is 2.76 MPa. Calculate:

(a)  thrust coefficient

(b)  momentum and pressure thrust

(c)  total thrust

(d)  exhaust velocity

(e)  specific impulse

(f)  mass flow rate

(g)  exit pressure

4.8 Assuming full-flowing nozzle at sea level recalculate the values for the nozzle in problem 4.7.

# Rocket Propellants

## 5.1   INTRODUCTION

Any substance that is used to propel a rocket i.e., generating thrust for a rocket is called propellant. It could be in solid or liquid or gas form or a combination of any of these. The types of chemical rockets are solid, liquid or hybrid. Propellant is the chemical mixture mainly containing an oxidizer and a fuel, burned in thrust chamber to produce thrust in rockets. A fuel is a substance that burns when combined with oxygen producing gasses for propulsion. An oxidizer is an agent to release oxygen for the chemical reaction with a fuel. The heat liberated in the chemical reaction transforms propellants (reactants) into hot gaseous reaction products, which in turn are thermodynamically expanded in a nozzle to produce thrust. High value of specific impulse, $I_{sp}$ is required for the better performance of the rocket engine. It is the product of characteristic velocity $c^*$ and thrust coefficient $C_F$ as shown in Eq. (4.45). The $c^*$ represents the thrust chamber performance i.e., how efficiently it is producing the high-pressure gases. It is given by the Eq. (4.37):

$$c^* = \frac{1}{\Gamma}\sqrt{RT_c} = \frac{1}{\Gamma}\sqrt{\frac{R_u T_c}{M}}$$

*So, to achieve high value of $c^*$ requires*:

1.  High value of combustion temperature $T_c$, i.e., high value combustion pressure $p_c$

2.  Low value of molecular mass of combustion gases $M$

3.  Low value of ratio of specific heat $\gamma$.

This demands that propellants(reactants) must release high heat per unit mass after chemical reaction, generate gases with low molecular mass and specific heat ratio. i.e., the change in enthalpy must be large.

## 5.2   MOLECULAR MASS OF COMBUSTION GASSES

To have a low molecular mass of the combustion gases, the atomic mass of the constituents of the propellant must be small. Figure 5.1 shows the periodic table arrangement of the first 17 elements. In the Periodic Table the chemical elements are arranged in order of increasing atomic number.

The atomic mass also increases with atomic number. Above atomic number 17, the atomic mass is higher, and the compounds formed with them are much heavier.

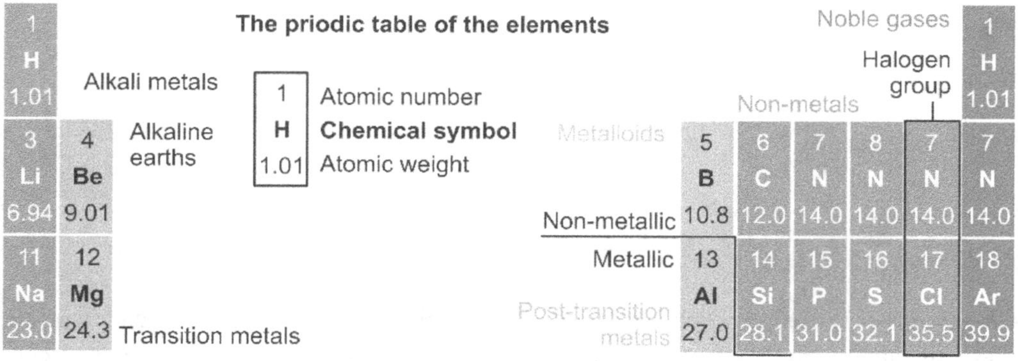

**Figure 5.1** Periodical Table arrangement for first 17 elements.

The elements can be classified and analyzed as below:

Fuel elements: H, Li, Be, B, C, Al

Oxidizer elements: O, Cl

Inert elements: He, N

Highly reactive elements: F, Na, Mg, P

The highly reactive elements are difficult to handle and so are not suitable as propellants. Though N is an inert element, many compounds formed with N like $N_2O_4$ are powerful oxidizers. So, propellants based on low atomic numbers like H, Li, Be, B, C, Al, O, Cl and N are suitable to have low molecular mass. The specific heat of mono-atomic gases is low and increases with complex combustion products. Thus, the combustion products must be simple like $CO_2$, $H_2O$ and $H_2$ to have a low specific heat value. The influence of specific heat ratio, $\gamma$ is not appreciable.

## 5.3 ALIPHATIC AND AROMATIC COMPOUNDS

An aliphatic compound is an organic **compound** containing carbon and hydrogen joined together in straight chains, branched chains, or non-aromatic rings. It is one of two broad classes of hydrocarbons, the other being aromatic compounds.

**Ex:** The simplest aliphatic compound is methane, $CH_4$. Others are ethylene, propane, butane etc.

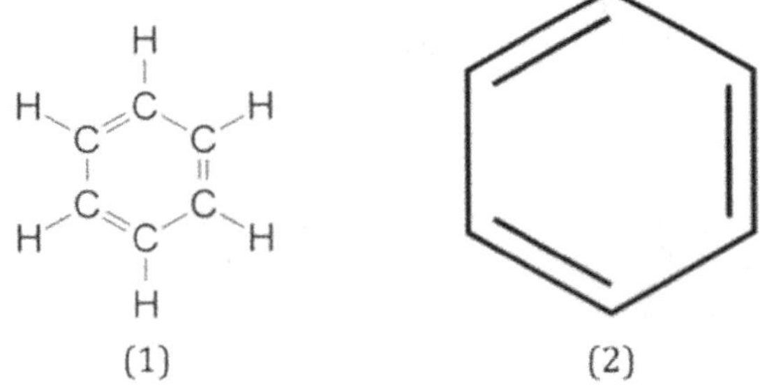

Aromatic compounds are substances that consist of one or more rings that contain alternating single and double bonds in its chemical structure.

Explosives are substances that have inbuilt oxygen. The fuel and oxygen are well mixed in it.

**Ex:** $H_2O_2$, $C_3H_5(ONO_2)_3$ Nitro-glycerin, $N_2H_4$ (Hydrazine) and $[C_6H_7(NO_2)_3O_5]_n$ Nitrocellulose.

Nitroglycerin is propane, $C_3H_8$ based compound. Substituting 3 hydrogen atoms with OH group $\rightarrow C_3H_5(OH)_3$ Propanetriol is formed. With further substituting OH group with nitro group $C_3H_5(ONO_2)_3$ Nitroglycerin is formed. $H_2O_2$, $C_3H_5(ONO_2)_3$, $HNO_3$ and Ammonium perchlorate $NH_4ClO_4$ are all oxidizers as they have excess oxygen after complete burning. $H_2O_2$ has one excess O after forming $H_2O$ and $NH4ClO4$ has two excess Os. Nitrocellulose $C_3H_5(ONO_2)_3$ is formed by nitrating cellulose $(C_6H_{10}O_5)_n$ and is fuel rich because oxygen atoms are insufficient to completely convert the fuel atoms C and H into $CO_2$ and $H_2O$.

## 5.4   MOLES AND MOLE FRACTION

It is convenient to work with chemical reactions in terms of moles and mole fractions rather than use of the mass of the substances. One mole of a substance is its molecular mass. Consider complete burning of butane $C_4H_{10}$ as shown in the following reaction.

$$2C_4H_{10} + 13O_2 = 8CO_2 + 10H_2O$$   .....(5.1)

i.e., 2 moles of butane gas react with 13 moles of oxygen to form 8 moles of carbon dioxide and 10 moles of water. As the molecular mass of butane is 58 gm, 1 mole of butane is equivalent to 58 gm on mass basis. 2 moles of butane are 2x58 gm of butane on a mass basis. Similarly, 13 moles of oxygen mean 13x32 gm of oxygen. In a mixture of several substances, the mole fraction, $x_i$ of a particular substance, say $i^{th}$ substance, is given by

$$x_i = \frac{n_i}{n}$$   .....(5.2)

where $n_i$ is the mole fraction of $i^{th}$ substance and $n$ is sum of the total number of moles in a mixture, i.e., $\sum n_i$

## 5.5   STOICHIOMETRY AND MIXTURE RATIO

A stoichiometric chemical reaction, which forms completely oxidized combustion products, releases a maximum amount of energy. When a chemical reaction goes to completion, that is, all the reactants are consumed and transformed into reaction products, the reactants are in *stoichiometric* proportions. For example, consider this reaction:

$$H_2 + \frac{1}{2} O_2 \rightarrow H_2O$$

All the hydrogen and oxygen are fully consumed to form the single product-- water vapor-- without any reactant residue of either hydrogen or oxygen. In this case it requires 1 mol of the $H_2$ and ½ mole of the $O_2$ to obtain 1 mol of $H_2O$. The mixer ratio is defined as

$$Mixer\ ratio\ (MR)_{st} = \frac{Mass\ of\ oxidiser}{Mass\ of\ fuel} = = \frac{\frac{36}{2}}{2} = \frac{8}{1}$$

The *stoichiometric* mixed ratio is 8:1.

The release of energy per unit mass of propellant mixture and the combustion temperature are highest at or near the stoichiometric mixture. Rocket propulsion systems usually do not operate with the proportion of their oxidizer and fuel in the stoichiometric mixture ratio. Instead, they usually operate fuel-rich because this allows light weight molecules such as hydrogen to remain unreacted; this reduces the average molecular mass of the reaction products, which in turn increases the specific impulse. For rockets using liquid hydrogen (LH$_2$) and liquid oxygen (LOX) propellants the best operating mixture mass ratio for high performance rocket engines is typically between 4.5 and 6.0, not at the stoichiometric value of 8.0.

## 5.6    HEAT OF FORMATION

The heat of formation or the enthalpy of a formation, $\Delta H_f\,^\circ$ is the change in enthalpy for the formation of one mole of a substance from the most stable form of its constituent elements at 1 bar (0.1MPa) and isothermally at 298.15 K or 25° C. The $\Delta$ implies that it is an energy change. The subscript f refers to formation and the superscript 0 means that each product or reactant substance is at its   standard state and at the reference pressure and temperature. Consider the formation of carbon dioxide from constituent stable elements carbon (solid) and oxygen (gas) at standard conditions of 1 bar and 25° C.

$$C + O_2 \rightarrow CO_2$$

One mole of carbon (solid) and 1 mole of oxygen (gas) react to give 1 mole of carbon dioxide (gas). During the reaction, an energy of 32,800 KJ/kg is released (exothermic reaction). The molecular mass of the carbon is 0.012 kg/mole and hence, the reaction releases energy of 0.012 $\times$ 32,800 = 394 kJ/mole. This energy increases the temperature of $CO_2$ gas. To bring back the temperature to standard reaction conditions of 25° C and one atmosphere pressure, it is essential to remove 394 kJ/mole of heat during reaction. Thus, the heat of formation of $CO_2$ from its elements $\Delta H_f^o = -394$ kJ/mole. So, the negative value of heat of formation indicates the exothermic reaction, whereas the positive value indicates the endothermic reaction. The above equation is now represented as

$$C + O_2 \rightarrow CO_2 - 394 \; kJ/mole$$

Similarly, for formation of water from its elements hydrogen (g) and oxygen (g) at standard conditions is

$$H_2 + \frac{1}{2}O_2 = H_2O - 286 \; kJ/mole$$

The heat of formation of $H_2O$ (l) is $\Delta H_f^o = -286$ kJ/mole. By convention, the heat of formation of the gaseous elements (e.g., $H_2$, $O_2$, Ar, Xe, etc.) is set to zero at these standard conditions of temperature and pressure. The heat of formation of some important pure substances is given in Table 5.1


## 5.7    HEAT OF REACTION

The heat of reaction $H_r\,^\circ$ is the energy released or absorbed when *products are formed from its reactants* at standard reference conditions, namely at 1 bar (0.1 MPa) and 25° C. The heat of reaction can be negative or positive, depending on whether the reaction is exothermic or endothermic. It is determined from the sum of the heats of formation of the products and the reactants.

**Table 5.1** Heat of Formation of Selected Substances at 298.15 K. (25°C) and 0.1 MPa (1 bar) (Source: JANAF Thermochemical Tables)

| Chemical formula | Species name | State | Standard heat of formation (kJ/mol) |
| --- | --- | --- | --- |
| $O_2$ | Oxygen | Gas | 0.0 |
| O | Element oxygen | Gas | 247.4 |
| $H_2$ | Hydrogen | Gas | 0.0 |
| H | Element hydrogen | Gas | 218.1 |
| OH | Hydroxyl | Gas | 42.3 |
| $H_2O$ | Water | Gas | −242.0 |
| $H_2O$ | Water | Liquid | −286.0 |
| $H_2O_2$ | Hydrogen peroxide | Liquid | −187.5 |
| $H_2O_2$ | Hydrogen peroxide | Gas | −133.2 |
| C | Graphite | Solid | 0.0 |
| CO | Carbon monoxide | Gas | −110.5 |
| $CO_2$ | Carbon dioxide | Gas | −394.0 |
| $CH_4$ | Methane | Gas | −74.5 |
| $C_2H_6$ | Ethane | Gas | −86.2 |
| $C_3H_8$ | Propane | Gas | −103.8 |
| $C_4H_{10}$ | Butane (n) | Gas | −124.7 |
| $C_4H_{10}$ | Butane (iso) | Gas | −131.8 |
| $C_2H_2$ | Acetylene | Gas | 226.9 |
| $CH_{1.842}$ | Kerosene | Liquid | −51.6 |
| $CH_3OH$ | Methyl alcohol | Gas | −201.0 |
| $CH_3OH$ | Methyl alcohol | Liquid | −238.6 |
| $N_2$ | Nitrogen | Gas | 0 |
| N | Element nitrogen | Gas | 471.8 |
| $N_2H_4$ | Hydrazine | Liquid | 50.4 |
| $HNO_3$ | Nitric acid | Liquid | −171.8 |
| $Cl_2$ | Chlorine | Gas | 0 |
| Cl | Chlorine atom | Gas | 121.4 |
| HCl | Hydrogen chloride | Gas | −92.1 |
| $NH_4NO_3$ | Ammonium nitrate | Solid | −365.3 |
| $NH_4ClO_4$ | Ammonium perchlorate | Solid | −290.5 |
| $NH_4Cl$ | Ammonium chloride | Liquid | −315.6 |
| $(CH_3)_2N_2H_2$ | UDMH | Liquid | 88.4 |
| $N_2O_4$ | Nitrogen tetroxide | Gas | 9.63 |
| $NO_2$ | Nitrogen dioxide | Gas | 33.9 |
| NO | Nitric acid | Gas | 90.4 |

$$H_r^0 = \sum n_i \left(\Delta H_f^0\right)_{products} - \sum n_i \left(\Delta H_f^0\right)_{reactants} \qquad \dots\dots(5.3)$$

where $n_i$ is the molar fraction of $i^{th}$ substance. The heat of reaction must be corrected to standard conditions when the reaction takes place at other temperatures and pressures in accordance with the change in enthalpy. Also, when phase change occurs (e.g., liquid to gas or gas to liquid) proper enthalpy changes to be considered in energy analysis.

If one of the reactants happens to be fuel and another one is the oxidizer, then combustion takes place, liberating a certain amount of heat, then the heat of reaction is known as the heat of combustion. The excess energy in the reactants above the product is the amount of heat available to increase the temperature of products. The value of heat of combustion depends on the temperature since the enthalpies of both reactants and products vary with temperature.

## Example 5.1

Determine the heat of combustion of one mole of methane with oxygen in stoichiometric ratio at standard conditions (1 atm and 298.15 K).

The stoichiometric reaction of methane is

$$CH_4(g) + 2O_2(g) \rightarrow CO_2(g) + 2H_2O(g)$$

The heat of reaction can be calculated by Eq. (5.3) and taking the heat of formation from Table 5.1

$$H_r^0 = \left(n_{CO_2} H_{f_{CO_2}}^0 + n_{H_2O} H_{f_{H_2O}}^0\right) - \left(n_{CH_4} H_{f_{CH_4}}^0 + n_{O_2} H_{f_{O_2}}^0\right)$$

$$H_r^0 = \left(1 \times (-394) + 2 \times (-242)\right) - \left(1 \times (-74.5) + 2 \times (0)\right) = -803.5 \frac{kJ}{mole}$$

Mass of fuel $CH_4 = (12+4) = 16$ gm/mole $= 16 \times 10^{-3}$ kg/mole

Heat combustion of 1 kg of fuel is

$$H_c^0 = -\frac{Heat\ of\ reaction \left(\frac{kJ}{mole}\right)}{Mass\ of\ fuel \left(\frac{kg}{mole}\right)} = \frac{803.5}{16 \times 10^{-3}} = 50.2 \times 10^3 \ kJ/kg$$

## 5.8 CALCULATION OF COMBUSTION TEMPERATURE AND PRODUCTS MOLECULAR MASS

The combustion process consists of a series of different chemical reactions that occur almost simultaneously. The chemical compounds breakdown into intermediate and final products subsequently. Consider a combustion process of a certain mixture ratio ($MR$) that leads to completion (equilibrium state) at constant pressure under adiabatic conditions. Then, the final temperature attained by the system is known as adiabatic flame temperature $T_c$. The adiabatic temperature of a combustion process depends on initial pressure, temperature, and composition of reactants. The theoretical analysis to calculate the combustion (adiabatic) temperature is only concerned with the initial and final conditions, before and after combustion. When the final compositions of the products are known, then the principles of the conservation of energy and the conservation of mass are sufficient to determine the combustion temperature.

To apply the conservation of energy, i.e., the first law of thermodynamics, to the combustion reaction, it is assumed that the chemical reaction occurs instantaneously but isothermally at the reference temperature, and the resulting energy release then heats the gases from this reference temperature to the final combustion temperature. As per the conservation of mass for the chemical reaction, the mass of any of the atomic species present in the reactants before the chemical reaction must be equal to the mass of the same species in the products. The propellants either solid or liquid, contain generally certain additive chemical compounds to improve the performance. Accordingly, the products of combustion contain many compounds or species. The number of nearly simultaneous chemical reactions that must be considered can easily exceed 150. Fortunately, many of these chemical species are present only in exceedingly small amounts and can usually be neglected. The final composition of the products must be known to calculate the combustion temperature. However, the final equilibrium composition depends on combustion temperature. Hence, one must resort to an iterative technique for determining both combustion temperature, $T_c$, and equilibrium composition. Also, to evaluate the enthalpy change of the products, the appropriate value of the specific heats must be chosen judiciously to get the correct combustion temperature. The specific heat of gasses at constant pressure varies with temperature as shown in Figure 5.2.

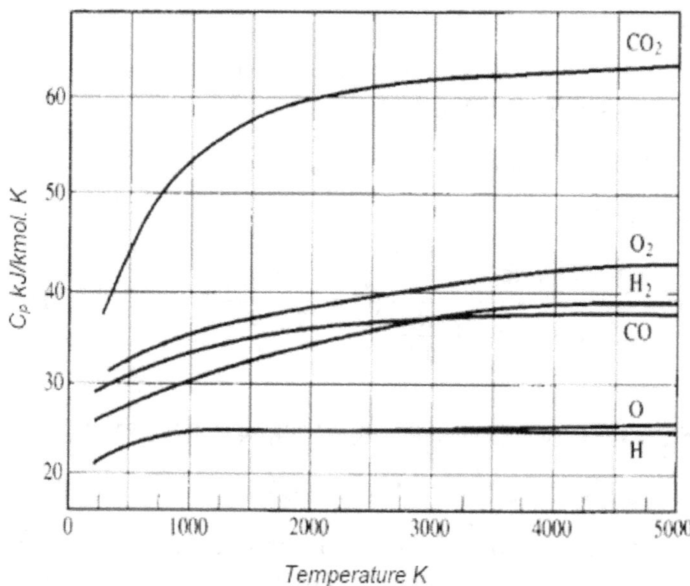


**Figure 5.2** Specific heat gases at constant pressure
(Source: JANAF Thermochemical Tables).

**Example 5.2**

Consider a rocket engine combustion with propellants liquid oxygen (LOX) and liquid hydrogen (LH$_2$). The LOX is at temperature 80 K and LH$_2$ at 20 K are burned in thrust chamber at stoichiometric ratio. The combustion temperature and molecular mass of the products will be calculated at stoichiometric conditions. The reaction equation is:

$$H_2 + \frac{1}{2} O_2 \rightarrow H_2 O$$

The heat of reaction is to be calculated at standard temperature 298 K. The reactants hydrogen and oxygen are brought to standard temperature. For example, to bring the liquid hydrogen at 20 K to standard temperature at 298 K, add sensible heat to bring it to boiling point at 22 K. add latent heat to bring it to gaseous state at 22 K and then add sensible heat to bring the gaseous hydrogen at 22 K to temperature 298 K. Similarly, product H$_2$O must be brought from liquid state at 298 K to combustion temperature $T_c$. Energy must be supplied to change the states. The heat generated by the reaction at standard state should be able to supply the required energy for the conversion of reactants to 298 K and to vaporize product H$_2$O to combustion temperature $T_c$. To summarize:

H$_2$ (Liquid @20 K) Sensible heat $\longrightarrow$ H$_2$ (Liquid @ boiling point 22 K) Latent heat h$_{fg}$   H$_2$

(gas @22K) sensible heat $\longrightarrow$ H$_2$ (Gas @298 K)

O$_2$ (Liquid @80 K) Sensible heat $\longrightarrow$ O$_2$ (Liquid @ boiling point 90 K) Latent heat h$_{fg}$   O$_2$

(gas @90K) sensible heat $\longrightarrow$ O$_2$ (Gas @298 K)

H$_2$O (Liquid @298 K) Sensible heat $\longrightarrow$ H$_2$O (Liquid @ boiling point 298 K) Latent heat h$_{fg}$ H$_2$O

(gas @298K) sensible heat $\longrightarrow$ H$_2$ O (Gas @ T$_c$ K combustion temperature)

The number of mols involved at standard temperature are:

*(1 mole of liquid hydrogen + $\frac{1}{2}$ mole of oxygen) $\rightarrow$ give 1mole of liquid water at standard temperature.*

Heat added for 1 mole of hydrogen:

Sensible heat required to raise 20 K to boiling point 22 K $= c_{p(l)}\Delta T = 20(22 - 20)40$

$$= 40 \text{ J/mole}$$

Latent heat of liquid Hydrogen $h_{fg} = 890$ J/mole

Sensible heat required to H$_2$ gas at 22 to bring to 298 k

$$= c_{p(g)} \Delta T = 30\left(\frac{J}{mole\ K}\right)(298 - 22) = 8280\,J/mole$$

Total heat required for hydrogen $= 40 + 890 + 8280 = 9210$ *J/mole*

For ½ mole of oxygen:

Heat required $= \frac{1}{2}\left[c_{p(l)}\Delta T + h_{fg} + c_{p(g)}\Delta T\right]$

Heat required $= \frac{1}{2}[29 \times 10 + 6800 + 35 \times 208] = 7185\, J/mole$

Total heat required for reactants (oxygen and hydrogen) = 9210 + 7185 = 16395 J/mole

For product 1 mole of $H_2O$:

Heat required = 90(J/mole K) *75 + 35000 (J/mole) + 58 (J/mole K) $\times (T_c - 373)$

$$= 41750 + 58*(T_c - 373)$$

Total heat required for reactants and products = 16395 + 41750 + 58 $\times (T_c - 373)$

$$= 58145 + 58 \times (T_c - 373)$$

The heat formation of $H_2O$ (g), from Table 5.1 is –242 kJ/mol, i.e., heat generated during reaction is 242,000 J/mole. Thus,

$$58145 + 58 \times (T_c - 373) = 242000$$

Therefore, the adiabatic temperature of combustion is,

$$T_c = 3170\ K$$

Molecular mass of the product $H_2O$, $M = 2+16 = 18$ gm/mole

Energy delivered = 242000 J/mole

Consider the reaction with more fuel, i.e., *fuel rich mass ratio*

$$H_2 + \frac{1}{4}\,O_2 \rightarrow \frac{1}{2}H_2O + \frac{1}{2}H_2$$

Mass ratio of mixture is $\dfrac{32/4}{2} = 4$

Molecular mass of the products $M = [\,(2+16)/2 + 2/2]/ (1/2+1/2) = 10$ gm/mole

The heat of formation of gases at standard state is zero and heat of formation of water is 242,000 J/mole.

For half mole of water, the heat formation = 121,000 J/mole

Energy released = 121,000 J/mole

Let consider an *oxidizer rich* propellant

$$H_2 + O_2 \rightarrow H_2O + \frac{1}{2}O_2$$

Mass ratio of mixture = 32/2 = 16

Molecular mass of the products $M = [18+32/2]/ (1+1/2) = 22.7$ gm/mole.

As the fuel is fully oxidized, the energy released is the same as stoichiometric reaction value= 242,000 J/mole. The values of energy released per unit mole of reactants for the cases considered above are tabulated in Table 5.2.

**Table 5.2** Energy released per unit mole of reactants with different mixture ratios.

| Item | Stoichiometric Mixture | Fuel Rich Mixture | Oxidizer Rich Mixture |
|---|---|---|---|
| Mixture Ratio | 8 | 4 | 16 |
| Molecular Mass (gm/mole) | 18 | 10 | 22.7 |
| Energy Released (J/mole) | 242,000 | 121,000 | 242,000 |
| Energy Released per unit mole of reactants (J/mole) | 161,300 | 96,800 | 121,000 |

The results are plotted against the mixture ratio as shown in Figure 5.3. Propellant combinations having fuel-rich mixture ratios form combustion products with lower molecular mass. This leads to also, lower value of specific ratios. The maximum temperature is obtained in the fuel-rich mixture near to the stoichiometric mixture ratio. As per the Eq. (4.14), the exhaust velocity $V_e$ is proportional to $\sqrt{T_c/M}$ . Maximum values of exhaust velocity i.e., specific impulse would, therefore, be obtained for fuel-rich mixture ratios instead of stoichiometric mixture ratios.

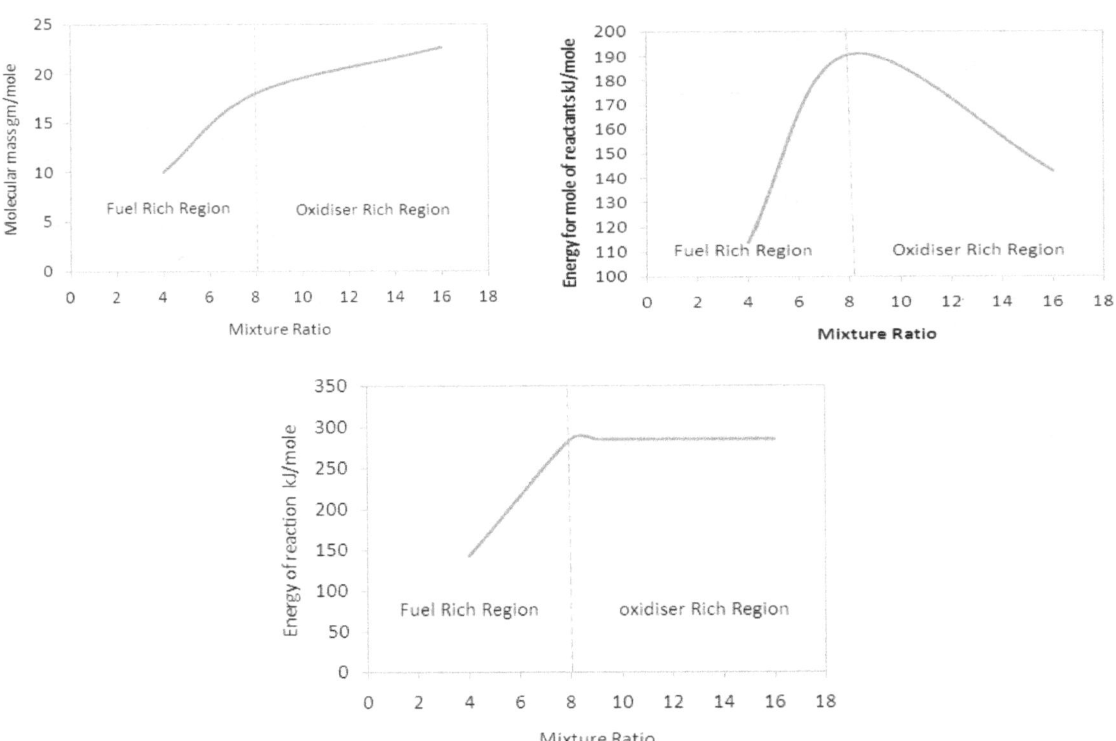


**Figure 5.3** Graphical Representation of Values in Table 5.2.

It is important to note that to find the correct adiabatic temperature, the final composition of the products must be known, and the appropriate value of specific heats must be used. The final equilibrium composition is dependent on the final combustion temperature. Also, the specific heat is a function of temperature. Hence, the final composition and combustion temperature are to be calculated using an iterative method.

The combustion temperature depends also on combustion pressure. The optimum mixture ratio that produces the maximum impulse is generally selected. Unlike other propellants, the optimum mixture ratio for LOX and LH$_2$ is not necessarily that which will produce the maximum specific impulse. Due to extremely low density of liquid hydrogen, the propellant volume decreases significantly at higher mixture ratios. Maximum specific impulse typically occurs at a mixture ratio of around 3.5, however by increasing the mixture ratio to, say, 5.5 the storage volume is reduced by one-fourth. This results in smaller propellant tanks, lower vehicle mass, and less drag, which generally offsets the loss in performance that comes with using the higher mixture ratio. In practice, most liquid oxygen/liquid hydrogen engines typically operate at mixture ratios from about 5 to 6.

## 5.9   CHEMICAL EQUILIBRIUM

Chemical equilibrium in a chemical reaction is the state in which both reactants and products are present in concentrations which have no further tendency to change with time. There will not be any observable changes in the properties of the system. In such state, the forward reaction proceeds at the same rate as the reverse reaction. Thus, there are no net changes in the concentrations of the reactant(s) and product(s). J.W. Gibbs, based on minimizing the free energy, developed necessary and sufficient thermodynamic condition for a chemical equilibrium. The free energy may be thought of as the chemical potential, the tendency or driving force for a chemical material to enter a chemical (or physical) change. When the chemical potential of the reactants is higher than that of the likely products, a chemical reaction can occur, and the chemical composition can change.

Combining the first and second laws of thermodynamics

$$dU + pdV - TdS \leq 0 \qquad \qquad \text{......(5.4)}$$

The Gibbs function is

$$G = H - TS = U + pV - TS \qquad \qquad \text{.....(5.5)}$$

where $U$ is internal energy, $p$ is pressure, $V$ is molar volume, $H$ is enthalpy, $T$ is temperature, and S is entropy. Differentiating the Gibbs function at constant temperature and pressure

$$(dG)_{T,P} = dU + pdV + Vdp - TdS - SdT = dU + pdV - TdS$$

Using the Eq. (5.4), the above equation can be written as

$$(dG)_{T,P} \leq 0 \qquad \qquad \text{......(5.6)}$$

The above equation confirms that a chemical reaction can only proceed in the direction of decreasing Gibbs function until it reaches its minimum value. In other words, chemical reaction cannot proceed in the direction of increasing Gibbs function as it violates the second law of thermodynamics. At the minimum value of Gibbs function no further change in chemical composition takes place. When a chemical reaction is in equilibrium, an equilibrium constant, $K_p$

has been devised which relates the partial pressures and the molar fractions of the species. Consider in the general reaction

$$aA + bB \rightleftharpoons cC + dD \qquad \dots\dots(5.7)$$

a, b, c, and d are the stoichiometric molar concentration coefficients of the chemical molecules (or atoms) A, B, C, and D. The equilibrium constant $K_p$, expressed in partial pressures, is a function of temperature.

$$K_p = \frac{p_C^c p_D^d}{p_A^a p_B^b} \qquad \dots\dots(5.8)$$

It is convenient to define the equilibrium constant $K_n$, in terms of mole fractions.

$$K_n = \frac{x_C^c x_D^d}{x_A^a x_B^b} \qquad \dots\dots(5.9)$$

The mole fraction of the species in a mixture can be expressed as

$$x_i = \frac{p_i}{p_m}$$

where $p_i$ is the partial pressure $i^{th}$ spice present in the mixture at a pressure $p_m$. According to the Gibbs-Dalton law, the mixture pressure is the sum of the partial pressure. Then,

$$\frac{p_i}{p_m} = \frac{n_i}{n_1 + n_2 + \cdots + n_n} \qquad \dots\dots(5.10)$$

where $n_i$ is the number of moles of constituent $i$ present in the mixture. The equilibrium constant for the chemical formation of a given species from its elements is $K_f$. The free energy and the equilibrium constant for the formation of a particular species at standard conditions from its atomic elements are related, namely

$$\Delta G^0 = -R_u T \ln K_f$$

The thermal properties like specific heat, heat of formation, Gibbs free energy and equilibrium constant at different temperatures of various spices that help to calculate the composition of products are available from JANAF tables (NIST-JANAF Thermochemical Tables). Table 5.3 gives Gibbs free energy and other thermodynamics values for $H_2O$.

**Table 5.3** Values of Gibbs Free Energy of Dissociation, Equilibrium Constant and Heat of Formation of $H_2O$.

$$\textbf{Gibbs } (\boldsymbol{H_2O \rightarrow H_2 + \tfrac{1}{2} O_2})$$

| | **J·K⁻¹ mol⁻¹** | | | | **kJ·mol⁻¹** | | |
|---|---|---|---|---|---|---|---|
| $T/K$ | $C_p°$ | $S°$ | $-[G°-H°(T_r)]/T$ | $H-H°(T_r)$ | $\Delta_f H°$ | $\Delta_f G°$ | log $K_f$ |
| 0 | 0. | 0. | INFINITE | -9.904 | -238.921 | -238.921 | INFINITE |
| 100 | 33.299 | 152.388 | 218.534 | -6.615 | -240.083 | -236.584 | 123.579 |
| 200 | 33.349 | 175.485 | 191.896 | -3.282 | -240.900 | -232.766 | 60.792 |

**Table *contd...***

| | | | | | | | |
|---|---|---|---|---|---|---|---|
| 298.15 | 33.590 | 188.834 | 188.834 | 0. | -241.826 | -228.582 | 40.047 |
| 300 | 33.596 | 189.042 | 188.835 | 0.062 | -241.844 | -228.500 | 39.785 |
| 400 | 34.262 | 198.788 | 190.159 | 3.452 | -242.846 | -223.901 | 29.238 |
| 500 | 35.226 | 206.534 | 192.685 | 6.925 | -243.826 | -219.051 | 22.884 |
| 600 | 36.325 | 213.052 | 195.550 | 10.501 | -244.758 | -214.007 | 18.631 |
| 700 | 37.495 | 218.739 | 198.465 | 14.192 | -245.632 | -208.812 | 15.582 |
| 800 | 38.721 | 223.825 | 201.322 | 18.002 | -246.443 | -203.496 | 13.287 |
| 900 | 39.987 | 228.459 | 204.084 | 21.938 | -247.185 | -198.083 | 11.496 |
| 1000 | 41.268 | 232.738 | 206.738 | 26.000 | -247.857 | -192.590 | 10.060 |
| 1100 | 42.536 | 236.731 | 209.285 | 30.191 | -248.460 | -187.033 | 8.881 |
| 1200 | 43.768 | 240.485 | 211.730 | 34.506 | -248.997 | -181.425 | 7.897 |
| 1300 | 44.945 | 244.035 | 214.080 | 38.942 | -249.473 | -175.774 | 7.063 |
| 1400 | 46.054 | 247.407 | 216.341 | 43.493 | -249.894 | -170.089 | 6.346 |
| 1500 | 47.090 | 250.620 | 218.520 | 48.151 | -250.265 | -164.376 | 5.724 |
| 1600 | 48.050 | 253.690 | 220.623 | 52.908 | -250.592 | -158.639 | 5.179 |
| 1700 | 48.935 | 256.630 | 222.655 | 57.758 | -250.881 | -152.883 | 4.698 |
| 1800 | 49.749 | 259.451 | 224.621 | 62.693 | -251.138 | -147.111 | 4.269 |
| 1900 | 50.496 | 262.161 | 226.526 | 67.706 | -251.368 | -141.325 | 3.885 |
| 2000 | 51.180 | 264.769 | 228.374 | 72.790 | -251.575 | -135.528 | 3.540 |
| 2100 | 51.823 | 267.282 | 230.167 | 77.941 | -251.762 | -129.721 | 3.227 |
| 2200 | 52.408 | 269.706 | 231.909 | 83.153 | -251.934 | -123.905 | 2.942 |
| 2300 | 52.947 | 272.048 | 233.604 | 88.421 | -252.092 | -118.082 | 2.682 |
| 2400 | 53.444 | 274.312 | 235.253 | 93.741 | -252.239 | -112.252 | 2.443 |
| 2500 | 53.904 | 276.503 | 236.860 | 99.108 | -252.379 | -106.416 | 2.223 |
| 2600 | 54.329 | 278.625 | 238.425 | 104.520 | -252.513 | -100.575 | 2.021 |
| 2700 | 54.723 | 280.683 | 239.952 | 109.973 | -252.643 | -94.729 | 1.833 |
| 2800 | 55.089 | 282.680 | 241.443 | 115.464 | -252.771 | -88.878 | 1.658 |
| 2900 | 55.430 | 284.619 | 242.899 | 120.990 | -252.897 | -83.023 | 1.495 |
| 3000 | 55.748 | 286.504 | 244.321 | 126.549 | -253.024 | -77.163 | 1.344 |

*Table contd...*

| 3100 | 56.044 | 288.337 | 245.711 | 132.139 | -253.152 | -71.298 | 1.201 |
| 3200 | 56.323 | 290.120 | 247.071 | 137.757 | -253.282 | -65.430 | 1.068 |
| 3300 | 56.583 | 291.858 | 248.402 | 143.403 | -253.416 | -59.558 | 0.943 |
| 3400 | 56.828 | 293.550 | 249.705 | 149.073 | -253.553 | -53.681 | 0.825 |
| 3500 | 57.058 | 295.201 | 250.982 | 154.768 | -253.696 | -47.801 | 0.713 |
| 3600 | 57.276 | 296.812 | 252.233 | 160.485 | -253.844 | -41.916 | 0.608 |
| 3700 | 57.480 | 298.384 | 253.459 | 166.222 | -253.997 | -36.027 | 0.509 |
| 3800 | 57.675 | 299.919 | 254.661 | 171.980 | -254.158 | -30.133 | 0.414 |
| 3900 | 57.859 | 301.420 | 255.841 | 177.757 | -254.326 | -24.236 | 0.325 |
| 4000 | 58.033 | 302.887 | 256.999 | 183.552 | -254.501 | -18.334 | 0.239 |
| 4100 | 58.199 | 304.322 | 258.136 | 189.363 | -254.684 | -12.427 | 0.158 |
| 4200 | 58.357 | 305.726 | 259.252 | 195.191 | -254.876 | -6.516 | 0.081 |
| 4300 | 58.507 | 307.101 | 260.349 | 201.034 | -255.078 | -0.600 | 0.007 |
| 4400 | 58.650 | 308.448 | 261.427 | 206.892 | -255.288 | 5.320 | -0.063 |
| 4500 | 58.787 | 309.767 | 262.486 | 212.764 | -255.508 | 11.245 | -0.131 |
| 4600 | 58.918 | 311.061 | 263.528 | 218.650 | -255.738 | 17.175 | -0.195 |
| 4700 | 59.044 | 312.329 | 264.553 | 224.548 | -255.978 | 23.111 | -0.257 |
| 4800 | 59.164 | 313.574 | 265.562 | 230.458 | -256.229 | 29.052 | -0.316 |
| 4900 | 59.275 | 314.795 | 266.554 | 236.380 | -256.491 | 34.998 | -0.373 |
| 5000 | 59.390 | 315.993 | 267.531 | 242.313 | -256.763 | 40.949 | -0.428 |

## 5.10  COMPOSITION OF PRODUCTS

To determine the combustion temperature and molecular mass of the mixture of gasses, the composition of the products of reaction is required. The composition of products depends on temperature and combustion pressure which in turn depends on composition. Hence, a trial-and-error procedure is required for the solution. Consider the stoichiometric reaction for kerosene and oxygen:

$$C_{12} H_{24} + 18 O_2 = 12 CO_2 + 12 H_2 O$$

At low temperatures, the nonstoichiometric reaction with excess oxygen completely oxidizes the products of combustion and the balanced oxygen is available in the products as given below.

$$C_{12}H_{24} + 20 O_2 = 12 CO_2 + 12 H_2 O + 2 O_2$$

In both cases it is easy to determine the composition by balancing the number of atoms of each element taking part in the reaction. At high combustion temperatures a portion of product compounds $CO_2$, $H_2O$ and $O_2$ dissociate into free atoms and radicals. Consider the combustion reaction of hydrogen-oxygen with possible product compounds at high temperature as shown below:

$$2H_2 + O_2 \rightarrow aH_2O + bO_2 + cH_2 + dO + eH + fOH \qquad .....(5.11)$$

Balancing the hydrogen and oxygen atoms

$$2a + 2c + e + f = 4$$

$$a + 2b + d + f = 2$$

The equilibrium reactions of product compounds and their equilibrium constants can be written as

$$2H_2O \leftrightharpoons 2H_2 + O_2, \qquad K_{p1} = \frac{p_{H_2}^2 p_{O_2}}{p_{H_2O}^2}$$

$$2H_2O \leftrightharpoons H_2 + 2OH, \qquad K_{p2} = \frac{p_{H_2} p_{OH}^2}{p_{H_2O}^2}$$

$$O_2 \leftrightharpoons 2O, \qquad K_{p3} = \frac{p_O^2}{p_{O_2}}$$

$$H \leftrightharpoons 2H, \qquad K_{p4} = \frac{p_H^2}{p_{H_2}}$$

Let $p_m$ and $T_m$ are mixture pressure and temperature are known. The above equations now can be transformed in terms of the number of moles using Eq. (5.10).

$$\frac{K_{p1}}{p_m} = \frac{c^2 b}{a^2(a + b + c + d + e + f)}$$

$$\frac{K_{p2}}{p_m} = \frac{cf^2}{a^2(a + b + c + d + e + f)}$$

$$\frac{K_{p3}}{p_m} = \frac{d^2}{b(a + b + c + d + e + f)}$$

$$\frac{K_{p4}}{p_m} = \frac{e^2}{c(a + b + c + d + e + f)}$$

The mixture composition can be found by solving the above equations for a, b...f. These equations are nonlinear and iterative solutions are difficult and time-consuming. The computer program called STANJAN, developed by Reynolds (Reynolds W.C., "STANJAN: The Element Potential Method for Chemical Equilibrium Analysis" Thermosciences Division, Department of Mechanical Engineering, Stanford University, Palo Alto, Calif., Jan. 1986), which may conveniently be used for such problems. Figure 5.4 shows the results of using STAJAN on the above reaction discussed. The mole fractions are shown as a function of combustion temperature ($K$) and pressure (atm). It can be observed at low temperatures, below 1500 K, $H_2$ and $O_2$ mole fractions are practically zero and the mole fraction of $H_2O$ tends to unity. Table 5.4 gives the mole fractions for the reaction shown in Eq. (5.11) at pressure 10 atm and temperature 4000 K from STANJAN Program.

**Table 5.4** STANJAN results for H/O=2, $T_m$=4000 K and $p_m$= 10 atm.

| $H_2$ | $O_2$ | $H_2O$ | OH | O | H |
|---|---|---|---|---|---|
| 0.1944 | 0.05521 | 0.2498 | 0.1695 | 0.1099 | 0.2213 |

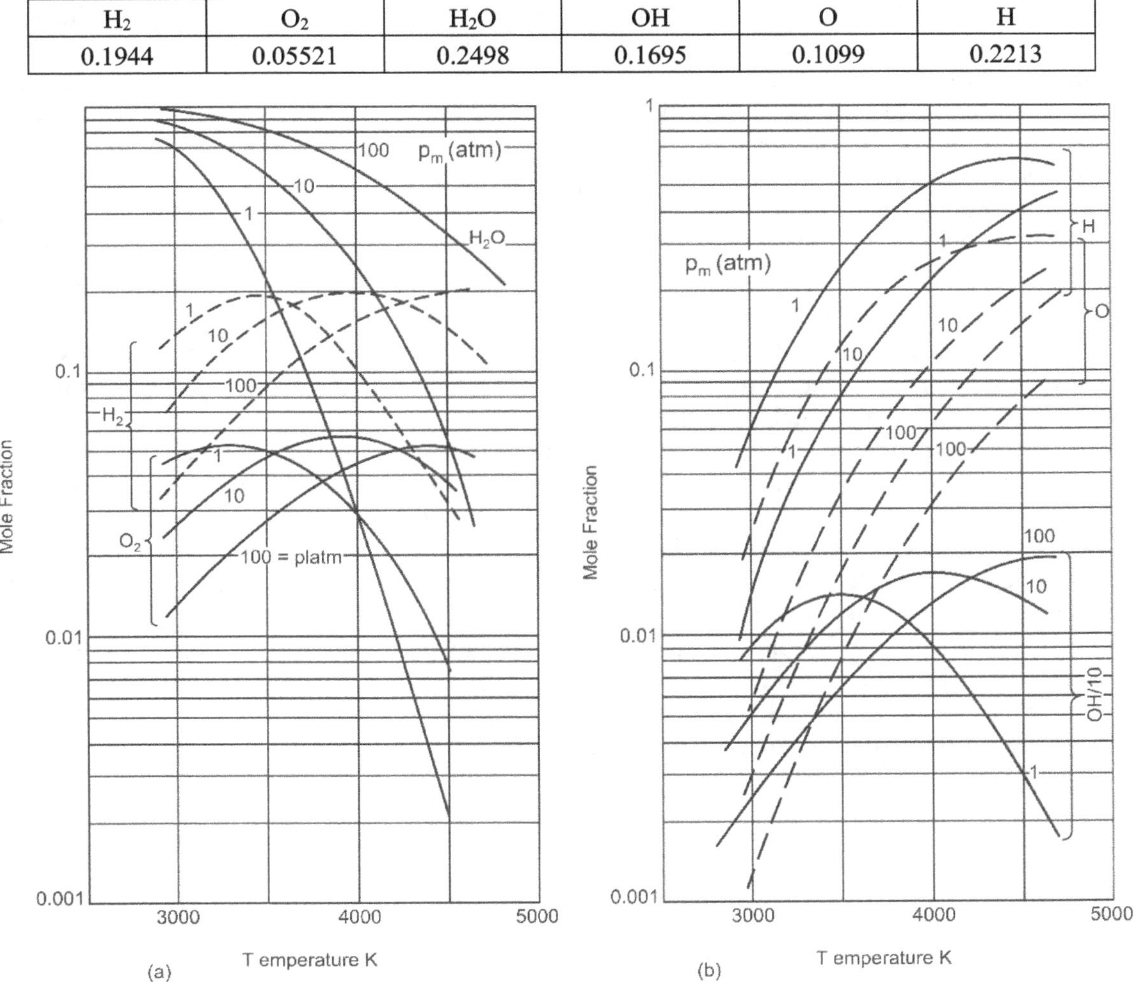


**Figure 5.4** (a) Equilibrium mole fractions of $H_2O$, $H_2$ and $O_2$ with H/O = 2; (b) equilibrium mole fractions of OH, O and H with H/O = 2.

## 5.11  FROZEN AND SHIFTING EQUILIBRIUM

In the nozzle flow analysis, it is assumed that the composition of combustion gasses is invariant throughout the nozzle sections. Thus, it is assumed the composition of gasses is "frozen" so that the nozzle exit is identical to that of its chamber condition, i.e., there are no chemical reactions or phase changes occurring during the flow in the nozzle. The results are known as frozen equilibrium rocket performance. It is a simple method, but the predicted values underestimate the performance, typically by 1 to 4%. However, as temperature of the hot gas reduces during the expansion process, the equilibrium constant changes and the composition would vary. At the high combustion temperatures, a small portion of the combustion gas molecules dissociate (split into simpler species) and as the process is endothermic, some energy is absorbed. As temperature reduces during the expansion process, the amount of the dissociated species would decrease and reassociation of species occurs releasing the energy. The equilibrium composition during the expansion process and the associated energy release can be determined by the minimization of Gibbs Free Energy method. So, as the expansion occurs in the nozzle, there would be a shift in equilibrium composition of gasses corresponding to the temperature at that section of the nozzle. The results so calculated are called *shifting equilibrium* performance. The analysis of nozzles is complex and overestimates the performance values typically by 1 to 4%.

Figure 5.5 shows the predicted performance values of a nozzle working with liquid oxygen (LOX) and RP-1 fuel under frozen and shifting equilibrium conditions. Because shifting equilibrium makes more enthalpy available for conversion to kinetic energy, it gives higher values of nozzle exit temperature and higher values of specific impulse ($I_{sp}$) and characteristic velocity ($c^*$). The influence of mixture ratio on chamber gas composition is evident from Figure 5.6. Breakdown into O, OH, or H and free $O_2$ occurs only at the higher temperatures or higher mixture ratios. As chemical reaction takes finite time and the flow velocities are high in nozzle, instantaneous achievement of chemical equilibrium is not possible. Therefore, the real performance values of the rocket nozzle would be in between the calculated values under frozen and shifting equilibrium assumptions. Due to subsonic flow in the convergent nozzle section, the residence time for a chemical reaction to take place is relatively larger compared to the divergent section of supersonic flow. So, in nozzle flow analysis shifting equilibrium assumption in convergent section and frozen equilibrium assumption in divergent section may give more reasonable estimates of performance parameters.

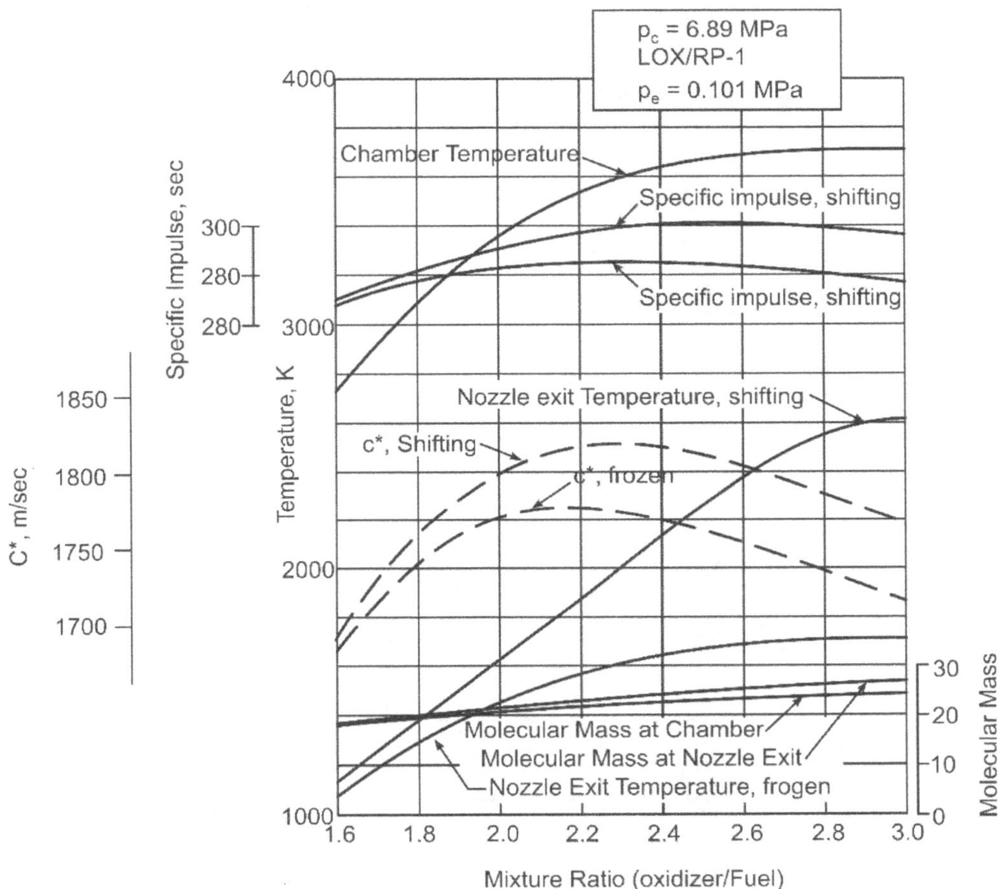


**Figure 5.5**  Calculated performance analysis of LOX and RP-1 as a function of mixture ratio (Source : George P Sutton).

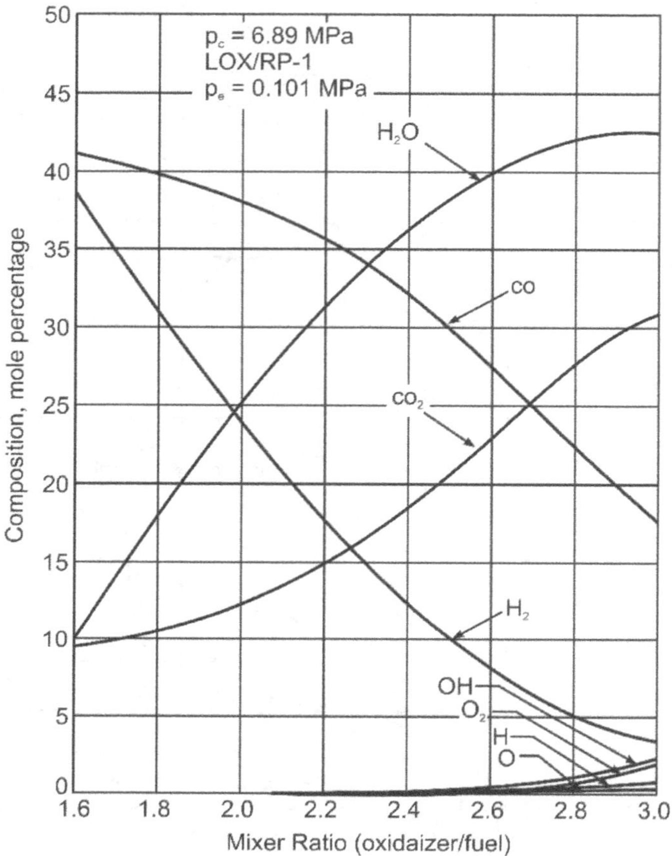



**Figure 5.6** Calculated chamber gas composition for LOX and RP-1 as a function of mixture ratio.

## Example 5.3

In a rocket combustion chamber, gaseous methane and gaseous oxygen are injected at 25° C at a mixture ratio of 3. The chamber pressure is 5 atm (0.5 MPa). A part of steam formed in the combustion products dissociates into hydrogen and oxygen. The boiling temperature of water at 0.5 MPa is 425 K, and the latent heat of vaporization is 37.95 kJ/mole. Taking appropriate values of specific heat and heat of formation (ref: Table 5.1, Table 5.3 & Figure 5.2) calculate the combustion temperature and characteristic velocity of the gases.

Consider 1 mole of $CH_4$. Let $x$ moles of oxygen react with methane. Now, the reactants are $(CH_4 + xO_2)$.

Given mixture ratio

$$MR = \frac{O}{F} = \frac{32x}{(12 + 4)} = 3$$

$$x = 1.5$$

Considering dissociation of a part of $H_2O$ at high temperatures as

$$H_2O \rightarrow H_2 + \frac{1}{2}O_2$$

The chemical reaction can be represented as

$$CH_4 + 1.5O_2 \rightarrow CO + xH_2O + yH_2 + zO_2$$

There are three unknowns, $x$, $y$ and $z$. Two equations can be obtained by balancing H and O atoms.

$$H: \quad x + y = 2$$

$$O: \quad x + 2z = 2$$

From above two equations the value of $x$ and $y$ can be expressed in terms of $z$ as

$$x = 2(1 - z)$$

$$y = 2z$$

The third equation can be obtained from the dissociation equation of $H_2O$. The equilibrium constant for the reaction in partial pressure terms is

$$K_p = p_{H_2}p_{O_2}^{0.5}/p_{H_2O}$$

As the solution is required in moles $x$, $y$ and $z$, the $K_p$ is to be expressed in mole fraction. For this purpose, the total number of moles in the products must be known. However, this is not known initially. As a first approximation, total number of moles can be assumed equals the moles in the product corresponding to a stoichiometric reaction:

$$CH_4 + 2O_2 \rightarrow CO_2 + 2H_2O$$

This gives the total number moles as 3. The chamber pressure, $p$ is given as 5 atm. The partial of the products now can be expressed in terms of moles:

$$p_{H_2O} = \frac{p}{3}x$$

$$p_{H_2} = \frac{p}{3}y$$

$$p_{O_2} = \frac{p}{3}z$$

Substituting in the equation for equilibrium constant:

$$K_p = \left(\frac{p}{3}\right)^{\frac{1}{2}}\frac{yz^{\frac{1}{2}}}{x} = \left(\frac{p}{3}\right)^{\frac{1}{2}}\frac{2z\,z^{\frac{1}{2}}}{2(1-z)} = \left(\frac{p}{3}\right)^{\frac{1}{2}}\frac{z^{\frac{3}{2}}}{(1-z)}$$

Substituting $p=5$ and rearranging:

$$\left(\frac{3}{5}\right)^{\frac{1}{2}}K_p = \frac{z^{\frac{3}{2}}}{(1-z)}$$

The value of $Kp$ depends on the temperature as given in Table 5.3. A temperature is initially assumed and checked for its validity taking energy balance. Hence, arriving at the correct combustion temperature is an iterative process. The value of $Kp$ can be taken Table 5.3 or calculated using equation

$$\Delta G^0 = -R_u T \ln K_f$$

Assuming the temperature of the chamber as 3200 K and taking the value from the table

$$\log K_p = -1.068$$

Therefore,

$$K_p = 0.0855$$

$$\frac{z^{\frac{3}{2}}}{(1-z)} = \left(\frac{3}{5}\right)^{\frac{1}{2}} \times 0.0855 = 0.0662$$

The above equation is to be solved iteratively. The values of $x, y$ and $z$ are:

$$x = 1.706$$

$$y = 0.294$$

$$z = 0.147$$

Now the chemical reaction with dissociation is:

$$CH_4 + 1.5O_2 \rightarrow CO + 1.706H_2O + 0.294H_2 + 0.147O_2$$

The heat of combustion for the above reaction$= -[\textit{Heat formation of products} - \textit{heat of formation of reactants}]$. Take the values from Table 5.1 at assumed temperature 3200 K.

$$= -[-110.5 + 1.706 \times (-286) - (-74.5)] = 523.916 \; kJ$$

Taking the energy balance, denoting the combustion temperature $T_c$

$$n_{CO}C_{p,CO}(T_c - 298) + n_{H_2O}\left[c_{pH_2O,(l)}(425 - 298) + LV + c_{p,H_2O}(T_c - 425)\right]$$
$$+ n_{H_2}C_{pH_2}(T_c - 298) + n_{O_2}C_{pO_2}(T_c - 298) = \textit{Heat of conbustion}$$

Substituting the proper values in the above equation

$$1 \times 0.037 \times (T_c - 298) + 1.706 \times [0.09 \times (425 - 298) + 37.5 + 0.058 \times (T_c - 425)]$$
$$+ 0.294 \times 0.40 \times (T_c - 298) + 0.147 \times 0.042 \times (T_c - 298) = 523.916$$

Simplifying and solving for $T_c$:

$$T_c = 3244 \; K$$

The calculated value of $T_c$ is close to the assumed value.

The molecular mass of the products, i.e., combustion gases is:

$$M = \frac{\textit{Molar masses of individual components}}{\textit{Number of moles}}$$

$$M = \frac{28 + 1.706 \times 18 + 0.294 \times 2 + 0.147 \times 32}{1 + 1.706 + 0.294 + 0.147} = 20.34 \frac{kg}{k\,mole}$$

The specific heat of the products at constant pressure $c_{p,product}$ is:

$$c_{p,product} = \frac{1 \times 0.037 + 1.706 \times 0.058 + 0.284 \times 0.40 + 0.147 \times 0.042}{1 + 1.706 + 0.294 + 0.147}$$
$$= 0.0489\ kJ/(mole\ K)$$

$$c_{v,product} = c_{p,product} - R_u = 0.0489 - 0.008314 = 0.0406 \frac{kJ}{mole\ K}$$

The ratio of specific heats of products $\gamma = \frac{0.0489}{0.0406} = 1.204$

The characteristic velocity is given by Eq. (4.37)

$$c^* = \frac{1}{\Gamma} \sqrt{\frac{R_u T_c}{M}} \ ; \qquad \Gamma = \sqrt{\gamma} \left(\frac{2}{\gamma + 1}\right)^{\frac{\gamma+1}{2(\gamma-1)}} = 0.6493$$

$$c^* = \frac{1}{0.6493} \sqrt{\frac{8.314 \times 3244}{20.35 \times 10^{-3}}} = 1773\ m/s$$

## 5.12 RELEVANT PROPERTIES FOR THE SELECTION OF PROPELLANTS

Some of the desirable properties in the selection of propellants are discussed below:

(a) Physical Properties: It is important to ensure that the physical properties like viscosity, boiling point, vapor pressure, freezing point and density are within the tolerable limits during storage and operation.

(b) Specific impulse and characteristic velocity: The combination of propellants (oxidizer and fuel) should generate high performance parameters like specific impulse, $I_{sp}$ and characteristic velocity $c^*$. These parameters depend on the mixture ratio, chamber temperature, pressure ratio, specific heat, and molecular weight.

(c) Specific Gravity: A higher specific gravity corresponds to a denser propellant. For a given mass of propellant, higher specific gravity results in lesser storage volume and in turn reduces the structural mass of the vehicle. The density specific impulse, defined as the product of the specific gravity of the propellant mixture times the specific impulse, is used to compare the propellant performance. Whereas the specific impulse judges the propellant combination for its impulse per unit mass, the impulse density allows to compare its impulse per unit volume. Thus, there is a relationship between the impulse density of propellant combination and the size of the vehicle to produce a certain impulse. A lower impulse density means a larger, lighter vehicle and vice versa.

(d) Heat Transfer Characteristics: It is an important property for a liquid rocket engine where the thrust chamber and nozzle side walls are circulated with propellant to keep the temperatures of the material within the limits to avoid damage due to high combustion temperatures. High thermal conductivity, high boiling point and decomposition temperature, high specific heat, and thermal stability are all desirable properties of propellant.

(e)  Material Compatibility: Selection of material for handling and use in the engine depends on the compatibility of the propellant with selected material. In case of use of cryogenic propellants like LOX and $LH_2$, the low temperatures can cause the mechanical properties of many materials to change considerably, notably making them more brittle and reducing their stress limits. The hydrazine-based propellants react with many materials. A certain type of stainless steel and aluminum as well as nickel are preferable. However, iron, copper (and its alloys such as brass and bronze), monel, magnesium, zinc, and even certain types of aluminum alloy are not suitable.

(f)  Aging characteristics and long life: They depend on the propellant's chemical and physical properties. The selected propellants should withstand high load and thermal cycles. Hydrogen peroxide decomposes by a few percent per year, even a clean storage tank. Cryogenic propellants, on the other hand, can evaporate by 5–20% per day.

(g)  Ignition: It is desirable to have predictable and reproducible ignition qualities. The ignition delay should be reasonably short to prevent unburned propellants from building up in the combustion chamber and then exploding.

(h)  Manufacturing: Simple, reproducible, safe, low-cost, controllable, and low-hazard manufacturing methods are preferred. Guaranteed availability of all raw materials and purchased components over the production and operating life of the propellant is required.

(i)  Safety and Environmental issues: They impose severe constraints in handling, storing and operation. While testing the engines or vehicle launching, the impact of the combustion exhaust products on nearby communities must be assessed. The exhaust should be non-toxic to avoid environmental damages. Care should be taken in handling cryogenic propellants and hazardous auto-ignite propellants like hydrogen peroxide and nitromethane.

(j)  In case of solid propellants, to have a minimum variation in thrust or chamber pressure, the pressure or burning rate exponent and the temperature coefficient should be small (will be discussed in the next chapter).


## 5.13  LIQUID PROPELLANTS

Liquid propellants used in rocketry can be classified into three types: hydrocarbons, cryogens, and hypergols. Hydrocarbon fuels are refined from crude oils and their properties depend upon the chemical processes used during the refinery phase. They are a mixture of complex compounds containing carbon and hydrogen. A specially refined petroleum fuel for rocket application is called RP-1. Another promising hydrocarbon fuel for future launch vehicles is methane with much more reproducible properties than RP-1. It can be cooled and liquified to increase its storage density. Cryogenic propellants are liquefied gases stored at extremely low temperatures, most frequently liquid hydrogen (LH2) as the fuel and liquid oxygen (LO2 or LOX) as the oxidizer. Hydrogen remains liquid at temperatures of $-253^0$ C and oxygen remains in a liquid state at temperatures of $-183^0$ C. Hypergolic propellants are fuels and oxidizers that ignite spontaneously on contact with each other and require no ignition source. As they are in liquid state at normal temperatures, storage and handling processes are simple and easy. The easy start and restart capability of hypergols make them ideal for spacecraft maneuvering systems. Hydrazine ($N_2H_4$), monomethyl hydrazine (MMH) and unsymmetrical dimethyl hydrazine (UDMH) are commonly used hypergolic fuels. The oxidizer is usually nitrogen tetroxide ($N_2O_4$) or nitric acid ($HNO_3$).

### 5.13.1    Liquid Oxidizers

Table 5.5 gives general properties of the liquid oxidizers used in rocket engines.

**Table 5.5** Properties of commonly used liquid oxidizers.

| Properties | Nitric Acid (99% pure) | Nitrogen tetroxide (NTO) | Liquid Oxygen | Hydrogen peroxide |
|---|---|---|---|---|
| Chemical formula | $HNO_3$ | $N_2O_4$ | $O_2$ | $H_2O_2$ |
| Molecular Mass (kg/kmole) | 63.02 | 92.016 | 32 | 34.02 |
| Melting Point (K) | 231.6 | 261.9 | 54.4 | 273.5 |
| Boiling Point (K) | 359 | 294.2 | 90.19 | 423.7 |
| Heat of Vaporization (kJ/mol) | 30.36 | 38.14 | 6.82 | 54.47 |
| Specific Heat (kJ/kg K) | 1.77 | 1.51 | 1.67 | 2.43 |
| Density (kg/m$^3$) | 1520(283 k) | 1450 (293) | 1141(91 K) | 1448 (203 K) |
| Heat of formation (kJ/mole) | -171.8 | -28.47 (Liq) | 0 | -187.5 |

(a)  Liquid Oxygen (LOX): Liquid Oxygen (LOX), a widely used nontoxic and noncorrosive oxidizer, has a boiling point of −183 °C at one atmosphere with the specific gravity of 1.145 and heat of vaporization of 213.5 kJ/ kg. It is used as an oxidizer in combination with fuels like alcohols, RP-1, and hydrogen. It burns with a bright white-yellow flame with most hydrocarbon fuels. The attainable performance is relatively high, and liquid oxygen is therefore a desirable and commonly used propellant in large rocket engines. Handling and storage are safe when contact materials are clean. Because of its ease of boiling, all feedlines and tanks must be well insulated. Care must be exercised by personnel handling LOX as prolonged contact with human skin, the cryogenic propellant causes severe burns.

(b)  Hydrogen Peroxide ($H_2O_2$): High concentrations of hydrogen peroxide in water (70–90%) are used in rocket applications. It is hypergolic with hydrazine and can burn efficiently with kerosene. The theoretical specific impulse of 90% hydrogen peroxide is 154 sec, when used as a monopropellant with a solid catalyst bed such as platinum, iron oxide, and solid manganese dioxide. It has not been used for a long time from 1965 due to its long-term storage stability, there has been some renewed interest in this dense oxidizer because it produces a nontoxic exhaust. Concentrated peroxide causes severe burns when in contact with human skin and may ignite and cause fires when in contact with wood, oils, and many other organic materials.

(c) Nitric Acid $HNO_3$: Presently the nitric acid is not used as an oxidizer. The most common type, red fuming nitric acid (RFNA), consists of concentrated nitric acid ($HNO_3$) that contains between 5 and 20% dissolved nitrogen dioxide. Nitric acid is highly corrosive. A small addition of fluorine ion (less than 1% of HF) inhibits the nitric acid, called inhibited red fuming nitric acid (IRFNA), causes a fluoride layer to form on the wall, and greatly reduces the corrosion with many metals.

(d) Nitrogen tetroxide ($N_2O_4$): Nitrogen tetroxide is the most used storable oxidizer by many space agencies with a fuel mixture consisting of hydrazine and unsymmetrical dimethylhydrazine. The temperature range for the liquid state is narrow, and it freezes or vaporizes easily. By adding a small amount of nitric oxide or NO the freezing point of $N_2O_4$ can be lowered but at the penalty of a higher vapor pressure. This mixture of NO and $N_2O_4$ is called mixed oxides of nitrogen (MON) and different grades have been 2 and 30% NO content. These are suitable for space storable and hence, used for control of spacecraft.

### 5.13.2 Liquid Fuels

Many different chemicals have been investigated and tested. Only a few have been used in production rocket engines. Table 5.6 gives the properties of commonly used fuels in rocket engines.

**Table 5.6** Properties of commonly used liquid fuels.

| Properties | Hydrazine | Hydrogen | Methane | MMH | RP-1 | UDMH |
|---|---|---|---|---|---|---|
| Chemical formula | $N_2H_4$ | $H_2$ (Liq) | $CH_4$ | $CH_3NHNH_2$ | $CH_{1.97}$ | $(CH3)_2N$ |
| Molecular Mass (kg/kmole) | 32.05 | 2.016 | 16.03 | 46.07 | 175 | 60.10 |
| Melting Point (K) | 274.7 | 13.96 | 90.5 | 220.7 | 225 | 215 |
| Boiling Point (K) | 386.7 | 20.4 | 111.6 | 360.6 | 460-560 | 336 |
| Heat of Vaporisation (kJ/mol) | 42.71 | 0.899 | 8.17 | 40.30 | 43.05 | 35.03 |
| Specific Heat (kJ/kg K) | 3.14 | 7.32 | 3.49 | 2.92 | 1.88 | 2.81 |
| Density (kg/m$^3$) | 1005 | 70.9 | 424 | 879 | 600-800 | 856 |
| Heat of formation (kJ/mole) | 50.4 | 0 | -74.50 | 53.09 | -51.6 | 88.4 |

(a) Rocket Propellant 1 (RP-1): It is a kerosene-based mixture of unsaturated and saturated hydrocarbons with a relatively narrow range of densities and vapor pressures. It is used mainly with LOX in rocket boosters. RP-1 is low in olefins and aromatics, which can cause carbonaceous deposits inside fuel cooling passages.

(b) Liquid Hydrogen ($LH_2$): Liquid hydrogen has the highest energy per unit mass among the fuels. Liquid hydrogen and LOX were commonly used in launch engines successfully because of the high specific impulse and their nontoxic characteristics. It also is an excellent regenerative coolant. The very low fuel density requires bulky fuel tanks, which necessitate very large vehicle volumes. The extremely low temperature makes the problem of choosing suitable tank and piping materials difficult, because many metals become brittle at low temperatures. Because almost all common liquids solidify in liquid hydrogen, it is critical that all passages are purged (e.g., flushed with helium) of any air or moisture before delivery of hydrogen to prevent solid particulate formation and its subsequent blockage of orifices and valves.

(c) Hydrazine ($N_2H_4$): Hydrazine is used as a bipropellant fuel as well as a monopropellant. Hydrazine and its related liquid organic compounds, mono methylhydrazine (MMH) and unsymmetrical dimethylhydrazine (UDMH), all have similar physical and thermochemical properties. Pure anhydrous hydrazine is a stable liquid; it has been safely heated above 530 K. It has been stored in sealed tanks for over 15 years.

(d) Unsymmetrical Dimethylhydrazine [$(CH3)_2NNH_2$]: A derivative of hydrazine, unsymmetrical dimethylhydrazine (UDMH), is often used instead of or in mixtures with hydrazine because it forms a more stable liquid, particularly at higher temperatures. Furthermore, it has a lower freezing point and a higher boiling point than hydrazine. The mixture of 50% hydrazine and 50% UDMH (known as A-50) was used in the Titan launch vehicles and many spacecraft engines. UDMH is being used in Russian, European, Indian, and Chinese rocket engines.

(e) Mono methylhydrazine ($CH_3NHNH_2$): Monomethyl hydrazine (MMH) has been used extensively as a fuel in spacecraft rocket engines, particularly in small attitude control engines, usually with $N_2O_4$ as the oxidizer. It is obtained by substituting one hydrogen with methyl radical.

(f) Methane ($CH_4$): Methane, a cryogenic liquid, is gaining significant interest in recent years. Both RP-1 and liquid methane are typically combusted with liquid oxygen. Current tendency in propellants research, particularly for use in boost engines, includes a transition from LOX/RP1 to LOX/Methane. Blue Origin BE-4 engine is a good example of a liquid oxygen/liquid methane engine. Methane is the simplest possible alkane, hydrocarbon (organic compound) made up of only carbon–carbon single bonds. It is non-toxic and inexpensive to produce. The atmosphere of Mars could also provide a means of producing methane in-situ. Its density is 0.422 gm/cm$^3$ is nearly six times heavier than liquid hydrogen (0.07gm/cm$^3$). Compared to other liquid hydrocarbon fuels, methane allows for a fuel-rich gas generator without soot formation, high cooling efficiency, and its gaseous state immediately downstream of rocket injectors lowers the risk of combustion instabilities. The advantages of liquid methane make it a preferable fuel for a different rocket application such as launch vehicles, space reaction control and maneuvering engines and descent and ascent engines for planetary missions.

*The most used liquid bipropellant combinations are*:

(i) the cryogenic oxygen-hydrogen propellant system, used in upper stages and sometimes booster stages of space launch vehicles

(ii) the liquid oxygen-hydrocarbon propellant combination, used for booster stages (and a few second stages) of space launch vehicles; its higher average density allows a more compact booster stage, when compared to the first combination

(iii) several storable propellant combinations like $N_2O_4$ and UDMH, used in large rocket engines for first and second stages and in almost all bipropellant low-thrust, auxiliary or reaction control rocket engines; they allow long-term storage and almost instant readiness to start without the delays and precautions that come with cryogenic propellants.

The Table 5.7 shows the theoretical performance of commonly used propellant combination, assuming ideal expansion in the nozzle, at chamber pressure of 6.895 MPa and nozzle exit pressure 0.101MPa. Adiabatic combustion and isentropic expansion of ideal gases are assumed in the calculations. The specific gravity is the average value calculated based on mixture ratio.

The mixture ratios mentioned in the Table 5.7 are for approximate maximum values of specific impulse.

## 5.14  SOLID PROPELLANTS

Solid rocket motors (SRM) are less complex compared to liquid rocket engines (LRE). They consist of a casing, usually steel, filled with a mixture of solid propellant compounds (fuel and oxidizer) that burn at a rapid rate, expelling hot gases from a nozzle to produce thrust. When ignited, a solid propellant burns in a controlled way to produce the desired thrust. Once ignited, they will burn until all the propellant is exhausted. A solid propellant grain is a combination of several chemical ingredients like fuel, oxidizer, binder, stabilizer, plasticizer, curing agent and cross-linking agent. The physical and chemical properties, combustion characteristics and performance depend upon the proportions of these ingredients. In addition to the relevant properties of a propellant discussed in section 5.12, the other desirable qualities of solid propellant grain are predictable and reproducible burning rate and mechanical properties over the intended range of operating conditions. Solid Propellants are classified as homogeneous or heterogeneous, according to their chemical composition and physical structure. The functions of different chemical ingredients added in a solid propellant are to improve the performance and to ease the manufacturing process. They are classified by their major functions like oxidizer, fuel, binder, plasticizer and curing agent. A small change in ingredient properties and containing impurities can cause measurable changes in physical properties, ballistic properties, aging and in ease of manufacturing.

**Table 5.7** Theoretical preference of some liquid rocket propellant combinations.

| Oxidizer | Fuel | Mixture Ratio | | Average Specific Gravity | Chamber Temp (K) | Chamber $c^*$ (m/sec) | $\mathcal{M}$ (kg/mol) | $I_s$ (sec) | | $\gamma$ |
| --- | --- | --- | --- | --- | --- | --- | --- | --- | --- | --- |
| | | By Mass | By Volume | | | | | Shifting | Frozen | |
| Oxygen | Methane | 3.20 | 1.19 | 0.81 | 3526 | 1835 | 20.3 | | 296 | 1.20 |
| | | 3.00 | 1.11 | 0.80 | 3526 | 1853 | | 311 | | |
| | Hydrazine | 0.74 | 0.66 | 1.06 | 3285 | 1871 | 18.3 | | 301 | 1.25 |
| | | 0.90 | 0.80 | 1.07 | 3404 | 1892 | 19.3 | 313 | | |
| | Hydrogen | 3.40 | 0.21 | 0.26 | 2959 | 2428 | 8.9 | | 386 | 1.26 |
| | | 4.02 | 0.25 | 0.28 | 2999 | 2432 | 10.0 | 389.5 | | |
| | RP-1 | 2.24 | 1.59 | 1.01 | 3571 | 1774 | 21.9 | 300 | 285.4 | 1.24 |
| | | 2.56 | 1.82 | 1.02 | 3677 | 1800 | 25.3 | | | |
| | UDMH | 1.39 | 0.96 | 0.96 | 3542 | 1835 | 19.8 | | 205 | 1.25 |
| | | 1.65 | 1.11 | 0.98 | 3594 | 1864 | 21.3 | 310 | | |
| Fluorine | Hydrazine | 1.83 | 1.22 | 1.29 | 4553 | 2128 | 18.5 | 334 | | 1.33 |
| | | 2.30 | 1.54 | 1.31 | 4713 | 2208 | 19.4 | | 365 | |
| | Hydrogen | 4.54 | 0.21 | 0.33 | 3080 | 2534 | 8.9 | | 389 | 1.33 |
| | | 7.60 | 0.35 | 0.45 | 3900 | 2549 | 11.8 | 410 | | |
| Nitrogen tetroxide | Hydrazine | 1.08 | 0.75 | 1.20 | 3258 | 1765 | 19.5 | | 283 | 1.25 |
| | | 1.34 | 0.93 | 1.22 | 3152 | 1782 | 20.9 | 292 | | |
| | 50% UDMH | 1.62 | 1.01 | 1.18 | 3242 | 1652 | 21.0 | | 278 | 1.24 |
| | 50% hydrazine | 2.00 | 1.24 | 1.21 | 3372 | 1711 | 22.6 | 289 | | |
| | RP-1 | 3.4 | 1.05 | 1.23 | 3290 | | 24.1 | | 297 | 1.23 |
| | MMH | 2.15 | 1.30 | 1.20 | 3396 | 1747 | 22.3 | 289 | | |
| | | 1.65 | 1.00 | 1.16 | 3200 | 1591 | 21.7 | | 278 | 1.23 |
| Red fuming nitric acid | RP-1 | 4.1 | 2.12 | 1.35 | 3175 | 1594 | 24.6 | | 258 | 1.22 |
| | | 4.8 | 2.48 | 1.33 | 3230 | 1609 | 25.8 | 269 | | |
| | 50% UDMH | 1.73 | 1.00 | 1.23 | 2997 | 1682 | 20.6 | | 272 | 1.22 |
| | 50% hydrazine | 2.20 | 1.26 | 1.27 | 3172 | 1701 | 22.4 | 279 | | |
| Hydrogen peroxide (90%) | RP-1 | 7.0 | 4.01 | 1.29 | 2760 | | 21.7 | | 297 | 1.19 |

(a) Oxidizer: They are the main ingredients that produce the high energy of combustion. The commonly used oxidizer is Ammonium perchlorate (AP) ($NH_4ClO_4$). It has more oxygen atoms and easily dissociates. It is not hygroscopic and can be stored for long periods. In propellant manufacture large crystals of AP (around 100 to 600 μm) are grinded to smaller sizes (down to 2 μm) just before they are incorporated into a propellant The other oxidizers are hydrazinium perchlorate ($N_2H_5ClO_4$) and hydrazinium nitroformate [$N_2H_5C(NO_2)_3$].

(b) Metal Fuels: Addition of metal fuel enhances the heat of combustion, propellant density, combustion temperature, and hence the specific impulse. Aluminum and boron are frequently added to propellant mixtures as fuel enhancers. Aluminum is one of the widely used metal additives in the form of micron (5-60 μm) size particles.

(c) Binders: They provide structurally a matrix in which solid granular ingredients are held together in a composite propellant. The binder impacts the mechanical and chemical properties, propellant processing, and aging of the propellant. The raw materials are liquid prepolymers or monomers. Composite propellants are often identified by the type of polymeric binder used. The two most common binders are polybutadiene acrylic acid acrylonitrile (PBAN) and hydroxy-terminator polybutadiene (HTPB). HTPB has been commonly used in recent years, as it allows higher solid fractions (total 88–90% of AP and Al) and relatively good physical properties.

(d) Burning Rate Modifiers or Catalyst: They are added to propellant to accelerate or decelerate combustion at the burning surface of the grain, i.e., they increase or decrease the propellant burning rate. In composite propellants, substances like iron oxide increase the burning rate whereas lithium fluoride decreases the burning rate. Recently, research is focused on nano-catalysts.

(e) Plasticizer: It is added to improve the processing properties of propellant like to reduce the viscosity of the slurry, therefore facilitating production and to get longer pot life of the mixed but uncured propellants. It is a relatively low-viscosity organic liquid, which also contributes to the thermal energy of oxidation. Some of the commonly used plasticizers are dioctyl phthalate (DOP), dioctyl adipate (DOA) and dimethyl phthalate (DMP).

(f) Curing Agents or Crosslinkers: They are used in composite propellants to cause the polymers to form longer chains of larger molecular mass and interlocks between chains. Although it is added in small amounts (0.2 to 3%), its presence impacts the propellant physical properties, manufacturability, and aging considerably. Some curing agents are isophorone diisocyanate (IPDI), hexamethylene diisocyanide (HMDI) and toluene-2,4-diisocyanate (TDI).

(g) Additives: There are many other chemicals added in small quantities to propellant to increase the performance or to improve ease of manufacturing. Opacifier is an additive to make the propellant opaque to prevent radiative heating at places other than burning surface. To improve the adhesion between the casing and propellant grain bonding agents are added. To improve the resistance to accidental ignition from unwanted energy stimulus, desensitizing agents are added. To moderately increase the performance of the propellant, energetic binders containing oxidizing species like azides or organic nitrates, are added.

### 5.14.1    Homogeneous Solid Propellants

In a homogeneous solid propellant, the ingredients are chemically linked at the molecular level and the resulting physical structure is homogeneous throughout. Depending on their chemical composition they are classified as

(i)   Single-base (SB) propellants that contain nitrocellulose, NC.

(ii)  Double-base (DB) propellants that contain nitrocellulose, NC and nitroglycerin, NG, or its mixture with other nitric acid esters like nitroglycol.

(iii) Triple-base (TB) propellants that are essentially DB propellants supplemented with 1-nitroguanidine

The single base propellant, nitrocellulose has both an oxidation and reduction capacity. It is a highly flammable compound formed by nitrating cellulose through exposure to nitric acid or another powerful nitrating agent. The nitric acid converts natural cellulose fibers from wood or cotton cellulose nitrate and water:

$$3HNO_3 + C_6H_{10}O_5 \rightarrow C_6H_7(NO_2)_3O_5 + 3H_2O$$

It forms a chain [ $C_6H_7(NO_2)_3O_5$ ]$_n$

Nitroglycerin (NG), is a dense, colorless, oily, explosive liquid most commonly produced by nitrating glycerol with white fuming nitric acid under conditions appropriate to the formation of the nitric acid ester. It is derived from propane

$$C_3H_8 \rightarrow C_3H_5(OH)_3 \ Propane \ triol \rightarrow C_3H_5(ONO_2)_3 \ nitroglycerin$$

Double base propellants usually consist of nitrocellulose and nitroglycerine, to which a plasticizer (like triacetin and diethyl phthalates) is added. The physicochemical properties of double base propellants like mechanical properties, energy density, stability and combustion characteristics depend on the proportions of NC, nitrate ester, stabilizers, plasticizers, and other catalysts. NC is fuel rich; NG is slightly oxygen rich and the combination forms good solid propellant. The energetic base of double-base propellants is mainly composed of nitrocellulose (40-70%) and nitroglycerine (15-41%).

Homogeneous propellants do not usually have specific impulses greater than about 210 seconds under normal conditions. Their main asset is that they do not produce traceable fumes and are, therefore, commonly used in tactical weapons. They are also often used to perform subsidiary functions such as jettisoning spent parts or separating one stage from another.

### 5.14.2    Heterogeneous or Composite Solid Propellants

Heterogeneous propellants are mixtures of crystalline oxidizers and fuel particles mixed physically. Composite propellants are made of a polymeric matrix, loaded with a solid powder oxidizer and possibly a metal powder that plays the role of a secondary fuel component. The commonly used oxidizers are ammonium perchlorate and ammonium nitrate and the fuel is aluminum. Table 5.8 shows the characteristics of some important oxidizers. Ammonium perchlorate is the most preferable oxidizer. It is dense, thermally stable and its decomposition produces only gases of which a large proportion is oxygen.

Prepolymers are the main element of the binder of composite propellants (70-80%). It is the prepolymer that confers on the binder its essential properties. The propellant is held together by a polymeric binder, usually polyurethane or polybutadienes, which is also consumed as fuel. Figure 5.8 shows the theoretical specific impulse of a few prepolymers as a function of ammonium perchlorate (AP). As mentioned, many additives are added so that the final product is rubber like substance with the consistency of a hard rubber eraser.

**Table 5.8** Some characteristics of the major oxidizers.

| Oxidizer | Chemical Formula | Molecular Mass | Oxygen content (%) | Density (kg/m$^3$) | Heat of Formation (kJ/mole) |
|---|---|---|---|---|---|
| Ammonium perchlorate | $NH_4ClO_4$ | 117.5 | 34 | 1950 | -290.5 |
| Ammonium nitrate | $NH_4NO_3$ | 80.0 | 20 | 1720 | -365.3 |
| Potassium perchlorate | $KClO_4$ | 138.5 | 46.2 | 2520 | -433.0 |
| HMX | $(CH_2N_2O_2)_4$ | 296 | -21.6 | 1910 | 84.0 |

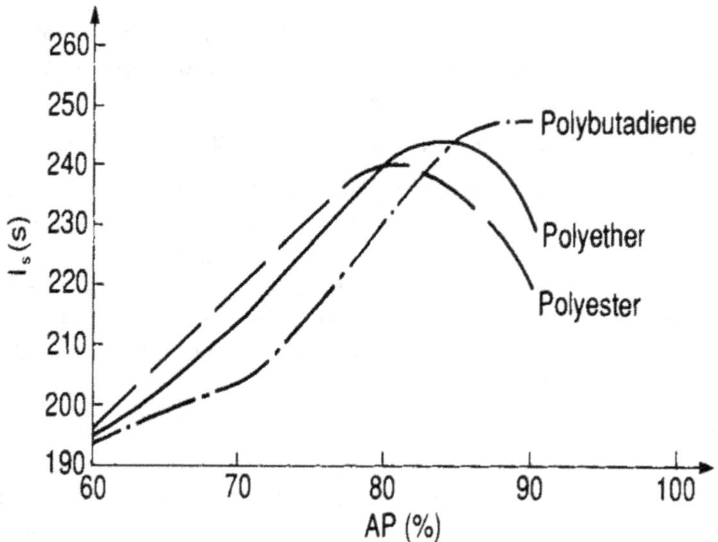


**Figure 5.7** Theoretical specific impulse as a function of AP concentration for three types of prepolymers.

Major plasticizers used in composite propellants are:

Diisoocytyl azelate

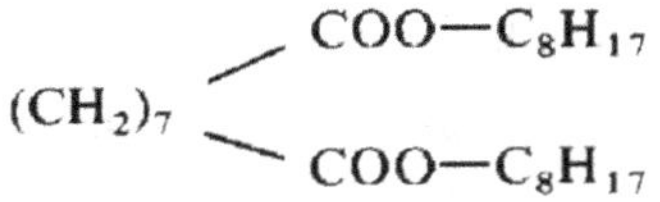

$$(CH_2)_7 \left\langle \begin{array}{l} COO-C_8H_{17} \\ COO-C_8H_{17} \end{array} \right.$$

Diisoctyl sebacate

$$(CH_2)_8 \left\langle \begin{array}{l} COO-C_8H_{17} \\ COO-C_8H_{17} \end{array} \right.$$

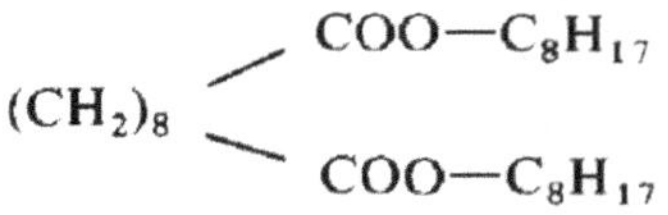

Isodecyl pelargonate     $CH_3-(CH_2)_7-COO-C_{10}H_{21}$

Polyisobutylene     $C_nH_{2n+2}$

Dioctyl phthalate

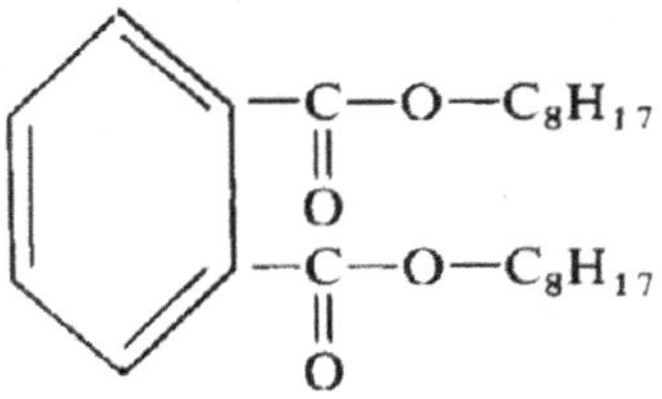

$$\begin{array}{l} -C-O-C_8H_{17} \\ \parallel \\ O \\ -C-O-C_8H_{17} \\ \parallel \\ O \end{array}$$

Composite propellants are often identified by the type of polymeric binder used. The two most common binders are polybutadiene acrylic acid acrylonitrile (PBAN) and hydroxy-terminator polybutadiene (HTPB). Butadiene ($C_4H_6$) is an aliphatic compound ($C = C$   $- C = C$). Polybutadiene is polymer of $(C = C - C = C)_m$. To make strong acrylic acid and acrylonitrile are added.

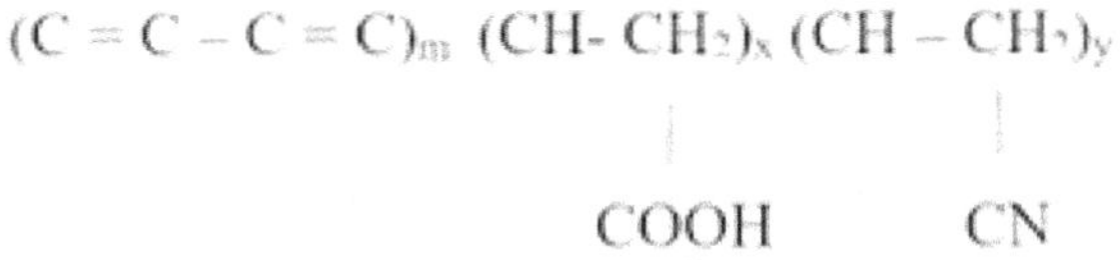

$$(C = C - C = C)_m \ (CH\text{-} CH_2)_x \ (CH - CH_2)_y$$
$$\qquad\qquad\qquad\quad | \qquad\qquad |$$
$$\qquad\qquad\qquad COOH \qquad CN$$

This is called as polybutadiene acrylic acid acrylonitrile (PBAN). If the polybutadiene is terminated with a hydroxyl group, it is known as hydroxy-terminator polybutadiene (HTPB).

$$OH \ (C = C - C = C)_m \ OH$$

PBAN formulations give a slightly higher specific impulse, density, and burn rate than equivalent formulations using HTPB. However, PBAN propellant is more difficult to mix and process. It requires an elevated curing temperature.

Presently, HTPB is the most widely used binder in composite propellant grains due to its good mechanical properties, adhesion properties and high heat of combustion. HTPB is also resistant to aging, having high oxidative and hydrolytic stability, while allowing for a high degree of loading with solids (up to 90% by weight). Composite propellants are cast from a mix of solid

(AP crystals) and liquid HTPB or HTPB ingredients. The propellant is hardened by crosslinking or curing the liquid binder polymer with a small amount of curing agent, and curing it in an oven, where it becomes hard and solid.

The conventional composite propellants usually contain between 60 and 72% ammonium perchlorate (AP) as crystalline oxidizer, up to 22% aluminum powder as a metal fuel, and 8 to 16% of elastomeric binder (organic polymer) including its plasticizer. Modified composite propellant where an energetic nitramine (HMX: His Master Explosive, cyclo tetra methyline tetra nitramine) or Research and Development explosive (RDX) is added for obtaining a little more performance and a somewhat higher density. RDX also stands for Royal Demolition explosive. It is also known as cyclonite or hexogen. The chemical name for RDX is 1,3,5-trinitro-1,3,5-triazine. RDX is an organic compound with the formula $(O_2NNCH_2)_3$. If a large amount of HMX is added, it can become a minimum smoke propellant with fair performance.

Small Solid propellant motors have a variety of uses like powering the final stage of a launch vehicle and to boost the payloads to higher orbits. Medium solids such as the Payload Assist Module (PAM) and the Inertial Upper Stage (IUS) provide the added boost to place satellites into geosynchronous orbit or on planetary trajectories. Large rocket launch vehicles like Ariane, PSLV, GSLV, Titan, and Space Shuttle use strap-on solid propellant rockets to provide added thrust at liftoff. The Space Shuttle uses the largest solid rocket motors ever built and flown. Each booster contains 500,000 kg of propellant and can produce up to 14,680,000 Newtons of thrust.

Recent research efforts are focused on development of green propellants to replace environmental harmful components. Many energetic systems have been proposed, nitrogen-rich organic cyclic species and their salts, as alternatives to ammonium perchlorate (AP). New energetic binders like Glycidyl azide polymer (GAP) and polymer of 3,3'-bis-azidomethyloxetane (PBAMO) are being studied in place of tried-and-true polymer HTPB. Such binders can improve the performance of propellants rather than only serving as an organic fuel and thermal ballast, as in the case of HTPB.

**Example 5.4**

A composite propellant contains ammonium perchlorate (AP) as oxidizer. The fuel combination is aluminum and hydroxy terminated polybutadiene (HTPB). Neglect other quantities like plasticizer, curator and burn rate additives in the propellant. The aluminum (Al) contains 10% and the solid loading is 90%. Determine the mixture ratio of the propellant. Also, find the molar composition of the propellant. Ascertain whether the composition is fuel-rich or oxidizer rich. Take the molecular mass of HTPB as 2734 kg/kmole.

Mixture Ratio:

Let the mass of AP, HTPB and aluminum in the propellant as $m_{AP}$, $m_B$ and $m_{Al}$ respectively.

$$Solid\ loading = \frac{m_{AP} + m_{Al}}{m_{AP} + m_{Al} + m_B} = 0.9$$

Aluminum contains 10% in the propellant, i.e.,

$$\frac{m_{Al}}{m_{AP} + m_{Al} + m_B} = 0.1$$

$$\frac{m_{AP}}{m_{AP} + m_{Al} + m_B} = 0.9 - 0.1 = 0.8$$

Therefore,

$$\frac{m_{AP}}{m_{Al}} = 8$$

Substituting in the expression for the aluminum fraction

$$\frac{m_{Al}}{9m_{Al} + m_B} = 0.1$$

$$\frac{m_B}{m_{Al}} = 1$$

Therefore, the mass ratio of the components in the propellant is

$$m_{AP} : m_{Al} : m_B = 8 : 1 : 1$$

The mixture ratio

$$MR = \frac{Oxidizer}{Fuel} = \frac{8}{1 + 1} = 4$$

Molar Composition:

The chemical formula for HTPB is

$$HO - (CH2 = CH - CH = CH2)n - OH$$

which can be written as $(C_4H_6)_n(OH)_2$.

The molecular mass of HTPB is given as 2734 kg/kmole. From the formula of HTPB molecular mass can be written as

$$(4 \times 12 + 6)n + (16 + 1)2 = 2734$$

$$n = 50$$

Therefore, the HTPB formula with molar mass 2734 kg/kmole is $C_{200}H_{302}O_2$.

The molar mass of AP ($NH_4ClO_4$) $= 14 + 4 + 35.5 + 64 = 117.5 \; gm/mole$

The molar mass of $Al = 27$ gm/mole

Let the moles of $AP$, $Al$ and binder by $n_{AP}$, $n_{Al}$ and $n_B$ respectively. Then,

$$n_{AP} : n_{Al} : n_B = \frac{m_{AP}}{M_{AP}} : \frac{m_{Al}}{M_{Al}} : \frac{m_B}{M_B} = \frac{8}{117.5} : \frac{1}{27} : \frac{1}{2734} = 0.068 : 0.037 : 0.00037$$

Therefore, the reactants can be expressed in terms of molar composition:

$$0.00037 C_{200}H_{302}O_2 + 0.068 NH_4ClO_4 + 0.037 Al$$

To find whether the propellant is oxygen or fuel rich, the stoichiometric reaction of the above composition of the reactants. The oxidized products of stoichiometric reaction are HCl, $H_2O$, $CO_2$ and $Al_2O_3$.

Total number of hydrogen atoms: $302 \times 0.00037 + 4 \times 0.068 = 0.3837$

H atom used to form HCL= 0.068

Remaining H atoms = $0.3837 - 0.068 = 0.3157$

O atoms required to oxidize to $H_2O$ =$0.3157/2 = 0.1578$

Number of C atoms = $0.00037 \times 200 = 0.074$

O atoms to oxidize C to $CO_2$ = $0.074 \times 2 = 0.148$

O atoms required to oxidize 0.037 moles Al = $0.037 \times \frac{3}{2} = 0.0555$

The total moles of $O_2$ required for stoichiometric combustion = $\dfrac{0.1578+0.148+0.0555}{2} = 0.181$

The number of moles of $O_2$ available in the reactants = $0.00037 + 2 \times 0.068 = 0.13637$

Therefore, the available $O_2$ moles are less than the required. Hence, the propellant is fuel rich.

## 5.15  HYBRID PROPELLANTS

In hybrid propellant engines, one of the propellants is solid, usually the fuel, while the other, usually the oxidizer, is liquid. The liquid is injected into the solid, whose fuel reservoir also serves as the combustion chamber. The main advantage of such engines is that they have high performance, like that of solid propellants, but the combustion can be moderated, stopped, or even restarted.

**Nomenclature**

$H_c^0$     heat of combustion per kg of fuel (kJ/kg)

$\Delta H_f\,^\circ$  heat of formation of substance at standard conditions (0.1 MPa, 298.15 K) (kJ/mole)

$H_r^0$     heat of reaction at standard conditions (0.1 MPa, 298.15 K) (kJ/mole)

$c^*$     characteristic velocity (m/s)

$c_p$     specific heat at constant pressure (J/mole K)

$G$     Gibbs function

$H$     enthalpy (kJ/mole)

$h_{fg}$     latent heat of liquid (J/mole)

$K_f$     equilibrium constant for the chemical formation of a given species from its elements

$K_n$     equilibrium constant in terms of mole fractions

$K_p$     equilibrium constant in terms of partial pressures

$M$     molecular mass (kg/mole)

$MR$   propellant mixture ratio

$n$     number of moles

$p_c$    combustion pressure ($N/m^2$)

$R$    specific gas constant (J/kg K)

$R_u$    universal gas constant (8.314 J/mole K)

$S$    entropy (kJ/K)

$T_c$    adiabatic combustion temperature (K)

$T_c$    Combustion temperature (K)

$U$    internal energy (J or kJ)

$V$    volume ($m^3$)

$x_i$    mole fraction of $i^{th}$ substance

$\gamma$    ratio of specific heats

## PROBLEMS

5.1   A rocket engine utilizes propellants LOX/RP 1 at a mixture ratio of 2.6. The combustion temperature is 3300 K. Assume complete combustion, calculate molecular mass, specific heat at constant pressure, specific heat ratio and characteristic velocity.

5.2   Many rocket engines commonly use nitrogen tetroxide ($N_2O_4$) and monomethyl hydrazine (MMH) as oxidizer and fuel, respectively. Assume that the combustion pressure is 20 atm. and mixture ratio 2.4. Take water vapor, CO, $CO_2$, $N_2$ and $O_2$ as combustion products (consider dissociation of carbon dioxide). Determine

(a) the adiabatic flame temperature assuming the reactants are gases at 298 K;

(b) the equilibrium composition of the product species.

(c) the molecular weight of the product gases.

(d) the mass fraction of each of the assumed product species; and

(e) the characteristic velocity obtained by this combustion process.

5.3   To transfer a satellite from LEO to geosynchronous orbit, a liquid rocket engine is used with propellant combination of MMH (density 865 $kg/m^3$) and $N_2O_4$ (density 1400 $kg/m^3$). Equal volumetric flow rates of MMH and $N_2O_4$ are supplied to the thrust chamber. Find the mixture ratio of the propellant combination used.

5.4   A cryogenic engine using sub-cooled liquid hydrogen and liquid oxygen at temperatures of 18 K and 80 K respectively at a mixture ratio of 6. Neglecting the dissociation of combustion products, find combustion temperature, specific heat ratio and characteristic velocity.

5.5   Assume in the following reaction the products are at pressure 10 atm and 1000 K. Determine the molar fractions of the product species.

$$CH_4 + 1.5\,O_2 \rightarrow aH_2O + bCO_2 + cO_2 + dH_2 + eCO$$

5.6 A Solid propellant grain has the following composition on percentage of mass

    Ammonium perchlorate, $NH_4ClO_4$            70%

    Aluminum, Al            18%

    HTPB $C_{7.3} H_{10.96} O_{0.06}$            12%

The propellant is combusted at a pressure of 35 atm and expanded in nozzle of area ratio 12. Find the characteristic velocity and specific impulse at sea level.

5.7 Determine the adiabatic flame temperature for the reaction containing hydroxyl radials:

$$2H_2 + O_2 \rightarrow aH_2O + bO_2 + cH_2 + dOH$$

Include the dissociation reaction for $H_2O$ and OH. Perform calculations assuming the reactants are initially at a temperature of 298 K for a pressure of 100 and 200 atmospheres. Write out the balanced chemical equations for both cases and explain differences in your results.

5.8 A nitramine propellant which consists of 85 per cent HMX [$C_4H_8N_4(NO_2)_4$] added to HTPB binder. The molecular mass of the HTPB binder can be taken as 2734 kg/kmole. The heat of formation of the binder is $-64$ kJ/mole. Neglecting the other small additives to the propellant and dissociation of the combustion products, find the heat released per kg of nitramine.

# CHAPTER 6 — Solid Propellant Motors

## 6.1 INTRODUCTION

The word "motor" is as common to solid rockets as the word "engine" is to liquid rockets. In the Solid Rocket Motor (SRM) only solid propellant is used to produce the thrust. In Liquid Rocket Engine (LRE) the oxidizer and fuel are in a liquid state. The unique characteristic of SRM that distinguishes it from LRE is the extremely simple way of controlling the energy release. This leads to highly reliable and low-cost devices to generate the thrust. SRMs can be designed and manufactured in many different types and sizes and vary in thrust from about 2 N to over 7000 kN. They can be more easily stored than liquid propellants. These features make SRM very attractive for space and defense applications. They find applications in large solid booster rocket motors, solid kick motors, ballistic missiles, sounding rockets, assisted take-off, air-launched missiles, antitank weapons, and retrorockets and as some special purpose motors.

Compared to liquid rocket engines, SRM offers a low specific impulse even using composite propellants. As in solid propellant, fuel and oxidizer are intimately mixed there is a danger for detonation. Hence, extensive safeguards are required during propellant manufacturing and launching processes. The solid rocket motor, once ignited it is not possible to stop the burning of the propellant grain and controllability of the burning and re-ignition, as obtained with liquid and hybrid propellant rockets is not feasible. Also, more recently, there is some concern about the effect of exhaust gasses on the environment.

With the ever-increasing need of payload weight, at the ground level for lift-off and to traverse the thick layers of atmosphere under very precise control conditions of acceleration, the rocket vehicle requires extremely high thrust and total impulse. Particularly, during the first 120 sec of flight, the vehicle may require delivering a thrust 10 times greater than the central rocket engine. This can be met by SRMs, and these are a basic complement to the central LRE launch vehicles.

## 6.2 COMPONENTS OF SRM

In SRM, the casing holding the propellant grain acts as the thrust chamber. The gasses generated by burning of propellant grain create the pressure and temperature. The gasses are then expanded in the nozzle to convert high pressure gasses to supersonic exhaust. Figure 6.1 shows the major components of a typical solid rocket motor.

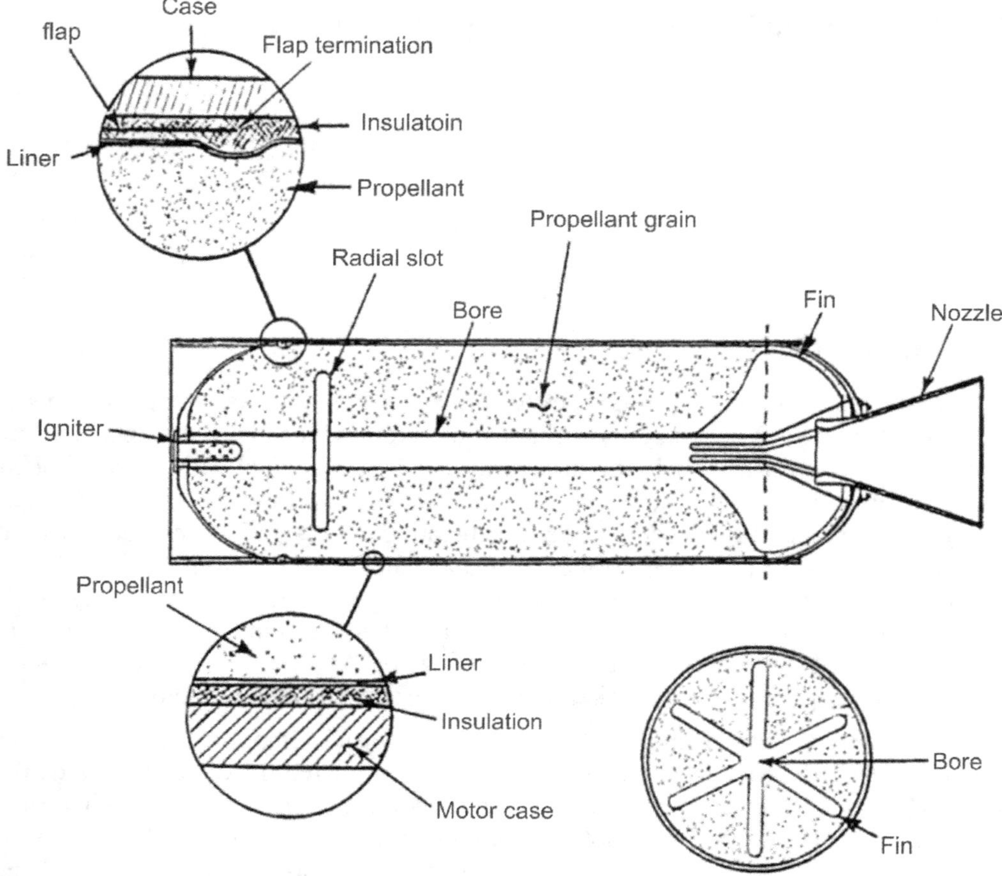



**Figure 6.1** Typical Solid Rocket Motor.

## 6.2.1   The SRM Case

The case must be designed considering it as a pressure vessel, capable of withstanding the combustion pressure with a sufficient safety coefficient. The value of safety margin ranges from 1.1 at yield for unmanned applications to 1.5 for manned vehicles. The SRM case is made of high strength metallic alloys or of a composite material such as fiber reinforced plastic (FRP). Generally, the bigger size motors the case is made of a low carbon steel such as 15CDV6 which combines a high yield strength with good toughness and weldability. It is readily welded with very little loss of properties during welding and without the need for further heat treatment. The advantages of metal case over filament-reinforced plastic or composite (FRP) case are:

(a)   metal case is rugged and can take rough handling

(b)   withstands higher temperatures

(c)   better aging characteristics.

The special steel used in booster and strap-on motors are 15 CDV 6 (French code) and maraging steel. Typical compositions of these steels are given below.

1. Composition of 15 CDV 6

| Carbon | : 0.1 to 0.16% |
|---|---|
| Chromium | : 1.25 to 1.45% |
| Molybdenum | : 08 to 09% |
| Vanadium | : 0.2 to 0.3% |
| Manganese | : 0.8 to 1.0% |

2. Composition of maraging steel

| Nickel | : 18% |
|---|---|
| Cobalt | : 8% |
| Molybdenum | : 5% |
| Aluminum | : 0.1% |
| Titanium | : 0.4% |

The metal case is used in booster stage, strap-on motors and also in second stage motors of launch vehicles.

The composite case is preferred for upper stage motors due to its light weight. The type of composite generally used are glass fiber reinforced epoxy and Kevlar fiber reinforced epoxy case. Both helical type filament winding, and polar winding machines are used to give desired contour. The Kevlar fiber reinforced epoxy case can withstand high temperatures up to 500° C. Table 6.1 gives comparative properties of the motor case.

**Table 6.1** Comparison of Motor Case Properties.

| Case Material | 15CDV6 | Maraging Steel | Glass Epoxy | Kevlar Epoxy |
|---|---|---|---|---|
| Tensile Strength (MPa) | 1660 | 1765 | 1230 | 1760 |
| % Elongation | 8 | 5 | - | - |
| Specific Strength (kN) | 21.7 | 23.1 | 52.5 | 123.7 |

For applications in smaller rockets, titanium alloys and aluminum alloys are used. The case has provisions for end covers, nozzle, and handling, etc. It is subjected to several quality assurance tests for strength, toughness, soundness of weld and pressure. Provisions are made in the rocket case to attach other components like motor igniter, nozzle, thrust skirt (connects motor case to vehicle structure to transfer thrust load to vehicle), staging connectors, thrust terminator, instrumentation systems and to handle ground & launch pad loading.

The rocket case is provided with thermal insulation on their inner surface to protect them from high temperatures of propellant grain combustion. Insulator is bonded to the inside of the motor case with suitable adhesive. The thickness of insulation will depend on the motor design, configuration, and exposure time. Generally, the insulation maximum at head end and nozzle end. The insulator is a highly flexible elastomer, having a coefficient of linear expansion around $0.7 \times 10^{-4} \frac{1}{°C}$, between that of the case and propellant can reduce the thermal stresses. The basic requirements of an insulator are low thermal conductivity, low erosion rate, and low density to reduce the dead weight. The types of rubber generally used are nitrile rubber and EPDM. The elastomers, with additional reinforcing fillers (silicate, Kevlar and carbon) either in the form of powder and fibers, are generally used due to their good insulating properties. The required rubber is masticated and then compounded with suitable fillers, vulcanizing agents, activators, and modifiers to be drawn to thin sheets of 1 to 3 mm thickness. By adding and bonding number of these sheets the required thickness of the insulation of the motor is obtained as shown in Figure 6.2. A thin layer of elastomeric polymer material of thickness 100-200 microns, known as liner, is sprayed uniformly on the insulator surface for proper bonding of propellant grain to insulator. A typical liner composition for HTPB propellant contains HTPB binder, tri-methylal propane crosslinker, carbon black filler, toluene di-isocyanate agent and solvent dichloromethane for spraying.

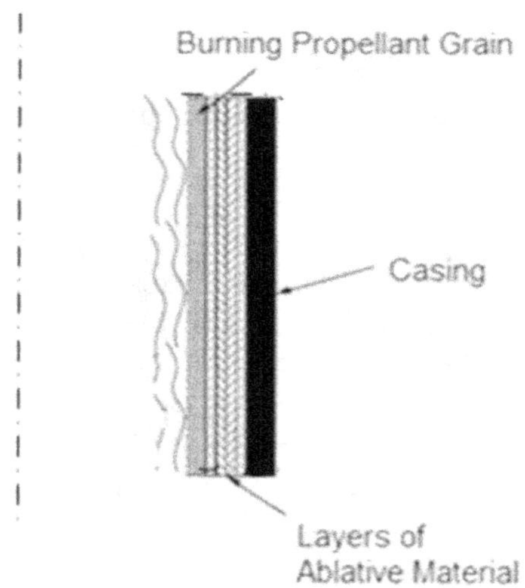


**Figure 6.2** Motor grain insulation.

## 6.2.2  Propellant Grain

The propellant grain is the solid body of the hardened propellant. Generally, it is made of composite propellant consisting of granular fuel and oxidizer in a rubber matrix binder. The grain accounts for 82 to 94% of the total motor mass. They are manufactured in two configurations: free-standing grain and case-bonded grain with various central port geometries. The free-standing grains are secured inside the case by various support elements such as wedges, springs or grids. Case-bonded grains are obtained by casting the propellant, before polymerization has occurred, directly into a case already provided with thermal insulation.

The propellant processing process is shown in Figure 6.3. In propellant processing, mixing is the most critical step. Horizontal sigma mixer or vertical change can mixer or continuous mixer is used to mix the solid and liquid ingredients to achieve homogeneous slurry product with viscosity ranging 3000 to 10000 poises. If the curing agent is not added at the premixing stage, the slurry can be stored for a few days. In the final mixing stage, the curing agent is added and mixed. This process of mixing is to cast large motors to avoid interruption in the casting process. The mixing process control parameters are mixing sequence, duration, temperature, and humidity. In the casting stage the viscous slurry is introduced carefully in the motor case without creating any porosity or blow holes. Suitable mandrels to achieve internal geometry of grain like star or cylindrical are used. The casting process parameters to be controlled during the process are vacuum level, valve/slit diameter, slit area, feed rate and slurry distribution pattern. During the curing process the liquid slurry converts into a solid mass. The curing is done at temperatures 50-60° C and duration varies on the size of the grain, from about 5 days for 50 kg grain to 15-20 days for 30 tons 3-meter diameter grain segment. Water or oil heated oven is used in curing small grains for safety reasons. In case of large grain segments oil or water is circulated through the mandrel to provide uniform heating. The cured motor is cooled to ambient temperature in a controlled cooling process to avoid debonding. After removing the mandrel, the ends are trimmed to required dimensions. To avoid combustion of certain specific areas of the propellant grain, like head end and nozzle side, inhibitor is casted layer by layer to achieve the required thickness. In cartridge loaded type grains are inhibited both ends and outside surface. Generally, an impregnated fabric with resin wound on the propellant surface to the required thickness.

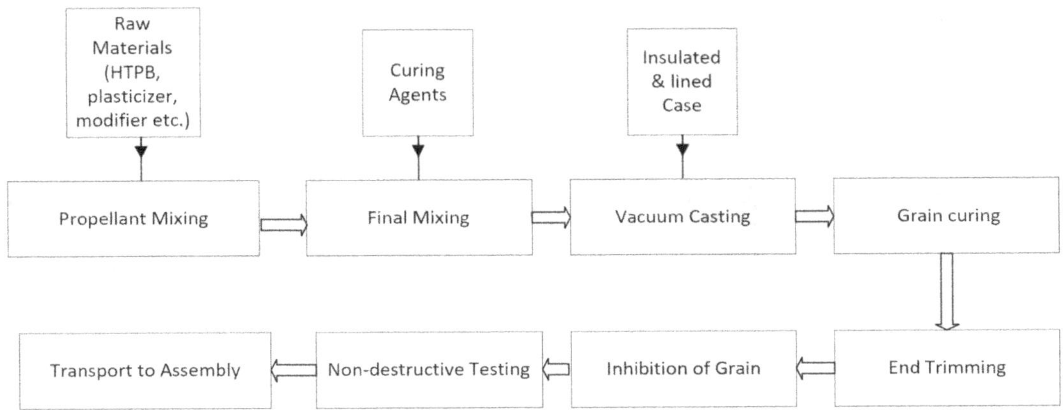


**Figure 6.3**  Propellant Processing Steps.

## 6.2.3    The Nozzle

The common type of nozzles used are shown in Figure 6.4. They are:

(a) Fixed Nozzle: This is the simple type used in tactical weapon propulsion systems, strap-on motors for space launch vehicles and spacecraft rocket motors for orbital transfer.

(b)  **Submerged Nozzle:** A significant portion of the nozzle structure is submerged within the combustion chamber. It reduces the overall length of the motor, in turn reduces the vehicle length.

(c)  **Movable Nozzle:** It provides thrust vector control (TVC) to a flight vehicle. Movable nozzles are typically submerged and have a ball-and-socket joint with two actuators 90° apart to achieve. They are able to make a rotation of around 7°. The nozzle is cooled by heat absorption in ablative materials. The shell is made of low carbon steel or high strength special alloys or aluminum. The different materials used to protect the erosion and to cool the nozzle are shown in Figure 6.5.

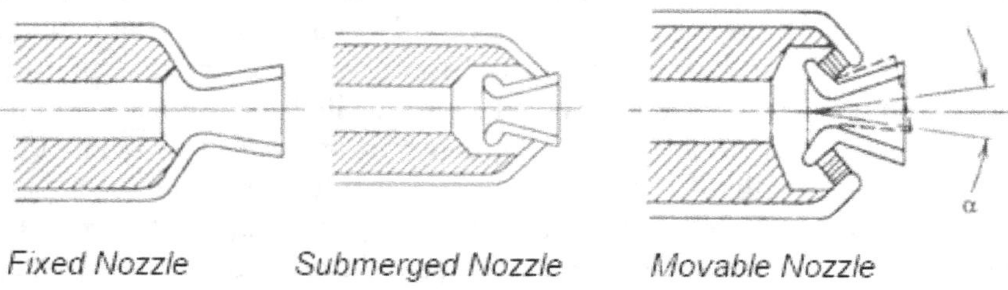


**Figure 6.4** Types of Nozzles Used in SRM.

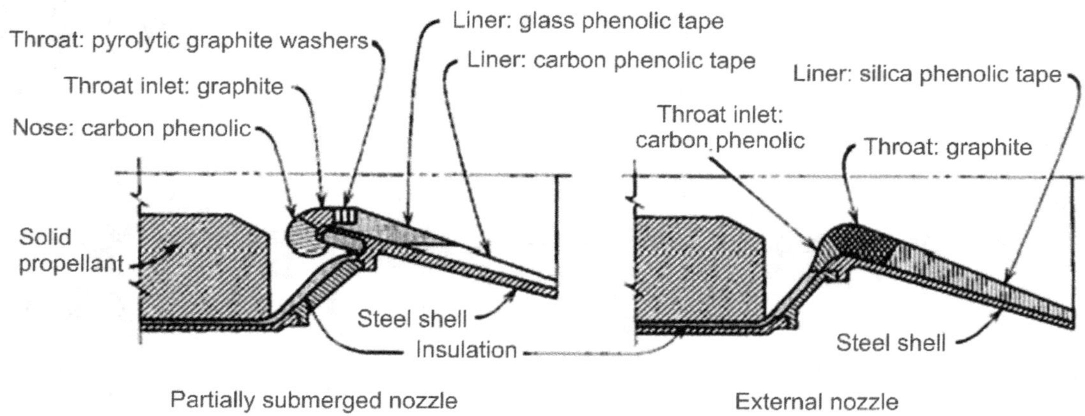


**Figure 6.5** Cooling Liners for SRM Nozzle (NASA SP-8115).

(d)  **The Ignitor System:** It provides the required amount of energy to the inner surface of the propellant grain to start the combustion process. The component is discussed in detail in section 6.10.

The booster rockets that develop high thrust to give sufficient acceleration during the first few minutes of launching are a basic complement to the classic liquid rocket engine. They are positioned on the periphery of the central liquid rocket engine. The Length/Diameter ratio of such SRM is high of the order of 10 to 12 and the propellant grain weight is over several hundred tons. It is difficult to manufacture such boosters in one single monolithic assembly. They are made into a few section segments called segments and then assembled. The three such big solid booster rockets are RSRM solid booster of Space Shuttle of USA, P230 solid booster of ARIANE-5 of ESA and the S200 solid booster for the Geosynchronous Satellite Launch Vehicle Mark III (GSLV-Mk III) of India. Ariane 5 has two solid boosters, each standing more than 30 meters tall with 237.8 metric tons of propellant. The boosters burn for 130 sec after the central Vulcain cryogenic engine has stabilized its thrust output. Figure 6.6 shows such a segmented motor with composite in a 14:18:68 HTPB: Al: AP mix and submerged movable nozzle hinged by a flexible bearing. The sea level specific impulse is 275.4 sec, motor diameter 3.05 m, length 27.34 m, total motor mass 269 tons and TVC with 6°.

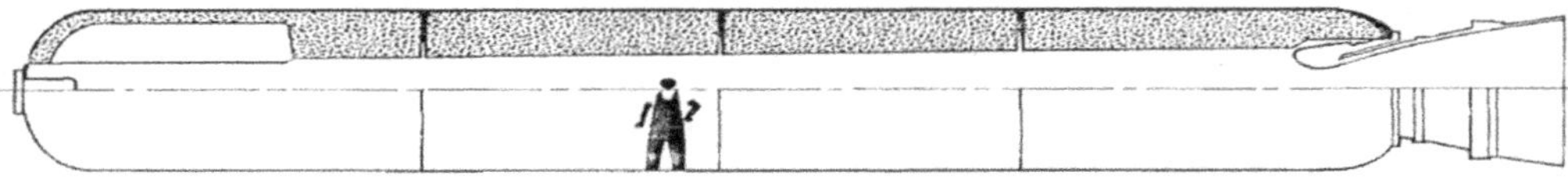

**Figure 6.6**  Ariane 5 booster configuration.

The solid rocket motors used for stage separation (called retro-motors) are of small size having a diameter around 0.20 m, length 1.0 m, propellant mass of 15 to 35 kg and have burning time 5 to 6 sec. These motors have canted nozzles, in the range of 6 to 25°, to ensure that the motor exhaust does not affect the main vehicle structure. The thrust developed will be in the range of 20 to 50 kN and average specific impulse around 235 sec.

Generally, rocket motors are used only once. To have major cost saving, the hardware must be recovered, cleaned, refurbished, and reloaded with propellant grain. The recovery and reusability make the design more complex. The reusable rocket hardware has a dominant effect on decreasing the cost of space travel. The SpaceX Falcon 9 and the Falcon Heavy lower rocket stages now can be recovered to reuse.

## 6.3  CLASSIFICATION OF SRM

SRM is classified based on grain composition, burning pattern, manufacturing method, geometry of grain and type of thrust developed as shown in Figure 6.7.

## 6.4  PERFORMANCE PARAMETERS

The performance of SRM depends on:

(a)  Propellant burn rate

(b)  Exposed propellant grain surface area

(c)  Grain Geometry

(d)  Amount of energy in propellant

The branch of applied science describing these is known as internal ballistics; it governs and controls the subsequent performance parameters of SRMs like thrust, exhaust velocity, flame temperature and duration of burning.

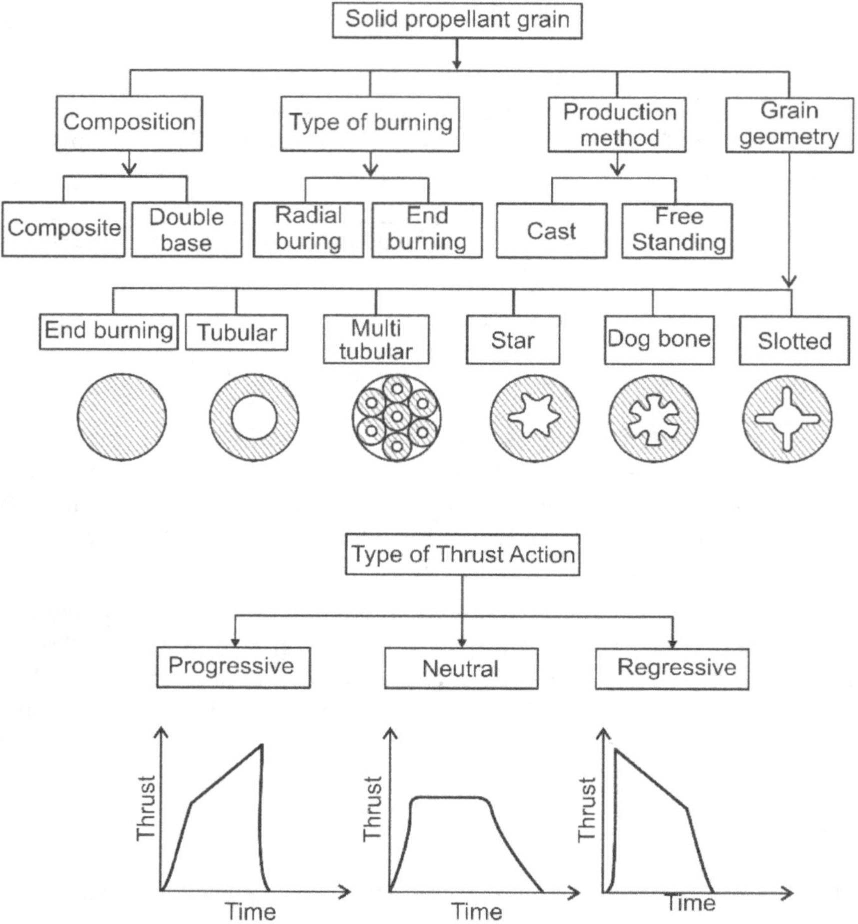



**Figure 6.7** Classification of SRM.

## 6.5 PROPELLANT BURN RATE

The propellant burn rate is the rate at which the exposed propellant surface is consumed. It is measured as distance normal to the surface consumed in a given time. The rate of regression or burn rate, $r$ usually expressed in cm/sec, mm/sec, or in./sec. The performance of SRM depends significantly on the burn rate behavior of the propellant grain under all operating conditions. The burn rate depends on the following conditions:

1. Initial temperature of solid propellant before start
2. Combustion gas chamber pressure

3. Combustion gas temperature
4. Combustion gasses flow velocities parallel to its burning surface
5. Rocket motor motions (acceleration and spin-induced grain stress)

### 6.5.1 Measurement of Burn Rate

The burn rate can be determined with the following experimental methods.

1. Standard strand burners, often called Crawford burners
2. Small-scale ballistic evaluation motors (BEM)
3. Full-scale rocket motors, properly instrumented

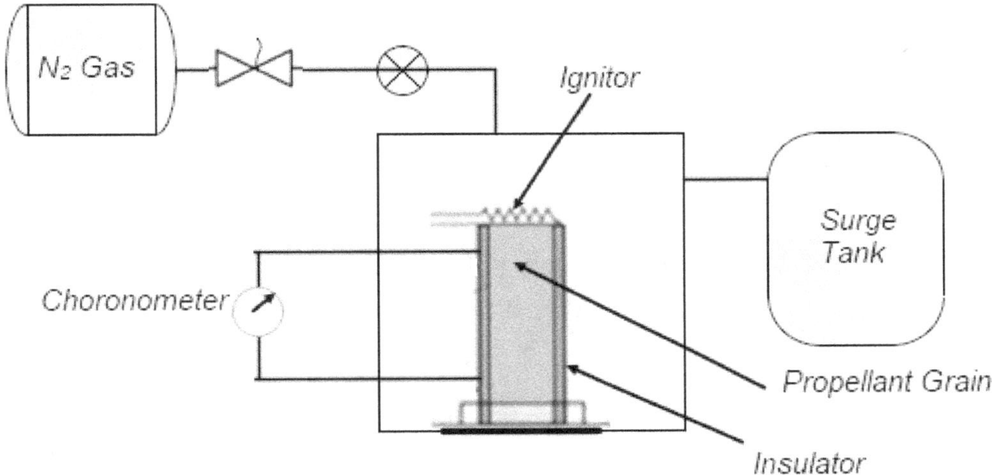


**Figure 6.8** Typical strand burner setup (*Crawford burner*).

The Figure 6.8 shows a schematic diagram of a typical strand burner. A small sample of size 5 to 10 mm diameter and approximately 170 mm long is fired in a bomb at constant pressure. They are inhibited along the whole length to ensure that combustion occurs perpendicular to the surface. An ignition wire is introduced on the surface and three to four lead wires are placed along the length which, by melting, allow the electrical detection of location of the flame front with time. The nitrogen gas is used to keep the test pressure constant during the combustion. The lead wires act as fuses to trigger the chronometer, a timekeeping device of great accuracy. The burn rate is calculated by knowing the burn distances and burning time. To find the burn rate at various temperatures glycol (low temperatures) and oil (higher temperatures) baths are used.

The BEM is basically a scaled-down model of a full-size rocket (1:10) and burn rates are inferred from the pressure–time plots. It is important to compare the burn rate data from the strand burner with those of the BEM. The burn rate measured on strand burners or small-scale ballistic motors are usually lower than that obtained from full-scale motor firing by 4 -10%, because it does not completely simulate the hot chamber environment.

A new or modified solid propellant is extensively characterized (i.e., tested for properties). These include:

(a) Burn rate testing (in several different ways) under different temperatures, pressures, and impurity conditions

(b) Measurements of physical and chemical properties, as well as manufacturing properties, ignitability, aging, and moisture absorption

(c) Compatibility with other materials like liners, insulators, and cases

For composite propellants, the burn rate can be increased by changing the propellant characteristics as follows:

1. Add a burn rate catalyst, often called burn rate modifier (0.1 to 3.0% of propellant) or increase percentage of existing catalyst

2. Decrease the oxidizer particle size

3. Increase oxidizer percentage

4. Increase the heat of combustion of the binder and/or the plasticizer.

5. Imbed wires or metal staples in the propellant

### 6.5.2 Effect of Chamber Pressure on Burn Rate

The mass flow rate $\dot{m}$, hot gases generated and flowing from the motor is given by:

$$\dot{m} = A_b r \rho_b \quad \left(m^2 \frac{m}{s} \frac{kg}{m^3}\right) \frac{kg}{s} \qquad \qquad .....(6.1)$$

where $A_b$ is the propellant grain burning area, $r$ the burning rate, and $\rho_b$ the solid propellant density prior to motor ignition. Propellant grains can also be designed for burning area $A_b$ to remain essentially constant.

The total effective mass $m$ of propellant burned is determined by integrating

$$m = \int \dot{m}\, dt = \rho_b \int A_b\, dt$$

*The operation of a typical rocket engine can be modeled by making the following assumptions:*

1. Uniform pressure across the entire combustion chamber

2. Combustion products can be treated as ideal gas

3. Burn rate remains constant over entire burning surface

4. The time for the mass to flow through the nozzle is negligible compared to the rate of pressure change within the combustion chamber even during transient state, shown in fig. 6.9.

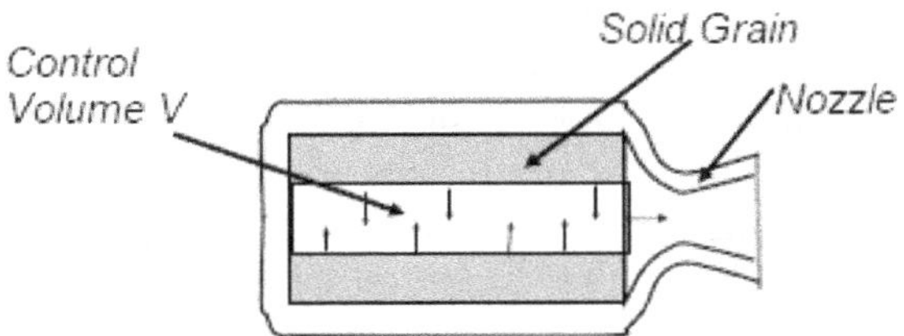


**Figure 6.9**  Control Volume for Burn Area.

Applying continuity equation to the control volume, as shown in Figure 6.9:

The gaseous propellant mass evolving from a burning surface per unit time

= the change in gaseous mass storage per unit time in the combustion chamber (due to increases in volume of the grain cavity) + the mass flowing out through the exhaust nozzle per unit time (Ref: Eq. 4.36)

$$A_b r \rho_b = \frac{d(\rho_g V)}{dt} + \frac{A_t p_c}{c^*}$$

The first term in RHS in the above equation is the rate of increase chamber cavity volume. The other terms are: $\rho_b$ is the solid propellant, $\rho_g$ is the chamber hot-gas density, $V$ is chamber gas volume, $A_t$ is the nozzle throat area, $p_c$ is the chamber pressure and $c^*$ is the characteristic velocity.

The chamber gas volume $V$ will measurably increase with burn time. But to fill this void requires relatively small amounts of gaseous propellant mass compared to what flows through the nozzle (the volume change of a unit of mass is about 1000 to 1 as the propellant gasifies during the combustion). As a result, the volume rate term can be neglected. Then, the above equation results

$$p_c = \frac{A_b}{A_t} \rho_b r\, c^* = K \rho_b r\, c^*$$

Over a limited operating range of chamber pressures, the burn rate, $r$ can be approximately expressed as function of chamber pressure $p_c$.

$$r = a p_c^n \qquad\qquad .....(6.2)$$

where $a$ is burn rate coefficient and $n$ is burn rate exponent. This equation is known as **St. Robert's Law**. The chamber pressure can be expressed as

$$p_c = K \rho_b a p_c^n c^*$$

$$K = \frac{p_c^{1-n}}{a \rho_b c^*}$$

Therefore, the chamber pressure, $p_c$ is

$$p_c = (K a \rho_b c^*)^{\frac{1}{1-n}} \qquad\qquad .....(6.3)$$

Figure 6.10 shows the experimental data of burning rate variation with chamber pressure for double base (DP) and composite modified double base propellants. The four samples are with varying percentage composition of oxidizer AP and its size. In general, it can be observed that the burning rate increases with chamber pressure $p_c$ with increasing concentration AP. The burning rate increases with the decreasing particle size of the oxidizer AP. The DB propellant has a higher burning rate compared to the CP for the same pressure. Hence, commercially available DB propellant has a higher combustion index $n$ in the range of 0.6–0.8 compared to that of 0.3-0.5 of the CP.

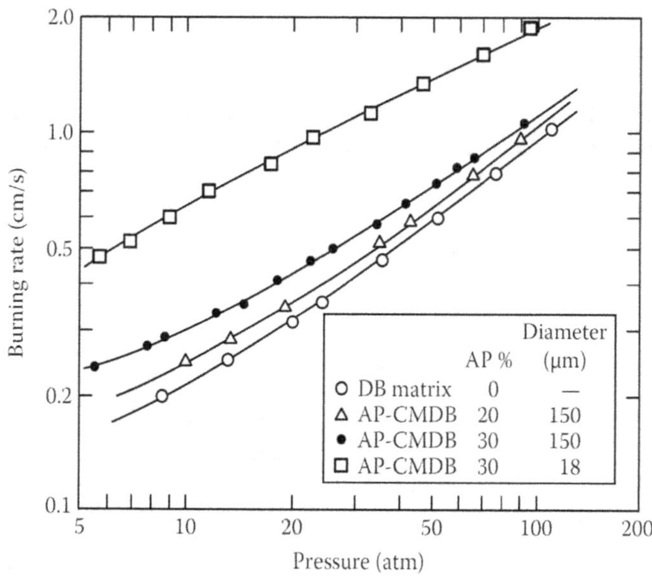


**Figure 6.10**  Variation of regression/burning rate $r$ with chamber pressure. ( Kubota N)

Assuming that the combustion temperature is constant, the mass flow rate through the fixed throat area nozzle is proportional to the chamber pressure $p_c$. This can be shown as a straight line as in Figure 6.11. The mass flow rate generated due to the burning of the propellant is proportional to $p_c^n$. The mass flow rates generated at different chamber pressures for $n < 1$ and $n > 1$ is shown in the same figure. The crossing point, S, between the mass flow rate generated and the mass flow rate through the nozzle corresponds to the stable operation of the rocket motor. This is because there would not be any accumulation of mass in the chamber i.e., the gas mass flow rate generated is equal to the mass flow rate passing through the nozzle, leading to a constant chamber pressure.

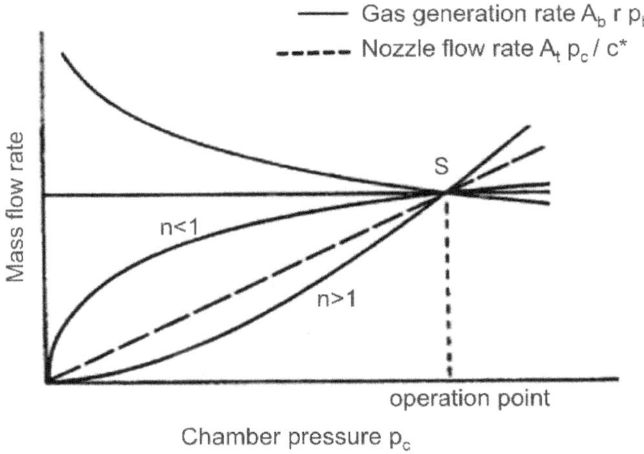


**Figure 6.11**  Balance of Mass Flow Rates.

Consider that there is a small change (perturbation) in chamber pressure by a small amount over its steady value. There will be two situations.

I.  Case 1 $n < 1$: If $p_c$ increases from stable chamber pressure $p_S$, the mass flow rate of gas generation rate is less than the nozzle flow rate. Hence the chamber pressure decreases and reaches the stable operating point. If $p_c$ decreases from $p_S$, the mass flow rate generated is higher than the nozzle flow rate. Hence chamber pressure builds up and reaches the stable point.

II. Case 2 $n > 1$: For $p_c > p_S$, the mass flow generation is higher than the nozzle flow. Hence chamber pressure further increases and goes away from stable pressure $p_s$ and drives the system to unstable condition. For $p_c < p_S$, the mass flow generation is less than the nozzle flow- chamber pressure further decreases and drives the system to unstable operation.

Hence, the hot gas mass flow rate is quite sensitive to the exponent $n$ and to have stable combustion process in the motor the burning rate exponent $n$ must be less than unity

**Example 6.1:**

With a strand burner the regression rates of a new propellant are found to be 18 mm/s and 35 mm/s at chamber pressures of 8 MPa and 20 MPa respectively. Determine the burn rate of Saint-Robert's. Also estimate the chamber pressure when the regression rates are 40 mm/s and 27 mm/s, assuming the law holds good in the range.

The Saint-Robert's Law is

$$r = a p_c^n$$

Given

At $p_c = 8$ MPa$=80 \times 0^6$ Pa,   $r = 18$ mm/s $= 0.018$ m/s

At $p_c = 20$ MPa $= 20 \times 10^6$,   $r = 35$ mm/s $= 0.035$ m/s

$$0.018 = a\,(8 \times 10^6)^n$$
$$0.035 = a\,(20 \times 10^6)^n$$

Dividing

$$\frac{18}{35} = \left(\frac{8}{20}\right)^n$$

$$0.5143 = (0.4)^n$$

$$\ln 0.5143 = n\,\ln 0.4$$

$$n = 0.726$$

$$a = \frac{0.018}{(8 \times 10^6)^{0.726}} = 1.752 \times 10^{-7}$$

The chamber $p_c$ is

$$p_c = \left(\frac{r}{a}\right)^{\frac{1}{n}}$$

For $r = 40$ mm/s $= 0.040$ m/s,

$$p_c = \left(\frac{0.040}{1.752 \times 10^{-7}}\right)^{\frac{1}{0.726}} = 24.03 \times 10^6\,Pa = 24.03\,MPa$$

For r = 27 mm/s $= 0.027$ m/s,

$$p_c = \left(\frac{0.027}{1.752 \times 10^{-7}}\right)^{\frac{1}{0.726}} = 13.99 \times 10^6\,Pa = 13.99\,MPa$$

### 6.5.3   Effect of Grain Temperature on Burn Rate

Temperature has a significant impact on the operation of SRM. It affects chemical reaction rates. The burning rate depends on the initial ambient temperature ($T_b$) of propellant grain. It changes noticeably the burning rate of the propellant. The sensitivity of the burning rate to propellant temperature can be expressed in the form of temperature coefficients, the two most common being are:

(i)   Temperature Sensitivity of Burn Rate ($\sigma_p$): The temperature sensitivity of burn rate, $\sigma_p$ expresses percent change of burn rate per degree change in propellant temperature at a particular chamber pressure.

$$\sigma_p = \left(\frac{\delta \ln r}{\delta T_b}\right)_{p_c} = \frac{1}{r}\left(\frac{\delta r}{\delta T_b}\right)_{p_c} \ {}^\circ C \qquad\qquad .....(6.4)$$

Typical value $\sigma_p$ of propellant lies in the range of *0.001–0.009 per °C*

For composite propellant $\sigma_p = 0.003\ \frac{1}{{}^\circ C}$

For double based propellant $\sigma_p = 0.005\ \frac{1}{{}^\circ C}$

(ii) **Temperature Sensitivity of Pressure ($\pi_K$):** The temperature sensitivity of pressure, $\pi_K$ expresses percent change of chamber pressure per degree change in propellant temperature at particular value of geometric function $K = A_b/A_t$.

$$\pi_K = \left(\frac{\delta ln p_c}{\delta T_b}\right)_K = \frac{1}{p_c}\left(\frac{\delta p_c}{\delta T_b}\right)_K \qquad .....(6.5)$$

Typical values of $\pi_K$ for propellant grain will be in the range of $0.00067$ and $0.0027$ *per* °C. Values of $\sigma_p$ and $\pi_K$ depend primarily on the nature of the propellant burning rate, the composition, and the combustion mechanisms of the propellant. Because variations in manufacturing tolerances act in addition to changes of ambient temperature, it is not simple to predict motor performance. These two sensitivity coefficients are used for setting temperature limits and maximum pressures for a given rocket motor.

### 6.5.3.1    Relation between $\sigma_p$ and $\pi_K$

The sensitivity coefficients are small numbers and can be written for convince in terms of differences. The temperature sensitivity of pressure, $\pi_K$ remains sufficiently constant over the temperature interval of operation. Eq. (6.5) can be integrated over from any reference temperature $T_{Rb}$ to ambient temperature $T_b$ to predict the chamber pressure $p_c$.

$$ln\, p_c = ln\, p_{Rc} + \pi_K(T_b - T_{Rb}) \qquad .....(6.6)$$

The burning rate equation Eq. (6.2) can be rewritten as

$$ln\, r = ln\, a + n\, ln p_c \qquad ......(6.7)$$

Differentiating the above equation with the assumption that the coefficient "*a*" depends only on $T_b$ and n is a constant.

$$\left(\frac{\delta(ln\, r)}{\delta T_b}\right)_{p_c} = \frac{\delta(ln a)}{\delta T_b} \qquad .....(6.8)$$

Using the above equation, the Eq. (6.4) for coefficient $\sigma_p$ can be written as

$$\sigma_p = \frac{\delta(ln\, a)}{\delta T_b} \qquad .....(6.9)$$

Expressing the chamber pressure Eq. (6.3) in logarithmic form and differentiating with respective ambient temperature $T_b$ considering K, $\rho_b,$ and $c^*$ *as constant*

$$ln\, p_c = \frac{1}{1-n}\left[ln\, a + ln\, (K\rho_b c^*)\right] \qquad .....(6.10)$$

$$\frac{1}{p_c}\frac{\delta(ln p_c)}{\delta T_b} = \frac{\delta(ln a)}{\delta T_b} \qquad .....(6.11)$$

Combining Eqs. (6.5), (6.9) and (6.11) gives a relation between the coefficients as

$$\pi_K = \frac{\sigma_p}{1-n} \qquad .....(6.12)$$

The above equations shows that $\pi_K$ and $\sigma_p$ are strong functions of the nature of propellant burn rate, composition, and combustion mechanism of propellant. Equation valid when the variables are constant over the chamber pressure and temperature range.

**Example 6.2:**

It is observed that for a SRM when the propellant unburnt temperature is 10°C and chamber pressure is 6 MPa, the regression rate of the propellant is 20 mm/s. It is planned to operate the SRM in desert area where the ambient temperature is 45°C. If the propellant grain is also soaked to the above desert temperature, estimate the new rate of regression for the SRM. Take the temperature sensitivity factor to be 0.006. If the temperature sensitivity pressure factor is 0.002, find the percentage change of chamber pressure due to increase in propellant unburnt temperature.

The temperature sensitivity coefficient is

$$\sigma_p = \left(\frac{\delta \ln r}{\delta T_b}\right)_{pc} = \frac{1}{r}\left(\frac{\delta r}{\delta T_b}\right)_{pc}$$

Integrating the equation between limits $r_f$ (reference burn rate) to $r$ and reference temperature $T_f$ to $T$

$$r = r_f\, exp\,\left[\sigma_p(T - T_f)\right]$$

$$r_{45} = 20 \times exp\,[0.006(45 - 10)] = 24.67 \frac{mm}{s}$$

The temperature sensitivity of pressure coefficient is

$$p_c = p_{0c}\, exp[(\pi_k(T_b - T_{0b})]$$

$$p_c = 6 \times exp\,[0.002 \times (45 - 10)] = 6.44\, MPa$$

The percentage increase of chamber pressure due to increase of unburnt temperature of propellant grain $= 0.44 \times \frac{100}{6} = 7.33\%$.

### 6.5.4    Effect of Internal Flows

It is generally assumed the velocity at the exit of the central grain port is negligible and the flow is accelerated in the convergent portion of the nozzle to the sonic velocity. However, the flow calculations have shown that the velocities at the exit of the grain port is of the order of 100-150 m/s and cannot be neglected in ballistic analysis of the propellant grain. Hence, the gas flow along the length of the grain generates following two types of activities to be considered in the ballistic analysis:

  (i)   pressure drops between forward and aft-end of the central port,

  (ii)  increase of burning rate due to erosive burning.

### 6.5.4.1   Pressure Drop

Pressure drops from grain head-end to grain aft-end due to energy losses inside the flow of combustion gasses and at the interface of the flow and propellant surface. The side injections from burning propellant surfaces also enhance the pressure losses. The critical operation step occurs just after ignition when port sections are having minimum area. The average pressure drop across the length of the grain is of the order 0.1 MPa. In some cases, for special configurations, pressure drops of more than 1 MPa have been observed. Computer programs are developed using Computational Fluid Dynamics (CFD) to map the pressure and velocity fields across different sections of the port.

### 6.5.4.2    Erosive Burning

The burning rate, $r$ defined without any consideration of combustion gasses flow along the length of the propellant grain. Enhancement of propellant burning rate due to tangential gas flow (compared to propellant burning rate without tangential flow) is known as erosive burning. This is due to an increase of heat transfer from the flame zone to the propellant surface. Figure 6.12 shows the variation of erosive burning ratio ($r_e/r$) with gas flow velocity $V$ for double- base propellants. It can be observed that the burning rate of the propellant remains constant below a certain velocity.

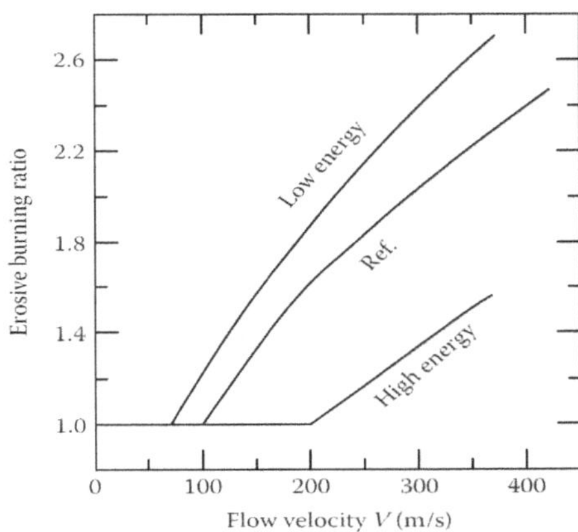


**Figure 6.12** Erosive burning ratio and threshold velocity for different double base propellants.

Once the gas velocity crosses threshold value, the erosive burning rate increases. For example, for a low-energy propellant the burning rate remains constant up to 70 m/s and beyond the threshold value the burning rate increases. Also, it can be noted that the high-energy propellant is less sensitive to convective flow velocity compared to a low-energy propellant. It has been observed that CPs exhibit less erosion than DB propellants. Extensive experimental and modeling studies have been conducted on erosive burning for both CP and DP by a number of investigators. The simple erosive law for modeling studies is

$$r_e = r[1 + \alpha(G - G_e)] \qquad\qquad .....(6.13)$$

where

$r_e$ = burning rate with erosive burning

$r$ = burning rate without erosive burning

G = mass flow rate in the given port section

$G_e$ = mass flow rate unit threshold (beyond which erosive burning occurs)

The values of $\alpha$ and $G_e$ are obtained empirically.

### 6.5.5    Effects of Acceleration on Burning Rate

The burning rate of the propellant grain is affected due to rocket vehicle longitudinal and lateral acceleration along its trajectory. The spinning of the vehicle for stabilization generates centrifugal forces. In radial burning grains, the centrifugal acceleration moves the flame nearer the surface.

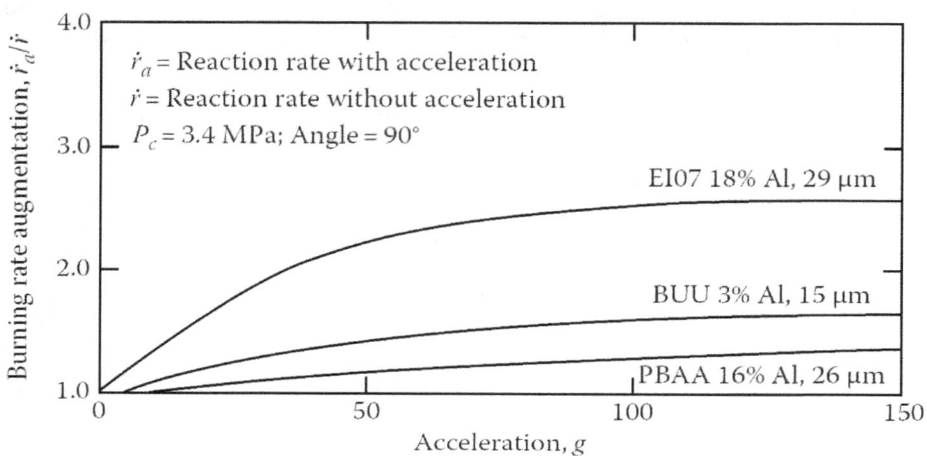


**Figure 6.13** Effect of acceleration on the burning rate for three different propellants.

This results in enhanced conduction heat transfer and an increase of the burning rate. The increase in burning rate depends on the composition of the grain, orientation of the acceleration vector with respect to the burning surface and the pressure level. When the acceleration vector forms an angle of 60°–90° with respect to the burning surface, the burning rate gets enhanced considerably. Generally, it must be greater than 10 $g$ to have a tangible effect on the burning rate. Also, due to the induced stress caused by its rapid acceleration or due to a sharp chamber pressure gradient, the propellant grain surface develops cracks. This increases the surface area and there will be an augmentation in the burning rate. Propellant grain designers depend on more experimental data, like as shown in Figure 6.13, as no satisfactory theory is developed to predict the effects of acceleration on the burning rate.

## 6.6    PROPELLANT GRAIN CONFIGURATION

Propellant grain, which looks like a hard plastic or rubber body, will burn on all its exposed surfaces forming hot gases that are then exhausted through a nozzle. The performance characteristics of the SRM depend upon material and geometrical configuration of the grain, operation modes of grain and burning rate. The propellant grain is designed to meet specific mission requirements. Its design, manufacturing, storage, and operation involve vast knowledge and numerous techniques related to different fields like chemistry, thermodynamics, geometry, combustion, fluid dynamics and mechanics of continuous media. Finite element analysis is used to predict the stress fields in the grain body and computational fluid dynamics methods are used to find the thermal and velocity field of the gas flow.

## 6.7    GRAIN ARCHITECTURE

As explained in section 6.2.2, there are two main types of grain architecture: free-standing grains and case-bonded grains. The free-standing grains are introduced into rocket cases (cartridge-loaded) after manufacture. In the second type the grains are bonded to the motor cases during the casting and curing steps of the manufacturing process. Generally, propellant grains having outer diameter 500 mm and above or weight of 300 kg and above are case-bonded. They have a central port, and the outer surface of the grain is bonded by a liner (and a thermal insulation) to the motor case. The final grain geometry is obtained during manufacture of the grain either by direct casting in the case around the mandrel or by machining the port after casting and curing have been completed. Figure 6.14 shows the casting pit for case bonded composite propellant and different shapes of mandrels generally used for central port configuration.

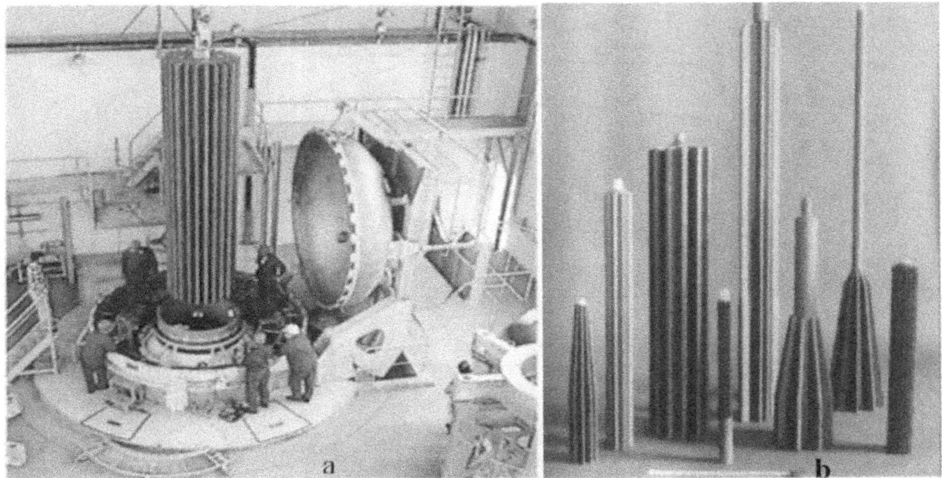


**Figure 6.14 (a)** Casting pit for case bonded composite propellant (b) Mandrels of different shapes for molding.

Free-standing grains are smaller in size and weight than case-bonded grains. Because they are not bonded to the case wall, they allow configurations which cannot be obtained with case-bonded grains (for instance an internal-external burning tube). To replace the grain independently of other motor components is ease, like stored missiles. The quality inspection of these grains is easier than in the case of case-bonded grains.

## 6.8   SELECTION OF GRAIN CONFIGURATION

The thrust versus time curve gives the variation of thrust with the progress of burning of grain. In turn, it provides the surface area of burning and pressure developed. There are three patterns of motor burning as shown in Figure 6.15:

(a)   Neutral burning: In this case the burning surface area remains constant. Hence, the thrust or pressure remains constant during burning. The variation of these parameters with time typically within about ±15%. The star shaped internal geometry motor and end burning motor are neutral burning motors.

(b)   Progressive Burning: In the case of a case bonded burning motor, the burning surface area increases with time and hence, the thrust or pressure increases with burning time.

(c)   Regressive Burning: In this case, the burning surface decreases with the time of burning and hence, the thrust or pressure decreases with time.

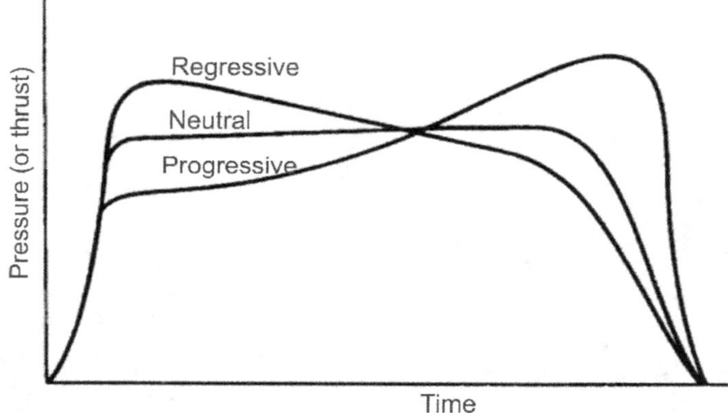


**Figure 6.15** Classification of grains according to their thrust–time characteristics.

*Some of the terms used in grain design are*:

*Perforation*: The central cavity port or flow passage of a propellant grain; its cross section may be a cylinder or a star shape or any similar geometrical shape.

*Sliver*: Unburned propellant residue or propellant lost i.e., expelled through the nozzle at the time of web burnout.

*Web thickness, b:* The minimum thickness of the grain from the initial internal burning surface to the insulated case wall; for an end-burning grain, $b$ equals the length of the grain.

*Web fraction, $b_f$ :* For a case-bonded internal burning grain, it is  the ratio of the web thickness $b$ to the outer radius of the grain.

*Volumetric loading fraction, $V_f$:* The ratio of propellant volume $V_p$ to the chamber volume $V_c$ (excluding nozzle) available for propellant, insulation, and restrictors.

$$V_f = \frac{V_p}{V_c}$$

### 6.8.1    Grain Requirements

The flight mission objectives determine the requirements of the motor. They include the total impulse, type of thrust time curve, total mass, available vehicle volume, storage and flight operation temperature and vehicle acceleration. The grain geometry with required burn surface area to meet the thrust time profile and to fit into the available volume is selected to meet the requirements. The propellant grain is designed to meet the ballistic properties, mechanical properties, structural integrity of the grain including the liner/insulator and easy manufacturing process to control the cost. Also, it should ensure proper combustion without any instabilities due to increase of port volume during burning.

### 6.8.2    Grain Configuration

Out of the many possible grain configurations, currently the configurations shown in Figure 6.16 are relatively popular since the requirements of a wide variety of solid rocket applications can be fulfilled. The grain configurations can be classified on their web fraction $b_f$, length to diameter ratio (L/D) and volumetric fraction $V_f$.

The end burning grain, also known as **cigarette burning grain**, is the simplest cylindrical grain. The combustion takes place across the transverse cross section of the grain and gives constant thrust during complete burning. It is restricted to small thrust and duration rockets because the chamber walls are continuously exposed to high temperature and pressure. This requires that to protect the casing from excessive heating thick wall chambers must be used. Moreover, the center of gravity of the propellant grain shifts while it is burning, making its motion unstable. The end burning grain has highest $V_f$, lowest grain port volume for a given thrust, low burning surface area and long burning time.

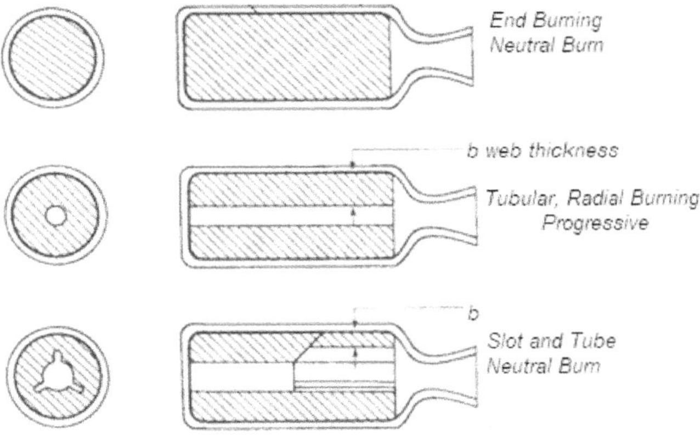


**Figure 6.16** Simplified diagrams of some grain configurations.

In the side burning grain, burning takes place along the length of the internal surface of the cylindrical grain and proceeds towards the case wall. Hence, the high pressure and temperature gasses are not exposed to the casing till the end of burning. Thus, the weight of the case is low compared to end burning. When larger values of thrust are required at the initial stage of take-off, such as in booster rockets, to increase the surface area of burning wrinkles or crumples are provided with shapes like star, wagon wheel and dendrites. In modern rocket motors, side burning propellant grains are used widely. The star configuration is suitable for web fractions 0.3 to 0.4 and is progressive above 0.4.

## 6.9   EVOLUTION OF BURNING SURFACE

To calculate the thrust, F of the rocket motor the evolution of the burning area with time is required.

It is equivalent to calculation of the burning perimeter, which will evidently vary with the geometry of the grain. The burning of any propellant grain proceeds in a direction along the local normal to the instantaneous burning surface. As per the Piobert's law, burning takes place by parallel layers where the surface of the grain regresses, layer by layer, normal to the surface at every point. The curved surface will propagate as a parallel curved surface as shown in figure. The burning of convex and concave surfaces is shown in Figure 6.17.

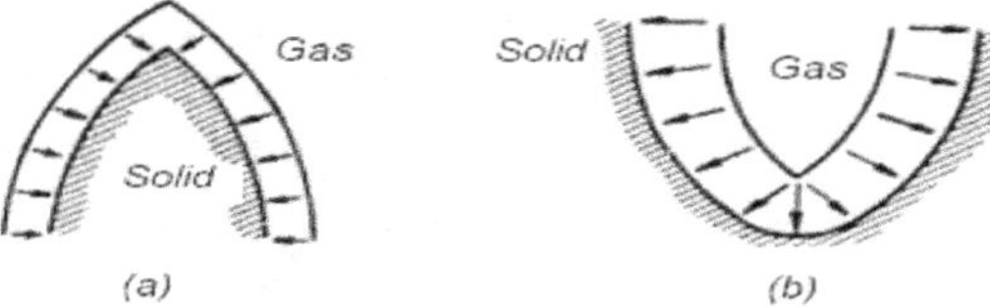


**Figure 6.17**  Effect of burning on (a) convex and (b) concave cusps.

(a)*End burning grain*: Consider an end burning cylindrical grain, as shown in Figure 6.18 of diameter D and length L in which combustion takes place across the transverse cross section of the chamber as shown in Figure. The thrust of the motor is given by Eq. (4.43),

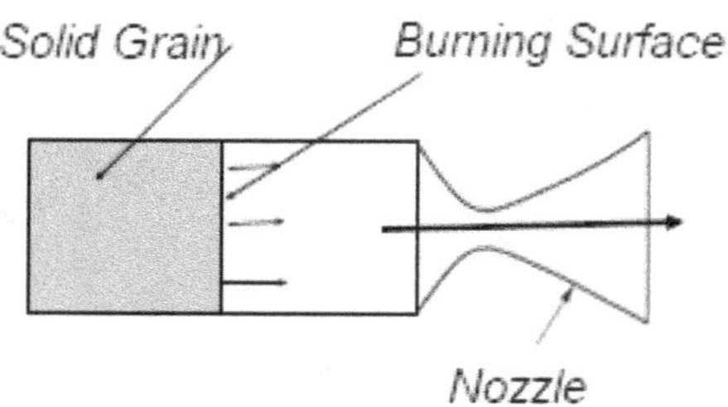


**Figure 6.18**  End Burning Cylindrical Grain.

$$F = C_F p_c A_t$$

Where $C_F$ is the thrust coefficient. The chamber pressure $p_c$ is given by Eq. (6.3).

$$p_c = (Ka\rho_b c^*)^{\frac{1}{1-n}} = \left(\frac{A_b a \rho_b c^*}{A_t}\right)^{\frac{1}{1-n}} \qquad .....(6.14)$$

Thus, the thrust $F$ is proportional to $A_b^{\frac{1}{1-n}}$. The burning area $A_b$ is constant. For large thrust we require a large diameter of rocket in case of end burning. The burning time of the motor with burn rate $r$ is

$$t_b = \frac{L}{r} \qquad\qquad .....(6.15)$$

(b) *Side burning grain*: In this case combustion takes place on the lateral surface of the grain as the end surfaces are insulated from burning. For the same length and diameter, the side burning grain covers higher burning surface area compared to end burning grain. The Figure 6.19 shows a circular grain of outside diameter $D$, inside diameter $d$ and length $L$.

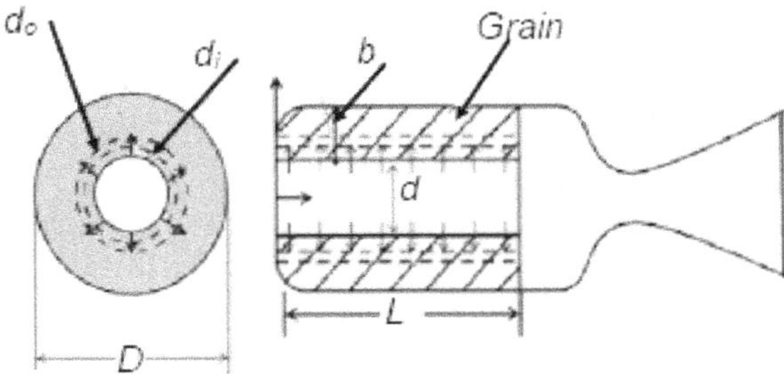


**Figure 6.19** Evolution of burning surface in cylindrical grain.

The initial burning area is

$$A_b = \pi dL$$

From Eq. (6.3), the initial chamber pressure $p_{ci}$ and final chamber pressure $p_{cf}$ are

$$p_{ci} = \left(\frac{\pi dLa\rho_b c^*}{A_t}\right)^{\frac{1}{1-n}}$$

$$p_{cf} = \left(\frac{\pi DLa\rho_b c^*}{A_t}\right)^{\frac{1}{1-n}}$$

The rocket thrust increases continuously as the burning area of the grain is increasing during the burning time of the motor. Consider in a small time $\delta t$, the radial burning of the grain takes from a diameter $d_i$ to $d_o$. Assuming the pressure $p_c$ is constant during the time and is given by

$$p_c = \left(\frac{\pi d_i La\rho_b c^*}{A_t}\right)^{\frac{1}{1-n}}$$

The burning rate $r$ is

$$r = ap_c^n = a\left(\frac{\pi d_i La\rho_b c^*}{A_t}\right)^{\frac{n}{1-n}} = aE(d_i)^{\frac{n}{1-n}} \qquad .....(6.16)$$

where $E$ is constant for a given grain and nozzle.

In time $\delta t$, a small thickness of the grain is burnt.

$$\frac{d_o - d_i}{2} = \frac{\delta d}{2}$$

The burning time is given by

$$\delta t = \frac{\delta d}{2r} = \frac{\delta d}{2aE(d_i)^{\frac{n}{1-n}}}$$

The time taken to burn the grain, $t_b$ from initial diameter d to final diameter D is given by integrating above equation

$$\int_0^{t_b} \delta t = \int_d^D \frac{\delta d}{2aE(d_i)^{\frac{n}{1-n}}}$$

Integrating and simplifying, the time burn is

$$t_b = \frac{(1-n)}{2aE(1-2n)}\left(D^{\frac{1-2n}{1-n}} - d^{\frac{1-2n}{1-n}}\right) \qquad .....(6.17)$$

For $n = 0.5$ the above equation is not valid. The integral becomes

$$\int_0^{t_b} \delta t = \int_d^D \frac{\delta d}{2aE(d_i)^1}$$

So, the time of burn for $n = 0.5$ is

$$t_b = \frac{1}{2aE} ln\frac{D}{d} \qquad .....(6.18)$$

(c) *Star Grain*: The star grain configuration is the most preferred radially burning cylindrical grain in solid motors. It has both a progressive burning tube and regressive burning star points. Hence, by varying these two aspects, the designer has more options of designing a suitable thrust pattern to meet the mission requirements. A neutral thrust pattern can be obtained with proper combination of both progressive and regressive surfaces in the grain. Figure 6.20 shows a view of a pointed star grain with 5-star points. It can be specified by independent variables: number of star points $n$, web thickness $b$, opening star point angle $\theta$, maximum inner diameter $d$, and outer diameter $D$. The web thickness for the star grain is the radial distance between the upper star point and the outer diameter. The burning surface area $A_b$ is to be evaluated for the calculation of thrust developed by the motor. A star propellant grain with $n$ vertices contains *2n* flat surfaces. The angle subtended by each of the *2n* surfaces of length $AB$ at the center would be π/n *(2 π/2n)* as shown in Figure 6.20. While burning the propellant surface AB regresses by distance $y$ normal to surface and evolves into surface $CB_1$ and the same time the star point regresses to arc $A_1C$

$$y = AA_1 = AC$$

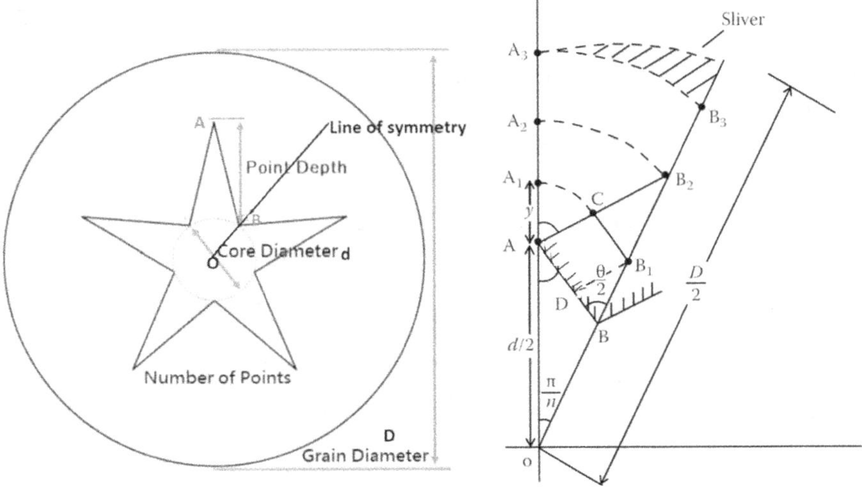



**Figure 6.20** Star Grain- Evolution of burning surface of star grain.

The new regress surface of the grain is $A_1CB_1$. This surface is repeated $2n$ times in this grain. The change in total burning surface area of this grain can be obtained by a net change in perimeter multiplied by depth $(L)$ and number of segments $(2n)$. Various angles from figure are

$$\angle AOB = \frac{\pi}{n}; \quad \angle OAB = \pi - \left(\frac{\pi}{n} + \pi - \frac{\theta}{2}\right) = -\frac{\pi}{n} + \frac{\theta}{2};$$

$$\angle A_1AC = \pi - \left(\frac{\pi}{2} - \frac{\pi}{n} + \frac{\theta}{2}\right) = \left(\frac{\pi}{2} + \frac{\pi}{n} - \frac{\theta}{2}\right)$$

$$CB_1 = AB - y\cot\frac{\theta}{2}$$

$$A_1C = y\left(\frac{\pi}{2} + \frac{\pi}{n} - \frac{\theta}{2}\right)$$

Then the net change of surface $(dS)$ in the initial segment AB is

$$dS = (A_1CB_1 - AB) = \left[\left(\frac{\pi}{2} + \frac{\pi}{n} - \frac{\theta}{2}\right) - \cot\frac{\theta}{2}\right]y$$

The net change in the burning surface area $A_b$ with regression of the propellant surface is

$$dA_b = 2nL\left[\left(\frac{\pi}{2} + \frac{\pi}{n} - \frac{\theta}{2}\right) - \cot\frac{\theta}{2}\right]y$$

If there is no net change in the burning surface area $A_b$, it can produce a neutral grain, i.e., $dA_b = 0$

$$\left[\left(\frac{\pi}{2} + \frac{\pi}{n} - \frac{\theta}{2}\right) - \cot\frac{\theta}{2}\right] = 0$$

The star grain angle $\theta$ is a function of the number of the star $n$. For a given star points the star angle $\theta_n$ for neutral thrust can be obtained by solving the above transcendental equation. For example, the neutral star grain angle $\theta_n$ is equal to 67° and 85° for $n = 6$ and 12, respectively. When the star grain angle $\theta$ is greater than $\theta_n$, it results in regressive burning. In contrast, propellant burning becomes progressive when $\theta$ is less than $\theta_n$. By choosing proper $\theta$, all three modes of burning can be obtained.

Table 6.2 gives values of characteristics generally observed in different solid propellant grain configurations.

**Table 6.2** Typical Characteristics of Some Grain Configuration.

| Configuration | Volumetric Loading Fraction | Burning Area | Burning Area Neutrality | Silver Fraction (%) | Web Fraction (d/w) |
|---|---|---|---|---|---|
| Star | $0.75 - 0.84$ | Intermediate | Good | 5 - 10 | $3.5 - 5.5$ |
| Slotted Tube | $0.75 - 0.85$ | Large | Good | 0 | 3 |
| Wagon Wheel | $0.50 - 0.70$ | Very Large | Excellent | 5 -10 | 6 - 12 |
| Dual Propellant | 0.90 | Intermediate | Excellent | < 5 | 2.5 - 3 |
| End Burning | $0.98 - 1.0$ | Small | Excellent | 0 | 1 |
| Internal Burning | $0.8 - 0.95$ | Intermediate | Good | 0 | 1.2 - 2 |
| Rod and Tube | $0.6 - 0.85$ | Large | Good | 0 | $2 - 2.5$ |
| Dog Bone | $0.7 - 0.8$ | Very Large | Excellent | 5 - 10 | 3.5 - 5 |

**Example 6.3:**

A SRM has a composite propellant grain of rectangular cross-section of 0.5 X 0.5 m, length 2 m with an outside diameter of 1.5 m. The burning of the grain starts from an initial port volume and proceeds towards the outer diameter. The density of the propellant is 1600 kg/m³.

(a) Estimate the sliver mass of the propellant grain and the percentage of the sliver mass compared to the initial propellant mass at the end of its operation. Comment on the grain design for its practical use.

(b) If the linear burn rate coefficient of propellant is $1.6 \times 10^{-7}$ and the index is 0.6, find initial and final chamber pressures. Take characteristic velocity as 1500 m/s and throat area as $3.0 \times 10^{-3} \ m^2$.

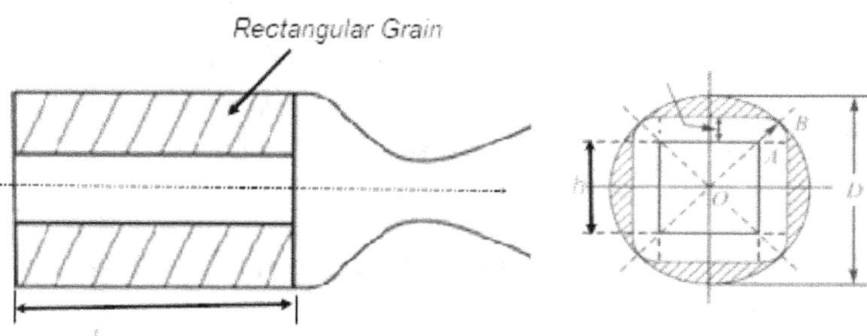


**Figure 6.21** Rectangular cross-section propellant grain.

(c)  As the burning proceeds normal to the surface, the corner points evolve as quadrants of a circle with its center at the initial corner point, while the flat surfaces progress parallel to the initial surface. The maximum burning surface area is obtained when the quadrant of the circle touches the outer diameter of the propellant grain. From the figure the web thickness of the grain (along the radius) is:

Web thickness

$$r_m = (OB - OA) = \frac{D}{2} - \sqrt{\left(\frac{b}{2}\right)^2 + \left(\frac{b}{2}\right)^2} = \frac{1}{2}(D - \sqrt{b^2 + b^2})$$

$$= \frac{1}{2}(1.5 - 0.707) = 0.397 \; m$$

Initial volume of the grain

$$Vi = \frac{\pi}{4}D^2L - b^2L = \left(\frac{\pi}{4}D^2 - b^2\right)L = \left(\pi \times \frac{1.5^2}{4} - 0.5^2\right) \times 2 = 3.034 \; m^3$$

(a) Initial mass of the grain $m_i = 1600 \times 3.034 = 4854.4 \; kg$

Volume of sliver is

$$V_s = \left(\frac{\pi}{4}D^2L - b^2L\right) - \left(4br_mL + 4 \cdot \frac{\pi}{4} r_m^2 L\right)$$

$$= 3.034 - (4 \times 0.5 \times 0.397 \times 2 + \pi \times 0.397^2 \times 2$$

$$= 0.456 \; m^3$$

Mass of the sliver

$$m_s = density \times V_s = 1600 \times 0.456 = 729.2 \; kg$$

The fraction of sliver mass to original mass

$$= \frac{729.2}{4854.4} = 0.15 \; or \; 15\%$$

This shows that 15 per cent of the propellant is burning in the tail-off region which is not desirable. Such grains are not used in practice.

(b) Initial burning surface area $A_{bi} = 4 \times 0.5 \times 2.0 = 4.0 \; m^2$

$$chamber \; pressure \; p_{ci} = (Ka\rho_b c^*)^{\frac{1}{1-n}}$$

$$K = \frac{A_{bi}}{A_t} = 4.0/ (3 \times 10^{-3}) = 1.33 \times 10^3$$

$$p_{ci} = (1.33 \times 10^3 \times 1.6 \times 10^{-7} \times 1600 \times 1500)^{\frac{1}{1-0.6}} = 5.9 \times 10^6 Pa$$

The perimeter of the grain corresponding to the maximum surface area is

$$= 4b + 4\left(\frac{\pi}{2} r_{max}\right) = 4b + 2\pi \; r_{max}$$

Maximum burning surface area $A_{bf} = L(4b + 2\pi\, r_{max})$

$$= 2 \times (4 \times 0.5 + 2\pi \times 0.397) = 8.99\ m^2$$

$$K = \frac{A_{bf}}{A_t} = 8.99/ (3\times 10^{-3}) = 2.99 \times 10^3$$

$$p_{cf} = (2.99 \times 10^3 \times 1.6 \times 10^{-7} \times 1600 \times 1500)^{\frac{1}{1-0.6}} = 44.7 \times 10^6 Pa = 44.7\ MPa$$

This maximum chamber pressure corresponds to the web burnout time. Thereafter, the burning surface area and the pressure progressively decrease with time. Figure 6.22 shows the evolution of burning area with time.

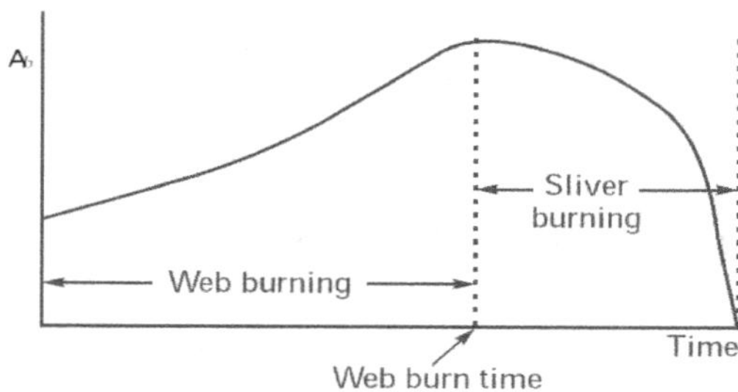


**Figure 6.22** Evolution of Burning Surface with time.

## 6.10  SOLID PROPELLANT GRAIN IGNITION

The ignition system of SRM supplies the energy to the surface of the propellant grain to start combustion. In case of large propellant grains used in space and ballistic missiles, there are three stages in the ignition system.

(a)  Initiator

(b)  Booster charge

(c)  Main charge

A pyrotechnic device used to transform the ignition signal (may be electrical, mechanical, or chemical, or a shock wave, a laser beam, or a combination of sources) into the steady of burning pyrotechnic material. The booster charge transmits the flame between the primer and the main grain. The main charge ignites the propellant grain and burns for a few tenths of a second, delivering a discharge approximately a tenth of the flow rate of the propellant grain. The booster and main charge contain a charge in the form powder, pellets, or propellant micro rocket grain.

### 6.10.1  Ignition Transient

The time interval between the first electric signal for the motor start-up and the attainment of the steady design operative condition of the engine is known as ignition transient. The different phase of the transient process is shown in Figure 6.23.

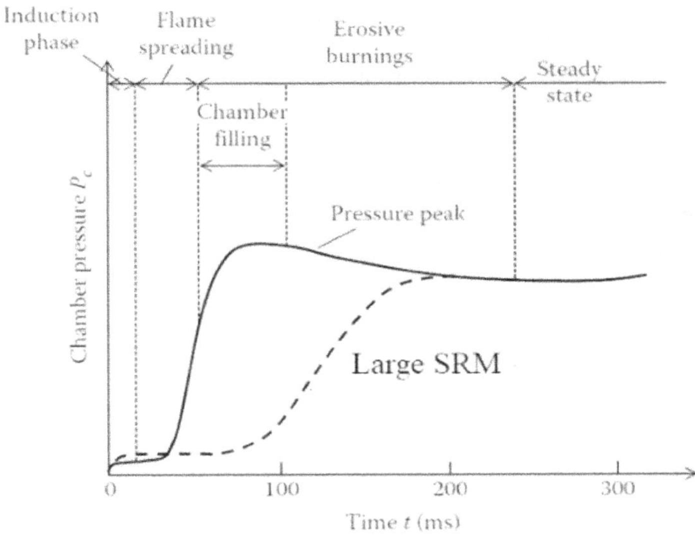



**Figure 6.23** Ignition process in a solid-propellant rocket engine.

(a) *Electric delay*: Time interval between the application of the electric signal to the start-up of the motor igniter and it is in the order of a few milliseconds.

(b) *Induction interval*: It is the interval between the start-up of the igniter and the very first revealing of the flame on the propellant surface and it is about 1/5 of the ignition transient.

(c) *Flame spreading*: It is the interval between the first detection of the flame on the propellant surface and the instant of the complete ignition of the grain and it is about 1/5 of the ignition transient.

(d) *Chamber filling*: It is between the complete ignition of the propellant grain and the reaching of the design equilibrium chamber pressure, characterizing the quasi-steady operative phase of the motor. This phase takes about 3/5 of the whole ignition transient.

Ignition is basically a complex physicochemical process. In composite propellant based on AP, the sublimation of AP is endothermic. It is followed by exothermic decomposition occurring either in gaseous phase in front of the propellant surface or through gas-solid reaction at the surface of the solid propellant. The ignition is dependent upon

(i)   the rate of heat flux
(ii)  gaseous composition in front of propellant surface
(iii) chamber pressure
(iv)  gas velocity cross grain surface area
(v)   surface condition and presence of AP particles.

## 6.10.2   Action Time and Burn Time

A typical variation of chamber pressure with time is shown in Figure 6.24.

(a) *Ignition delay $t_d$*: In the Figure 6.24 zero time refers to the time at which ignition voltage is applied. The chamber pressure changes very slowly over a small period of time and reaches a point A, 10% of maximum chamber pressure $p_{c,max}$. 0-1 marked in the figure is known as ignition delay period $t_d$.

(b) *Ignition rise time*: Beyond the delay time, the chamber pressure rises rapidly due to the generation of combustion gases because of combustion of propellants in both igniter and main propellant grain. The time duration 1-2, corresponding to 75% of maximum chamber pressure, is called the ignition rise time

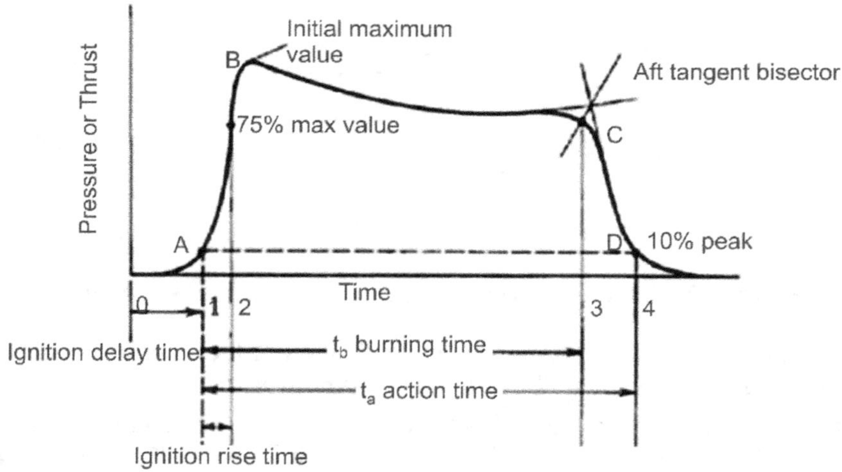


**Figure 6.24** Typical Pressure-Time curve for SRM.

(c) *Pressure-Thrust build up*: It is the time from 75% of $p_{c,max}$ to the time of attaining equilibrium pressure or thrust. After reaching the steady pressure, it remains constant (or follows the design values) during the major portion of the motor operation. Here, the rate of gas production is exactly equal to the rate of discharge.

(d) *Web burn time $t_b$:* The chamber pressure slowly drops to ambient pressure towards the end of propellant burning as all the propellant is almost burnt out except the residual portion known as sliver, which cannot be utilized for developing any thrust. This tail-off phase, CD, indicates the end of the grain burning. The intersection point between the tangent to grain burning and the tangent to the tail-off phase is, C indicates the end of the burning phase. The burning time 1-3 in the figure, denoted by $t_b$ is defined web burn time. It is used for calculating motor burn rate.

*Action time $t_a$*: The action time $t_a$, represented by 1-4 in the figure, is defined as the time duration (AD) between the ignition delay time (point A) and the time corresponding to 10% of the maximum pressure on the tail-off curve (point D). It is used to calculate the specific impulse.

### 6.10.3    Ignition System

The function of the igniter is to induce the combustion reaction at a stipulated rate in a controlled and predictable manner. The main components of solid rocket igniters are

(i)   an initiation system

(ii)   an energy release system

(iii)  the hardware to contain the above 1 & 2 systems and to provide for mounting them in or to the rocket motor.

*The ignition of motor is characterized by*:

(a)   Ignition peak is below the Maximum Expected Operating Pressure (MEOP) of the motor

(b)   Ignition delay and ignition time should meet the time specification fixed from mission consideration

(c)   Rate of pressure rise, i.e., *dp/dt,* in the motor chamber during ignition transient should not be large to induce shock which could damage the motor grain.

*The other requirements of ignition system are*:

1.  It should be compatible with the motor and its propellant.
2.  Perform under all specified conditions
3.  The heat generation rate is more than the rate of heat loss to the surroundings
4.  The heat flux provided by the igniter should be sufficient to raise the propellant grain surface temperature from ambient to its auto-ignition temperature.
5.  During combustion, igniter container fragments should not damage the propellant grain or choke the nozzle flow.
6.  Should have a long storage life and be easy to produce and assemble.

An electro-explosive device, referred to as an "initiator" or "squib", is a commonly used method to initiate motor grain combustion. Protection against misfire or inadvertent ignition is important to avoid hazards. The main hazards to electrical initiators are

(a)   induced current from electromagnetic radiation

(b)   static electrical discharge

(c)   spurious signal pickup

(d)   heat, vibrations, and shocks

The energy release system of the igniter provides the heat efflux necessary to ignite the propellant and raise it to a self-sustaining combustion level. Commonly used types of igniters are pyrotechnics or pyrogen igniters or some variation or combination of them.

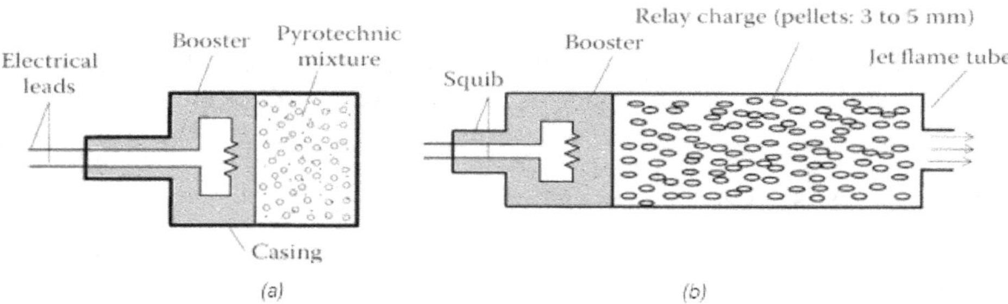


**Figure 6.25** Schematic of pyrotechnic igniters: (a) basket igniter; (b) jet flame igniter.

Two types of pyrotechnic igniters, basket igniter and jet flame igniter, as shown in Figure 6.25, are used for both small and medium sized rocket engines. The electrical initiator squib uses a fine nichrome wire of diameter 0.12 to 0.0125 mm attached to two electrical terminals. It is coated with a readily ignitable substance like potassium nitrate with charcoal powder and sulfur, generally in the ratio of 0.75: 0.15 : 0.10. On the passage of current through the squibs, it gets sufficiently heated due to resistive heating, to cause the deflagration of the heat-sensitive charge in contact with the wire. The electrical current passing through the squib is stipulated by fixing "no fire" and "all fire" current levels. The "no fire" is the safe current level that can be passed through the bridge wire to check continuity and circuit connection. The "all fire" is the current that is to be supplied through the squib for successful initiation of ignition.

The energy released by the squib charge is transported to the relay charge which is an intimate mixture of pyrophoric metal powders or granules like boron, magnesium, aluminum, and oxidizers like potassium perchlorate. The charge is either in powder form or number of pellets of size 5 mm to 10 mm depending upon the motor size and requirement. The hot gas produced by ignition is ejected from circumference and end cover holes forming as a jet that can spread over the entire propellant grain easily.

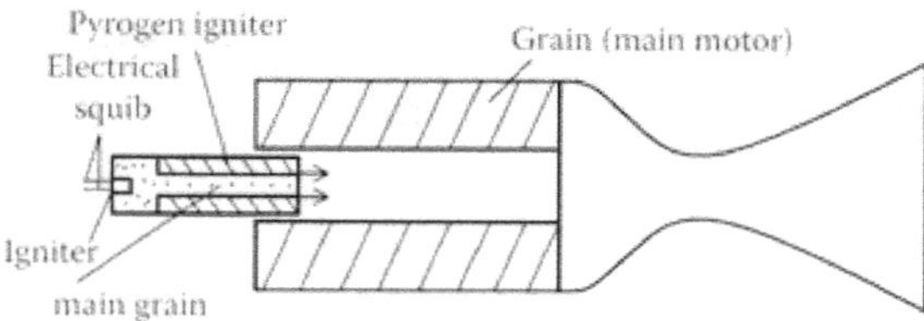


**Figure 6.26** Schematic of Pyrogen Igniter.

In case of a large solid propellant rocket the amount of igniter charge required for ignition goes up significantly. Controlled burning of a large amount of charge is difficult and the practice is to substitute a small solid propellant rocket for the igniter known as Pyrogen Ignitor as shown in Figure 6.26. The boost charge, usually a readily ignited pelleted pyrotechnic, propagates the ignition train from the initiator to the pyrogen propellant grain.

*The basic requirements of ignitor propellant are*:

1. Fast high heat release and gas evolution per unit ignitor propellant to allow rapid filling of the grain cavity with hot gas and partial pressurization of the chamber.
2. Rapid initiation of ignitor burning and low ignition delays.
3. Low burning rate exponent and its sensitivity to ambient temperature should be low.
4. Operation over required ambient temperature range of lunching.
5. Good aging characteristics and storage life.
6. Easy to manufacture and low cost.

The basic difference between the motor and ignitor propellant formulation is the percentage of metallic fuel in it. The ignitor propellant contains a low percentage of metallic fuel of the order 1- 2% only. The specific impulse is of the order of 210 s.

## Nomenclature

$a$      burn rate coefficient

$A_b$      solid propellant grain burning area ($m^2$)

$A_t$      nozzle throat area ($m^2$)

$b$      web thickness (mm)

$b_f$      web fraction

$c*$      characteristic velocity (m/s)

D, d      diameter of the solid propellant grain (m)

$G$      mass flow rate in the given port section

$G_e$      mass flow rate unit threshold (beyond which erosive burning occurs)

$\dot{m}$      mass flow rate (kg/s)

$n$      burn rate exponent

$n$      number star points of the star configuration grain

$p_c$      chamber pressure (MPa)

$r$      burn rate (mm/s or m/s)

$r_e$      erosive burn rate (mm/s or m/s)

$t_b$      burn time of motor (s)

$V$      gas volume ($m^3$)

$V_f$      volumetric loading fraction

$\pi_K$      temperature sensitivity of pressure

$\rho_b$      the solid propellant density (kg/$m^3$)

$\sigma_p$      temperature sensitivity of burn rate

$\rho_g$      hot-gas density (kg/$m^3$)

## PROBLEMS

6.1  The test results of regression rates of solid propellant samples at chamber pressures of 7 and 12 MPa are reported as 25 and 34 mm/s, respectively. Using Saint-Robert's law, find the chamber pressure for the regression rate 30 mm/s.

6.2  A SRM hollow cylindrical grain, radially burning, has inner diameter 25 mm and outer diameter 120 mm. The density of the propellant is 1675 kg/m$^3$. The length of the grain is 750 mm. The throat diameter is 15 mm. The characteristic velocity is 1250 m/s and thrust coefficient 1.12. The burn rate follows the Saint Robert's law. Take the burn rate coefficient as 0.02, the exponent 0.4 and the chamber pressure is in Pa.  Find the initial thrust, burn out thrust and duration of burning. What is time taken for the chamber pressure to fall from maximum to 10% of it during tail-off.

6.3  A SRM cylindrical grain has the following physical measurements:

(i)   Length = 2000 mm

(ii)  Outside diameter = 500 mm

(iii) Core diameter = 200 mm

(iv)  Nozzle throat = 80 mm

During a test firing of this motor, the initial and final chamber pressures were 2.8 and 7.0 MPa, respectively. The propellant used in this test has a density of 1700 kg/m$^3$ and a characteristic velocity of 1250 m/s. Assuming the propellant burns according to St. Robert's law determine the burning rate coefficient and exponent.

6.4  A rectangular propellant having cross section 0.4 × 0.4 with an outer diameter of 1.2 m and length of 2.0 m is proposed to be used in a SRM. The density of the propellant is 1600 kg/m$^3$. Find the percentage of sliver mass compared to the initial propellant mass at the end of its operation,

6.5  At an unburnt temperature 15°C and chamber pressure of 7 MPa, the burn rate 26 mm of the propellant grain is 26 mm as per Saint–Robert's law. The temperature sensitivity of the burning rate is 0.005 per °C the burn rate exponent is 0.58. Determine the new burning rate at an unburnt temperature of 40°C.

6.6  Consider the problem 6.4, determine the chamber pressure and thrust at the unburnt temperature of 40°C. Take thrust coefficient 1.10 and characteristic velocity 1700 m/s. and assume it remains constant. Take the burning surface area to throat area ratio 45 and throat area as $360 \times 10^{-6}$ m$^2$. The density of the propellant is 1600 kg/m$^3$.

6.7  The test SRM has the following characteristics:

Characteristic velocity = 1550 m/s

Density of propellant = 1675 kg/m$^3$

Specific heat ratio = 1.2

Chamber Pressure = 6 MPa

Burning time = 12 s

Total Impulse = 100,000 N s (vacuum)

Expansion ratio = 25

The outside diameter of the cylindrical grain = 300 mm

The burn rate follows the law $r = 0.62\, p_c^{0.38}$ ($r$ in cm/s and $p_c$ in MPa)

Find:

(a)  specific impulse (vacuum)
(b)  throat diameter
(c)  propellant mass required to maintain the total impulse requirements
(d)  the grain length and inside diameter.

6.8  A SRM with characteristic velocity of 1500 m/s, thrust coefficient of 1.2, and chamber pressure of 7 MPa is used to carry a payload of 200 kg. The initial total mass of the rocket engine happens to be 10,000 kg. The propellant with a density of 1700 kg/m$^3$ follows Saint-Robert's law. Take the burn rate coefficient as 0.02, the exponent 0.4 and the chamber pressure is in Pa.  If it is subjected to an initial acceleration of 1.6g, determine the throat diameter of the booster rocket and the initial burning area.

# Liquid Rocket Engines

## 7.1  INTRODUCTION

Liquid rocket engine (LRE) is the most common thrust source in the launch vehicle technology in use today. It is also a preferred propulsion unit in interplanetary missions. Compared to solid rocket motors, the liquid rocket engines are more flexible and have the advantage of throttling capability to permit thrust changes during flight trajectory. It can provide thrust levels ranging from a few Newtons to several hundred Newton. The characteristics of LRE, high specific impulse, longer burning time and high thrust-to-weight ratio, will lead to lower cost per kg of payload. The liquid rocket engine systems are more complex than the solid rocket systems. The other disadvantage is the engine has a long pre-launch check-list that takes considerable launch preparation time. Hence, for military applications, LRE is not preferred. The liquid rocket engines are of two types based on the number of propellants used, monopropellant and other bipropellant systems. In a monopropellant system, a single liquid propellant is used and in the presence of a catalyst it decomposes to create combustion gases. They find applications in propulsion systems of spacecraft and satellite control systems where the thrust requirements are not large. These are generally referred to as thrusters and are very simple systems compared to bipropellant systems used in launch vehicles. Chapter 13, Chemical Thrusters for Spacectaft Maneuvers, covers the working and analyses of these thrusters. This chapter deals with the bipropellant LRE, engine cycles and describes various components. The liquid propellants, engine performance and thermochemistry are dealt in with the previous chapters.

Typical bi-propellant engines use either earth storable propellants, generally combinations of nitrogen tetroxide ($N_2O_4$), asymmetric di-methylhydrazine (UDMH), or mono-methyl-hydrazine (MMH), mixtures of storable and cryogenic propellants like liquid oxygen (LOX) and kerosene or fully cryogenic, liquid oxygen (LOX) and liquid hydrogen ($LH_2$). Table 7.1 gives characteristic data of these propellants for quick reference. They can be classified depending on usage and position in the launch vehicle as booster, core (main) stage and upper stage engines and satellite propulsion and attitude control systems. The amount of heat released in the combustion process of such systems exceeds by far typical values of nuclear power plants ($\sim$ 3-4 GW) and is of order of 10 to 30 GW. Obviously, such power levels are only possible with high combustion chamber pressures and even higher propellant mass flow rates which may exceed 1000 kg/s. The present space launching systems can be divided into two major types: (a) launchers with large strap-on boosters with core and upper stage(s) like European Space Agency ARIANE, Indian PSLV or GSLV and USA SpaceX Falcon (b) launchers with a booster, an optional sustainer and an upper stage like American ATLAS and Delta, the Russian PROTON, the Chinese Long March. Launch booster engines deliver a high lift-off thrust of order 4000 – 9000 kN with a short burning time of around 150 s. Engines with small thrust levels are clustered together to provide the necessary lift-off thrust. Core engines provide a comparable lower lift-off thrust of order 1000 to 2000 kN but with a higher burning time in the range of 200-500 s. The upper stages engines typically operate at lower thrust, at higher specific impulse and longer duration. Recent trend is to use booster engines using kerosene/LOX propellants and the most impressive examples are SpaceX Falcon Merlin 1D and ATLAS (RD-180) engines.

**Table 7.1** Characteristic Data of Common Liquid Propellant Combination.

| Oxidizer | Fuel | Density $(Kg.m^3)$ | Mixture Ratio | Specific Impulse (s) |
|---|---|---|---|---|
| | Kerosene | 820 | 2.77 | 358 |
| LOX | $LH_2$ | 700 | 4.83 | 455 |
| | $LCH_4$ | 430 | 3.45 | 369 |
| | UDMH | 791 | 1.95 | 342 |
| $N_2O_4$ | MMH | 880 | 2.37 | 341 |
| | UM25 | 850 | 2.15 | 340 |

Recent popular rocket launch vehicles characteristic data is summarized in Table 7.2 (a), (b) and (c). Ariane 5 is an expendable European Space Agency (ESA) launch system designed to deliver payloads into LEO and GTO. SpaceX Falcon Heavy is a partially reusable launch vehicle consisting of the Falcon 9 first stage as the core with two additional Falcon 9 like first stages as strap-on boosters. Nine SpaceX Merlin engines power the Falcon 9 first stage with up to 854 kN thrust per engine at sea level, for a total thrust of 7,686 kN at liftoff. The first-stage engines are configured in a circular pattern, with eight engines surrounding a center engine. Twenty-seven SpaceX Merlin engines power the Falcon Heavy first stages for a total thrust of 22,900 kN at liftoff. Figure 7.1 shows the arrangement for the center core and side boosters of the launch vehicle.

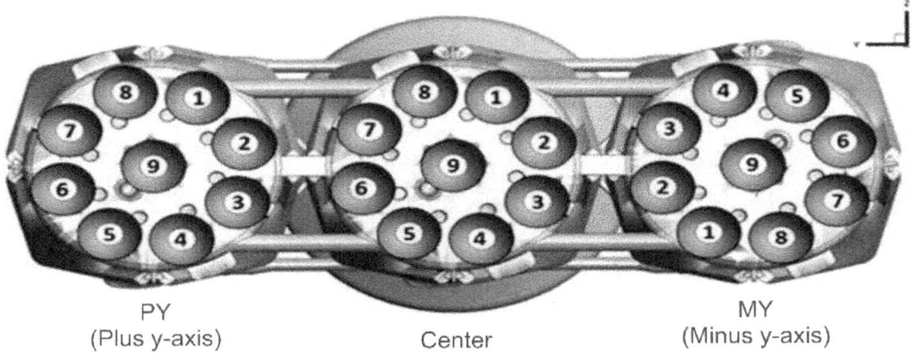


**Figure 7.1** Falcon Heavy booster stage layout.

The PSLV of Indian Space Research Organization (ISRO) has four stages using solid and liquid propulsion systems alternately. The first (core) stage, one of the largest solid rocket boosters in the world, carries 138 tons of HTPB propellant. The first stage thrust is augmented by six strap-on solid boosters.

**Table 7.2(a)** Characteristic Data of ESA Ariane 5 Launch Vehicle.

| | Boosters | Core Stage | 2nd Stage | Total Height |
|---|---|---|---|---|
| Length m | 31.6 | 23.8 | 4.71 | =48-52 m |
| Diameter m | 3.06 | 5.4 | 5.4 | Total Mass =777,000 kg |
| Empty Mass kg | 33000 | 14,700 | 4,540 | Payload |
| Gross Mass kg | 273,000 | 184,700 | 19,440 | LEO Over 20,000 kg |
| Engines | 2 boosters P241 | Vulcain 2 | HM 7B | GTO =10,865 kg |
| Thrust kN | 7,080 per Booster | 960 (sl) 1,390(vacuum) | 67 | |
| Specific Impulse sec | | 310(sl) 432 (Vacuum)) | 446 | |
| Burn Time sec | 140 | 540 | 945 | |
| Fuel | AP,Al,HTPB | LH2/LOX | LH2/LOX | |

**Table 7.2(b)** Characteristic Data of SpaceX Falcon 9 Launch Vehicle.

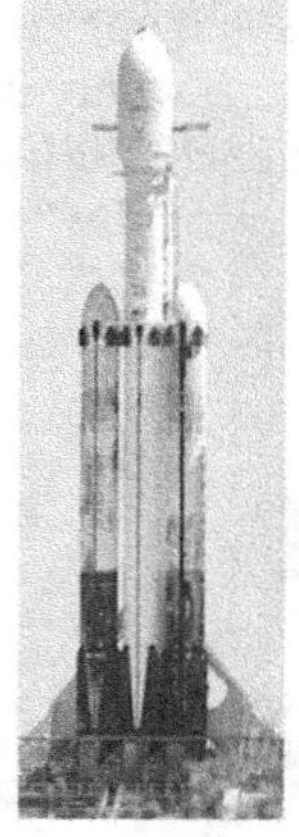

| | Booster (2) | Core stage | 2nd stage | Total height |
|---|---|---|---|---|
| Length m | 42.6 | 42.6 | 12.6 | = 70 m |
| Diameter m | 3.66 | 3.66 | 3.66 | Total Mass |
| Engine | 9 X Merlin 1D | 9 X Merlin 1D | 1 Merlin 1D | =1,420,788 kg |
| Propellant | RP1/LOX | RP1/LOX | RP1/LOX | Payload |
| Thrust kN | 2 x 9 x 845 | 9 x 845 | 934 | =63800 kg (LEO) |
| Specific Impulse sec | 282 | 282 | 348 | =26700 kg (GTO) |
| Burn time sec | 154 | 187 | 397 | |

**Table 7.2(c)** Characteristic Data of ISRO PSLV Launch Vehicle.

| | Boosters | 1st Stage | 2nd Stage | 3rd Stage | 4th stage | Total Height |
|---|---|---|---|---|---|---|
| Length m | 12 | 20 | 12.8 | 3.6 | 3 | =44 m <br> Total Mass |
| Diameter m | 1 | 2.8 | 2.8 | 2 | 1.3 | =320,000 kg |
| Propellant Mass | 12,200 each | 138,200 | 42,000 | 7600 | 2500 | Payload <br> LEO <br> Over 3,8000 |
| Gross Mass kg | - | | | | | kg <br> GTO |
| Engines | 6 Boosters | S139 | 1 Vikas | S 7 | 2xL25 | =1,200 kg |
| Thrust kN | 703.5 each | 4846 | 803.7 | 240 | 14.66 | |
| Specific Impulse sec | 262 | 237 <br> 269(V) | 293 | 295 | 308 | |
| Burn Time sec | 70 | 110 | 113 | 83 | 425 | |
| Fuel | HTPB | HTPB | $N_2O_4$/UDMH | HTPB | MMH/MON | |

## 7.2   BASIC CONFIGURATION

*The basic subsystems of a bipropellant liquid rocket engine are*:

(a) *Thrust chambers*: It is a metal structure where propellants, oxidizer and fuel, are injected through an injector to take place combustion. It is designed to contain the high-pressure combustion gases. The thrust chamber is cooled through a regenerative cooling system where the fuel is circulated through a network tube before being injected to the combustion chamber.

(b) *Feed mechanism*: It is a system to feed the propellants from storage tanks to the thrust chamber at the required pressure, i.e., higher than the combustion gases pressure. There are two basic configurations of feed systems, pressure-fed or blowdown system and pump-fed system. Inert gas at high pressure is used to force propellants into the thrust chamber. Applications that require large thrust are pump-fed systems in which pumps run by a gas turbine force propellant into the combustion chamber. Suitable plumbing or piping to transfer the liquid propellants under pressure is a part of the feed mechanism.

(c) *Power source*: In pump-fed systems, to run the propellant pumps to feed the thrust chamber at high pressure requires power. An impulse gas turbine is used to supply the power to pumps. A gas generator unit to supply the gas to the turbine is a part of the power system.

(d) *Control devices*: To load propellant to tanks, to start and stop the engine and to control the propellant flow require several various types of valves and instruments.

The size of the LRE depends upon the total propellant flow rate, $\dot{m}$ which can be calculated from thrust and specific impulse requirements. For a given mixture ratio, $MR\ (= \dot{m}_{ox}/\dot{m}_f)$ the flow rates of oxidizer and fuel are

$$\dot{m}_{ox} = \frac{(MR)\dot{m}}{1+MR} \qquad\qquad\qquad .....(7.1)$$

$$\dot{m}_f = \frac{\dot{m}}{1+MR} \qquad\qquad\qquad .....(7.2)$$

The propellant bulk density, $\rho_b$ is the ratio of the total mass of the propellant and the total volume it occupies. It can be written as

$$\rho_b = \frac{1+MR}{\frac{1}{\rho_f}+\frac{MR}{\rho_{ox}}} \qquad\qquad\qquad .....(7.3)$$

High bulk density is preferred for smaller tankage volume for a given total mass of the propellant. This is the primary problem with cryogenic fuel liquid hydrogen (LH$_2$) whose density is 71 kg/m$^3$. For a cryogenic LOX/LH$_2$ bipropellant engine with a mixture ratio 6 the bulk density is

$$\rho_b = \frac{1+6}{\frac{1}{71}+\frac{6}{1140}} = 361.8\,\frac{kg}{m^3}$$

Compared to water it is only around one-third. Hence, large tanks are required to house this cryogenic combination of propellants.


## 7.3 PRESSURE-FED SYSTEM

Figure 7.2 shows the basic features of a pressure-fed LRE. A high-pressure gas (helium or nitrogen) stored in a separate tank forces the propellants into the thrust chamber. A pressure regulator is used to regulate the pressure in the propellant tanks. In another type, known as blow-down mode, the gas or vapor column above the liquid surface in the tank, known as ullage volume, is directly pressurized. The mass or volume of the pressurant gas is proportional to the thrust requirement. Hence, the systems get quite large and heavy if the thrust is substantial. These systems are used for small burning time and thrust levels that require small quantities of propellant. The engine operates at low thrust chamber pressures. The main requirements of pressurant gas are low molecular weight of the gas, high gas density at storage conditions, minimum residual-gas weight, and high allowable stress-to-density ratio of the pressurant-tank material. Helium fits these requirements and finds its use in as pressurant gas invariably in all pressure-fed systems.

A good example of a pressure-fed system is the Aestus engine that powers the Ariane 5 ES version bipropellant upper stage for insertion of payloads into LEO, SSO and GTO. Figure 7.3 shows the detailed layout of the system. It includes propellant tanks, pressurant gas tanks, feed line and control valves. The fuel MMH is used to cool the nozzle and thrust chamber. Table 7.3 gives the main specification of the Aestus engine.

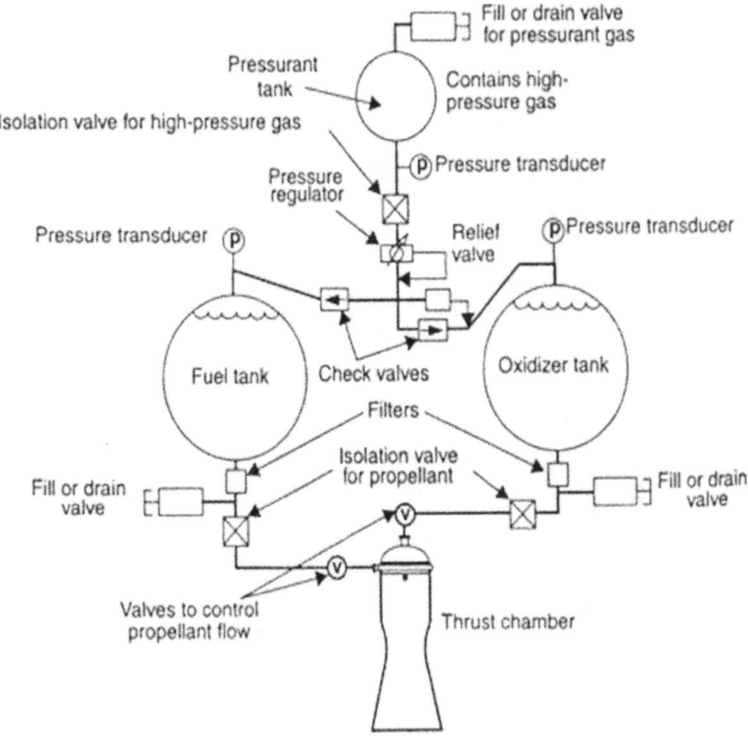


**Figure 7.2** Schematic of Pressure-fed System.

**Table 7.3** Ariane 5 upper stage Aestus engine Specifications.

| | |
|---|---|
| Engine Cycle | : Pressure-fed |
| Propellant | : $N_2O_4$/MMH |
| Thrust | : 29.6 kN |
| Chamber pressure | : 11 bar |
| Specific speed (vac) | : 3178 m/s |
| Burn time | : 1100 s |

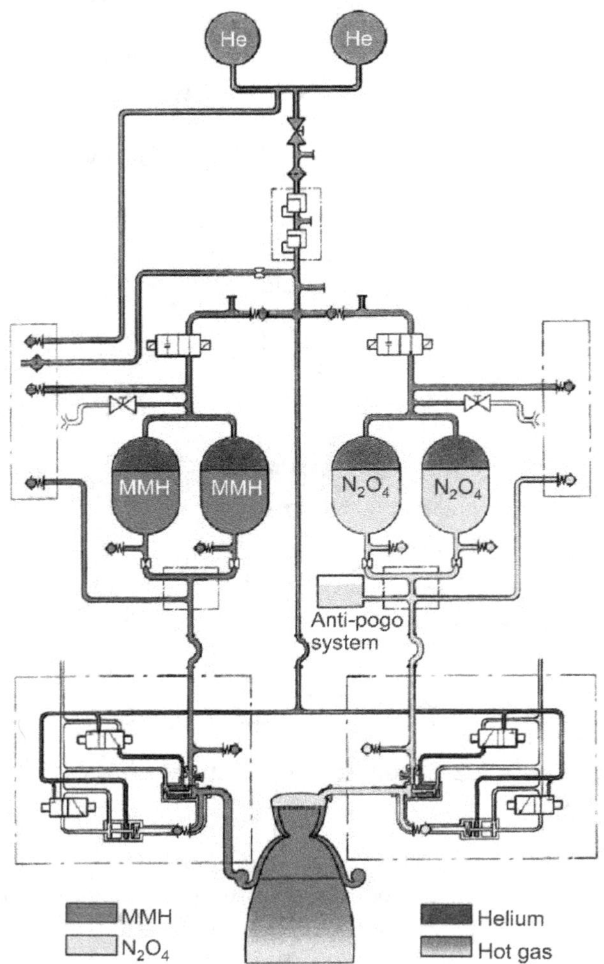



**Figure 7.3** Ariane 5 Aestus engine layout (Source: Turner, M.J.L.).

## 7.4　TURBO PUMP-FED SYSTEMS

The term turbopump is used to describe a pump and a turbine that are integrated into a single unit. Turbo pump-fed liquid rocket engines are useful to meet the high thrust and long duration operation requirements of launch vehicles. These systems lower the overall weight and increase the performance as compared to pressure-fed systems. It is an essential configuration of liquid rocket engine to **overall trend towards higher chamber pressure for better performance of the engines.** Figure 7.4 shows a schematic layout of a turbo pump fed liquid rocket engine. It requires relatively low inlet pressure at the pump inlet and thus requires low propellant tank pressure. **This saves considerable tank weight, particularly in large vehicles.** The pumps raise the required pressure of propellants received from propellant tanks and deliver them to the thrust chamber.

The feed line ducts and valves control the required propellant flow rates and pressure commensurate with rated engine operation. The main components of turbo pump feed system are:

Propellant pumps

Turbine(s) to drive the pump

Gas source for the turbine(s)

Speed-reduction gear train, if required

Bearing, seals and lubrication system

Pump inlet and outlet ducts

Turbo pump mounts

A Schematic layout of a turbopump-fed liquid rocket engine is shown in Figure 7.4. The various components of the pump-fed system are illustrated in the Figure 7.5.

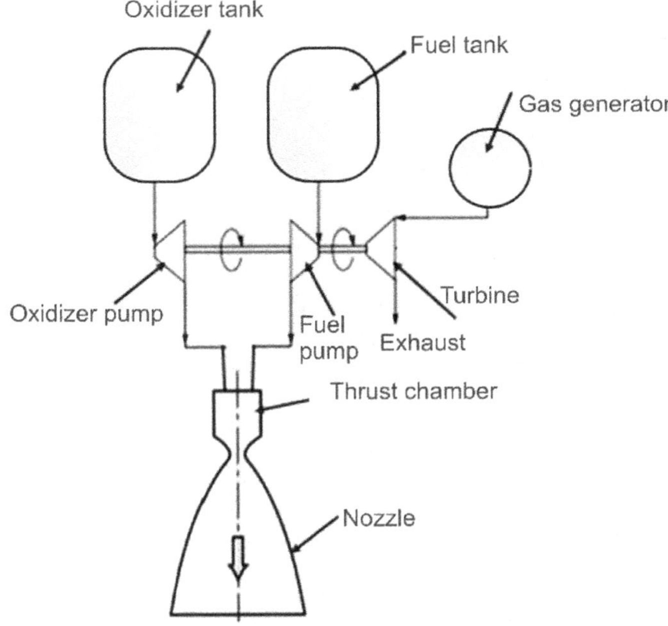


**Figure 7.4** Schematic of Turbo pump-fed.

## 7.4.1    Turbo Pump

Turbo pump assembly is the most complex and key unit for the performance of LRE. Figure 7.5 and 7.6 shows the cross section and main components of a typical LOX/Kerosene turbo pump. The rotating assembly of pump impellers and turbine are housed in a stationary casing structure fixed to the engine. The main design requirements are pump discharge pressure and flow rate, pump suction pressure, turbine drive cycle and efficiency, fluid properties, throttling range, life, reliability, weight and size. The design of turbo pump requires the interaction of many disciplines, such as hydrodynamics, aerodynamics, mechanics, materials and processes, structural mechanics and analysis, rotor dynamics and thermodynamics. In the entire field of rotating machinery, the rocket turbo pump system has the highest power-to-weight ratio. This is possible by multifold increasing of the rotational speed of the unit. The turbo pumps operate at very high speed of range 20,000 to 60,000 RPM and transmit power as high as 750 MW.

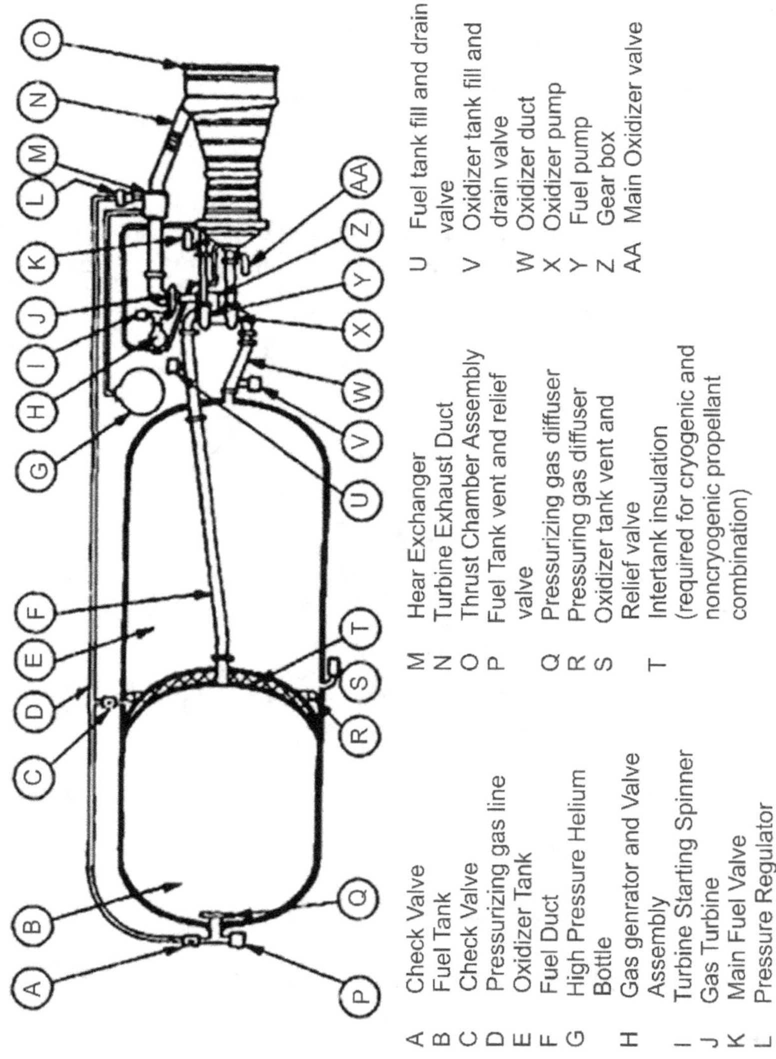


**Figure 7.5** Typical turbo pump-feed Liquid Rocket Engine (Source: Dieter K. Huzel and David H. Huang).

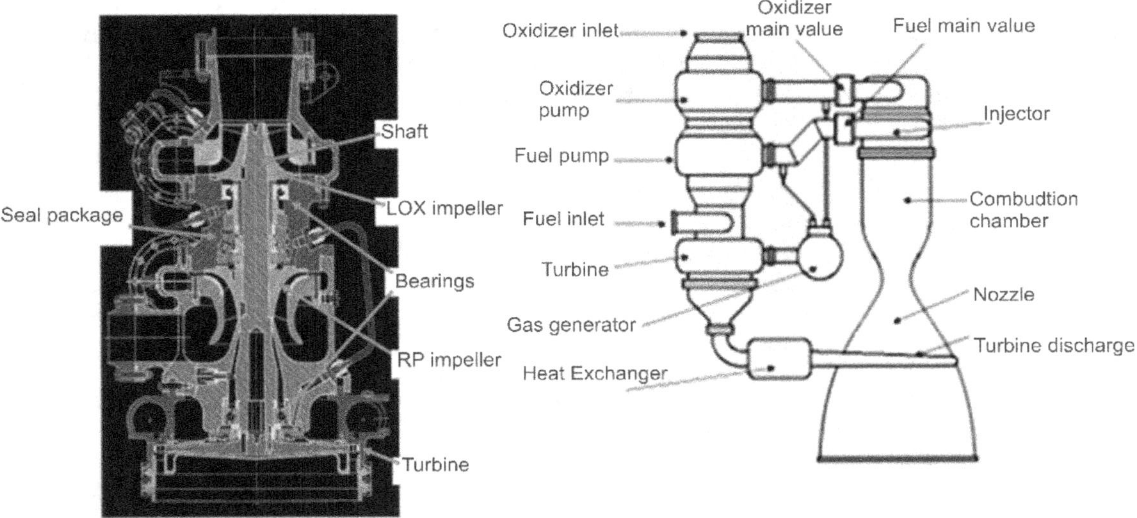


**Figure 7.6** Components of a typical LOX/RP Turbo pump (Source: Ballard Richard).

The power required to run the propellant feed pumps is enormous. An order of magnitude of the power can be estimated from the following example. Consider a liquid propellant rocket engine (Vulcain 2 of ESA), having a chamber pressure of 11.6 MPa, generating a thrust of 940 kN with a specific impulse of 3140 Ns/kg. Assuming ideal condition and neglecting all losses of flow to the thrust chamber, the combined flow of propellant can be determined as

$$Propellant\ flow\ rate = \frac{Thrust\ N}{Specific\ impulse\ \frac{Ns}{kg}} = \frac{940 \times 10^3}{3140} = 299\ \frac{kg}{s}$$

Assume the average density of propellants is 985 kg/m³.

$$The\ volume\ flow\ rate\ of\ propellant = \frac{299}{985} = 0.304\ \frac{m^3}{s}$$

$$The\ power\ required\ to\ run\ the\ pumps$$
$$= volume\ flow\ rate\ of\ propellants\ \times\ pressure\ rise$$

$$= 0.304 \times 11.6 \times 10^6 = 3.526\ \times 10^6\ \frac{J}{s}\ (W) = 3.526\ MW$$

This is a huge amount of power and the figures are comparable with a small electrical power plant. Figure 7.7 shows a typical static pressure profile at various points of the engine components. The propellant tank pressure is maintained at a low level to achieve vehicle weight reduction due to thinner tank walls. Low pressure at pump inlet and high speed of pump can create low pressure on moving blades lower than the liquid vapor pressure resulting in cavitation. This results in reduction in pump performance. This problem is more predominant in cryogenic propellants. An inducer, basically a small axial pump, is provided upstream of a main centrifugal impeller to avoid unacceptable cavitation, improve the suction performance, and reduce the propellant tank pressure and weight. The design of the inducer is focused on obtaining sufficient cavitation margin rather than high efficiency and the head rise is just sufficient to suppress the cavitation. Hence, many inducers will operate under slightly cavitation conditions.

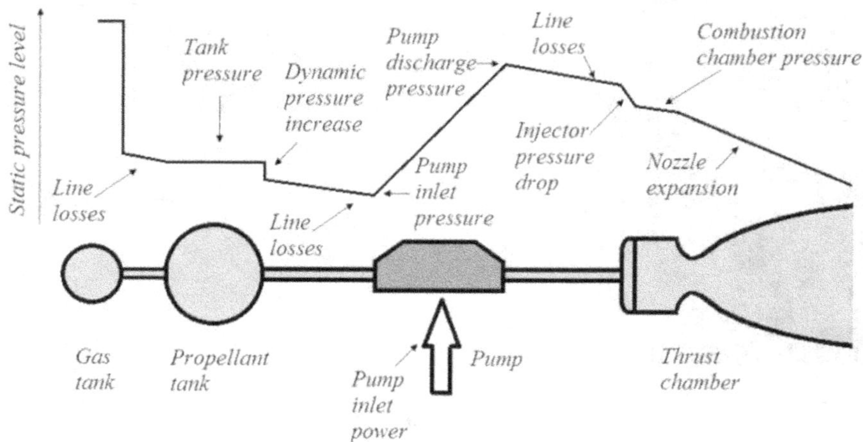


**Figure 7.7** Static pressure profile of typical arrangement of LRE.

A gas turbine provides shaft power to run propellant pumps. The energy to run the turbine is derived from high pressure and temperature gases produced by gas generators or pre-burner or heat exchangers heated by the main thrust-chamber combustion products, or gases tapped directly from the main combustion chamber. There are two types of turbines, impulse and reaction turbines. Impulse turbines have been used in rocketry as they are suitable for high pressure and low gas flow rates. The turbine includes one or more stages to develop required power. The turbine stage begins with a stationary row of blades called stators, or more commonly turbine nozzles followed by a moving blades row known as rotor. All the enthalpy change, i.e., pressure change occurs in the stator and the pressure is constant in the rotor and only the direction of flow changes. Hence, they are referred to as constant pressure turbines.

The type of coupling between turbine and pumps depends on the propellants being pumped. Also, it depends on overall engine design and configuration. Figure 7.8 shows schematically various turbo pump drive arrangements, direct drive, geared drive and dual shafts. Rocket engines using oxidizer and fuel having similar density, such as RP-1/LOX, will have similar discharge pressure requirements and run at the same speed. This permits the fuel and oxidizer pumps to be placed on a common shaft and driven by a common turbine. The turbine can be located either on the shaft end or between pumps. Liquid engines Merlin 1D, RS-27 or the RD-180 are examples for single shaft turbo pump arrangement. Placing the pumps back-to-back helps to cancel some of the axial thrust loads and packaging issues may favor placing the turbine on the end of the stack or in the middle between the two pumps. Figure 7.9 shows the SpaceX Merlin engine turbo pump components and assembly. Cryogenic propellant combinations like LOX/LH$_2$, due to the large density difference and the pressure drop required for regenerative cooling usually have two separate pumps since the hydrogen pump must operate at much higher speed to provide the necessary high exit pressures. Figure 7.10 shows the geared turbo pump used in the Saturn 1 engine. Geared turbo pump has several shortcomings since the gearbox assembly requires a separate lubrication system, creates some frictional losses and adds weight to the engine. However, the pumps can be designed to operate near optimal efficiency and thus reduce size and weight of the components. The modern rocket engine designers are preferring direct drive configuration. Dual-shaft turbo pump arrangements with pump and turbine for each propellant on separate shafts include two gas turbines in series, with the discharge gas from the first turbine driving the second turbine, and two gas turbines in parallel, both receiving gasses directly from the gas power source. Figure 7.11 shows the SSME (Space Shuttle Main Engine) turbo pumps, one for fuel and another for oxidizer.

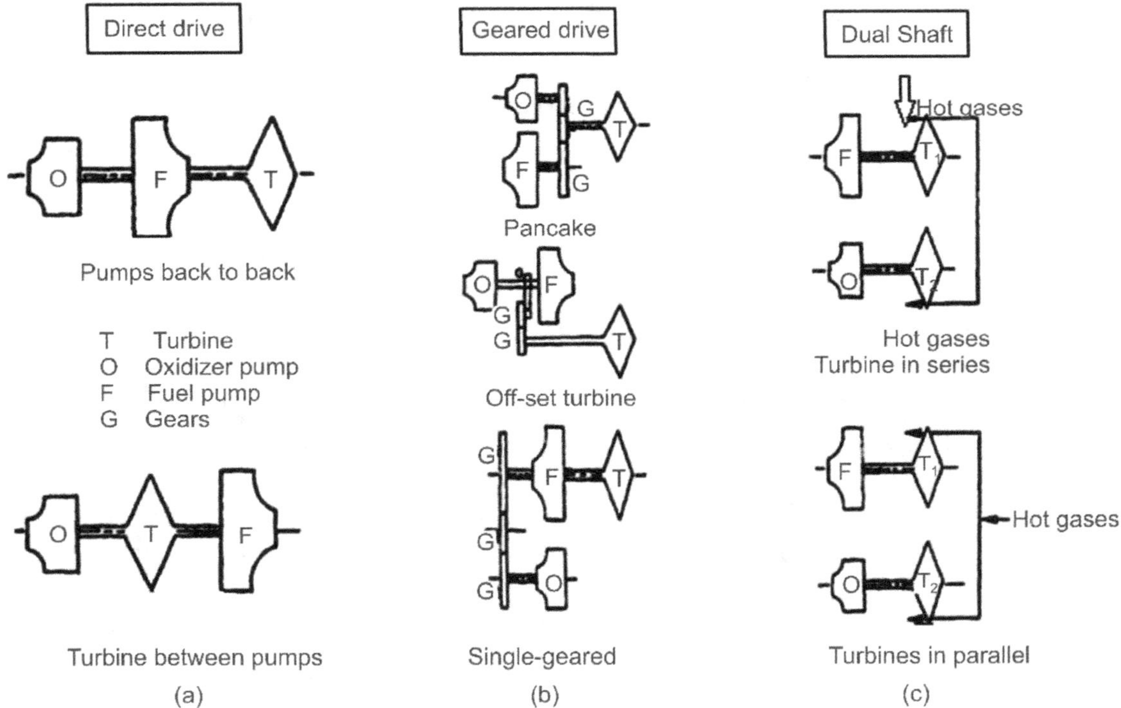


**Figure 7.8** Schematic Representation of Turbo pump Configuration
(a) Direct Drive (b) Geared (c) Dual Shaft.

**Figure 7.9** SpaceX Merlin LOX/RP-1 Turbo pump.
(Source: https://barber-nichols.com)

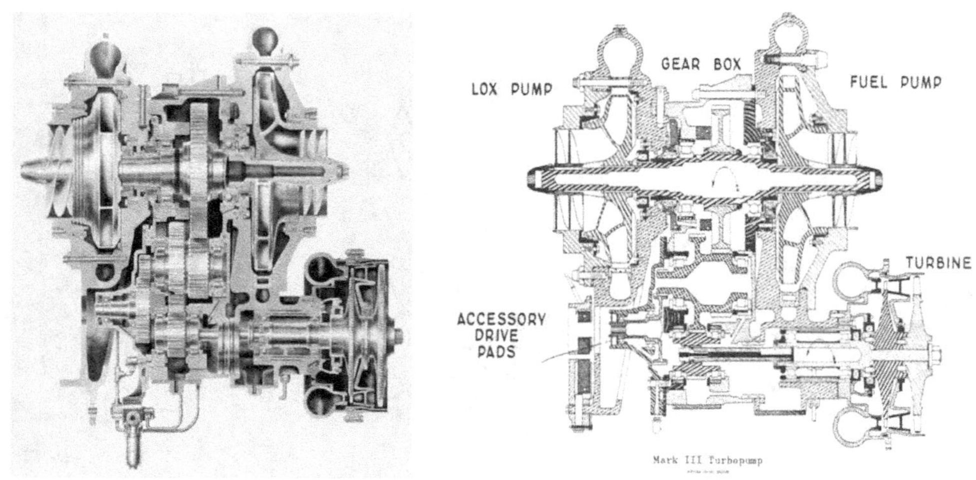


**Figure 7.10** H 1 Engine Mark III Turbo pump assembly (Source: NASA SP-8100).

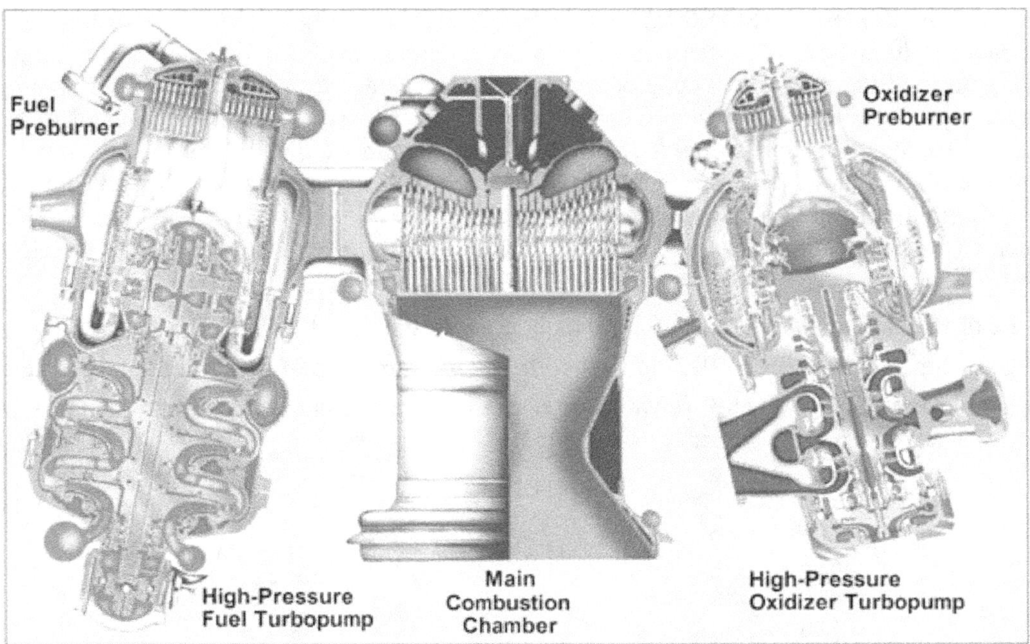


**Figure 7.11** SSME Powerhead Component Arrangement (Source: Cannon, J.L.).

## 7.5    ENGINE CYCLES

Engine cycle is the power cycle that the rocket engine uses to run the turbo pumps to deliver high pressure propellants to the thrust chamber. They are classified depending upon the turbine fluid source and discharge location after expansion in the turbine. For large thrust requirements, bipropellant engines are used with turbo pump fed systems. In such systems the walls of the thrust chamber and nozzle are regeneratively cooled. The fuel is circulated in specially designed flow channels around the thrust chamber and nozzle and acts as a coolant. It is an efficient way to recover the energy lost through the walls and preheats the fuel. Oxidizer is not generally used to avoid the potential ignition hazards. The most common cycles used to run the turbo pump fed systems are gas generator, expander and staged cycles.

### 7.5.1  Bipropellant Gas Generator (BGG) Cycle

A gas generator is a device to generate hot gas to run the turbine(s) either from the chemical combustion of the engine main propellants or through a separate solid or liquid source. Figure 7.12 shows a schematic representation of a bipropellant gas generator (BGG) with single shaft and dual shaft with parallel or series turbo pump arrangements. The fuel flows through a regenerative cooling jacket prior to entering the injector. The gas generator is operated to a fuel rich mixture ratio to limit the temperature at the turbine inlet to avoid the damage of the turbine blades. The fuel rich mixture runs with fuel like RP1 (kerosene), unburnt fuel form soot. This soot can block the injector holes. This cycle works well in dual shaft arrangement for propellants with different fluid properties (LOX/LH) that require different pump speeds. The turbine flow is subsequently dumped in a separate nozzle (known as open cycle) or introduced to some low-pressure point of the nozzle exit cone (known as closed cycle). Closed cycle engines have higher efficiency over the open cycle engines in that turbine exhaust gases are expanded to higher velocities. Hence, the specific impulse of a closed cycle engine will be higher than the open cycle alternative if all other design parameters are kept fixed. In some applications the turbine exhaust exits through a moveable nozzle that can provide vehicle roll control. The gas generator engine is much easier to develop and operate. It does have a performance penalty, however, due to three facts:

1. the turbine-drive gases are not fully expanded as in the thrust chamber,

2. the turbine gases are not fully combusted to the mixture ratio of the thrust chamber, and

3. there is a mixture-ratio shift (away from optimum) created in the main thrust chamber when flow is taken from the system to supply the gas generator.

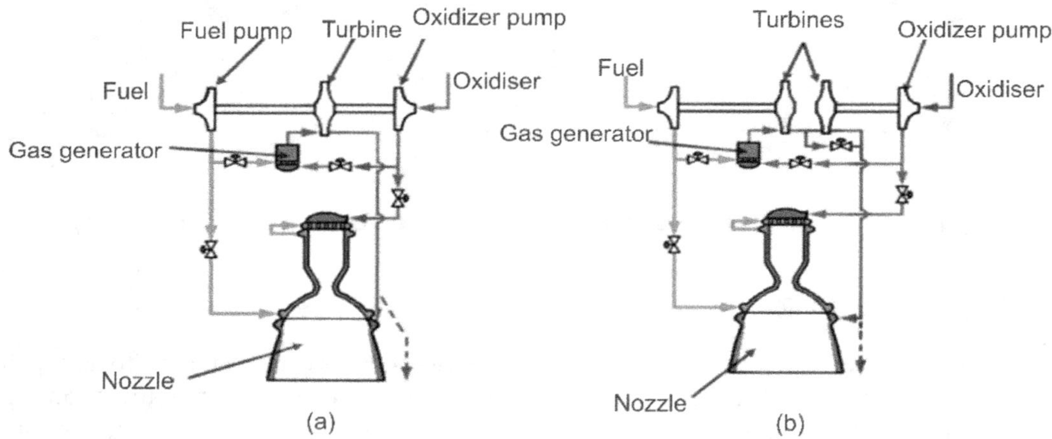


**Figure 7.12** Bipropellant gas generator cycle (BGG) (a) with direct drive turbo pump arrangement (b) with dual shaft turbo pump arrangement.

The inherent simplicity of this type of engine makes it a prime candidate for both booster and space engines. The J-2 engine used in the upper stage of the Saturn launch vehicle with dual turbine in series and the Vulcan engine used in the Ariane launch vehicle with turbine in parallel are examples of BGG cycle. The Merlin engine of the SpaceX vehicle and RS-68 engine used on the Delta IV (first stage) rocket vehicle are also BGG cycle engines. Figure 7.13 gives the layout of the RS-68 engine. The thrust chamber/nozzle assembly consists of a combustion chamber and ablative nozzle. It uses separate turbo pumps for oxidizer (LOX) and fuel (LH). Turbo pumps are single-shaft with direct drive turbines. It also supplies pressurization gasses to vehicle fuel and oxidizer propellant tanks through a heat exchanger located at the exhaust gases of the oxidizer turbine line. It has the fuel turbine exhaust roll control nozzle.

The initial design of the engine cycle requires the power balance i.e., the power developed by the turbine equals to the pumps power demand. Figure 7.14 shows the flow process in a BGG. The turbine exit gases are expanded in a separate nozzle. The subscripts *"ox," "f"* and *"gg"* represent oxidizer, fuel and gas generator, respectively. The temperature in GG is $T_{gg}$, is equal to the inlet turbine temperature $T_i$. The outlet turbine temperature after expansion is $T_o$. The flow in turbines and pumps are assumed to be isentropic.

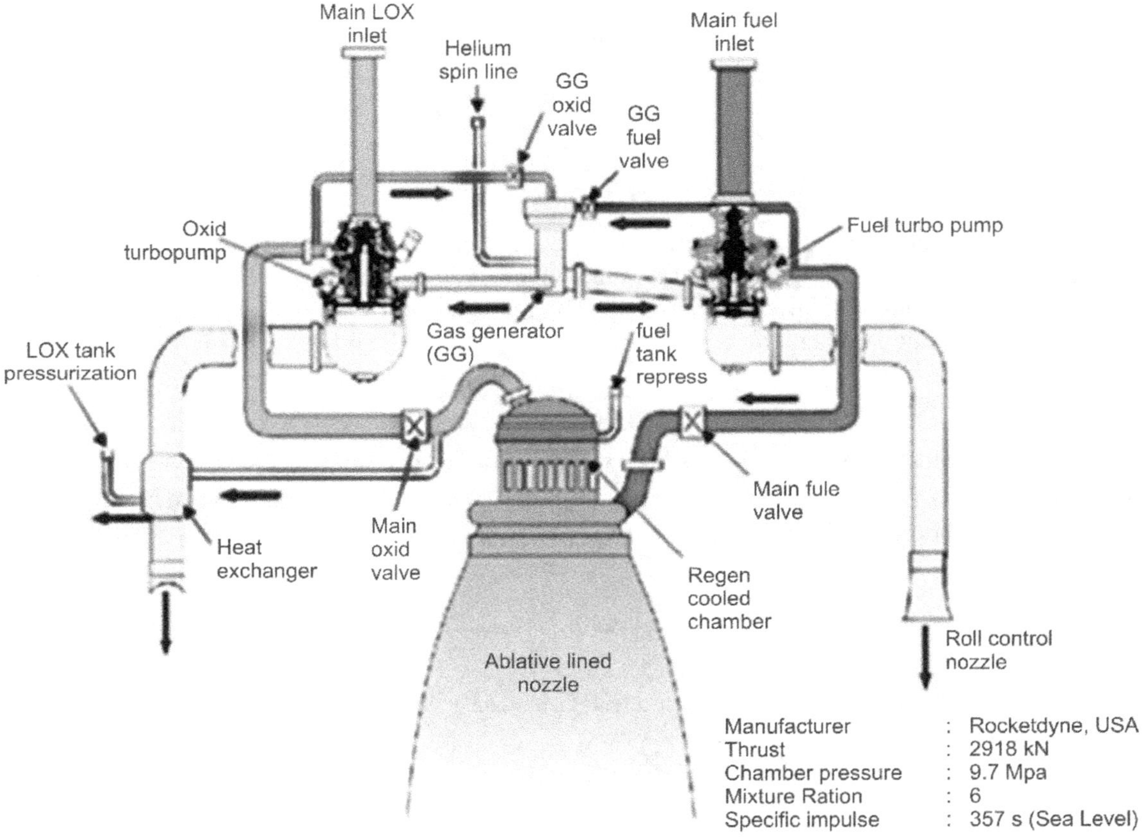

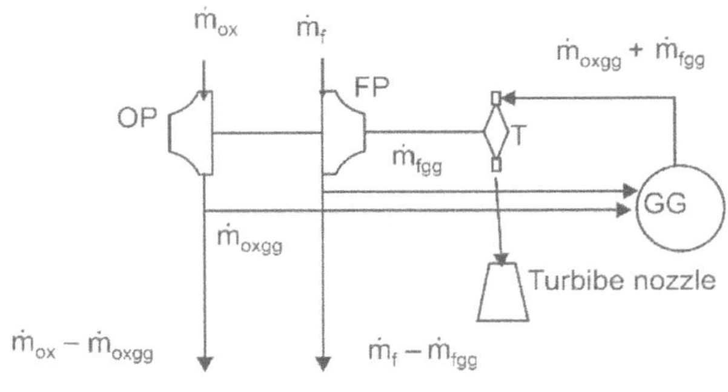


**Figure 7.13** RS-68 Schematic layout (Source: Byron K. Wood).



**Figure 7.14** Flow process in BGG.

*The mass flow rate into GG = mass flow rate of oxidiser + mass flow rate of fuel*

$$\dot{m}_{gg} = \dot{m}_{oxgg} + \dot{m}_{fgg}$$

The power developed by the turbine, $P_t$ is

$$P_t = \dot{m}_{gg}c_p(T_i - T_o)\eta_t = \dot{m}_{gg}c_pT_i\left(1 - \frac{T_o}{T_i}\right)\eta_t$$

Using isentropic relation, the above equation can be expressed in terms of turbine pressure ratio.

$$P_t = \dot{m}_{gg}c_p\eta_tT_i\left[1 - \left(\frac{p_o}{p_{gg}}\right)^{\frac{\gamma-1}{\gamma}}\right] \qquad .....(7.4)$$

$\eta_t$ is the turbine efficiency, $c_p$ is the average specific heat of the gases, $p_o$ is the turbine outlet pressure and $p_{gg}$ is the gas generator pressure which is equal to inlet pressure of the turbine neglecting power losses. The above power is supplied to the liquid propellant pumps, where the pressure of the propellant is increased. The power required by the pumps is

$$P_p = \frac{\dot{Q}}{\eta_p}\Delta p = \frac{\dot{m}_{ox}}{\eta_{oxp}\rho_{ox}}\Delta p_{ox} + \frac{\dot{m}_f}{\eta_{fp}\rho_f}\Delta p_f \qquad ......(7.5)$$

$\dot{Q}$ is the volumetric flow rate, $\eta_{oxp}$ is the oxidizer pump efficiency, $\rho_{ox}$ is the oxidizer density, $\eta_{fp}$ is fuel pump efficiency, $\rho_f$ is the fuel density, $\Delta p_{ox}$ is the oxidizer pump pressure rise, $\Delta p_f$ is the fuel pump pressure rise. Equating power developed with power required, for the power balance

$$\dot{m}_{gg}c_p\eta_tT_i\left[1 - \left(\frac{p_o}{p_{gg}}\right)^{\frac{\gamma-1}{\gamma}}\right] = \frac{\dot{m}_{ox}}{\eta_{oxp}\rho_{ox}}\Delta p_o + \frac{\dot{m}_f}{\eta_{fp}\rho_f} \qquad .....(7.6)$$

The turbine exit temperature can be related to GG pressure ratio using turbine efficiency.

$$\eta_t = \frac{\dfrac{(T_i - T_o)}{T_i}}{1 - \left(\dfrac{p_o}{p_{gg}}\right)^{\frac{\gamma-1}{\gamma}}}$$

The above equation can be rearranged as

$$\frac{T_o}{T_i} = 1 - \eta_t\left[1 - \left(\frac{p_o}{p_{gg}}\right)^{\frac{\gamma-1}{\gamma}}\right] \qquad .....(7.7)$$

To select the operating GG pressure, $p_{gg}$, the minimum of the two exit pump pressures is to be considered accounting the losses due to valves, feed lines and loss across the gas generator injector, i.e.,

$$p_{gg} = min(p_{ox}, p_f) - \Delta p_{val} - \Delta p_{line} - \Delta p_{injgg}$$

The gas generator mixture ratio ($\dot{m}_{oxgg}/\dot{m}_{fgg}$) is kept significantly fuel-rich such that the turbine inlet temperature ($T_i$) is less than about 900 K. For LOX/LH2, the GG mixture ratio is about 0.6 and for LOX/RP1 it is about 0.4. The pressure ratio across the turbine is to be selected so that the exit pressure of the turbine can create a choked flow in the turbine nozzle. The resulting

$I_{sp}$ from the turbine nozzle can be calculated from the theoretical exit velocity and thrust as explained in chapter 4. With the given $p_{gg}$ and GG mixture ratio using Eqs. (7.6) & (7.8) the GG flow rate and turbine exit temperature can be calculated. An engine level performance can be estimated based on the total thrust and total flow as follows. The fraction of propellant flow rate through the gas generator, f is

$$f = \frac{\dot{m}_{oxgg} + \dot{m}_{fgg}}{\dot{m}_{ox} + \dot{m}_f}$$

The engine specific impulse is combination of specific impulse due to main combustion chamber, $I_{spcc}$ and specific impulse due to turbine nozzle $I_{spt}$.

$$I_{speng} = (1 - f)I_{spcc} + f I_{spt}$$

## 7.5.2 Tap-off Cycle

In this cycle, the hot gasses in the thrust chamber are used as the energy source to run turbo pumps as shown in Figure 7.15. The main advantages of this approach include the elimination of the gas generator and the wide range of throttling capability afforded by the tap-off features. The hot gases at temperature suitable to the turbine (around 900 deg C) will be off directly from the main combustion chamber to drive the turbine. Sometimes, to reduce the temperature of the gases fuel is mixed. The J-2S was developed using a tap-off cycle as a part of the J-2 series which was designed as a replacement for the upper stages of the Saturn vehicleJ-2S is a technology demonstrator and has never been flown before but has been tested extensively. The BE-3 engine developed by Blue Origin with LOX/LH2 is the first flight tested engine using the tap-off cycle. The engine is being used on the New Shepard suborbital rocket, for which test flights began in 2015 and a first crewed flight occurred in 2021. It provides 490 kN of thrust and can throttle down to 89 kN which allows it to make vertical landings and be reused multiple times suitable to take space tourists up to 100 km altitude.

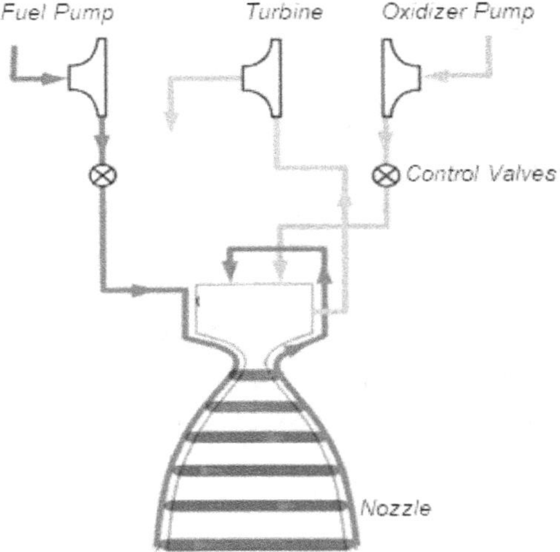


**Figure 7.15** Tap-off Cycle.

### 7.5.3 Expander Cycle

In the expander cycle the turbine is placed in series with the thrust chamber and exhausts the expanded gases in the turbine into the chamber as shown in Figure 7.16. Thus, it avoids combustion upstream of the turbine. Cryogenic fuels like liquid hydrogen are suitable in these cycles. The fuel after it has been heated in the thrust chamber cooling jacket expands in the turbine. This limits the turbine inlet temperature and the power developed to run the pumps. Hence, the chamber pressure in these cycles is limited and is less than 6 MPa. The expander engine is highly suitable as an upper stage engine where it exhausts in vacuum, consequently, have a very high nozzle area ratio despite its lower chamber pressure.

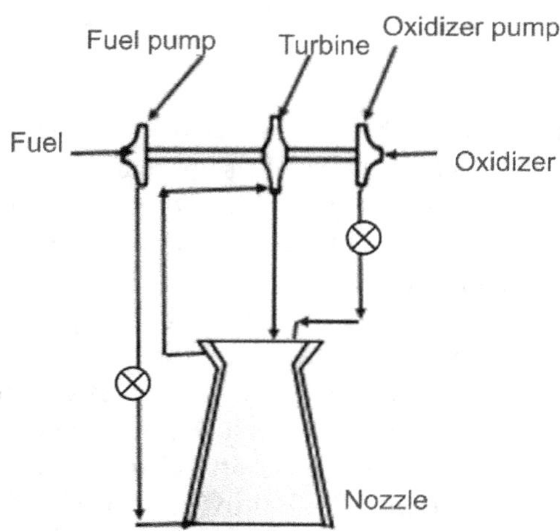


**Figure 7.16** Expander Cycle.

The RL-10 versions, developed by Aerojet Rocketdyne, are expander cycle engines, working with LOX/LH2, as upper stages engines of launch vehicles Atlas, Titan IV, Delta and future space vehicles. Modern versions produce up to 110 kN of **thrust** per engine in vacuum. Figure 7.17 gives the schematic of RL10A-3-3A version. It is based on an expander cycle, in which the fuel is used to cool the main combustion chamber and the thermal energy added to the fuel drives the turbo pumps. The fuel turbine drives the fuel pump (31571 RPM) and oxidizer pump (12675 RPM) through a gear train. The system normally operates at a chamber pressure of 3.28 MPa, a mixture ratio (O/F) of 5.0, and a thrust of 73.4 kN.

The VINCI engine is a 180 kN restartable upper stage expander cycle cryogenic engine of ESA used in ARIANE 5+ programs. Its specific impulse is 457 s. It has two turbines in series, one for fuel LH2 and other for oxidizer LOX. Figure 7.18 shows the schematic layout of the propellant flow in the engine.

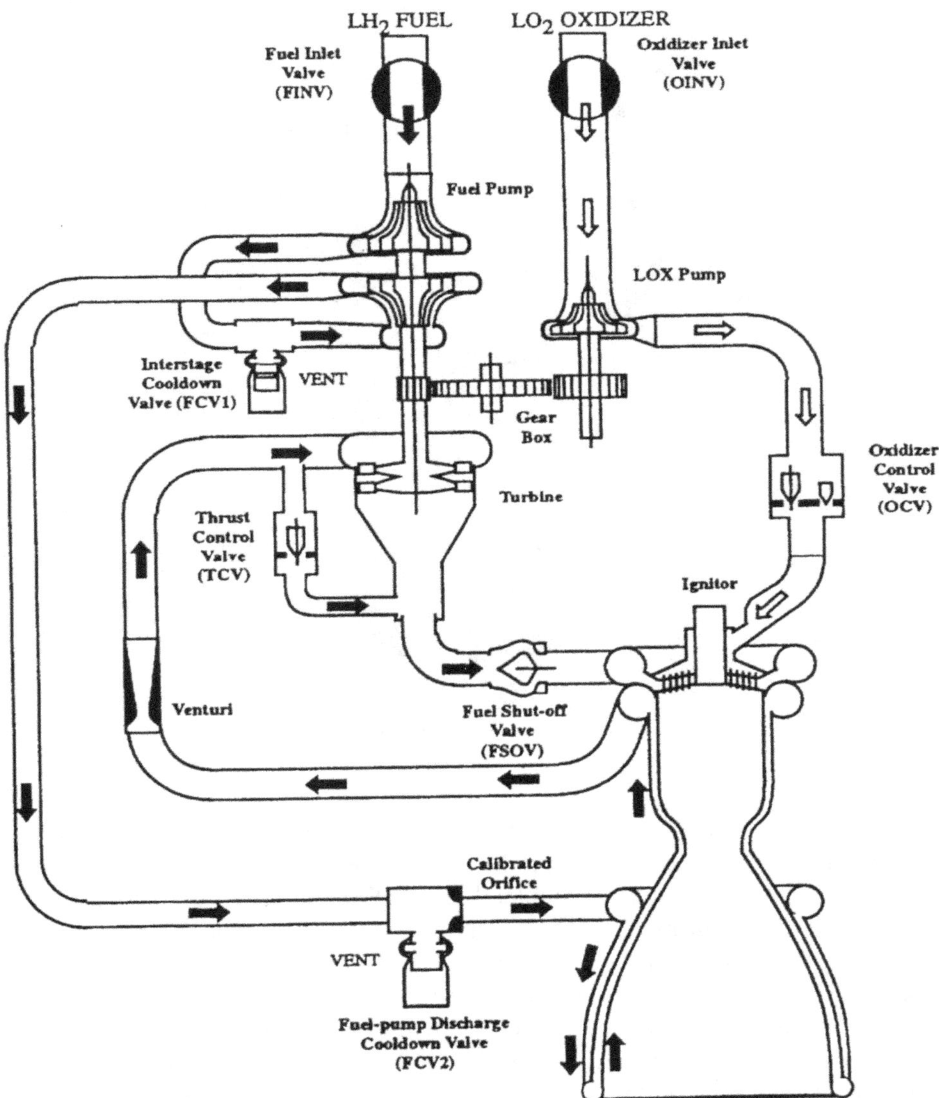


**Figure 7.17** RL10A-3-3A Engine System Schematic ( Source: Michael Binder et al).

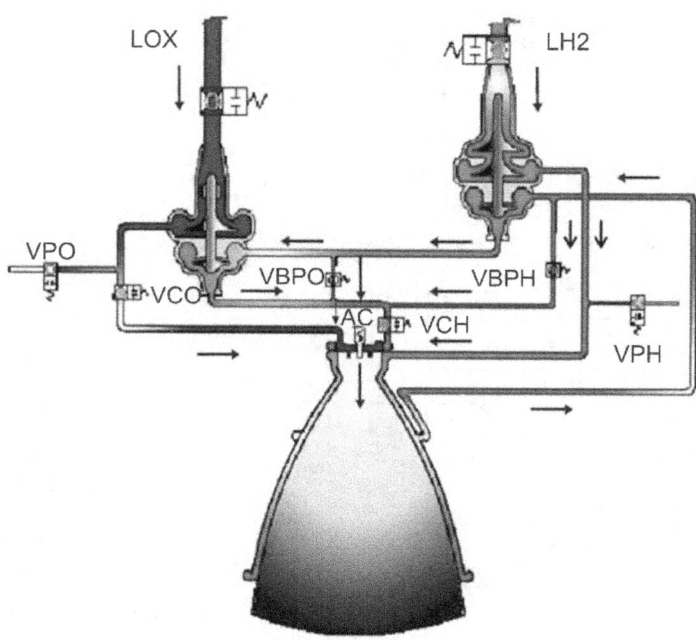


**Figure 7.18** VINCI flow schematic (Source: Alliot P et al).

### 7.5.4   Staged Cycle

The staged combustion cycle shown in Figure 7.19 is the most advanced of the closed cycle. It has combustion upstream of the turbine and the exhaust gases from the turbine directly enter the thrust chamber. One of the propellants completely flows through the preburner and combusts with a small portion of the other propellant. In the Figure 7.19 all the fuel (LH2) enters the preburner after passing through the regenerative cooling system and combusts with a portion of oxidizer (LOX). The fuel rich combustion gases give the energy to the turbine and the exhaust from the turbine is fed to the thrust chamber. As the preburner flow is much higher than from a typical gas generator and the inlet turbine temperature is also high, the turbine can generate a huge amount of power. This results in high pump pressure rises and high chamber pressures and consequently, allows larger nozzle area ratio. For this reason, the staged combustion alternative provides the highest performance for booster applications. Like in fuel rich staged combustion (FRSC), in oxidizer rich staged combustion (ORSC) all the oxidizer is combusted in the preburner with a small amount of fuel. The cycle is not suitable for using kerosene fuels due to the potential for coking within the heavily fuel-rich preburner exhaust stream. Though relative to other cycles, the staged combustion cycle has the highest performance, it is more complex and introduces engineering challenges and costs associated with containing the high pressures in difficult oxidizer- or fuel-rich combustion gas environments.

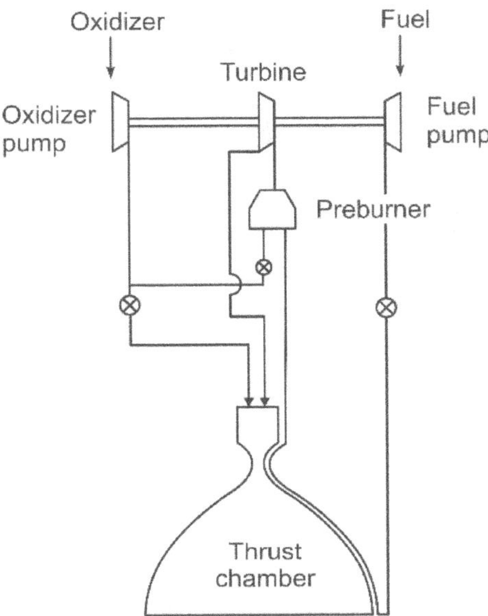


**Figure 7.19** Staged combustion cycle.

In bipropellant engines the oxidizer amount is much higher as the mixture ratio (O/F) is greater than 2. The ORSC cycle develops substantially more turbine power and hence chamber pressure compared to the FRSC approach. However, the major challenge lies in working with hot oxidizer gases produced by the preburner since they are highly reactive and can flow path walls as "fuel." This requires development of oxidizer resistant metal alloys or coatings that inhibit flammability at the aggressive conditions. In 1960-70 the Soviet Union overcame this problem with the new alloys and developed engines like RD-170, RD-180 and RD-181. Currently, the Russian RD-180 is used as first-stage propulsion on the Atlas V launch vehicle. US firms like SpaceX, Blue Origin and Aerojet Rocketdyne are being active in the development of ORSC approach engine development.

Figure 7.20 shows the Full Flow Staged Combustion (FFSC) cycle having two separate preburners one for oxidizer and other for fuel. The fuel side preburner generates fuel rich gasses and oxidizer side preburner oxidizer generates rich gases to drive the respective fuel and oxidizer pumps. Also, the dual shaft arrangement eliminates the danger of mixing the propellants from the inter-propellant seal of the turbo pump. The FFSC cycle is supposed to be the highest efficient bipropellant engine cycle as the total flow of each of the propellants is producing turbine power and hence pressure rise in the respective pumps. Hence, the main combustor operates efficiently with gas/gas injection and is believed to be more stable than combinations involving a liquid propellant.

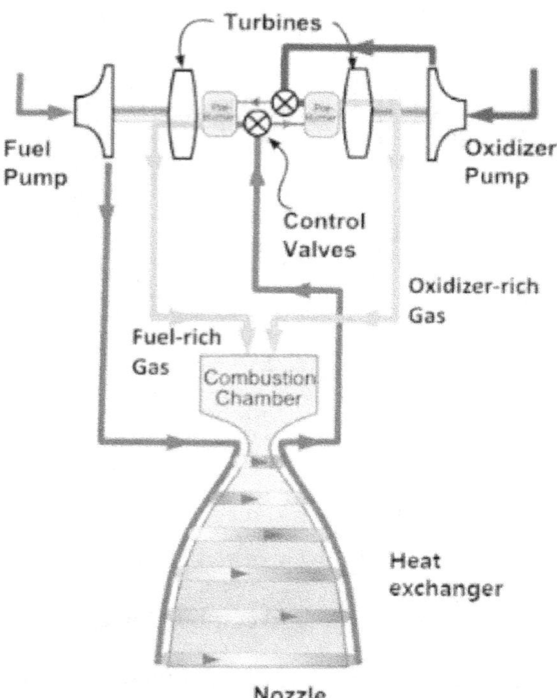


**Figure 7.20** Full Flow Staged Combustion (FFSC) cycle (Source: https://blogs.nasa.gov/J2X/tag/closed-cycle-engine/).

The Space Shuttle Main Engine (SSME), now known as RS-25, is a cryogenic (LOX/LH2) liquid rocket engine working with fuel rich staged combustion (FRSC) cycle. Figure 7.21 shows the functional diagram showing the flow of propellant of RS-25. It has separate preburners for oxidizer and fuel. The low-pressure axial flow oxidizer pump (5,150 rpm) is driven by a six-stage turbine powered by high-pressure liquid oxygen from the high-pressure oxidizer turbo pump to boost oxygen pressure from 0.7 to 2.9 MPa. The pressure boost helps to run the high-pressure oxidizer pump at high speed (28,190 rpm) without cavitation. In a similar configuration, the low-pressure axial flow fuel pump (16,185 rpm) driven by a two-stage turbine powered by gaseous hydrogen boosts the liquid hydrogen pressure from 0.2 to 1.9 MPA. With this inlet pressure, the three-stage centrifugal high pressure fuel pump runs at 35,360 rpm without cavitation to raise the pressure to 45 MPa. This engine is the most complicated, reusable and efficient engine developed as of now. NASA is continuing to use the RS-25 on the Space Shuttle's successor, the **Space Launch System** (SLS), a super heavy-lift launch vehicle that provides the foundation for human exploration beyond Earth's orbit. With its unprecedented power and capabilities, SLS is the only rocket that can send Orion spacecraft, astronauts, and cargo directly to the Moon on a single mission. The core stage of SLS is powered by 4 RS-25 engines.

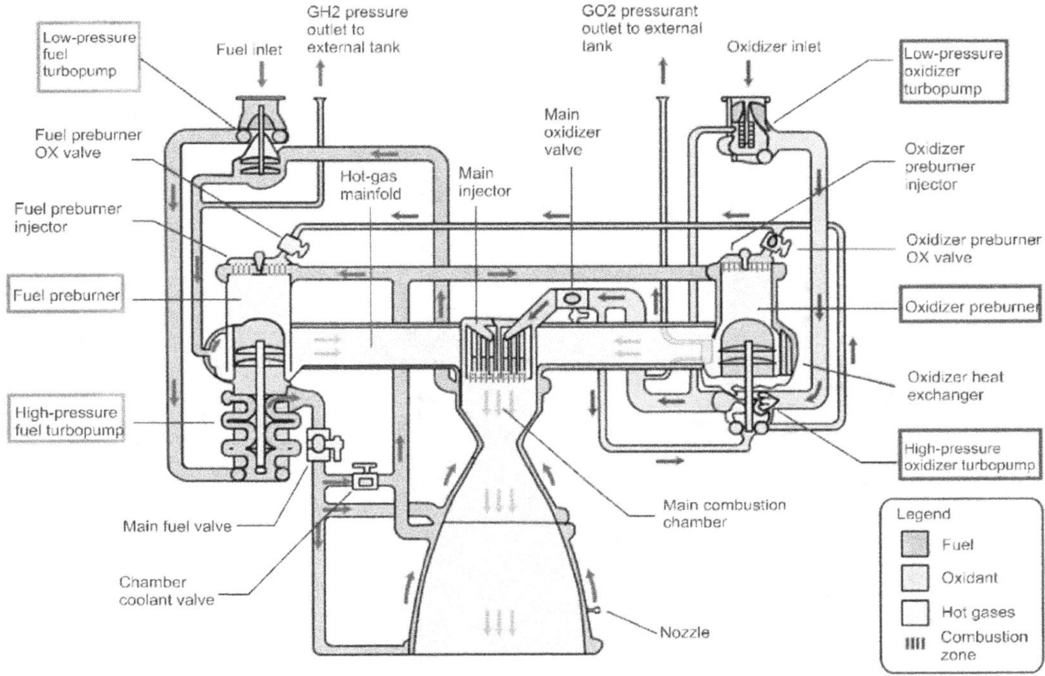


**Figure 7.21** Schematic of SSME (RS-25)
(Source: https://en.wikipedia.org › wiki › RS-25).

SpaceX Raptor is a highly reusable full flow staged combustion (FFSC) engine working with liquid methane and liquid oxygen (LOX/CH4). It is being developed to power the Interplanetary Transport System (ITS) to transport cargo (around 450 tons) and crew between Earth and Mars in future. It has a deep throttling capability down to 20% of the rated thrust to enable propulsive landing. Figure 7.22 shows the schematic layout of the Raptor engine. It employs boost pumps (not shown in figure) on both the fuel and oxidizer sides that operate at a lower speed than the main pumps and create an engine inlet pressure sufficient for the operation of the turbo pumps. The engine is regeneratively cooled using methane fuel. Both propellants reach the engine injector in gaseous phase. The engine is being developed in two versions – one for use on the first stage booster of the ITS launch vehicle and one optimized for operation in vacuum for operation outside Earth's atmosphere for the interplanetary insertion and in the ambient Martian atmosphere for retro-propulsion ahead of landing. Methane as fuel is selected in the Raptor engine as it prevents a build-up of deposits in the engine compared to other fuels like kerosene, a process known as coking, while its higher performance allows for lower costs. Also, the other major consideration is the possibility of generating methane with the resources present on Mars. The ITS booster stage employs a cluster of 42 engines (outering 21 engines, inner ring 14 engines and 7 center cluster engines for vehicle control) generating a liftoff thrust 127.8 MN. Blue Origin, headed by Amazon CEO Jeff Bezos, is also developing its own LNG/LOX engine called Blue Engine 4 (BE-4). Two

BE-4 engines will power the first stage of United Launch Alliance (ULA) new rocket, Vulcan Centaur. Table 7.4 summarizes the important specifications of present rocket engines.

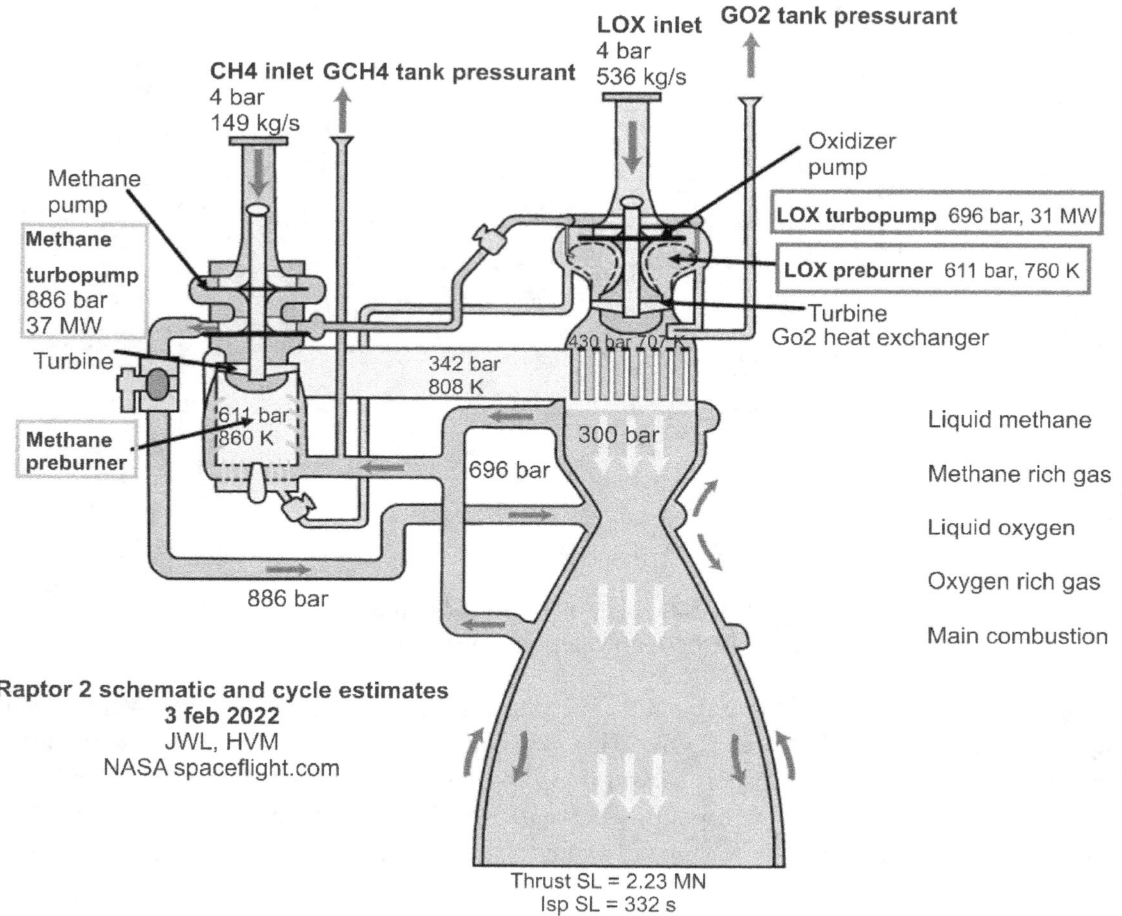


**Figure 7.22** SpaceX Raptor engine schematic flow diagram (Source: https://en.wikipedia.org/wiki/SpaceX_Raptor).

**Table 7.4** Summary of important rocket engine specifications.

| Parameter | Merlin 1D | RD-180 | RS-25 | Raptor (Booster Ver) | Raptor (Space Ver) | BE-4 |
|---|---|---|---|---|---|---|
| Company | SpaceX | USSR | Rocketdyne | SpaceX | SpaceX | Blue Origin |
| Propellant | LOX/RP1 | LOX/RP1 | LOX/LH2 | LOX/CH4 | LOX/CH4 | LOX/LNG |
| Cycle | GG | ORSC | FRSC | FFSC | FFSC | ORSC |
| Thrust (Sea Level) kN | 845 | 3830 | 1860 | 3050 | - | 2445 |
| Thrust (Vacuum) kN | 915 | 4150 | 2279 | 3297 | 3500 | - |
| Chamber Pressure (MPa) | 9.7 | 26.7 (Two chambers) | 20.64 | 30 | 30 | 13.44 |
| Sp. Impulse $I_{sp}$(SL) s | 282 | 311 | 366 | 334 | - | 311 |
| Sp. Impulse $I_{sp}$(Vac) s | 311 | 338 | 452 | 361 | 382 | 338 |
| Nozzle Ratio | 16 | 37 | 78 | 40 | 200 | - |
| TWR Thrust(SL)/Weight | 183 | 78 | 60 | - | - | - |
| Mixer Ratio | 2.34 | 2.72 | 6.03 | 3.8 | 3.8 | - |
| Mass flow kg/s | 236 | 1250 | 514.5 | 931 | 931 | - |
| Throttle range | 57-100% | 47-100% | 67-109% | 20-100% | 20-100% | Deep Throttle |

The schematics of the various engine cycle shown in this section are only a simple overview. Obviously, a large engineering effort required to develop the plumbing, valving, and mechanical systems required to make a functional device. While the performance of the staged combustion alternative is attractive, this performance gain must be weighed against the complexity. the additional development cost and time required as compared to simpler cycles. So, it is not a foregone conclusion that any one cycle is superior in any given instance.

**Table 7. 5** Gas Generator/Pre-burner Characteristic Data of Selected Engines.

| | Vulcain 2 | LE - 7 | SSME | RD-0120 | RD - 180 |
|---|---|---|---|---|---|
| Propellant | LOX/LH2 | LOX/LH2 | LOX/LH2 | LOX/LH2 | LOX/RP1 |
| Temperature (K) | 875 | 810 | 940/870 | 846 | 820 |
| Pressure (MPa) | 10.1 | 21.0 | 35/36 | 42.4 | 55.6 |
| Mixture Ratio | 0.9 | 0.55 | 0.89/0.8 | 0.81 | 54 |
| Mass flow rate (kg/s) | 9.7 | 53 | 80/30 | 78.6 | 887 |
| Power MW | 5/14 | 4.5/19 | 56/21 | 62 | 93.5 |

Table 7.5 gives the characteristic data of gas generator/preburner of cryogenic engines of Vulcain 2 (ESA), LE – 7 (Japan), SSME (USA), RD – 0120 (USSR) working in fuel rich mode and RD – 180 (USSR) operating in oxidizer rich mode. The mixture ratio of fuel rich operation mode is between 0.8 to 0.9. The oxidizer-rich operation of RD -180 allows for a sufficient reduction of the turbine entry temperature since the high mass flow rates are available. In oxidizer rich cycles, like RD -180, all the component materials coming in contact with oxidizer rich gases, (pre-burner, piping system, turbine and injector head) should be compactible for a high temperature (>800 K) working environment. It can be observed from the table that the pre-burner works at high pressure and high mixture ratio. A fuel-rich operation will have problems with soot formation in the gas generator and subsequently soot deposition along the flow lines. The only explanation might be that for the relatively short operation times of less than 200 s and low pressures in the gas generator soot deposition in the turbine might still be tolerable compared to the complexity and cost of development of oxidizer rich cycle engine.

## 7.6 THRUST CHAMBER ASSEMBLY

The function of the thrust chamber assembly is to generate thrust efficiently by the combustion of the propellants in the combustion chamber, followed by acceleration of the hot gases through a convergent divergent nozzle. The main elements of the thruster chamber assembly consist of injector head which holds the propellant distribution manifolds, ignition system (for nonhypergolic propellants), the combustion chamber liner including the throat section and the first part of the nozzle to which the inlet manifold for the coolant is welded and suitable surfaces and structures for component mounting and to carry the thrust forces to the vehicle. Figure 7.23 illustrates the Vulcain thrust chamber and its parts. For low thrust engines the combustion chamber and the nozzle may be one piece, but for high thrust large engines the nozzle is a separate component attached to the thrust chamber body at a small distance beyond the throat as shown in Figure 7.23b. The injector head distributes propellants to the injectors, holds the ignition system and the propellant injectors that decouples the combustion chamber from the fluid system. The liquid oxygen enters directly from the turbo pump while the hydrogen enters after passing through the regenerative cooling channels. All the components of the thrust chamber assembly are closely coupled and therefore must be designed in parallel.

### 7.6.1 Thrust Chamber

The propellants at proper mixture ratio are injected into the combustion chamber and atomized into droplets. These droplets get vaporized due to heat transfer from surrounding hot gases. Thus, the size and the velocity of the droplets change continuously, mix rapidly and promptly react, during their entrainment increasing the temperature and gas flow rate. This gas phase reaction is aided by high-speed diffusion of molecules and atoms. This process of combustion is essentially completed upstream of the chamber throat. The combustion chamber is basically a casing to retain the propellants for a sufficient period, called stay time, to ensure complete mixing and combustion. With the choked throat condition, the gases are expanded in the divergent nozzle to produce the thrust.

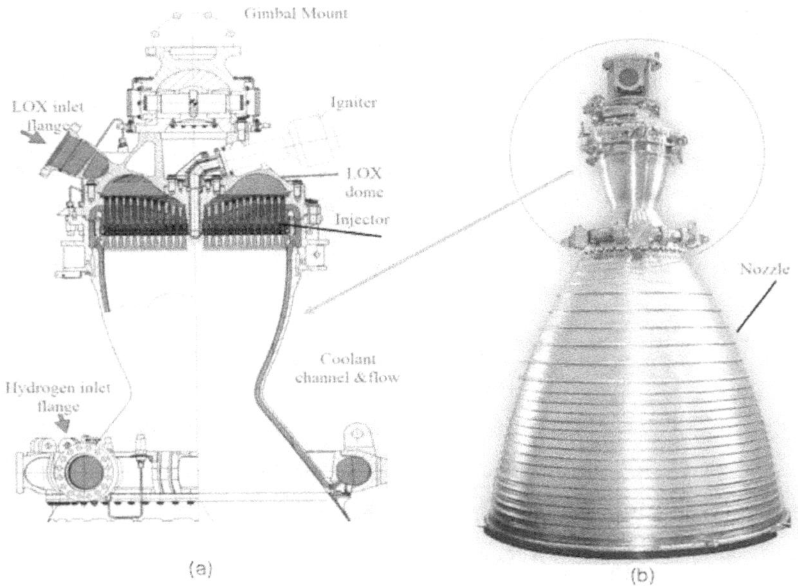

**Figure 7.23** Vulcain Thrust Chamber (a) without nozzle (b) with nozzle (Source: Haidn, O.J.).

The stay time is the function of many parameters like propellant combination, the injected condition of propellants, injector design, drop sizes, mixing time, kinetic rates of reaction, turbulence level in the chamber and speed (Mach number) of gases in the chamber. Thus, the combustor volume has a definite effect on combustion efficiency. A longer stay time is required for higher thermodynamic and combustion efficiencies. But it will increase the cooling requirements and chamber weight. A long chamber with a small cross-sectional area result in high pressure losses and space limitation to accommodate the required number of injection elements. A short chamber with a large cross-sectional area the propellant atomization and vaporization zone occupies a large portion of the chamber volume compared to mixing and combustion zone that affects the combustion efficiency. The exact shape is determined considering other factors like heat transfer, combustion stability, weight, ease of manufacturing and designer experience. Figure 7.24 gives frequently used shapes of the thrust chamber. The commonly used shape is cylindrical combined with a converging section to the throat.

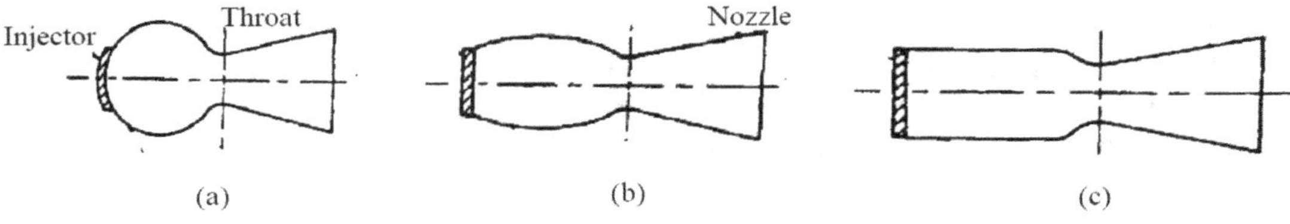

**Figure 7.24** Shapes of Thrust Chambers (a) Spherical (b) Near Spherical (c) Cylindrical.

The specific impulse, the overall quality of the thrust chamber, is the product of characteristic velocity $c^*$ and thrust coefficient $C_F$ as shown in Eq. (4.45). The $c^*$ represents the thrust chamber performance i.e., how efficiently it is producing the high-pressure gases. The $C_F$ represents the nozzle performance i.e., how efficiently it is converting high pressure gases into high velocity gases. The characteristic velocity $c^*$ is a function of specific heat ratio $\gamma$, gas constant $R$ and the temperature of combustion. For a selected combination of propellant and mixture ratio, the values of $\gamma$ and $R$ are known and thus, the $c^*$ entirely depends on the combustion temperature $T_c$ of gases and hence the combustion pressure $p_c$. The chamber length should be enough to complete combustion and attain the required temperature and pressure of gases. The stay time, $t_s$ can be expressed as

$$t_s = \frac{length\ of\ chamber}{velocity\ of\ gases} = \frac{l_c}{v_c} = \frac{\rho_c(A_c l_c)}{\dot{m}} \qquad \text{.....(7.8)}$$

where $\rho_c$ is the density of gases, $A_c$ is the cross-sectional area of the chamber and $\dot{m}$ mass flow rate of gases. From Eq. (4.36), $\dot{m} = \frac{p_c A_t}{c^*}$. Treating the combustion gases as ideal gas, $\rho_c = \frac{p_c}{RT_c}$. Substituting these values in Eq.(7.8) and rearrange the terms

$$t_s = \frac{c^*}{RT_c} \frac{A_c l_c}{A_t} \qquad \text{.....(7.9)}$$

The fraction $\frac{c^*}{RT_c}$ depends on the propellant combination and is constant for a given propellant. The term $\frac{A_c l_c}{A_t}$ is the ratio of chamber volume and throat area of the nozzle and is therefore a unit of length. So, the stay time now can be written as

$$t_s = constant \times L^* \qquad \text{.....(7.10)}$$

$L^*$ is called the characteristic length of the chamber. It is the length that a chamber of the same volume would have if it were a straight tube whose diameter is the nozzle throat diameter, that is, if it had no converging nozzle section. The ratio of chamber cross sectional area to nozzle throat area is called contraction ratio, $CR$. It is established with experiments that the value of $c^*$ will increase to an asymptotic maximum. Increasing $L^*$ beyond a certain value tends to decrease the overall engine performance due to increase in thrust chamber volume and weight, cooling load and frictional losses in the chamber. The selection of the value of $L^*$ is based on extensive experimentation with different chamber lengths $(L^*)$ for the chosen combination of propellant and on experience of the designer with similar propellants and engine size. Table 7.6 gives typical values of $L^*$ for common propellant combinations. It can be observed from Table 7.6 that the hydrogen fuel combinations have the lowest value of $L^*$ whereas kerosene fuel combinations have the highest value. This is due to the high evaporation rate and kinetic reaction. speed of combustion of cryogenic propellants and longer combust time of heavier hydrocarbon molecules contained in kerosene. With throat area and minimum $L^*$ value selected the chamber volume and the length can be estimated with the relation,

**Table 7.6** Characteristic chamber length (L*) values for selected propellants.

| Propellant Combination | Characteristic Length $L^*$ (m) |
|---|---|
| Nitrogen tetroxide/hydrazine base fuel | 0.76-0.89 |
| Liquid oxygen//liquid hydrogen (LH$_2$ injection) | 0.76-1.02 |
| Liquid oxygen//liquid hydrogen (GH$_2$ injection) | 0.56-0.71 |
| Liquid oxygen/RP-1 | 1.02-1.27 |
| Nitrogen peroxide/RP-1 (including catalyst bed) | 1.52-1.78 |

$$l_c = \frac{A_c L^*}{A_t} \qquad \qquad .....(7.11)$$

Sizing the thrust chamber for specified thrust and chamber pressure is still an arbitrary process depending on designer experience and data available for previous engines of the same class. The combustion process is not scalable. A good start for the design of the thrust chamber is to determine the throat diameter. The chamber contraction factor, *CR*, is another important parameter and is basically related to the mean Mach number of combustion gases. The value of *CR* ranges from 1.3-10. Generally, large engines use low *CR* and comparatively long length; and smaller engines use a large *CR* with shorter length. Selection of *CR* involves conflicting criteria. Increasing CR increases the chamber diameter and decreases the acoustic frequencies in the chamber, making them more susceptible to excitation leading to combustion instability. It also increases the chamber diameter and hence the chamber weight. Reducing *CR* increases the chamber Mach number, and Rayleigh losses become more pronounced. Figure 7.25 shows the historical relation of CR with throat diameter.

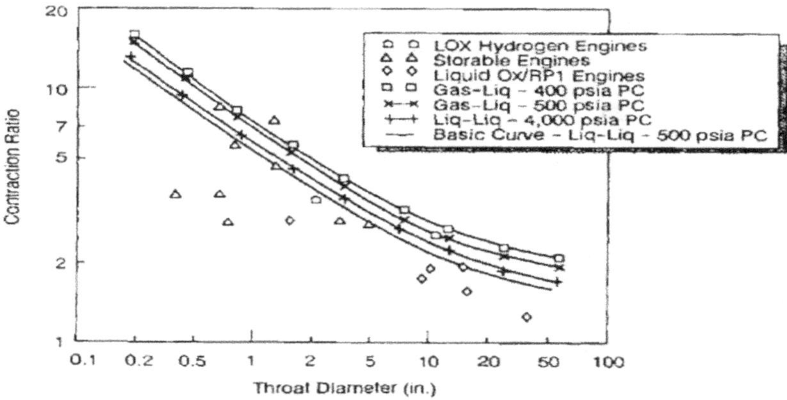


**Figure 7.25** Data of Contraction Ratio in relation with Throat Diameter of previous successful Engines (Source: Dieter K Huzel and David H Huang).

### Example 7.1

Calculate major dimension of thrust chamber along 80% bell nozzle to suit for the following specifications:

Propellant combination: LOX/RP-1

Thrust = 3300 kN

Chamber pressure = 6.89 MPa

Area ratio $\varepsilon$ =14

Mixture ratio 2.24

**Solution:**

The chamber temperature and other parameters can be taken from Table 5.7 or from Figure 5.5 or can calculated from relevant equations given chapter 4 for the propellant combination and mixture ratio.

Chamber temperature = 3571 K

$c^* = 1774$ m/s

Molecular mass $M = 21.9 \, kg/mole$

$\gamma = 1.24$

$I_{sp} = 285.4$ s

$$C_F = \frac{I_{sp}}{c^*} = 285.4 \times 9.81/1774 = 1.578 \text{ Eq (4.45)}$$

The throat area can be calculated from Eq. (4.42)

$$F = C_F A_t p_c$$

$$A_t = \frac{3.3 \times 10^6}{1.578 \times 6.89 \times 10^6} = 0.3035 \, m^2$$

Throat diameter is

$$D_t = \sqrt{\frac{4 \times 0.3035}{\pi}} = 0.622 \, m = 662 \, mm$$

Throat radius $R_t = 331 \, mm = 0.331 \, m$

Given area ratio $\epsilon = 14$

So, the exit diameter of the nozzle

$$D_e = \sqrt{14} \times 662 = 2477 \, mm \cong 2480 \, mm$$

Exit Radius $R_e = 1240 \, mm = 1.240 \, m$

Referring to Table 7.6, select a characteristic length of the thrust chamber $L^* = 1.150$ m =1150 mm

Chamber volume $V_c = A_t \times L^* = 0.3035 \times 1.15 = 0.349\ m^2$

Referring Figure 4.23, the general shape of bell nozzle contour:

Take nozzle convergent half angle $\alpha = 20°$.

The radius of circular arc of converging cone $R = 1.5 R_t = 1.5 \times 331 = 497\ mm$

The thrust level of the engine is high and used for the booster stage. The contraction ratio, CR is small. From Figure 7.25 a CR value of 1.6 can be selected. Then,

The chamber diameter $D_c = \sqrt{CR} \times D_t = \sqrt{1.6} \times 662 = 837\ mm \cong 840\ mm$

The chamber Radius $R_c = 420\ mm = 0.420\ m$

The convergent cone length is given by

$$L_{con} = \frac{R_t\left(\sqrt{CR} - 1\right) + R(\sec\alpha - 1)}{\tan\alpha}$$

$$L_{con} = \frac{331\left(\sqrt{1.6} - 1\right) + 497(\sec\sec 20 - 1)}{\tan\tan 20} = 329\ mm$$

The approximate convergent cone volume can be calculate using cone-frustum volume equation:

$$V_{con} = \frac{\pi}{3} \times L_n[(R_c^2 + R_t^2) + R_c R_t]$$

$$V_{con} = \frac{\pi}{3} \times 0.329(0.42^2 + 0.331^2 + 0.42 \times 0.331) = 0.1464\ m^3$$

The required volume of cylindrical chamber section

$$V_{cyl} = V_c - V_{con} = 0.349 - 0.1464 = 0.2026\ m^3$$

The length of the cylindrical chamber

$$l_{cyl} = \frac{V_{cyl}}{CR \times A_t} = \frac{0.2026}{1.6 \times 0.3035} = 0.417\ m = 417\ mm$$

The length of the throat from injector face $= 417 + 329 = 746\ mm \cong 750\ mm$

The design of 80% bell nozzle:

Downstream of the throat the nozzle contour radius $R_{ds} = 0.382 R_t = 0.382 \times 331 = 126\ mm$

The bell nozzle length is 80% of the equivalent 15° half angle cone nozzle and is given by:

$$L_{bn} = 0.8 \left[\frac{R_t\left(\sqrt{\epsilon} - 1\right) + R_{ds}(\sec 15 - 1)}{\tan 15}\right]$$

$$L_{bn} = 0.8 \times \frac{331\left(\sqrt{14} - 1\right) + 126(\sec 15 - 1)}{\tan 15} = 2723\ mm \cong 2.75\ m$$

The parabolic contour angles can be selected from Figure 4.23(b) for 80% long bell nozzle.

$$\theta_N = 27° ;\ \theta_e = 10°$$

The position of the point N (Fig.4.23) on the nozzle profile:

With respect to throat $N_t = 0.382R_t\ Sin\theta_N = 0.382 \times 331 \times sin27 = 57.4\ mm$

With respect to nozzle axis $N_a = R_t + 0.382R_t(1 - cos\theta_N)$

$$N_a = 331 + 0.382 \times 331 \times 0.11 = 345\ mm$$

Similarly, the position point E on the contour is given by

$$E_t = L_{bn} = 2750\ mm;\ \ E_a = R_e = 1240\ mm$$

The divergent contour of the nozzle can be achieved accurately by method of characteristic using a computer program.

The important dimensions are shown in the Figure 7.26.

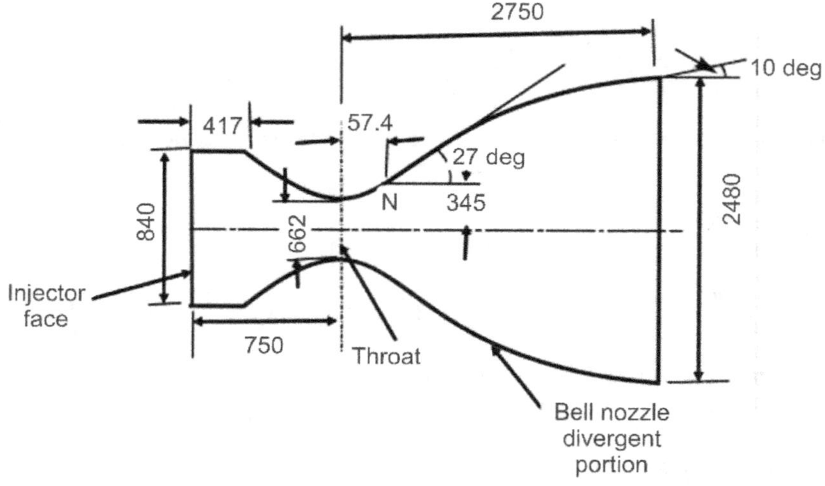


**Figure 7.26** Thrust Chamber and Nozzle Details-Example 7.1.

## 7.7  INJECTOR

The injector injects the propellants into the combustion chamber in correct mixture ratio, with the uniform distribution and right condition for fine atomization to yield efficient and stable combustion process. As the specific impulse $I_{sp}$, is directly proportional the combustion efficiency represented by $c^*$, each percentage point loss of it means a loss of the same magnitude in overall propulsive efficiency. The injector must function adequately during engine start, normal design thrust level operation and throttling levels as per the demand of the mission. The placement of injection elements is important in achieving good mixing interaction between propellant jets, uniform mass distribution to minimize combustor wall-heating. The engine stable operation will mainly depend on the design of the injector along with provision for damping any oscillatory combustion phenomena. Also, injection elements have a significant pressure loss, about 15-20% of the combustion pressure, work as damping elements and at least partially decoupled dynamically the propellant feed system from the combustion chamber. In the design of injector, the designer's experience plays a major role as two-phase flow and its jet atomization characteristics are less understood, hence, its design remains largely as an art and experimental techniques.

Achieving mass distribution uniformly across the injector face is difficult due to mechanical space adjustment for the perimeter of the manifolds. This pushes the outer row injector elements towards the inside away from the chamber wall creating a mass deficient zone around the injector perimeter. The initial gas generated, centered inward, causes an outward flow to fill the chamber and thus hitting against the chamber wall. Another major concern is oxidizer-rich gases mixed with cryogenic droplets may get in contact with the combustion chamber walls. With most propellant combinations, the oxidizer vaporizes more rapidly than the fuel. This oxidizer rich hot gases scrub the walls of the chamber, increasing local heating. The result of this process, a combined physical and chemical attack, burns on the surface of the chamber and is known as blanching. To cope with this problem, the outermost row of the injector operates with a smaller mixture ratio to operate at low temperature and avoid oxidizer rich gas hitting the surface. Another method is to provide an excess fuel rich stream to impinge against the chamber wall at shallow angles to form a coolant film and to reduce the wall temperature. These methods involve a performance penalty since the fuel mainly will be unavailable for effective combustion.

### 7.7.1  Atomization Process

Figure 7.27 shows the process of atomization. Initially, the bulk liquid is converted into a liquid sheet or jet of continuum body. The various internal and external forces competing on the surface of this liquid column create oscillations and perturbations of the surface. The forces include surface tension, viscosity and aerodynamics. The surface tension force tends to pull the liquid in the form of a sphere as it has minimum surface energy. The viscosity of the liquid affects the atomization process adversely as it opposes any change in the shape and geometry of the liquid. The aerodynamic forces disrupt liquid ligaments into droplets. In other words, it can be stated that when the sum of disruptive forces exceeds the overall surface tension force, the liquid is disintegrated into ligaments and subsequently gets converted into smaller droplets. The three most important physical parameters related to spray breakup are Weber number of the jet $We_j$, Reynold number $Re$ and Ohnesorge number of the jet $Oh_j$. The Weber number represents the ratio of aerodynamic force (inertia force) to the force of surface tension, given by

$$We_j = \frac{\rho V_j^2 d_j}{\sigma} \qquad\qquad .....(7.12)$$

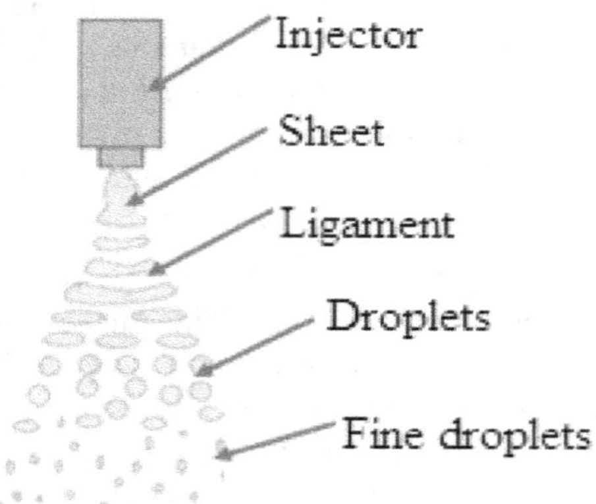


**Figure 7.27** Atomization Process.

$V_j$ is jet the velocity (m/s)

$d_j$ is the diameter of undisturbed liquid jet (m)

σ is the surface tension of the liquid in contact with the surrounding gas (N/m)

Only when the aerodynamic force is much greater than that of the surface tension force ($We_j > 1$), a liquid jet breaks up into droplets. The jet Reynolds number, a ratio of inertial forces to viscous forces, that plays an important role in the formation of droplets is defined as

$$Re = \frac{\rho V_j d_j}{\mu} \qquad \qquad .....(7.13)$$

$\mu$ is the dynamic viscosity of liquid (Pa s)

The third relevant number is the Ohnesorge number, describes the jet disintegration comprehensively, representing the ratio of liquid viscous forces to surface tension forces and is defined as

$$Oh_j = \frac{\sqrt{We_j}}{Re} = \frac{\mu}{\sqrt{\sigma d_j \rho}} = \frac{Viscous\ force}{\sqrt{surface\ tension * inertia\ force}} \qquad .....(7.14)$$

The variation of $Oh_j$ with Reynolds number is shown in Figure 7.28. Four zones can be observed in this figure. The physical effects and mechanisms discussed below are all present in each zone in a certain extent; however, it is their dominance that is changing because of changes in physical circumstances.

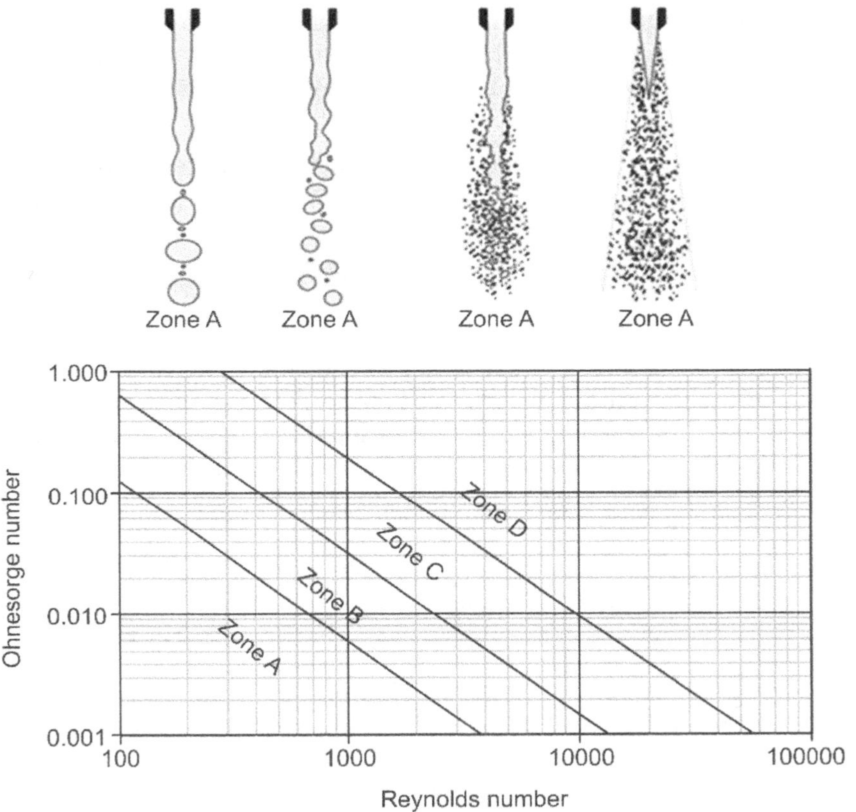


**Figure 7.28** The regimes of liquid breakup as a function of Reynolds and Ohnesorge numbers.

**Zone A: Rayleigh Breakup:** At low jet velocities (low Reynolds number) jet breakup is because of surface tension forces. It promotes contractions but hinders flare of the liquid column when the displacement of liquid from the contracted region into neighboring sections occurs. Any small perturbation led to axial-symmetric oscillations that breakup the liquid column resulting in droplets. These droplets are larger than the diameter of the original jet.

**Zone B: First wind-induced breakup:** This zone marks the beginning of the influence of the ambient gas. With increase in Reynolds number (increase of relative jet velocity), the aerodynamic forces on the surface of the liquid column gain influence. The pressure distribution resulting from the flow around the surface perturbations contribute to the jet disintegration. The resulting droplets are approximately the same in diameter as the original jet.

**Zone C: Second Wind-induced Breakup:** The processes of the first wind-induced breakup continue to a larger extent and more rapidly in this zone with the further increase of Reynolds number. The droplet size is smaller than the original diameter of the jet. Besides that, in a process referred to as surface stripping, short surface waves (Kelvin-Helmholtz instabilities) also distract small droplets from the jet surface.

**Zone D: Atomization:** At higher Reynolds number due to the high interaction of the jet with ambient gas, high aerodynamic drag and flow turbulence the liquid jet disintegrates within a short distance from orifice forming a fine cloud of droplets. The size of the droplet is much less than the jet diameter.

The chart shown Figure 7.28 does not take into consideration the density of ambient gas, which is otherwise a very important factor related to breakup. Figure 7.29 shows the regimes based on Reynolds number and gas density for various gas Weber numbers $We_g$. The dashed line of $Weg=13$ represents the beginning of the atomization regime.

### 7.7.2 Injector Types

Injectors can be largely divided into two categories: (1) non impinging and (2) impinging. Parallel injection in form of a showerhead, shear coax (coaxial) injector and swirl injector are non-impinging types of injectors. The impinging type of injectors are classified into two types, like-on-like and like-on-unlike. A combination of these concepts like a swirl coax injection (HM-7B engine) and or a swirl and impinging (AESTUS engine) can be found. Though many types of

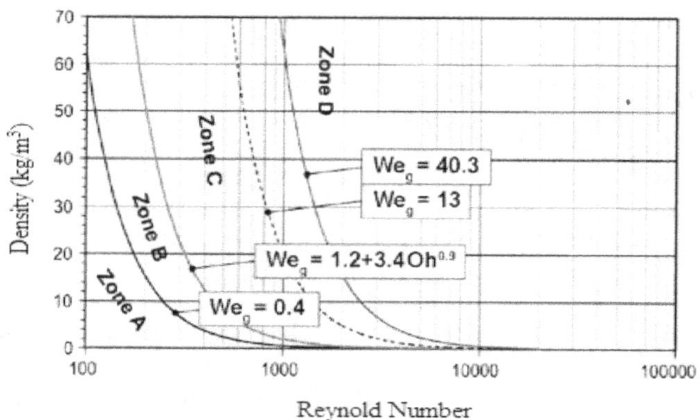


**Figure 7.29** The regimes of liquid breakup as a function of Reynolds number and ambient gas density.

injectors are developed, only a few types of injector designs are presently in use in bipropellant liquid rocket engines. The number of injectors in an engine depends on the size and engine cycle selected. The SSME has three injectors: main engine injector and one injector each for the fuel and oxidizer preburners.

### 7.7.2.1 Non impinging Injectors

Showerhead injector is the oldest type of injector used in rocket engines and was abandoned early in the industry. As shown in Figure 7.30 alternating rows of fuel and oxidizer rings feed axial holes in each ring to form the spray cones that interact with each other to stimulate atomization and mixing. The process of atomization and mixing is inefficient and hence, requires a longer thrust chamber for complete combustion. However, it is effective for chamber wall cooling as the flow is axial near the wall.

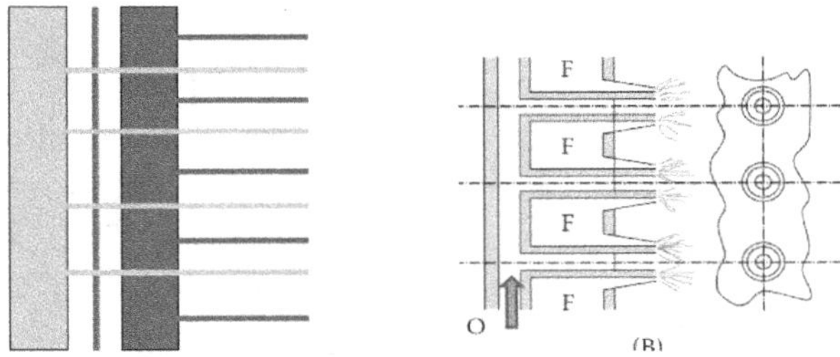

**Figure 7.30** Nonimpinging Type Injector Schematic.

Coaxial injector (shear coax injector) is a non-impinging type frequently used in nonhypergolic propellant combination. It is a preferred injector for semi-cryogenic engines. It consists of two concentric flow passages as shown in Figure 7.31. In the central tube, generally liquid propellant (oxygen) is admitted at a lower velocity (10 to 25 m/s), while in the concentric outer tube fuel is admitted at a high velocity (250 to 300 m/s). The high velocity gaseous fuel breaks the liquid surface into ligaments and subsequently into fine droplets. As the oxidizer is surrounded by fuel, it shields the combustion process and hence restricts the heat transfer to the chamber wall. Also, it mitigates to great extent the combustion instability. Coaxial injectors with or without swirl are the element of choice when one of the propellants is in its liquid phase. Before leaving the oxygen liquid, some gaseous hydrogen is mixed to improve the throttling capability due to increase in pressure drop. Also, it improves the mixing and atomization process.

The injector element developed by the USSR in the early 1980s for ORSC (oxygen rich staged combustion) engine has been the recent research topic in many research institutes connected with space science. It is a variant of coaxial injection element where the oxidizer rich gases from pre-burner flow in the inner of the injector. Fuel is admitted as a swirling film in the outer tube. Coaxial injectors are almost exclusively used in cryogenic engines using liquid hydrogen/liquid methane/natural gas fuels. The fuel after heated in a regenerative cooling jacket enters the element as a gas or a low-density supercritical fluid. Either of the fluids may be swirled as required for better performance. Swirl injectors, introduce a tangential velocity component to the propellants. This rotational momentum provides centrifugal forces that cause the liquid film leaves the orifice to spread outward in the shape of a diverging bell-shaped sheet. As it progresses forward, it becomes thinner and the growth of disturbances on it breaks it up into droplets.

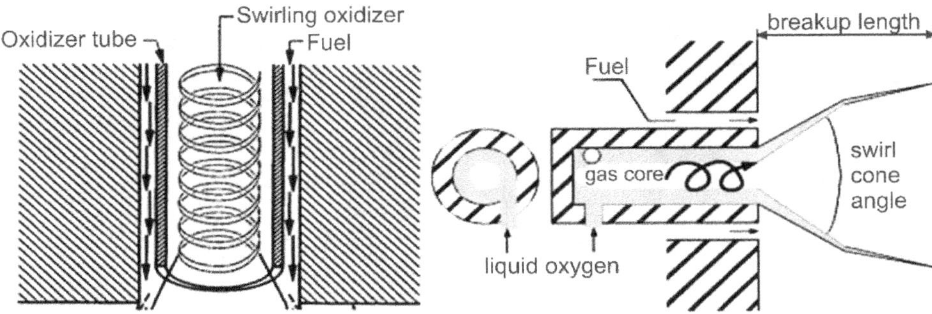


**Figure 7.31** Schematic of a shear coax injector element (Source: Nan Zong and Vigor Yang).

### 7.7.2.2 Impinging Injectors

In an impinging type of injectors, two/three streams of propellant jets are impinged on each other so that the dynamic pressure in the jet helps to decompose it into smaller fragments and create atomization. If propellants of the same kind (fuel-fuel or oxidizer -oxidizer) impinge on one another, it is called like-on-like (like doublet) as shown in Figure 7.32. If fuel and oxidizer jets impinge each other then it is known as like-on-unlike (unlike doublet). These type injectors are generally used because of their better performance, simplicity, and lower cost. Typical impingement angles vary from 45° to 60° in many designs. The propellant combination and injection condition strongly influence the choice of injection element. For hypergolic propellants like-on-unlike (unlike doublet), a single jet impinging a single oxidizer jet, or two oxidizer jets impinging a single fuel jet (unlike triplet), is an attractive choice. For liquid/liquid propellant combinations having similar volume flow rates (i.e., the ratio of volume flow rate $V_o/V_f$ is nearly equal to one), like NTO/UDMH, the choice of unlike doublet type injector is a good candidate. A like-on-like configuration has been employed with LOX/kerosene large booster engines. With hypergolic propellant gases will begin to generate in the contact region that can blow the two streams apart and reduce the mixing effectiveness. To avoid this effect high impingement velocities are needed to essentially keep the breakup time below that of gas evolution time. In case of non-hypergolic propellants, like-on-like configuration is preferred as overheating of the injector face can be avoided by maintaining sufficient distance between the surface and mixing/flame region.

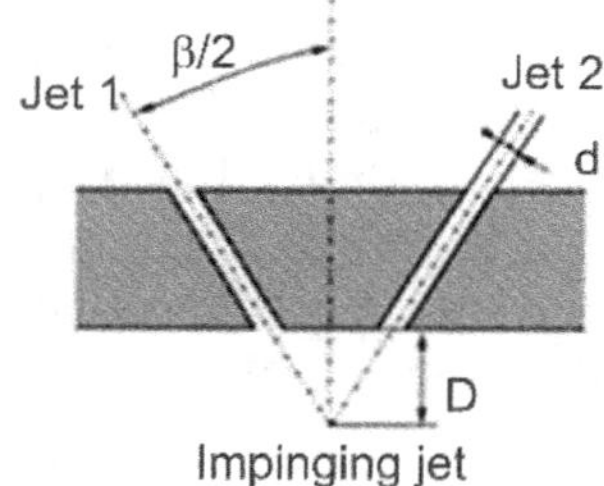


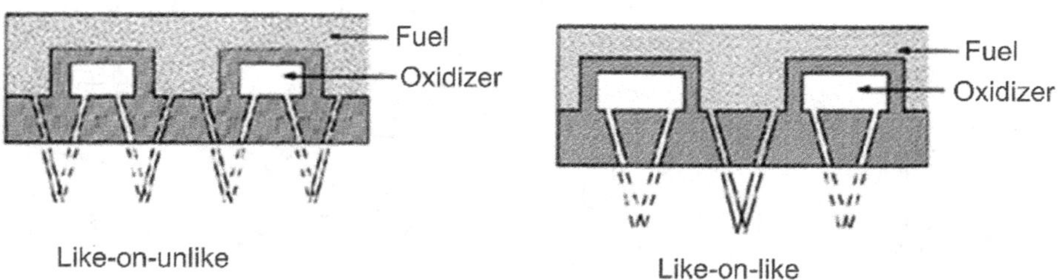


**Figure 7.32** Schematic of Impinging Jet Types.

The pintle injector is the main injector type in use in modern rocket engines where throttling capability is essential. The pintle injector design can deliver high combustion efficiency (typically

96–99%) and enables implementing some unique operating features, such as deep throttling and injector face shutoff. As shown in Figure 7.33 one of the propellants (oxidizer) enters the device from the upper part, flows down inside the pintle and is ejected through a set of calibrated orifices located at the pintle tip to form a uniform radial flow. The other propellant (fuel) is fed through outer injector circumferential annular flow passages and exit the injector as an axially flowing annular sheet. Collisions between the radial and the axial flow stream result in the appropriated mixing and atomization. As shown in figure multiple rows of holes might be included to better cover the perimeter of the injector and improve mixing of the two streams.

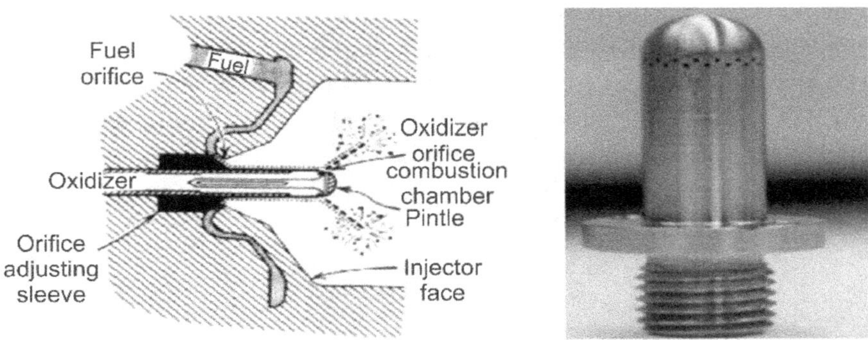


**Figure 7.33** Schematic of pintle injector.

The central injector sleeve, shown in the figure, is easily designed to be movable to provide a convenient and reliable means of throttling unit. This will maintain nearly constant injection velocities across a wide range of injected propellant flow rates. Figure 7.34 shows another variant of pintle injector and is known as Face Shutoff-only (FSO) injector. With a proper design the sleeve can be made to fully shut off both propellants at the injector face. This feature is advantageous for applications demanding intermittent engine usage.

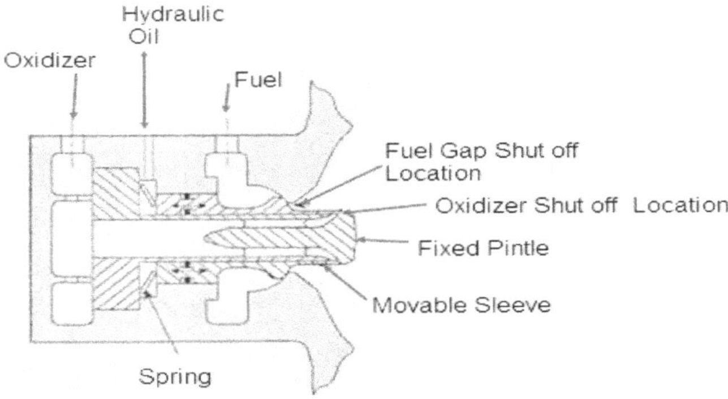


**Figure 7.34** Face Shut Off Only (FSO) Pintle (Gordon A. Dressler and J. Martin Bauer).

The Apollo Lunar Module Descent Engine (LMDE) developed to land Apollo astronauts on the moon is a most familiar example of pintle type injector usage. The Merlin 1D engine of SpaceX uses this pintle type injector. Compared to conventional rocket engines, pintle rocket engines are generally longer in physical length and higher in chamber contraction ratio (both being required to support the chamber's major recirculation zones) operating at the same chamber pressure and thrust level. Table 7.7 summarizes the key performance characteristics of the rocket engines using conventional and pintle injectors.

### 7.7.3 Injector Design Considerations

The thrust chamber and injector design process can be started with the thrust requirements, selection of propellant combination and injector topology selection. The selected cycle analysis begins with the assumed values of pressure drop and efficiency based on team experience with similar engines. For the given thrust level and certain exhaust velocity (specific impulse, $I_{sp}$) to be achieved, the overall propellant mass flow rate can be calculated from Eq. (3.22), $\dot{m} = F/V_e$. For the selected mixture ratio (MR= Oxidizer flow rate/fuel flow rate) Eqs. (7.1) & (7.2) give mass flow rates of oxidizer and fuel. The oxidizer flow rate.

$$\dot{m}_{ox} = \frac{(MR)\dot{m}}{1+MR}$$

The mass flow rate of fuel

$$\dot{m}_f = \frac{\dot{m}}{1+MR}$$

Assuming that the velocity of propellant in manifold is negligible, applying Bernoulli equation for incompressible flow through circular hole orifices, propellant velocity and mass flow rate can be expressed as

$$V = C_d\sqrt{\frac{2\Delta p}{\rho}} \quad ; \quad \dot{m} = C_d A_i\sqrt{2\rho\Delta p}$$

where

$V$ is the velocity of propellant through the orifice, m/s

$C_d$ is the discharge coefficient of the orifice

$A_i$ is the injection (orifice) area, m$^2$

**Table 7.7** Comparison of Typical Performance characteristics of Rocket Engines Using Conventional Injector and Pintle Injector (Source: Gordon A. Dressler and J. Martin Bauer).

| Parameter | Chamber flow pattern in a typical liquid rocket | Chamber flow pattern in a typical in TRW Pintle Rocket |
|---|---|---|
| Propellant injection | Distributed across injector face | Only at central location |
| Fuel and oxidizer injection geometry | Multiple intersection or shearing propellant streams; intersecting streams are of like or unlike propellants | Single annular outer sheet of one propellant impinges on (a) multiple radial spokes of other propellant or (b) thin radial fan of other propellant |
| Fuel and oxidizer collision geometry | In plane immediately adjacent to injector face | In significantly offset from injector face |
| Droplet trajectories | Approximately axial down to chamber | Initially at large angle to chamber axis |
| Chamber circulation | None | Two major circulation zones in chamber |
| Droplet vaporisation and combustion | Proceed in planar fashion to down chamber length | Proceeds along axially symmetric, but highly non-planar, contours in chamber |
| Secondary droplet breakup | Comparatively small due to axial flow and homogeneous distribution | Comparatively large due to wall impingement and recirculation zones |
| Energy release zone geometry | Uniform and planar across chamber diameter (facilitates acoustically-coupled combustion instability) | Radially varying and canted down cross chamber-together with stable zones having different gas properties (O/F, molecular mass, $\gamma$, T) Serve to prevent acoustic instability. |
| Chamber for optimum performance | Is relatively short and small contraction ratio | Is relatively long and high contraction ratio |
| Wall film cooling | Established by separate injection ports | Established by pintle injector "tuning", eliminates need for separate ports |
| Injection metering orifice | Relatively small and contamination sensitive. | Relatively large and insensitive to contamination |

The value of the discharge coefficient, $C_d$, depends on many parameters like propellant properties, geometry of orifice/injector, pressure drop and chamber pressure. The geometric parameter of cylindrical injector, $l/d$ (length/diameter) and orifice entry shape also affect the value $C_d$. Higher the length more friction loss in the injector and the value of $C_d$ includes these losses. Generally, the $C_d$ value is determined experimentally and varies between 0.65 to 0.98. The pressure across the injector orifices imparts high injection velocity to the streams. This provides requisite energy for atomization, leading to controlled mass distribution and better mixing between fuel and oxidizer propellants in the chamber. Also, the pressure drop across the injector provides isolation between chamber pressure disturbance and propellant flow rate. General thumb rule for the starting point of design for pressure drop is 20% of combustion chamber pressure $p_c$. However, some applications such as engines require throttling or those with stability sensitivity may require higher than 20% values of pressure drop across the injector. The total injector area can be calculated as

$$A_i = \frac{\dot{m}}{C_d\sqrt{2\rho\Delta p}} \qquad\qquad .....(7.15)$$

If $N$ is the number of injector elements, the diameter $d_i$ of the orifice is given by

$$d_i = \sqrt{\frac{4\dot{m}}{\pi N C_d\sqrt{2\rho\Delta p}}} \qquad\qquad .....(7.16)$$

The orifice sizing is a complex process. For better atomization and efficient mixing would be the maximum number of injector elements or smaller diameter orifice for fine spray. However, from the manufacturing point it is quite difficult and costly to fabricate tiny-holed orifices. Also, it is equally difficult to maintain as they are more prone to blockages due to plugging of contaminations like soot particles, during its operation. Moreover, fine spray produced by small diameter orifices are more prone to combustion instabilities due to higher level of premixing mixture in the combustion chamber. Fortunately, large rocket engines generally have more stability concerns, tend to have high stay (residence) time and do not require fine spray with small orifices. Small engines require compact combustion chambers, are less prone to stability problems and can use fine injection patterns. Most injector diameters are in the range 0.5 to 2 mm. For the F-1 booster engine the orifice diameter is 7 mm. To accommodate all the injector elements sufficient chamber face area must be designed. With chamber cross-section area fixed, chamber length $l_c$ required for acceptable efficiency can be determined as explained in section 7.6.2. From Eq. (4.36), $\dot{m} = \frac{p_c A_t}{c^*}$ the throat can be calculated.

Figure 7.35 shows the model a fuel and an oxidizer jet impinging on each other (doublet) resulting in a single stream making an angle $\theta$ with a chamber axis. The fuel and oxidizer streams with the velocity of $V_f$ and $V_{ox}$ make an angle $\alpha_f$ and $\alpha_{ox}$ respectively with the chamber axis. To have the spray fan perpendicular to injector face, the tangential momentum of the two streams must cancel each other, i.e.,

$$\dot{m}_{ox}V_{ox}\sin\alpha_{ox} = \dot{m}_f V_f \sin\alpha_f$$

The impingement point depends on impingement angles; too steep may lead to splash back onto the injector face and angles that are too shallow can lead to premature aerodynamic distortion or breakup of the liquid jets. Typical values of $\alpha_f + \alpha_{ox}$ would lie in the 45-60° range. Generally, the cross-sectional area of the oxidizer jet is greater than that of fuel jet.

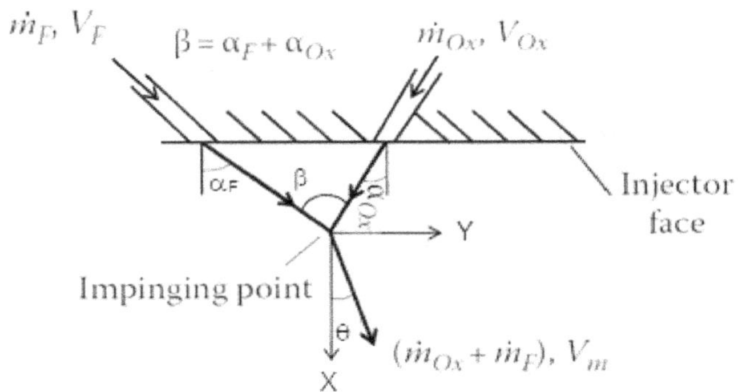



**Figure 7.35** Doublet impinging injector element.

**Example 7.2**

A rocket engine with LOX/LH$_2$ propellant combination has the following specifications:

Thrust $F$ : 660 kN

Chamber pressure : 30 MPa

Mixture ratio : 1.2

Characteristic velocity $c^*$ : 1700 m/s

Thrust coefficient : $C_F$ 1.858

Density of LOX : 1450 kg/m$^3$

Density of Gaseous hydrogen : 12 kg/m$^3$ (at 100 K)

Determine the sizes of coaxial injection elements used in the engine, injection velocities and velocity ratio.

***Solution:***

The specific impulse

$$I_{sp} = \frac{C_F c^*}{g} = \frac{1.895 \times 2280}{9.81} = 440 \ s$$

The mass flow rate of propellant

$$\dot{m} = \frac{F}{I_{sp} \ g} = \frac{660 \times 10^3}{440 \times 9.81} = 152.9 \ kg/s$$

Mass flow rate of oxidizer can be calculated from Eq.(7.1)

$$\dot{m}_{ox} = \frac{(MR)\dot{m}}{1 + MR} = \frac{5 \times 152.9}{1 + 5} = 127.4 \frac{kg}{s}$$

Mass flow rate of fuel

$$\dot{m}_f = 152.9 - 127.4 = 25.5 \frac{kg}{s}$$

The throat area can be calculated from Eq. (4.42)

$$F = C_F \, A_t \, p_c$$

$$A_t = \frac{0.66 \times 10^6}{1.895 \times 6.0 \times 10^6} = 0.0581 \; m^2$$

Throat diameter is

$$D_t = \sqrt{\frac{4 \times 0.0581}{\pi}} = 0.272 \; m \cong 275 \; mm$$

From Figure 7.25 a CR value of 1.6 can be selected. Then,

Chamber diameter $D_c = \sqrt{CR} \times D_t = \sqrt{1.6} \times 275 = 348 \; mm \cong 350 \; mm$

The surface of the injector

$$A_{in} = \frac{\pi}{4} \times D_c^2 = 0.0957 \; m^2$$

Consider 300 injector elements (typically 2 elements per square inch injector face area is used).

Take a discharge coefficient of 0.65. Consider a pressure drop of 15% of chamber pressure, i.e., 0.9 MPa.

Oxidizer injector element ID can be calculated by Eq. (7.18)

$$d_{iox} = \sqrt{\frac{4\dot{m}_{ox}}{\pi N C_d \; \sqrt{2\rho\Delta p}}}$$

$$d_{iox} = \sqrt{\frac{4 \times 127.4}{\pi \times 300 \times 0.65\sqrt{2 \times 1140 \times 0.9 \times 10^6}}} = 4.3 \times 10^{-3} \; m \cong 4.5 \; mm$$

The velocity of the oxidizer jet is

$$V_{ox} = \frac{127.4 \times 4}{300 \times 1140 \times (4.5 \times 10^{-3})^2} = 23.4 \frac{m}{s}$$

Take the thickness of the oxidizer element = 0.5 mm

The outer diameter of the oxidizer injector element = 4.5+ 0.5+0.5 = 5.5 mm

The total fuel injector elements area is calculated from Eq. (7.17)

$$A_{f,in} = \frac{\dot{m}_f}{C_d\sqrt{2\rho_f \Delta p}} = \frac{25.5}{0.65 \times \sqrt{2 \times 12 \times 0.9 \times 10^6}} = 0.00844 \ m^2$$

Fuel element area per element $= \dfrac{0.00844}{300} = 2.813 \times 10^{-5} \ m^2$

Total injector element area = oxidizer element area + fuel element area

$$= \frac{\pi}{4}(4.5 \times 10^{-3})^2 + 2.813 \times 10^{-5} = 4.403 \times 10^{-5} \ m^2$$

Internal diameter of the element

$$D_{element} = \sqrt{\frac{4 \times 4.403 \times 10^{-5}}{\pi}} = 7.5 \times 10^{-3} \ m = 7.5 \ mm$$

$$\text{Fuel gap} = \frac{7.5 - 5.5}{2} = 1 \ mm$$

Velocity of fuel jet

$$V_f = \frac{\dot{m}}{\rho_f A_{f,in}} = \frac{25.5}{12 \times 0.00844} = 251.8 \frac{m}{s}$$

Velocity ratio = 251.8/23.4 = 10.8

For good atomization and mixing, the velocity ratio will be in the range of 10 to 20.


## 7.8   ENGINE START METHOD

Two methods, tank-head start and spin start, are generally used in starting the liquid rocket engine until the main power source takes over. The main factors affecting the selection of start methods are the type of fuel used and the engine cycle. A tank-head start uses propellants out of the main tanks for initial turbine power. As the storable fuels, like RP-1 and hydrazine, do not gasify a low pressure, relatively high flow rates can be obtained under tank head conditions. Usually for the cryogenic expander cycle and staged combustion cycle a tank-head start method is suggested. In the spin start method an auxiliary power source is used to provide the initial turbine speed buildup. Spin starts use a stored high-pressure gas, a solid-propellant gas generator, or bipropellant start tanks generate the initial turbine power. A spin start will usually be used for hydrogen-fueled engines with a gas generator.

### Nomenclature

$\dot{m}$     mass flow rate (kg/s)

$\dot{m}_{ox}$   mass flow rate of oxidizer (kg/s)

$\dot{m}_f$   mass flow rate of fuel (kg/s)

$\dot{m}_{gg}$   mass flow rate of the propellant in gas generator (kg/s)

$\rho_b$     bulk density (total mass of propellant /volume it occupies) (kg/m$^3$)

| | |
|---|---|
| $MR$ | mixture ratio of the propellant |
| $\dot{Q}$ | volumetric flow rate (m$^3$/s) |
| $p_{gg}$ | pressure in gas generator (MPa) |
| $\eta_p$ | pump efficiency |
| $\eta_t$ | turbine efficiency |
| $c*$ | characteristic velocity (m/s) |
| $p_c$ | chamber pressure (MPa) |
| $A_c$ | cross sectional area of thrust chamber (m$^2$) |
| $C_F$ | thrust coefficient |
| $T_c$ | temperature of combustion gases (K) |
| $\rho_c$ | density of gases (kg/m$^3$) |
| $l_c$ | length of the thrust chamber (m) |
| $R$ | gas constant (J/kg K) |
| $t_s$ | stay time (s) |
| $L^*$ | characteristic length of the thrust chamber (m) |
| $W_{ej}$ | Weber number of the jet |
| $Oh_j$ | Ohnesorge number of the jet |
| $V_j$ | jet the velocity (m/s) |
| $d_j$ | diameter of undisturbed liquid jet (m) |
| $\sigma$ | surface tension of the liquid in contact with the surrounding gas (N/m) |
| $\mu$ | dynamic viscosity of liquid (Pa s) |
| $Re$ | Reynolds number |

## PROBLEMS

7.1    For a rocket engine having the following specification, calculate the thrust chamber dimensions. Consider an 80% bell nozzle for the expansion of the hot gases.

Propellant combination: LOX/LH$_2$

Thrust = 660 kN

Chamber pressure = 6 MPa

Area ratio $\varepsilon$ =40

Mixture ratio 5

7.2    A rocket engine with $N_2H_4/N_2O_4$ propellant combination has the following specifications:

| | | |
|---|---|---|
| Thrust $F$ | : | 30 kN |
| Chamber pressure | : | 0.75 MPa |
| Mixture ratio | : | 1.2 |
| Characteristic velocity $c^*$ | : | 1700 m/s |
| Thrust coefficient $C_F$ | : | 1.858 |
| Density of Oxidizer | : | 1450 kg/m$^3$ |
| Density of Fuel | : | 1005 kg/m$^3$ |

Determine the sizes of like-on-unlike injection elements used in the engine, injection velocities and velocity ratio.

7.3    For the problem in 7.2 find the size of the thrust chamber with 70% bell nozzle. Take the expansion ratio equal to 35.

7.4    A liquid engine with hydrazine and nitrogen tetroxide propellant combination is used to develop a thrust of 5 kN with characteristic velocity of 1900 m/s at chamber pressure of 3 MPa and mixture ratio of 1.4. If its thrust coefficient $C_F$ and characteristic length happen to be 1.485 and 1.45 m, respectively, determine combustion chamber volume and volumetric flow rate of each propellant by considering oxidizer and fuel densities to be 1450 and 856 kg/m$^3$.

7.5    A liquid rocket engine using $N_2O_4/MMH$ propellant combination develops a thrust of 1.1 kN with characteristic velocity of 1900 m/s at chamber pressure of 0.9 MPa and mixture ratio of 1.5. Determine throat of nozzle, and propellant flow rates of oxidizer and fuel. Take the thrust coefficient as 1.832. Take oxidizer and fuel densities to be 1450 and 856 kg/m$^3$. Considering 10 element doublet impinging injection systems determine the size and velocities of injection.

7.6    Consider the liquid rocket engine using LOX and RP-1 propellants that deliver a characteristic velocity of 1800 m/s at $\gamma = 1.2$. The desired vacuum thrust level is 130 kN and a nozzle expansion ratio of 45 with a chamber pressure of 8 MPa. Take the chamber characteristic length as 1.1 m. Find the size of the thrust chamber and the bell nozzle.

CHAPTER # Hybrid Rocket Engines

# CHAPTER 8 — Hybrid Rocket Engines

## 8.1  INTRODUCTION

The Hybrid Rocket Engine (HRE), also mentioned as Hybrid Rocket Motor (HRM), contains one liquid oxidizer and one solid fuel as shown in Figure 8.1. The opposite configuration, liquid fuel and solid oxidizer, is called a reversed hybrid. The fuel, usually a polymer, having a ring or star cross section shape (like in SRM) is stored in a metal case. Liquid oxidizer, through a pressurizer or turbo pump system, is injected into the forward section of the solid fuel chamber in which combustion occurs. The inner fuel surface melts and vaporizes because of the convection and radiation heat transfer. If the propellant combination is not hypergolic, a separate ignition device is provided. By varying oxidizer flow through a control valve, throttling capability can be achieved. The mixture formation and combustion process take place in the close area of the fuel inner surface. As there is not enough time and space in the inner port for complete combustion, an additional space is provided called a post combustion chamber. The combustion gases expand in a convergent divergent nozzle to produce thrust.

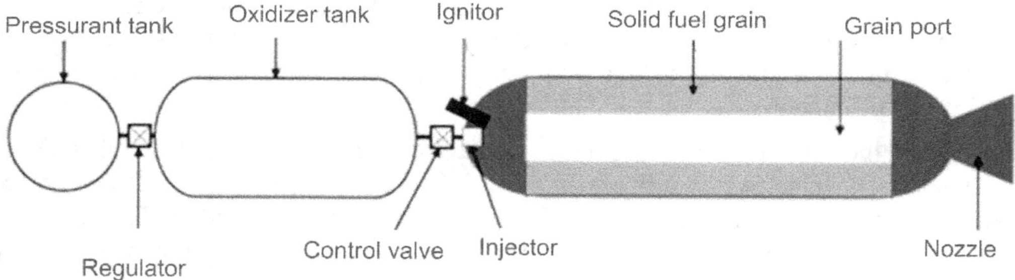


**Figure 8.1** Schematic of Hybrid Rocket Engine (HRE).

In 1933 a hybrid motor was first time test-flown in the Soviet Union built by Tikhonarov and Korolev and it was a breakthrough in rocket technology. The propellant was LOX and gelled gasoline. The interest in hybrid technology was not continued as the space scientist's community was involved in the development of launch vehicles based on solid (SRM) and liquid (LRE) propellants. During the time frame 1960-1980 extensive experimental and modelling studies were conducted in hybrid rocket technology. Karabeyoglu (Source: A Hybrid rocket propulsion for future space launch. Department of Aeronautics and Astronautics, Stanford University, Aero/Astro 50th Year Anniversary (2008)) mentioned this period as "Era of Enlightenment ").

Efforts were focused to understand the performance and combustion phenomena of HRE using conventional propellants. American Rocket Company (AMROC), later acquired by SpaceDev, tested hybrid motors up to 324 kN. In 1995, under Hybrid Propulsion Demonstration Program (HPDP), NASA and DARPA several versions of the Hyperion rocket were built based on HTPB/$N_2O$ technology. Lockheed launched the HYSR hybrid rocket in 2002. Table 8.1 contains key data of hybrid rocket systems flight-tested between 1980 and 2005.

**Table 8.1** Hybrid Rocket Vehicles Flight Tested During 1980-2005.

| Name | Dolphin | Hyperion | HYSR | SpaceShipOne |
| --- | --- | --- | --- | --- |
| Entity | Starstruck | eAc | Lockheed Martin | Scaled Composite |
| year | 1984 | 1996 | 2002 | 2004 |
| Oxidizer | LOX | $N_2O$ | LOX | $N_2O_4$ |
| Fuel | PB | HTPB | HTPB +Al | HTPB |
| Grain Type | Multiport (8) | Slotted single port | Multiport (11) | Multiport |
| Diameter (m) | 1.07 | 0.28 | n/a | 8.05 (wingspan) |
| Length (m) | 15.54 | 5.7 | n/a | 8.53 |
| Burn Duration (s) | n/a | 17.0 | 33.4 | 87 |
| Max Thrust (kN) | 155.7 | 6.67 | 267 | 74 |
| Max Altitude (km) | 201 (454kg payload) | 36.5 | 42 | 112 |

## 8.2  PROPELLANTS

The most common fuel used in HRE is hydroxyl-terminated polybutadiene (HTPB). Hydrocarbons might be also used as hybrid fuels. Paraffin, a hydrocarbon, is solid even in the ambient temperature and simplifies the storage process. Liquid oxidizers are like those used in liquid propellant rockets. LOX or a combination of LOX and liquid fluorine viz., FLOX are energetic oxidizers. GOX, $N_2O_4$, $H_2O_2$ and the red fuming nitric acid (RFNA) are also viable oxidizers. Table 8 2 presents performance possible to reach by several hybrid propellant combinations in comparison with liquid oxygen and kerosene. The main disadvantage with LOX is the requirement of a pyrotechnic ignition device. That can be used only once, i.e., the hybrid motor cannot be restarted. Hence, for restart capability of the motor, the liquid oxygen should be vaporised before the mixture forms. It is observed that incomplete oxygen vaporisation before the combustion port causes the intensification of the low-frequency combustion instabilities. The alternative for liquid oxygen is nitrous oxide ($N_2O$). It is also a cryogenic liquid. At atmospheric pressure, it boils at a temperature of $-59°$ C. At $20°$ C the $N_2O$ vapor pressure is 5.85 MPa. If stored at high pressure as a liquid it could be directly fed into a solid fuel chamber as a gas. This feature makes it possible to eliminate additional pressurization devices in the oxidizer feed system. It has also been used in the hybrid rockets developed by Scaled Composites and Virgin Galactic for space tourism.

**Table 8.2** Hybrid Propellant Performance Compared LOX/RP-1 Combination.

| Oxidizer | Fuel | O/F | Specific Impulse (m/s) | Density impulse $\times 10^6$ (kg/m$^2$ s) |
|---|---|---|---|---|
| LOX | RP-1 | 2.5 | 3450 | 3.55 |
| $H_2O_2$ 98% | HTPB | 6.6 | 3200 | 4.34 |
| $H_2O_2$ 98% | PB | 7.2 | 3100 | 4.21 |
| LOX | HTPB | 2.5 | 3450 | 3.63 |
| $N_2O$ | HTPB | 8.5 | 3050 | 2.54 |

The other important oxidizer for hybrid motors is 80-98% $H_2O_2$ (also known as High Test Peroxide). Its decomposition temperature 900 -1100 K depending on concentration is much above ignition temperatures of polymeric hybrid fuels and thus, eliminates the need for an ignition device. The other advantage is the decompose products of HTP are environmentally friendly.

## 8.3 HYBRID ROCKET CHARACTERISTICS

The characteristic features of hybrid rocket systems are briefly mentioned below. The advantages include the following:

(a) As oxidizer and fuel are separated by phase, distance and the solid fuel grain is classically inert, hybrids have almost no explosion hazard and very few failure modes. The fluid oxidiser can reach and react with the fuel surface in gaseous phase and at low pressures. The combustion process in HRE is insensitive to solid fuel grain imperfections, pores and cracks to cause any transition to detonation.

(b) HREs may deliver a higher specific impulse than SRMs. They may have higher density specific impulse than LREs.

(c) With only the oxidiser in liquid phase, the feeding system hardware is simplified, increasing the reliability.

(d) Wide choice of propellant is available. Additives for different purposes can be embedded in the fuel grain.

(e) Throttling, reignition and motor shutdown can be easily achievable with oxidizer flow control.

(f) The combustion products are environmentally friendly.

(g) The cost of handling hybrid rockets is low. They pose almost no explosion hazard during manufacture, transport, ground test, and storage. Recurring costs are also low due to high level safety and minimum failure modes.

*The major disadvantages of hybrid rocket are*:

(a) They cannot reach the specific impulse of cryogenic LREs. They have lower density specific impulse compared to SRMs.

(b)  The major concern of hybrid rockets is its very low regression rates. Hence, a large fuel surface is needed to produce the required thrust level. This results in large length-to diameter ratios, poor fuel loading and low thrust densities.

(c)  During the combustion of side burning grain, the port area increases. Also, the burning area increases. The regression rate primarily depends on the mass flux (the ratio of mass-flow and port area). These two effects lead to variation of fuel mass flow with constant oxidizer flow. Thus, the mixture ratio shifts during operation affecting the performance.

(d)  The combustion efficiency is lower than SRMs and LREs. A part of the fuel under the flame at the grain port may exit through the nozzle without mixing and reacting with oxidizer.

(e)  Slow transient/response to throttling.

(f)  High residual fuel mass. To obtain a very low and reproducible level of residuals on multiple ports will be a hard point.

## 8.4    REGRESSION RATE AND MODELLING

Figure 8.2 shows a combustion flow field model close-up view near the fuel surface within a typical HRE fuel port. The pure liquid oxidizer is injected as a spray of droplets or else as a gaseous oxidizer in the port volume of the solid fuel grain. The gaseous oxidizer moves at high velocities over the solid fuel surface and begins to mix and combust near the fuel surface. Below the flame zone, a mixture of fuel and product gases exist. Above the flame zone a mixture of oxidizer and product gases is present. The blowing of the fuel gases departing the surface increases the boundary layer thickness. The flame height depends on energy from the flame vaporizing the fuel, and this turbulent boundary layer development along the fuel port. The blowing of the fuel vapor from the fuel surface hinders the heat transfer from flame zone to the fuel surface and thus reduces the vaporization of the fuel. This results in the regression of solid fuel at low levels, lower than solid propellant in which fuel and oxidizer are premixed. As a result, the fuel mass flux (G) in a hybrid per surface area is low. The fuel regression rate is expressed as

$$\dot{r} = aG^n x^m = a\left(\frac{\dot{m}_{ox}}{A_P}\right)^n x^m \qquad\qquad .....(8.1)$$

where $G$ is oxidizer mass flux (kg/m²s), $\dot{m}_{ox}$ is oxidizer mass flow rate (kg/s) and $A_P$ is the port area (m²). Ballistic coefficients $a$, $n$ and $m$ are determined from experiments. For pure polymer fuels these parameters depend only on the fuel-oxidizer composition, not the chamber pressure. Pressure dependency appears only in case of metal and solid oxidizer addition. In general, the axial coordinate factor is neglected. The above equation is simplified to

$$\dot{r} = aG^n = a\left(\frac{\dot{m}_{ox}}{A_P}\right)^n \qquad\qquad .....(8.2)$$

The high $G$ values will minimize the port area of the solid grain for a given desired propellant flow and hence the fuel section volume. However, with high $G$ value excessive velocities exist in the fuel port that may cause "blowing the flame out". Because the fuel flow is not controlled, it is observed that the mixture ratio continually varies along the axis as the regression proceeds. The drastic variations of mixture ratio in the combustion volume would lead to poor burning and hence performance. The values of $a$ and $n$ may be dependent on many factors such as $G$ range, fuel formulation, grain production method, engine dimensions (scale effect), port/grain geometry, injector geometry, and flow features. Experimentally determined ballistic coefficients and fuel regression rates for nitrous oxide are shown in Table 8.3. Regression rate is in mm/s and oxidiser mass flux in g/cm$^2$/s.

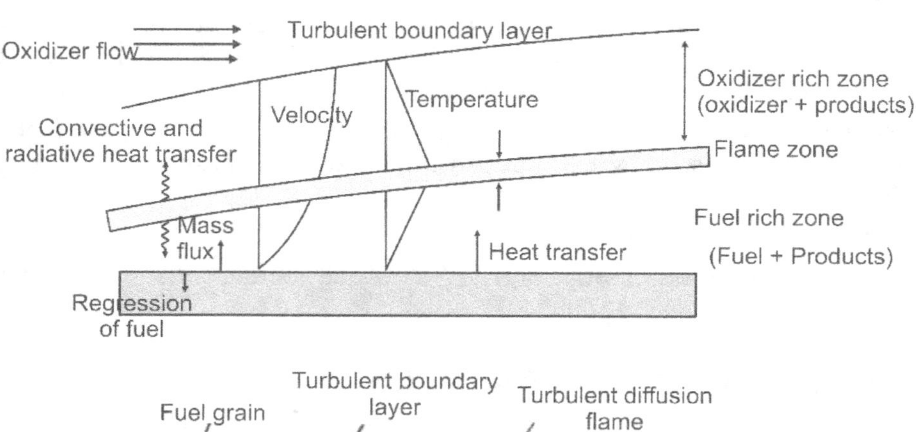


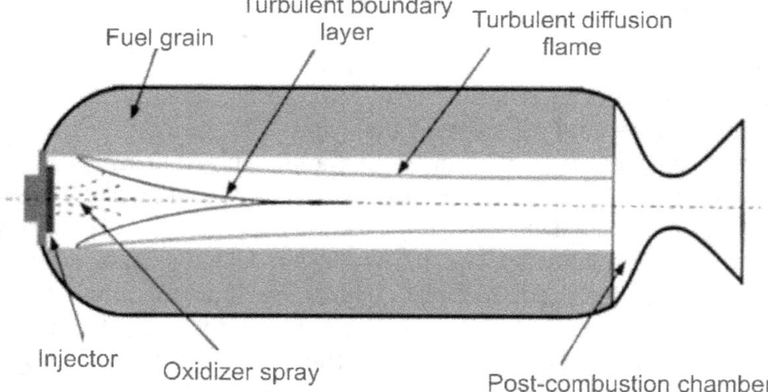


**Figure 8.2** Schematic view of HRE combustion near burning of the solid fuel surface.

**Table 8.3** Ballistic Coefficients for Polymer fuels and Nitrous Oxide. (Source: Lohner K et al)

| Fuel | $a$ | $n$ | $\dot{r}$  $mm/s$ |
|------|------|------|------|
| Polyethylene | 0.104 | 0.352 | 0.35 - 0.7 |
| Plexiglas | 0.111 | 0.377 | 0.4 − 0.8 |
| HTPB | 0.198 | 0.310 | 0.6 -1.0 |
| Paraffin | 0.488 | 0.620 | 0.5 -5.4 |

From Eq. (8.2), the mass flow rate of fuel can be expressed as

$$\dot{m}_f = a\rho_f \left(\frac{\dot{m}_{ox}}{A_P}\right)^n A_b \qquad\qquad .....(8.3)$$

Where $\rho_f$ is the density of the fuel and $A_b$ is burning surface area.

The mixture ratio, $R$ is

$$R = \frac{\dot{m}_{ox}}{\dot{m}_f} = \frac{\dot{m}_{ox}^{1-n}}{a\rho_f} \frac{A_p^n}{A_b} \qquad \qquad .....(8.4)$$

The area ratio $\frac{A_p^n}{A_b}$ is a function time. Hence, the mixture ratio R changes even if the oxidizer flow rate $\dot{m}_{ox}$ is kept constant. This characteristic behaviour of hybrid rockets is known as mixture-ratio shifting.

For low melting temperature hydrocarbon fuels, especially paraffin, Stanford University (B. Cantwell, A. Karabeyoglu, G. Zilliac: Recent Advances in Hybrid Propulsion, Stanford University, Department of Aeronautics and Astronautics, 2007) developed the "Liquid Layer

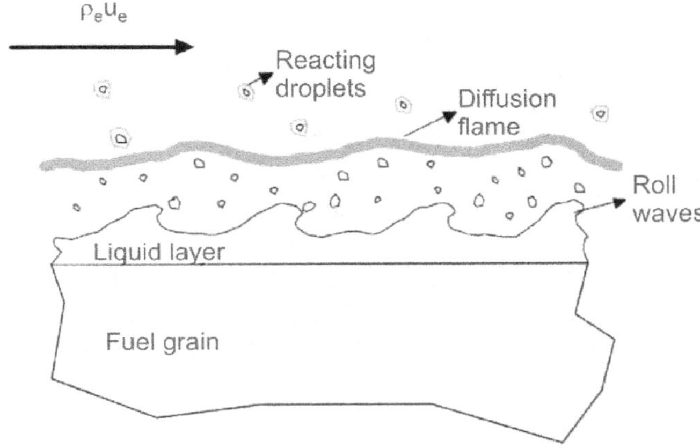


**Figure 8.3** Schematic of Paraffin wax liquid entertainment mechanism.

Hybrid Combustion Theory". While combustion occurs, the fuel melts and creates a liquid layer over the fuel grain with low viscosity and low surface tension as shown in Figure 8.3. It consists of liquified paraffin fuel droplets. This layer merges with the oxidizer flow and becomes hydrodynamically unstable in the fuel port. The high-velocity oxidizer is injected at one end of solid fuel, which destabilizes this liquid layer. The melting layer instability generates liquid fuel droplets and rolling waves. The instability created lift-offs paraffin fuel droplets from the grain surface that further increased the mass transfer rate of fuel into the oxidizer gas flow. This is called the "liquid entrainment mass transfer mechanism". This additional mass transfer mechanism increases the regression rate of hybrid motor combustion. Experiments performed on paraffin with melting point 69° C under gaseous oxygen as oxidizer reported an increase of three times higher than HTPB regression rate. The researchers also validated their predicted entrainment theory with a scale-up test conducted on the large motor.

The paraffin-based fuel has poor mechanical properties. It is a complicated manufacturing process to cast sizable paraffin grains to withstand combustion pressure, inertial flight stresses and thermal loads. It is reported that adding 20% of ethylene-vinyl acetate (EVA) improve the tensile strength by 50%. Addition of polyethylene (PE) can improve mechanical performance, thermal stability, and combustion efficiency significantly.

## 8.5   REGRESSION RATE ENHANCEMENT TECHNIQUES

For the trust levels used in space applications the low regression rate of conventional hybrid rocket fuels is unattractive. Intensive research efforts have been on-going by many research teams in recent years. The fuel regression rate is low due to non-uniform diffusion into the combustion zone resulting in poor mixing of fuel and oxidizer. It is affected by combustion chamber fluid dynamics, grain geometry, and chemical properties of solid fuel. Accordingly, the regression enhancement techniques can be classified into two groups: fluid dynamics and chemical methods. In the fluid dynamics methods turbulence is boosted to improve the mixing of fuel and oxidizer and heat transfer rate. The injection system is modified to create swirl or vertex flow and/or mechanical device is embedded in solid fuel to increase the turbulence in the chamber as shown in Figure 8.4. Swirl injectors provide rotational momentum to the oxidizer vapor and better mixing with the fuel vapor. However, maintenance of swirl over the entire length of the fuel grain is a complex design. Though use of multi-port grain geometry can solve the problem of low solid fuel regression rate there are many disadvantages like excessive unburned mass fraction (sliver), complex design/manufacturing process, low structural integrity, uneven burning of individual ports and combustion instability. Multiport design reduces solid fuel volumetric loading and increases the launch vehicle diameter to accommodate the required fuel mass for a given mission. In a chemical method, to improve the combustion kinetics, heat release and flame temperature, the solid fuel is doped with energetic metal additives or light metal hydrides. Combustion of metals has inherent advantages as they help to increase both specific impulse and density specific impulse. Table 8.4 summarizes some regression rate enhancement techniques tested by various research teams.

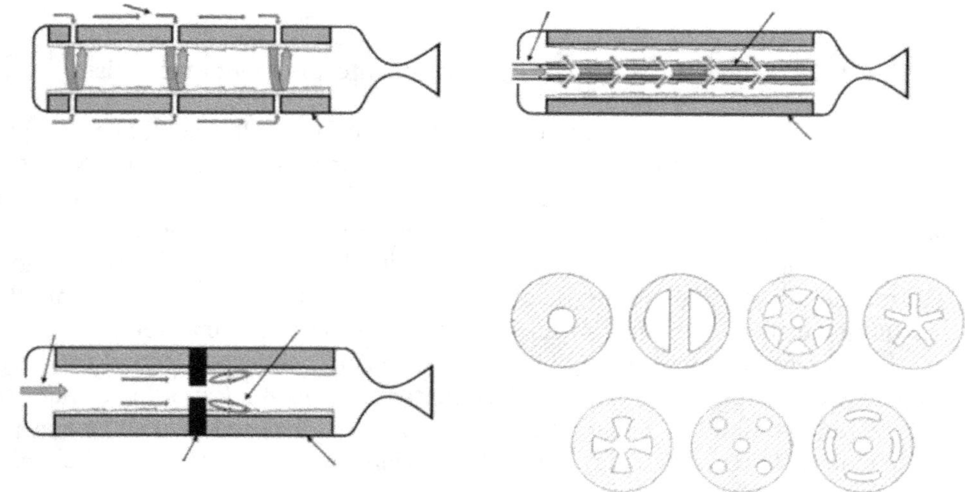

**Figure 8.4** Regression Rate Enhancement Methods (a) multi-section swirl (b) double tube (c) diaphragm and (d) multiport grain (Source: Christopher Glaser et al).

The increasing space accessibility and its applications call for reduction in development, production and overall transport cost with reference to one kilogram of payload. These measures are required for satellite manoeuvring and orbit transfer, space probe and lander propulsion and also for increasing enthusiasm in a very interesting new transport area called space tourism. The main advantages of hybrid rocket engines are: simplicity, safety, stop and restart ability, throttling ability and use of non-toxic storable propellants. There is a revival of interest in hybrid propellant rockets due to the increasing research activities and opportunities arising from additive manufacturing of fuel grains. New players declare ambitious deadlines for orbital operations, while experienced entities including Virgin Galactic and Nammo Space have been developing their hybrid propulsion competences for years to reach altitudes close to the Von Karman Line.

**Table 8.4** Techniques Overviews for Regression Rate Enhancement.

| Name | Description |
|---|---|
| Nested helix | Swirl inducing structure in paraffin |
| Armoured grain | Reinforcing structure in paraffin |
| Aluminium additives | Higher temperature and radiation |
| Embedded metal wires | Promote heat transfer to grain |
| Transition metal additives | Decrease thermal stability |
| Carbon black powder | Opacifying agent |
| Carbon nano-tubes | Increase heat conduction process |
| AP + Ferric oxide | Acts as catalyst |
| GAP | Energetic binder material |
| Conical axial injection | Formation of recirculation zone |
| Hollow cone injector | Bidirectional injection |
| Pressure swirl injector | Annular sheet injection |
| Vortex injection | Rotating injection |
| Swirl injection | Swirl injection head end |
| Multi-section swirl | Swirl injection along grain |
| Double tube | Injector tube inside fuel port |
| Multi-port grain | Increase burning surface |
| Diaphragm | Induce turbulence and mixing |
| Bluff body | Alter recirculation zones |
| 3D printed mould | Complex port geometries |

## Nomenclature

$a, n$    ballistic coefficients

$A_b$    burning surface area (m$^2$)

$A_P$    solid fuel grain port area (m$^2$)

$G$    oxidizer mass flux (kg/m$^2$s),

$\dot{m}_{ox}$    oxidizer mass flow rate (kg/s)

$\dot{r}$    regression rate of fuel grain (mm/s)

$\rho_f$    density of the fuel (kg/m$^3$)

# CHAPTER 9

# Thrust Chamber Cooling

## 9.1 INTRODUCTION

The thrust chamber cooling is a major design consideration due to high combustion temperatures (3000 to 3500 K) and the high heat transfer rates (of the order of 100 MW/m$^2$) from the hot gases to the chamber wall. The fluxes in the rocket thrust chamber are more in nuclear reactor. The thermal management task is the most difficult among all manmade devices. The temperature levels are much higher than the melting temperatures of metal alloys used in engines. Also, the material strength decreases at high temperatures. For short-duration operation (up to a few seconds) uncooled chambers may sometimes be used. In this case, a thick chamber wall will absorb the heat acting as a heat sink, until the wall temperature approaches the failure level. For long duration flight operations, an efficient and steady state cooling system is used to lower the wall temperature of combustion chamber and nozzle. To meet the cooling requirements, several chamber cooling techniques have been utilized. This chapter describes different thrust chamber methods available, but many topics will be introduced in a brief manner. It introduces some of the relevant approaches to solve rocket propulsion heat transfer problems. It is assumed that the reader has fundamental knowledge of heat transfer and analysis methods.

Table 9.1 gives common variables and non-dimensional numbers used in analysis of heat transfer for ready reference. The heat flux, $\dot{q}$ is the important quantity that gives the amount of heat flow per unit area of surface exposed to said heat and it allows to relate temperature at various points in the system. The quantity $\rho c_p$ in thermal diffusivity term $\alpha$, measures the amount of energy the material can store in a certain volume and involves in finding the rate at which wall surfaces heat up when exposed to a given hot gas source. The convective heat transfer coefficient $h$, associated with fluid flow is helpful to find the amount of convective heat transfer that occurs for a given temperature difference. In rocket cooling problems this is the chief heat transfer mechanism. It depends upon flow characteristics of gas or liquid and is related to the nature of boundary layers near heated/cooled walls. The convective heat transfer coefficient, $h$ is expressed in terms of dimensionless number $Nu$. In convective heat transfer problems usually, the Nu is expressed in terms of other dimensionless fluid flow parameters such as Re and Pr. Reynolds number, $Re$ is a key variable for consideration of boundary layers in fluid flow and it is the ratio of inertial and viscous forces. Prandtl number, $Pr$ is the ratio of the thickness of the momentum boundary layer to the thermal boundary layer and hence represents the relative ability of the fluid to transport momentum and heat via molecular collisions in viscous/thermal layers. The following sections describe common approaches for cooling of the thrust chamber.

**Table 9.1** Common Variables and Dimensionless Numbers in Heat Transfer Problems.

| Variable/Non-dimensional number | Variable/Number | Units |
|---|---|---|
| Total heat amount | $Q$ | Joules |
| Heat transfer rate $\dot{Q}$ | $\dot{Q} = dQ/dt$ | Watts |
| Heat flux rate $\dot{q}$ | $\dot{q} = \dot{Q}/A$ | Watts/m$^2$ |
| Thermal diffusivity | $\alpha = k/(\rho c_p)$ | m$^2$/s |
| Heat transfer coefficient $h$ | $h = \dot{q}/\Delta T$ | W/(m$^2$ K) |
| Nusselt number $Nu$ | $Nu = hL/k$ | - |
| Reynolds number $Re$ | $Re = \rho VL/\mu$ | - |
| Prandtl number $Pr$ | $\mu c_p/k$ | - |
| Stanton number $St$ | $St = Nu/(Re \times Pr)$ | - |

## 9.2 REGENERATIVE COOLING

Regenerative cooling is the most widely used method of cooling a combustion chamber. The liquid fuel/oxidizer with high velocity is allowed to pass through the passages placed outside of the combustion chamber and nozzle. The coolant with the heat input from cooling the chamber and nozzle is then discharged into the injector and utilized as a propellant as shown in Figure 9.1. The space industry has acquired a large experience base of regenerative cooling techniques as several rocket engines (like H-1, J-2, F-1 and RS-27) used this method. In tubular wall combustion chamber design, an array of suitably shaped tubes is brazed to the walls of the chamber as shown in Figure 9.2 (a) and (b).  Due to structural integrity this kind of chamber the cooling system is limited to around 10 MPa pressure. The best solution to date is the channel design as shown in Figure 9.2 (c) and (d) where the coolant passes through the rectangular channels machined or formed into a hot gas liner fabricated from a high-conductivity material.

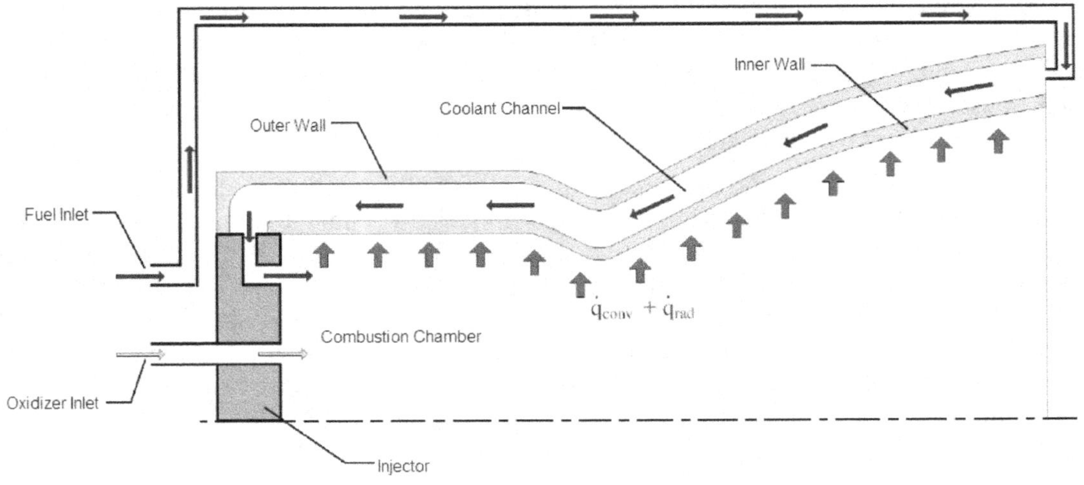


**Figure 9.1** Schematic of Regenerative Cooling System.

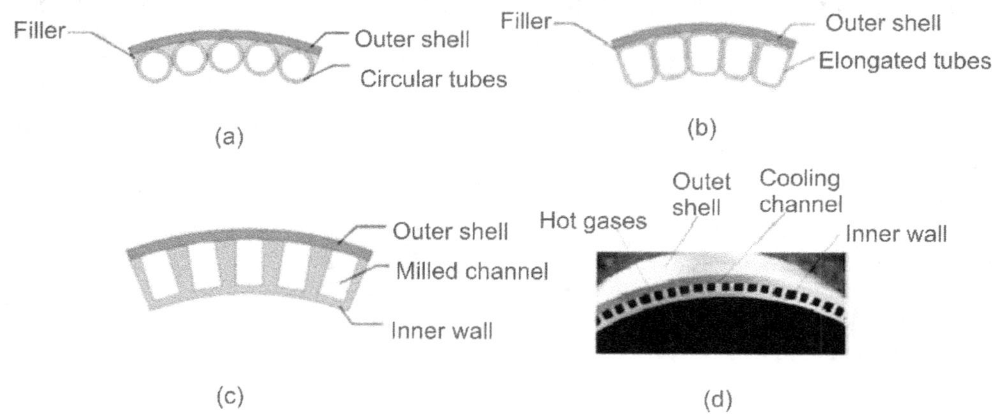


**Figure 9.2** Schematic of cooling jacket sections in regeneratively cooled thrust chambers: (a) circular tubes (b) elongated tubes (c) milled channel (d) typical cross section of milled channels with copper alloy inner liner and applied nickel alloy outer jacket.

To determine the rate of heat transfer from combustion gases to chamber and nozzle walls, the boundary layer thickness along hot gases is required. To estimate the boundary layer thickness is difficult as it depends on many factors like type of flow, fluid, wall curvature, pressure and temperature gradients. In the convergent portion of the nozzle as the flow is accelerating, the boundary layer thickness decreases along the flow direction and attains a minimum value at the throat. It increases marginally in the divergent portion. Hence, the heat transfer is maximum at the throat. To accurately predict the heat transfer, the Navier-Stokes equation along with the energy equation must be solved using CFD (Computational Fluid Dynamics) techniques. The quasi-one-dimensional heat transfer analysis is explained here. From a thermal point of view, regenerative cooling consists of the steady flow of heat from a hot gas through a solid wall to a cool fluid. The combustion heat from gases conducts in the radial direction in the walls and is carried away by the flowing fluid in the channel by convection. This model ignores the heat conducted through the channel ribs. The steady-state model the heat flow at any point in the system must be the same, i.e., the heat conducted through the wall is the same as the heat transferred to coolant by convection.

The hot gas with velocity $V_g$ and temperature $T_g$ is flowing over the inner wall of the thrust chamber. Figure 9.3 shows the one-dimensional model: (a) coolant channel (b) temperature gradients and (c) gas temperature $T_g$ and stagnation temperature $T_{0g}$. The adiabatic wall temperature, $T_{wa}$ is the temperature that would be attained at the wall in case of adiabatic condition. The gas flows in the chamber with high velocity and attains sonic velocity at throat. In such high velocity flow, the driving potential is not based on the free stream temperature $T_g$ but on the adiabatic wall temperature $T_{wa}$ as shown in Figure 9.4. Its deviation from the stagnation temperature of the free stream is evaluated by the recovery factor RF

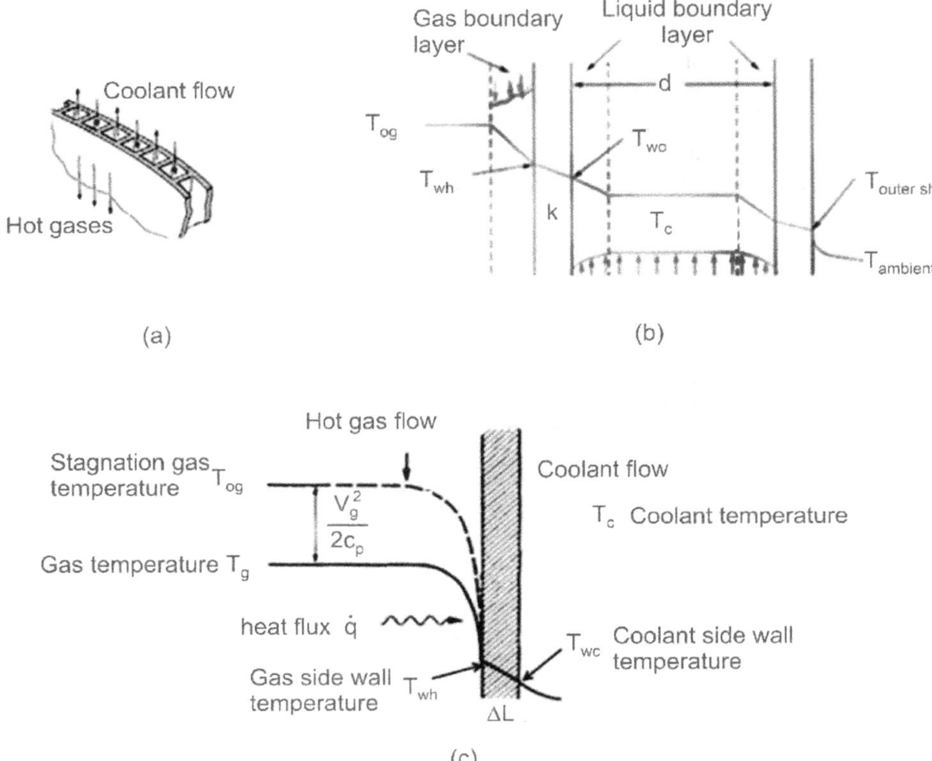



**Figure 9.3** One-dimensional thermal model of regenerative cooling.

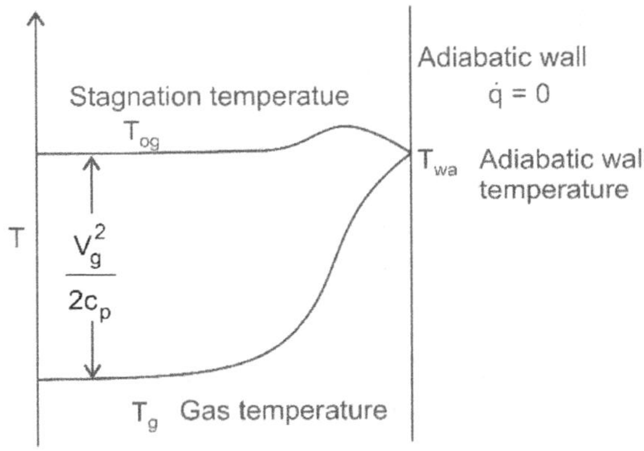



**Figure 9.4** Adiabatic wall or recovery temperature (Source: Hill, P.G. and Peterson, C.R).

$$RF = \frac{T_{wa} - T_g}{T_{0g} - T_g}$$

$T_g$ is the free stream temperature

$T_{0g}$ is the free stream stagnation temperature

$T_{wa}$ is the adiabatic wall temperature

For subsonic flow, RF is equal to 1.0, while for supersonic flow, RF will be less than 1.0. The typical value of R is around 0.9. For turbulent flows it is approximated as:

$$RF = (Pr)^{\frac{1}{3}}$$

where is $Pr$ is Prandtl number.

Assuming the combustion gas obeys the perfect gas laws, the stagnation temperature is:

$$\frac{T_{0g}}{T_g} = 1 + \left(\frac{\gamma - 1}{2}\right) M^2$$

Hence,

$$\frac{T_{wa}}{T_g} = 1 + (RF)\left(\frac{\gamma - 1}{2}\right) M^2$$

Therefore,

$$\frac{T_{wa}}{T_{0g}} = \frac{1 + (RF)\left(\frac{\gamma - 1}{2}\right) M^2}{1 + \left(\frac{\gamma - 1}{2}\right) M^2}$$

For relatively high values of RF (near to one), the adiabatic wall temperature is essentially equal to the free stream stagnation temperature (i.e., the combustion temperature).

### 9.2.1  Heat Transfer Coefficient on Hot Gas Side

Referring to Figure 9.3b, to find the heat transfer coefficient, $h_g$ on the hot gas side, many correlations for the Nusselt number are presented in the literature. These relations are based on the assumption that $Nu \propto Re^{0.8}$. These correlations determine the convective heat transfer coefficient ($h_g$) values within the range of $\pm 20\%$. The fluid properties are evaluated at local bulk temperature or the average value between the wall and the free stream temperature. The general correlation for turbulent flow heat transfer suggested by Colburn is a well-known equation used for estimation of hot gas side convective heat transfer coefficient.

$$\frac{h_g D}{k_0} = 0.026 \left(\frac{\dot{m}D}{\mu_0 A}\right)^{0.8} \left(\frac{\mu c_p}{k}\right)_0^{0.4} \left(\frac{\rho_{am}}{\rho_f}\right) \left(\frac{\mu_{am}}{\mu_0}\right)^{0.2} \tag{9.1}$$

In the above equation the subscript "0" refers to properties evaluated at the combustion temperature and "am" refers to properties evaluated at the average between the wall and the free stream temperature. $\rho_f$ is the free stream local gas density.

Since,

$$\frac{h_g D}{k_0} \propto \left(\frac{\dot{m}D}{\mu_0 A}\right)^{0.8} \propto \left(\frac{\dot{m}}{D}\right)^{0.8} \quad (\,Substituting\ A = \frac{\pi}{4}D^2)$$

Hence,

$$h_g \propto \frac{(\dot{m})^{0.8}}{D^{1.8}}$$

For above it can be concluded that the maximum heat transfer coefficient occurs in the nozzle where the diameter value is the minimum, i.e., at the throat. However, it is observed that the maximum value of $h_g$ occurs slightly upstream of the throat because of the multidimensional flow field. Figure 9.5 shows the variation of $h_g$ for the Space Shuttle main combustion chamber. The value of $h_g$ increases with the increase of mass flow rate of the propellant, i.e., increases with thrust chamber pressure.

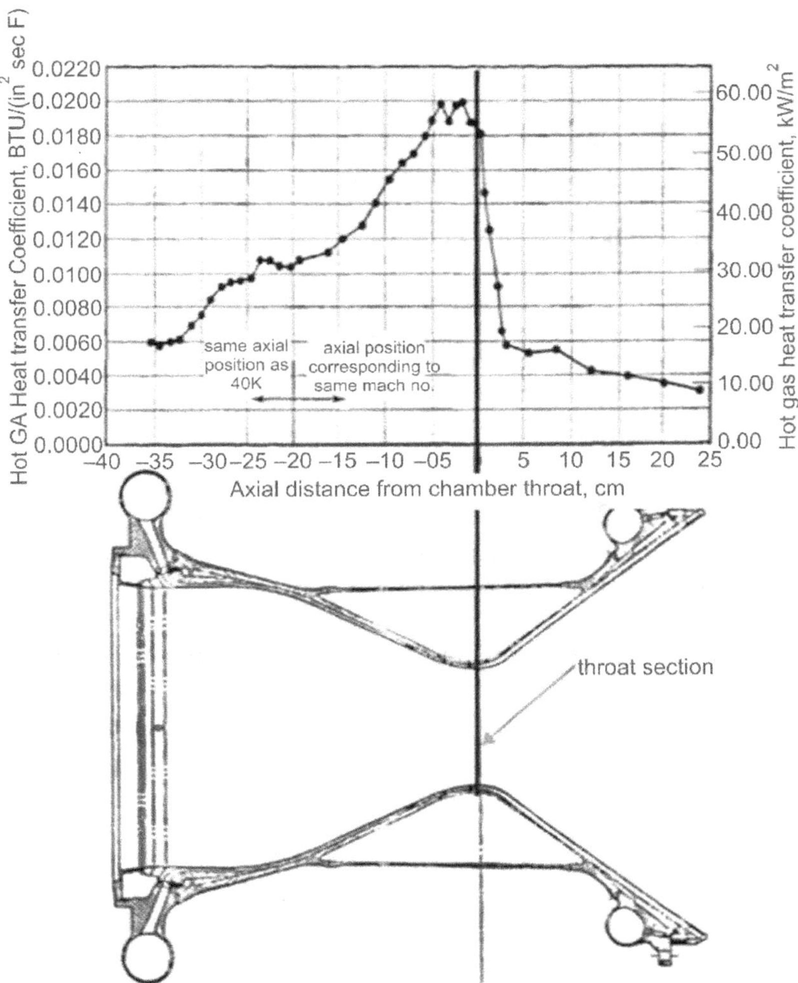


**Figure 9.5** Variation of hot gas side heat transfer coefficient along the nozzle of Main Combustion Chamber (MCC) of Space Shuttle Main Engine (SSME) (Source: Dexter C E et al).

Now, considering one-dimensional steady state flow heat transfer, the convention heat flux rate, $\dot{q}_g$ on hot gas side can be written as

$$\dot{q}_g = h_g(\dot{T}_{wa} - T_{wh}) \qquad \text{.....(9.2)}$$

$T_{wa}$ is the adiabatic wall temperature

$T_{wh}$ is the hot gas side wall temperature

For rocket nozzle application, Bartz developed the following relation for convective heat transfer coefficient on hot gas side modifying Eq.(9.1).

$$h_g = \frac{0.026}{D_t^{0.2}} \left(\frac{\mu^{0.2} c_p}{Pr^{0.6}}\right) \left(\frac{p_c}{c^*}\right)^{0.8} \left(\frac{D_t}{R}\right)^{0.1} \left(\frac{A_t}{A}\right)^{0.9} \lambda \qquad \text{.....(9.3)}$$

where $\lambda$ is the correction factor for property variations across the boundary layer, $R$ is the nozzle radius of curvature at the throat and $A$ is the cross-sectional area along the nozzle axis.

The dimensionless factor $\lambda$ is expressed in terms of local wall temperature of hot gas side $T_{wh}$, stagnation temperature $T_{0g}$ and local Mach number as follows:

$$\lambda = \left[0.5 \frac{T_{wh}}{T_{0g}} \left(1 + \frac{\gamma-1}{2} M^2\right) + 0.5\right]^{-0.68} \left(1 + \frac{\gamma-1}{2} M^2\right)^{-0.12} \qquad \text{.....(9.4)}$$

Prandtl number and viscosity $\mu$ was approximately calculated by:

$$Pr = \frac{4\gamma}{9\gamma - 5}$$

$$\mu = 1.184 \times 10^{-6} \mathcal{M}^{0.5} T^{0.6}$$

where $T$ is the temperature hot gas mixture and $\mathcal{M}$ is molecular mass. Bartz computed the value of $\lambda$ as a function of $T_{wh}/T_{0g}$ and $\gamma$ and presented it as shown in Figure 9.6.

The calculated value of $h_g$ from above equation would be lower if the dissociation of and subsequent association of combustion gases components occur near the wall, the gases contain high radiant components and due to high frequency flow instabilities. On the other hand, the calculated values would be higher because of incomplete combustion and carbon deposits on the chamber walls.

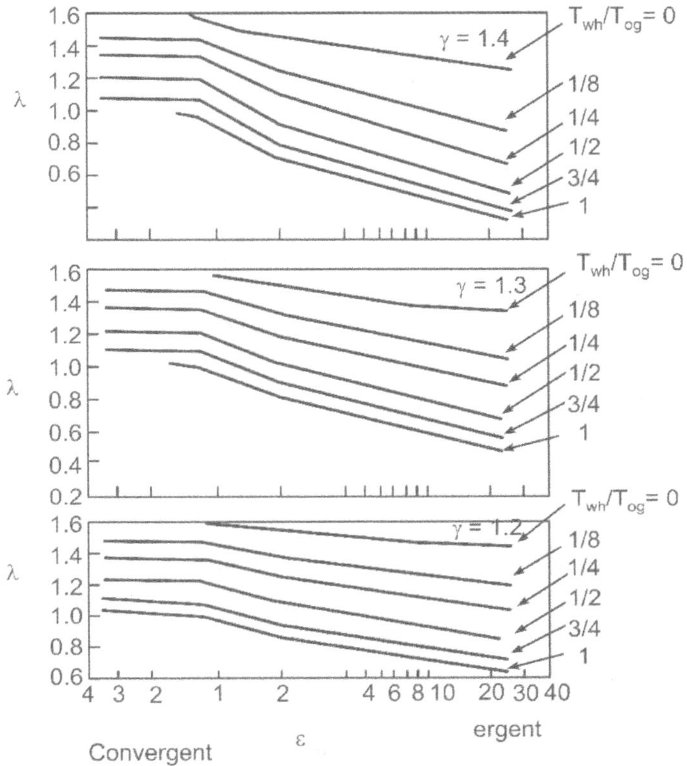



**Figure 9.6** Bartz equation correction factor ($\lambda$) for different $T_{wh}/T_{0g}$ and $\gamma$ ratios at axial location of the nozzle $\varepsilon$, the ratio of the local area of cross-section to the throat area (Source: Dieter K Huzel and David H Huang).

### 9.2.2    Radiation Heat Transfer

The hot gases in the thrust chamber also radiate energy. This radiation energy augments the heat transfer rate to the thrust chamber walls. Combustion gases like $CO_2$, $H_2O$ and $CO$ emits energy over discrete bands of wavelengths (infrared region). In general, polyatomic gases radiate more strongly than diatomic gases. Also, these gases absorb radiation as they radiate it. If the gases contain solid or liquid particles, these may appreciably contribute to the emitted radiant energy. These particles radiate over a continuous spectrum of the wavelengths.

The chamber wall is assumed to absorb all the incident radiation on it. The reradiation of the wall is negligible as its temperature is much less than the combustion gases. As the radiation heat flux is proportional to $T^4$, it is more appreciable in the combustion chamber than in nozzle. Generally, radiative heat transfer in rocket thrust chamber is modest and it forms 5% to 30% of the total heat transfer. The radiative heat flux can be written as:

$$\dot{q}_r = \epsilon_g \sigma T_g^4 \qquad\qquad\qquad .....(9.5)$$

$\epsilon_g$ is the gas emissivity (<1)

$\sigma$ is the Stefan Boltzmann constant ($= 5.6697 \times 10^{-8}$ W/m² K⁴)

### 9.2.3    Coolant Side Heat Transfer Coefficient

As mentioned, the coolant flows through several channels in either circular or different rectangular shapes. For non-circular cross-section, the hydraulic diameter ($d_h = 4 \times Perimeter/Area\ of\ cross\ section$) is used in the calculation of heat transfer. To evaluate the non-dimensional heat transfer numbers (*Re, Pr and Nu*), the fluid properties like viscosity, specific heat, density and thermal conductivity are used at average fluid temperature. Many empirical relations are available in literature with an uncertainty range of around 20%. The coolant heat transfer coefficient can be calculated using the Dittus-Boelter relation as given below.

$$\frac{h_L d}{k_L} = 0.023\,(Re)^{0.8}(Pr)^{0.33} = 0.023 \left(\frac{\rho V d}{\mu}\right)_L^{0.8} \left(\frac{\mu c_p}{k}\right)_l^{0.33} \qquad \dots\dots(9.6)$$

Here, $V$ is the velocity of coolant in the pipe. The coolants function under different thermodynamic regimes that influence the cooling performance. Strong variations in properties occur near the critical point; for example, specific heat values tend to peak in a narrow region near the critical temperature. Table 9.2 gives the values of critical pressure, $p_c$ and temperature $T_c$ for selected coolants. Figure 9.7 shows the typical phase diagram and heat transfer regimes against $T_{wc} - T_L$. Depending on the critical pressure, $p_c$ and boiling temperature, $T_b$ the different heat transfers regimes can be recognized.

(i)   $p < p_c$ and $T_{wc} < T_b$ or $p > p_c$ : Forced convection and it is a single-phase heat transfer.

(ii)   $p < p_c$ and $T_{wc} - T_b \leq 50K$ : Nucleate boiling. Local turbulence increases and convective heat transfer coefficient values increase due to the small vapor bubbles.

(iii) $p < p_c$ and $T_{wc} - T_b \geq 50K$ : Film boiling.

To achieve a good heat-absorbing capacity of the coolant, the coolant flow velocity is selected so that boiling is permitted locally at the wall but the bulk of the coolant does not reach boiling condition.

**Table 9.2** Critical pressure and temperature for typical propellants.

| Propellant | Critical Pressure | Critical Temperature |
|---|---|---|
| | $(p_c)$ bar | $(T_c)$ K |
| Hydrazine | 147 | 652 |
| Methane | 46 | 190 |
| Kerosene | 20 | 678 |
| Hydrogen | 13 | 33 |
| Oxygen | 50 | 155 |

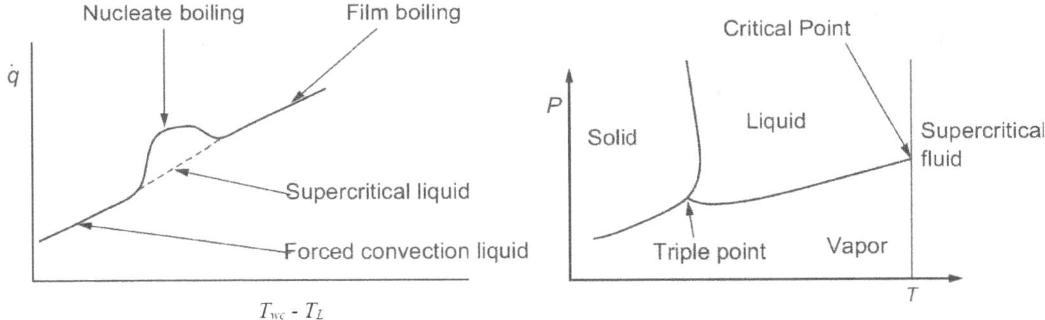


**Figure 9.7** Heat Transfer Regimes in Coolant.

The coolant pressure drop must be properly regulated because a higher pressure drop allows a higher coolant velocity in the cooling channel (and thus a better cooling). But it lowers the engine specific impulse and requires a heavier feed system that increases the engine mass. The pressure drops $\Delta p$ in a channel of length $l$, diameter $d$ can be evaluated as follows:

$$\Delta p = \frac{f l V^2 \rho}{2d} \qquad \qquad .....(9.7)$$

where $f$ is the friction factor and $\rho$ is the density of the coolant. The friction factor is a function of the Reynolds number and the relative roughness of the wall. The relative roughness is a dimensionless ratio of the wall surface height variation to the passage diameter. The narrow passages of coolant and manufacturing methods like laser melting/sintering produce high roughness. The cryogenic coolants create at critical temperature compressibility effects as the Mach number becomes appreciable as the fluid is heated effecting the values of friction factors. The good starting point for the hydraulic design is to use Moody diagram for selecting the friction factor to estimate the pressure drop the cooling channel. The convective heat flux at the coolant side can be written as

$$\dot{q} = h_L(T_{wc} - T_c) \qquad \qquad .....(9.8)$$

### 9.2.4  Chamber Wall Heat Transfer

As shown in Figure 9.3b, the heat transfer in the chamber wall of thickness $\Delta L$ is due to conduction. The heat flux rate due to conduction across the wall can be expressed as

$$\dot{q}_{wl} = \frac{k_w}{\Delta L}(T_{wh} - T_{wc}) \qquad \ldots\ldots(9.9)$$

where

$k_w$: thermal conductivity of the wall material

$\Delta L$: wall thickness

$T_{wh}$: is the hot gas side wall temperature

$T_{wc}$: coolant side wall temperature

### 9.2.5  Coupled Heat Transfer in the Thrust Chamber

By considering steady one-dimensional heat transfer model for regenerative cooling, heat balance across combustion chamber is:

$$(Convective\ hot\ gas\ heat\ flux + hot\ gas\ radiative\ heat\ flus)$$
$$= chamber\ wall\ conductive\ heat\ flux$$
$$= coolant\ side\ convective\ heat\ flux$$

i.e.,

$$\dot{q} = \dot{q}_g + \dot{q}_r = \dot{q}_w = \dot{q}_L \qquad \ldots\ldots(9.10)$$

$$\left[h_g(T_{wa} - T_{wh}) + \epsilon_g \sigma T_g^4\right] = \frac{k_w}{\Delta L}(T_{wh} - T_{wc}) = h_L(T_{wc} - T_L)$$

It can be observed from above equation, if the radiative heat transfer contribution is low and is neglected, then $\dot{q}_g \sim \dot{q}_L$. Hence, $h_g \ll h_L$ because the thermal potential $(T_{wa} - T_{wh}) \gg (T_{wc} - T_L)$. In order to keep the hot gas side temperature at lower side, it is desirable to achieve lower value of $h_g$ and higher values of $h_L$ and $k_w/\Delta L$ within the permissible limits of mechanical design (wall thermal stresses) and coolant properties (vaporization or decomposition). However, to increase the heat transfer rate to the coolant as in case of expander cycle, the hot side heat transfer coefficient is kept at higher side. This can be achieved by using higher chamber pressure and increasing wall roughness. The heat transfer rate also increased by providing longitudinal ribs (i.e., increasing the effective heat transfer area) as shown in Figure 9.8 or with a longer cylindrical part of the combustion chamber. To increase the value of coolant side heat transfer coefficient $h_L$, increase the coolant velocity, tube wall roughness and use high-aspect ratio cooling channels. High aspect ratio channels increase the number of channels and fin length and both effects will increase the cooling heat transfer area. This could result in an excessive coolant pressure drop. To increase the wall conduction heat transfer, it is desirable to use high conductivity material and thin structure.

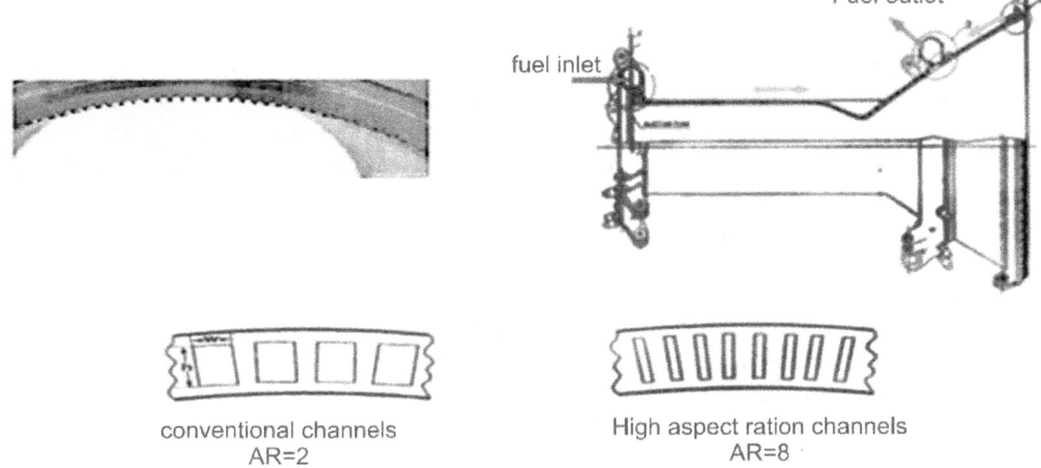


**Figure 9.8** Methods to increase the effective heat transfer area of the hot gas side.

The adiabatic wall temperature $T_{wa}$ will increase due to the radiation heat transfer of the gas. The enhanced adiabatic wall temperature $T_{wa}^r$ can be expressed as

$$T_{wa}^r = T_{wa} + \frac{q_r}{h_g} \qquad (9.11)$$

The heat flux through the wall is

$$\dot{q} = h(T_{wa}^r - T_L) \qquad (9.12)$$

where $h$ is the overall heat transfer coefficient given by

$$\frac{1}{h} = \frac{1}{h_g} + \frac{1}{\left(\frac{k_w}{\Delta L}\right)} + \frac{1}{h_L} \qquad (9.13)$$

In the above analysis of heat transfer in the thrust chamber one dimensional model is assumed. However, the heat transfer in the thrust chamber is complex and multidimensional. Heat is transferred also through the ribs that separate the coolant passages and act as cooling fins. Figure 9.9 shows the direction of heat flux and isotherms in cooling channel ribs. The above one-dimensional steady analysis can be extended for complex configuration by considering the effective area on both the hot gas and coolant side, which can be higher than the actual area. Equation (9.8) can be modified as

$$(\dot{q}_g + \dot{q}_r)A_g = \dot{q}_w A_w = \dot{q}_L A_L$$

where

$A_g$ is gas side area

$A_w$ is effective wall area

$A_L$ is effective coolant side area

The combustion process is more energetic than the convective heat transfer. Hence, the coolant can absorb a few percentages of the heat of combustion. The design of the cooling system is relatively easier in large thrust engines than small engines. The heat energy absorbed by the coolant is not wasted. In the expander cycle engine, it is used to power the turbine. If the coolant is directly injected in the combustion chamber, it augments the initial energy content of the propellant prior to injection increasing marginally the specific impulse ($\sim 1\%$). The maximum permissible coolant temperature depends upon its properties. In case of kerosene, carbon deposition on the channel wall limits its coolant maximum temperature and the safety consideration limits maximum temperature for hydrazine. Kerosene coking occurs at temperatures larger than 450 K, while pure methane does not present this problem when operated as rocket coolant. Hydrazine ($N_2H_4$) and mono-methyl-hydrazine (MMH) can exothermically decompose at temperatures as low as 370 K. Unsymmetrical-di-methyl-hydrazine (UDMH) decomposes at temperature 490 K. Under some conditions these decompositions can be a violent detonation. Hydrogen, having a high specific heat and is chemically stable, is an excellent coolant. It leaves no residues.

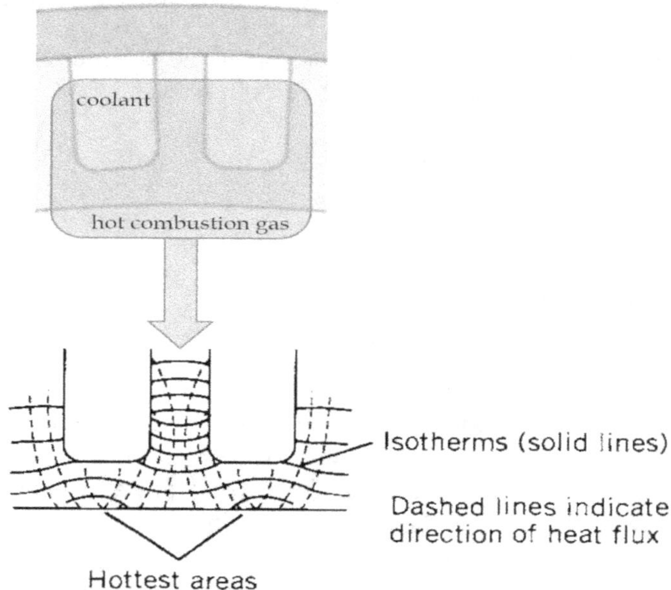


**Figure 9.9** Heat transfer in cooling channel ribs. (Source: Cornelisse J.W et al).

Some of the studies (NASA-CR-165621) carried on 90% bell shape nozzles using fuel as coolant are summarized below:

Coolant inlet temperature:

Liquid hydrogen 21 K

Kerosene RP-1 298 K

Liquid Methane 112 K

**Regenerative coolant discharge or film coolant inlet pressure:**

    Liquid    1.176 times chamber pressure (minimum)

    Gas      1.087 times chamber pressure (minimum)

**Maximum coolant velocity (Regenerative):**

    Liquid    61 m/s

    Gas      Mach 0.3

Coking Limit:

    RP-1    561 K

    $LCH_4$    978 K

**Dimensional Limits:**

    Tubular construction

    Minimum wall thickness    0.25 mm

    Non tubular construction

    Minimum slot width    0.76 mm

    Minimum slot depth/width    4 to 1

    Minimum web thickness    0.76 mm

    Minimum wall thickness    0.64 mm

**Example 9.1**

A rocket engine with $LOX/LH_2$ propellant combination has the following specifications:

Thrust $F$ :             660 kN

Chamber pressure:        6.0 MPa

Mixture ratio:           5

Characteristic velocity $c^*$:    2279 m/s

Thrust coefficient $C_F$:    1.895

Density of LOX:          $1140 \ kg/m^3$

Density of Gaseous hydrogen : $12 \ kg/m^3$ (at 100 K)

Stagnation temperature:      3200 K

Molecular mass:          5.44 kg/mole

Calculate the hot gas side and cooling liquid side heat transfer coefficients.

Refer to the example 7.2 in chapter 7.

Throat diameter 275 mm

The average nozzle radius of the throat section = (1.5+0.382) /2 x 275/2 = 130 mm

Molecular mass $= 12$ kg/mole

Given stagnation temperature $= 3200$ K

Due to friction effects, heat transfer, nonaxial flow, nonuniform flow distribution and nonperfect gas behaviour the real stagnation temperature is less than the ideal stagnation temperature.

Take stagnation recovery factor $= 0.91$

$T_{aw} = 3200 \times 0.91 = 2912$ K. Take it as 2900 K

$$c_p = \frac{\gamma R_u}{M(\gamma - 1)} = \frac{1.21 \times 8.3144}{12 \times 0.21} = 3.992 \frac{kJ}{kg\ K}$$

$$Pr = \frac{4\gamma}{9\gamma - 5} = \frac{4 \times 1.21}{9 \times 1.21 - 5} = 0.822$$

$$\mu = 1.184 \times 10^{-6} M^{0.5} T^{0.6} = 1.184 \times 10^{-6} \times 12^{0.5} \times 3200^{0.6} = 520 \times 10^{-6} \frac{kg}{m\ s}$$

$$h_g = \frac{0.026}{D_t^{0.2}} \left( \frac{\mu^{0.2} c_p}{Pr^{0.6}} \right) \left( \frac{p_c}{c^*} \right)^{0.8} \left( \frac{D_t}{R} \right)^{0.1} \left( \frac{A_t}{A} \right)^{0.9} \lambda$$

$$h_g = \frac{0.026}{(275 \times 10^{-3})^{0.2}} \frac{(520 \times 10^{-6})^{0.2} \times 3.992 \times 10^3}{0.822^{0.6}} \left( \frac{6 \times 10^6}{2279} \right)^{0.8} \left( \frac{275}{130} \right)^{0.1} \left( \frac{A_t}{A} \right)^{0.9} \lambda$$

$$h_g = 0.0337 \times 990.4 \times 544.9 \times 1.078 \times \left( \frac{A_t}{A} \right)^{0.9} \lambda$$

$$h_g = 19.57 \times 10^3 \times \left( \frac{A_t}{A} \right)^{0.9} \lambda$$

Assume $T_{wh}$ as 900° K

The ratio

$$\frac{T_{wh}}{T_c} = \frac{1000}{3200} = 0.31$$

At the combustion chamber

$$\left( \frac{A_t}{A} \right)^{0.9} = \left( \frac{1}{1.6} \right)^{0.9} = 0.655$$

The value of $\lambda$ from Figure

$$\sigma = 1.30$$

$$h_g = 19.57 \times 10^3 \times 0.655 \times 1.30 = 16.66 \times 10^3\ W/(m^2\ K)$$

At throat the area ratio $A_t/A$ is equal to 1. The value of $\lambda = 1.33$ from

$$h_g = 19.57 \times 10^3 \times 1 \times 1.33 = 26.02 \times 10^3\ \frac{W}{m^2 K}$$

The adiabatic temperature = 2900 K

The heat flux at throat

$$\dot{q}_g = h_g(T_{aw} - T_{wh}) = 26.02 \times 10^3 \times (2900 - 1000) = 49440 \times 10^3 \; W/m^2$$

Select Chromium nickel alloy (14-17% Cr) ASTM B637 for construction of the nozzle

Tensile strength 1325 MPa

Thermal conductivity   20 w/m K

$$T_{wc} = T_{wh} - \frac{\dot{q}_g \Delta L}{k_w} = 1000 - \frac{49440 \times 10^3 \times .25 \times 10^{-3}}{20} = 382 \; K$$

The average value of the tube

$$T_{avtube} = \frac{1000 + 382}{2} = 691 \; K$$

The above temperature is a safe wall temperature of the tube.

Assuming the bulk temperature of coolant 75 K

$$h_L = \frac{\dot{q}}{T_{wc} - T_L} = \frac{49440 \times 10^3}{382 - 75} = 161 \times 10^3 \; \frac{W}{m^2 K}$$

## 9.3   FILM COOLING

Film cooling, another efficient method of chamber cooling, in combination with regenerative cooling is a very effective method to reduce the wall thermal stresses. This is an important consideration because thermal stresses may establish the feasibility limits of conventional regenerative cooling. The coolant, fuel or fuel-rich propellants far from stoichiometric condition, is injected through orifices around the injector periphery or through manifold orifices in the chamber wall near the injector and sometimes in several more planes toward the throat as shown in Figure 9.10. A thin film of coolant formed on the inner side wall of the chamber acts as an insulating layer. Flow of the coolant along the walls absorbs the heat transferred from the hot combustion gases through sensible and latent heat. The film layer travels downstream toward the nozzle and beyond and eventually vanishes.

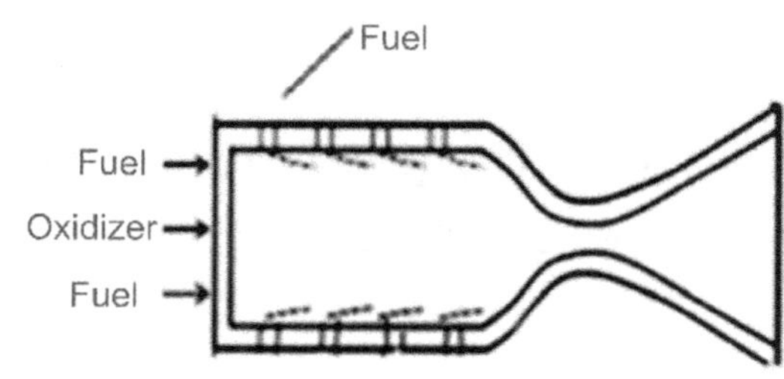


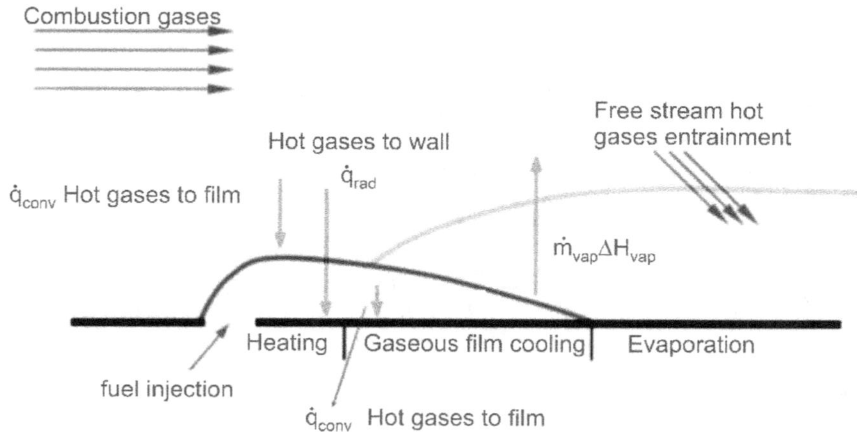


**Figure 9.10** Film cooling process.

Liquid film cooling has an advantage over gaseous film cooling because the utilization of the latent heat of vaporization of a liquid significantly improves the cooling method. In the case where a liquid film layer exists, the temperature of the wall would be at maximum equal to the vapor saturation temperature of the coolant. As shown in Figure 9.10 the combustion gases increase the sensible enthalpy of the film by radiation, convection and conduction. When saturation temperature is reached, heat is used to vaporize the coolant. This is an effective way to protect the wall of the thrust chamber from hot gases by directing the cooling air into the boundary layer to provide a protective cool film along the surface. Typically, film-coolant flow is approximately 3 to 10% of the propellant. Fuel rich propellant is preferred to oxidizer rich since they give lower temperatures. Also, the oxidizer-rich gases lead to oxidation of the metal walls and to be avoided.

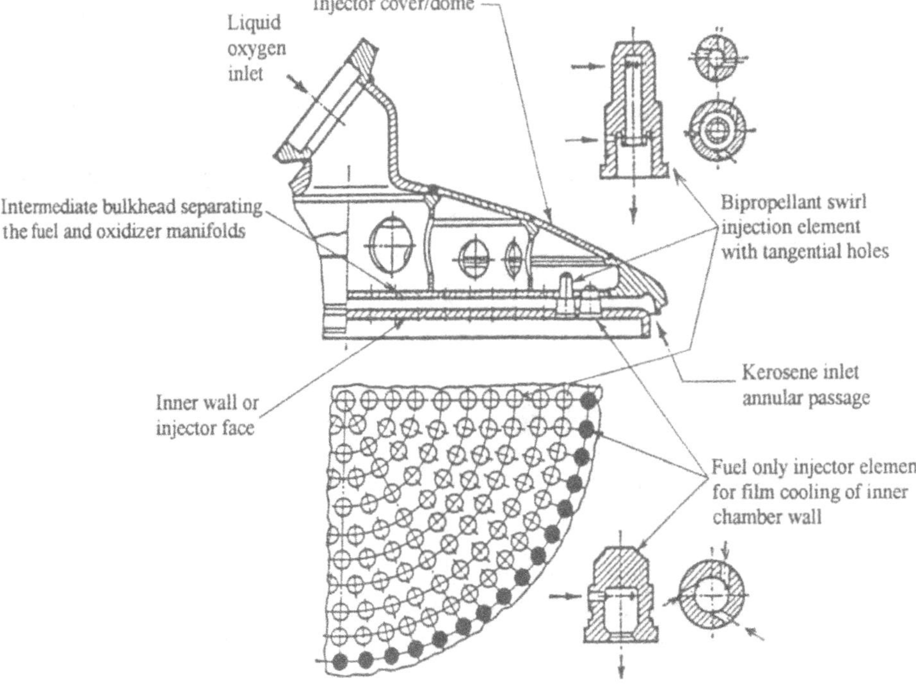


**Figure 9.11** RD 107 Rocket Engine Injector (Source: George P Sutton 2005).

Figure 9.11 shows the RD 107 engine injector details. There are 337 injection elements arranged in 10 rings. The outer ring spray only fuel on the inner chamber wall for film cooling and keep the wall temperature below 650 K. Proper injection of liquid coolant is required to provide uniform coverage and to minimize liquid entrainment by the combustion products. In some injector designs small orifices are provided along the outer ring of injector near to the chamber wall. It is observed by experience that the orifices diameter around 0.25 to 0.50 mm and maximum center-to-center orifices spacing of 6 to 12 mm will provide a sufficiently uniform film-coolant coverage.

## 9.4   ABLATIVE COOLING

In ablative cooling systems, insulating material that melts and vaporizes at high temperature (thermal erosion) is used on the combustion gas side walls of the thrust chamber. The thermal erosion is called ablation. Due to phase change of sacrificial liner, the temperature of the material remains constant and keeps the chamber wall at much lower temperature than the combustion gases. Ablative liners are lightweight, relatively simple to implement, and inexpensive. Also, being a good thermal insulator allows a low rate of heat transfer to the outer structure. Ablative materials can be categorized into two types, pyrolyzing and non-pyrolyzing. They are made of resins such as phenolic resins, epoxies and unsaturated polyesters and strengthened using carbon and silica fibers. They are also known as composite materials. The carbon phenolic composites resin erodes and undergoes an endothermic degradation (pyrolysis) when exposed to high velocity and temperature combustion gases. This causes material to char. Similarly, the silica phenolic composites initially form a melt layer of silica. Non-pyrolyzing materials, such as carbon and graphite, melt or sublimate causing thermochemical erosion at the surface. The charring and melting temperatures of ablative materials (around 1400 K) are low enough to keep the temperature of the chamber wall much below melting point. Of these two composite materials, the silica phenolic ablative is a better insulator compared to carbon phenolic, since silica has lower thermal conductivity than carbon. Also, it can also withstand oxidation unlike the carbon phenolic composite. At the throat, the heat flux is highest and to avoid the erosion and change of throat diameter, graphite (carbon) and sintered metals are used in the throat region.

Figure 9.12 typical experimental results that shows the effect of peripheral-zone combustion temperature on ablative throat erosion for a rocket engine nozzle with throat diameter 76,2 mm and operating at a chamber pressure of 690 kPa with propellants $N_2O_4$/UDMH. It can be observed that the ablative erosion rate decreases almost linearly, and the onset of erosion is delayed at the lower combustion temperatures. Erosion is also driven by the chamber pressure and gas velocity at the throat, which influence the heat transfer coefficient and heat flux. The use of ablative cooling method is aptly fit in the ever-increasing interest to develop low-cost launches of small satellites to low Earth orbit. To meet this demand, one approach includes the utilization of relatively inexpensive propellants such as RP-1 fuel and liquid oxygen (LOX) in a low chamber pressure, pressure-fed engine, with an uncooled combustion chamber and nozzle. This eliminates complex high-pressure turbopumps, and avoidance of cryogenic fuels with insulated storage tanks and transfer lines. Thus, simplifies the entire system, increasing reliability and lowering costs.

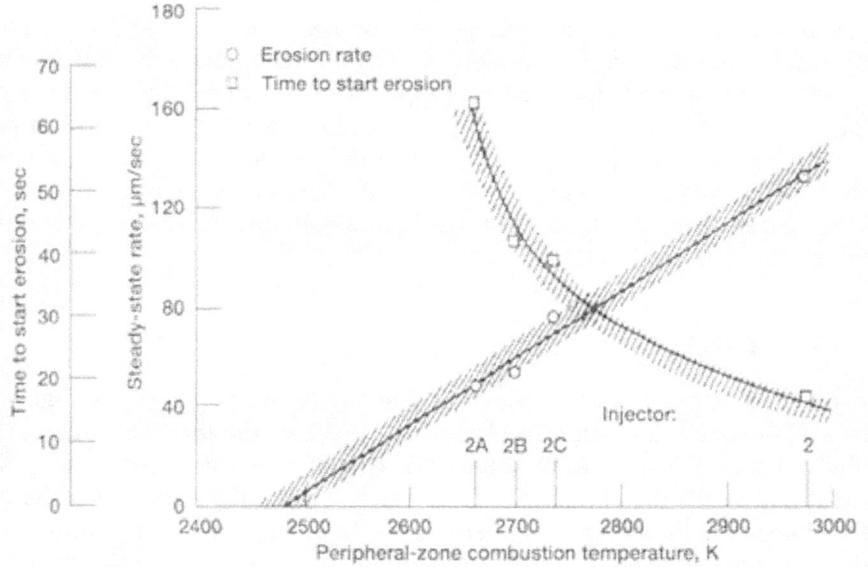


**Figure 9.12** Effect of peripheral-zone combustion temperature on ablative throat erosion (Source: Winter, Jerry M et al).

## 9.5    RADIATION COOLING

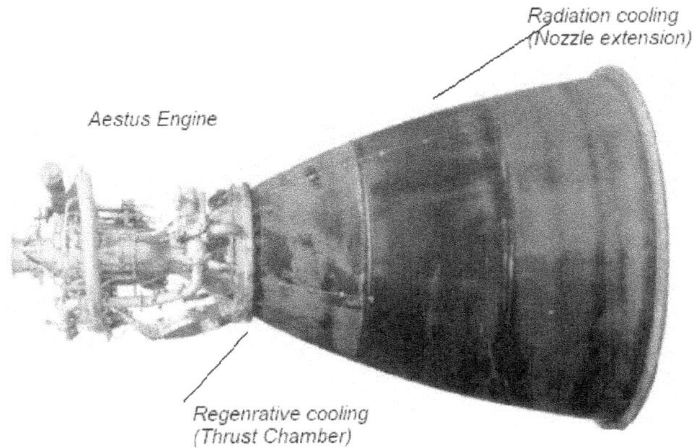


**Figure 9.13**  Radiatively cooled Aestus Engine Nozzle.

The heated surface radiates heat to the surroundings. Radiation heat transfer effectively occurs in a vacuum and when the surroundings are relatively much cooler. Materials, such as, columbium and carbon-carbon are used for radiation-cooled liquid propellant rockets. The method is useful for nozzle cooling of upper stage engines and space transfer vehicles. The radiatively cooled nozzle sections can rapidly reach thermal equilibrium, thus can be used for unlimited duration. Figure 9.13 shows the Aestus engine (Ariane 5 upper stage) nozzle using radiation cooling. The Space Shuttle small thrusters used radiation cooling with refractory metals.

## Nomenclature

$\dot{Q}$      heat transfer rate (W)

$\dot{m}$      mass flow rate (kg/s)

$\dot{q}$      heat flux (W/m$^2$)

A      cross sectional area (m$^2$)

$c^*$      characteristic velocity (m/s)

$c_p$      specific heat at constant pressure (J/mole K)

D,d      diameter (mm)

$f$      flow friction factor

$h$      convective heat transfer coefficient (W/m$^2$ K)

$l$      length of the pipe (m)

$M$      Mach number

$Nu$      Nusselt number

$p_c$      chamber pressure (MPa)

$Pr$      Prandtl number

$R$      radius of nozzle near throat (mm)

$Re$      Reynolds number

$St$      Stanton number

$T_{0g}$      stagnation temperature of hot gases (K)

$T_g$      free stream temperature combustion gases (K)

$T_{wa}$      adiabatic wall temperature (K)

$T_{wh}$      hot gas side wall temperature (K)

$V_g$      velocity of combustion gases in thrust chamber (m/s)

$\gamma$      ratio of specific heats

$\rho$      density (kg/m$^3$)

$T_{wc}$      coolant side wall temperature (K)

$k_w$      thermal conductivity of the chamber wall material (W/m K)

$\mu$      dynamic viscosity of liquid (Pa s)

$\sigma$      Stefan Boltzmann constant ($= 5.6697 \times 10^{-8}$ W/m$^2$ K$^4$)

$\epsilon_g$      gas emissivity

# Combustion Instabilities

## 10.1 INTRODUCTION

Although combustion instability can occur in many thermal devices when unsteady release of heat is coupled with pressure oscillations, it is particularly challenging for rocket engine combustors because of their exceedingly high energy release rates and power density of the order of $10^3$ kW/cm$^3$ in the combustion chamber. The combustion instability is due to the presence of self-sustaining pressure oscillations in the combustion chamber. The low acoustic damping and thin thermo-structural design margins also aggregate the problem in rocket vehicles. The energy needed by these oscillations is a very small part of the chemical energy made available by the propellants. Hence, the average or steady state thrust of the engine is not affected by the oscillations. However, the oscillating pressure in the combustion chamber and thrust oscillations can excite the structural vibrations to hazardous levels. Also, the heat transfer rates between the hot gas and the internal surface of the combustion chamber can be highly increased causing the damage of the injector surface and chamber wall. The severity of the problem can well be understood by the efforts devoted by the Project 1 team of NASA for more than 4 years to solve combustion instabilities encountered during the F-1 engine development used in Saturn V vehicle in manned Apollo mission. It was fixed by a laborious and expensive process of trial and error, taking over 2000 tests and 108 different injector configurations to finally arrive with a stable design. The present design and practices to suppress the instabilities in rocket engines have their source more this vast knowledge gained during that program. In Europe, after combustion instability caused a failure of a Viking engine a comprehensive study was started at French Aerospace Lab ONERA. Significant amount of research and development work was conducted in the Soviet Union to understand and suppress the combustion instabilities in their launch vehicles. Thus, the instabilities in rocket engines, particularly in liquid rocket engines, the spontaneous excitation of high frequency combustion instabilities, has been a design issue to overcome throughout their development history. Mathematical treatment of rocket combustion instabilities is not in the scope of this book. The physical situation that creates the instabilities in the operation of the rocket engine and the methods used to suppress them for smooth operation are discussed briefly.

## 10.2  WAVE MOTION

Combustion instabilities occur in different forms and arise from the interaction between the acoustic, combustion and hydrodynamic processes in the engine. Also, system induced behaviors such as feed system and structure interactions can also result in combustion instability. The combustion chamber of LRE is designed for steady state operation. However, an unstable behavior may occur as a result of small perturbations caused due to sudden change in pressure, heat rate, propellant flow rate or temperature that cause self-sustaining oscillations of pressure in the burning gas. These perturbations travel in the chamber in wave form. The frequencies of these oscillations vary in a wide range. Low-frequency instabilities are present at frequencies lower than the acoustic modes of the chamber and high-frequency instabilities are usually associated with the acoustic modes of the combustion chamber.

The rocket combustion chamber may be considered as a closed chamber having certain complicating features like a through-flow, two-phase flow with combustion and droplet resistance. One end is the injector and the other end is the nozzle throat. The rapid acceleration of the gasses at the throat region of the nozzle causes large density gradients and acts as if it were a closed end. The walls and injector face of the combustion chamber may be considered as acoustic absorption devices. Any disturbance generated in a cavity at some local zone, such as the flame zone, will travel in the cavity. Figure 10.1 show a sinusoidal pressure wave travelling along the axial length $x$ of a cavity with velocity $a$ having the maximum pressure amplitude, $A$ and wavelength $\lambda$. In a time t, the wave moves a distance $at$ and the wave position is shown in dotted position. The pressure fluctuations in the travelling wave can be represented by the equation

$$p' = A \sin \frac{2\pi}{\lambda} (x - at) \qquad\qquad .....(10.1)$$

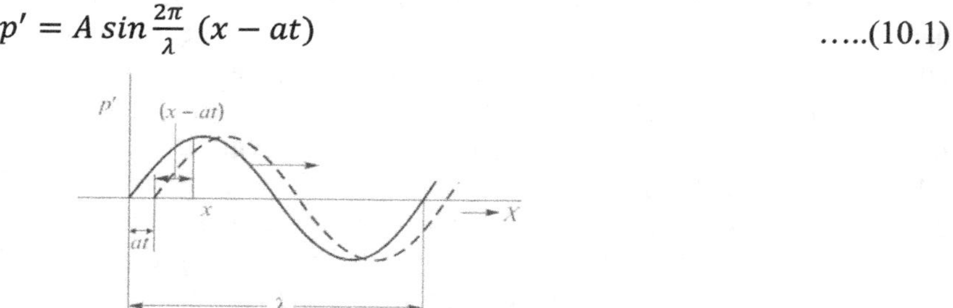


**Figure 10.1**  Sinusoidal Travelling Pressure Wave

The period, i.e., the time taken to travel one wavelength $\lambda$ is

$$T = \frac{\lambda}{a}$$

The frequency, $f$ of the wave is

$$f = \frac{1}{T} = \frac{a}{\lambda} \qquad\qquad .....(10.2)$$

The circular frequency, $\omega = 2\pi f$. Therefore, the wave speed can be expressed as

$$a = \frac{\lambda}{2\pi} \omega \qquad\qquad .....(10.3)$$

The wave number is defined as

$$K = \frac{2\pi}{\lambda}$$ .....(10.4)

The wave reflects from the nozzle throat as it acts as a solid body. Consider that the wave reflects without any attenuation, then the equation of the reflected wave is given by

$$p' = A \, sin\frac{2\pi}{\lambda} \, (-x - at), i.\,e.,\, p' = -A \, sin\frac{2\pi}{\lambda} \, (x + at)$$ .....(10.5)

The interaction of the incident and reflected wave gives the resulting wave and its equation is given by adding Eq. (10.1) & (10.5):

$$p' = A \, sin\frac{2\pi}{\lambda} \, (x - at) - A \, sin\frac{2\pi}{\lambda} \, (x + at)$$

Using Eq. (9.2) & (9.3) and simplifying

$$p' = A \, sin(Kx - \omega t) - A \, sin(Kx + \omega t)$$

$$p' = -2A \, cos \, Kx \, sin \, \omega t$$ .....(10.6)

The above equation represents the resulting wave and is stationary as there is no travelling component $at$.

If $cos \, Kx = 1$, the pressure perturbation is maximum, known as the pressure antinode.

If $cos \, Kx = 0$, the pressure perturbation is zero, known as the pressure node.

Thus, the resulting wave has fixed locations of maximum and minimum amplitudes and is called a stationary wave or standing wave. Consider the length of the chamber up to nozzle throat as L. Then at $x = L$

$$cos \, KL = 1$$

Therefore,

$$KL = n\pi \quad where \, n = 1, 2, 3, ... ...$$

Using Eq. (10.4), the definition of the wave number, substituting the value of $K$ in above relation

$$\lambda = \frac{2L}{n}$$ .....(10.7)

Substituting the value of wavelength, $\lambda$ in Eq. (9.2), the frequency of the wave can be expressed as

$$f = \frac{na}{2L}$$ .....(10.8)

$n = 1$ corresponds to the fundamental frequency of oscillation of the stationary wave. The wave length for fundamental or basic frequency is $\lambda = 2L$. The harmonic frequencies of the wave are when $n \geq 2$ . The first harmonic of the wave is when $n = 2$ with frequency $f = \frac{a}{L}$ and wave length $\lambda = L$ . The second harmonic is for $n = 3$ with a frequency of $f = 3a/2L$ and wave length

$\lambda = \frac{2}{3}L$. Figure 10.2 shows the waveform for a fundamental and first and second harmonic frequencies in a rectangular cavity of length L.

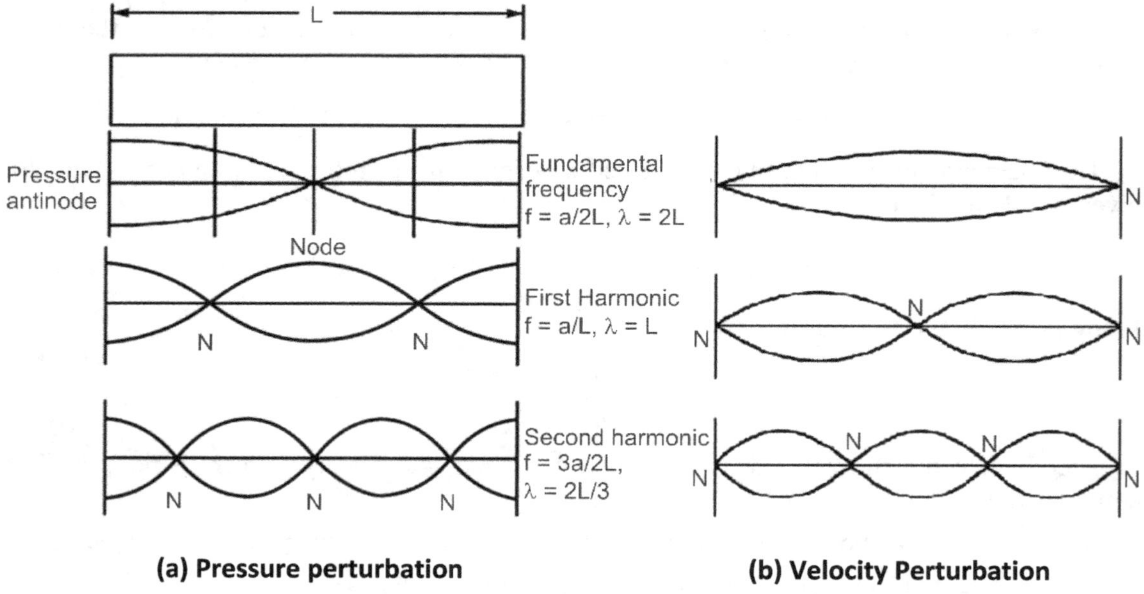


**Figure 10.2** Stationary Wave Forms for the Fundamental and 1st & 2nd Harmonic Frequencies.

The velocity perturbations, $v'$ due to pressure perturbations in the chamber can be derived from the momentum equation (rate of change of momentum, i.e., mass times acceleration, is equal to applied force):

$$\rho \frac{dv'}{dt} = -\frac{dp'}{dx}$$

Differentiating Eq. (10.6) w.r.t $x$ and integrating with respective to time $t$ give the velocity perturbations in the form:

$$v' = -B\sin Kx \cos \omega t = B\sin Kx \cos\left(\omega t - \frac{\pi}{2}\right) \qquad .....(10.9)$$

B is the maximum amplitude of the velocity perturbation. The velocity perturbation is seen to lag the pressure perturbations by $\frac{\pi}{2}$. The wave form velocity perturbations in the cavity of length L are shown in Figure 10.2 b.

## 10.3 CLASSIFICATION OF COMBUSTION INSTABILITIES

Instabilities in the combustion chamber exist as high amplitude damaging pressure oscillations. Generally, on the basis of frequency of oscillations they are classified as low-frequency or high-frequency instabilities. Low-frequency instabilities are present at frequencies lower than the acoustic modes of the chamber. The combustion in a liquid rocket engine is never smooth. The fluctuations of pressure, temperature and velocity are always present in the thrust chamber. When these fluctuations are in phase with chamber acoustic or natural frequencies of the propellant feed system large periodic superimposed oscillations develop that are harmful to the engine or the mission performance. As a conventional practice in the liquid rocket engine industry, when the ratio of the amplitude of the chamber pressure fluctuation to the mean chamber pressure during operation, as shown in Figure 10.3, do not exceed $\pm 5\%$ the combustion process is considered smooth. If the periodic oscillations of pressure within the combustor are exceed the limit of $\pm 5\%$, combustion instabilities are present and the combustion is considered as rough and necessitates dedicated tests on the engine stability. Once a small disturbance/perturbation in the steady state combustion occurs, an unsteady heat release is associated with the combustion. It can be thought of as an acoustic source which propagates sound disturbances throughout the combustor from the source. These pressure waves reflect back towards the injector once they encounter the acoustic boundary, i.e., the nozzle throat. These waves create acoustic pressure and velocity fluctuations in the vicinity of the injector plate, which alter the incoming mass flow rate of the propellant and cause local perturbations to the unsteady heat release rate. When the unsteady heat release and the acoustic fluctuations are in phase, then even a small perturbation will amplify, i.e., the oscillations in a combustor are self-sustained when the rate of heat release and the pressure fluctuations are in phase. If this acoustic energy generated is greater than the acoustic losses produced (dampening power of combustor chamber), combustion instabilities can propagate within the combustor. Thus, the more acoustic dampening that exists within the chamber, the less likely combustion instabilities will be generated. However, in a typical rocket engine operating state of combustion chamber the acoustic losses tend to be less than the acoustic energy present within the system. This creates an environment favorable for the acoustic modes to become excited, as well as spontaneous combustion instabilities. Lord Rayleigh in 1878 proposed the general theory of thermoacoustic instability and it is most widely quoted in the field of combustion instability:

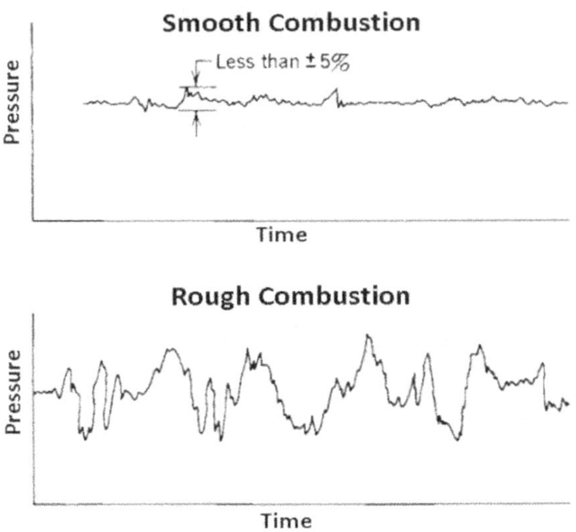


**Figure 10.3** Pressure Oscillations Comparison for Smooth and Rough Combustion.

"If heat be periodically communicated to, and abstracted from, a mass of air vibrating (for example) in a cylinder bounded by a piston, the effect produced will depend upon the phase of the vibration at which the transfer of heat takes place. If heat be given to the air at the moment of greatest condensation, or be taken from it at the moment of greatest rarefaction, the vibration is encouraged. On the other hand, if heat is given at the moment of greatest rarefaction, or abstracted at the moment of greatest condensation, the vibration is discouraged."

The essence of the principle is: (1) Acoustic perturbations are enhanced when the total energy gained through unsteady combustion is larger than the total energy dissipated through the system boundaries (2) heat addition tends most strongly to drive acoustic waves if the energy is added in the region of space where the oscillating pressure reaches greatest amplitude and is in phase.

Combustion instabilities are generally classified on frequency ranges, though there are no sharp limiting lines, known as low-frequency and high-frequency instabilities. The low-frequency instabilities in liquid rocket engine are called "chug" or $L^*$ instabilities and "chuffing" in solid motors. They are bulk mode oscillations since the entire volume or bulk of the gases in the combustion chamber oscillates. They are dependent on flow residence time and a representative time scale of combustion, with typical frequencies less than 300 Hz. In other words, the length of these waves, $\lambda$ are much larger than the characteristic length $L^*$ of the chamber ($\lambda \gg L^*$). Such instabilities occur due to a coupling between the combustion process and the feed system. They are eliminated by increasing the pressure drop in the injector (injection pressure drop is proportional to the square of the flow rate) or by increasing the length-to-diameter ratio in the injector, or by decreasing the volume of the combustion chamber. High-frequency instabilities, referred also as screech, the most damaging of instability types are of the order of frequency 1000 to 10000 Hz. They are distinguished by very large acoustic pressure and velocity fluctuations at high frequencies, which are highly localized throughout the combustor. Their frequencies are close to the acoustic modes of the chamber, and are typically classified according to the associated modes like longitudinal and transverse instabilities. The acoustic fluctuations increase the localized heat release rates through faster mixing of the propellants, and if the localized heat release rate becomes too large, it can ultimately cause harm to the combustor.

Combustion instabilities can also be classified according to whether they are linear or nonlinear. Linear instabilities grow exponentially in time from low-level noise and have waveforms that are mostly sinusoidal. Nonlinear instabilities include those that appear very suddenly and reach a limit cycle (amplitude) in a few periods of oscillation. To understand and analyse these instabilities it is required to distinguish linear and nonlinear behaviour of acoustic waves. In linear instability analysis, it is assumed that the perturbation magnitudes are infinitesimally small. This assumption allows the mathematical development of control differential equations in linear perturbation terms, neglecting terms containing powers or products of perturbations. To study the oscillatory performance of the system, the dependent variable (ex., pressure) is expressed into its study part and an oscillatory perturbation, the behaviour of which is to be determined and provides the key to the problem of instability. NASA SP-194 report "Liquid Propellant Rocket Combustion Instability" discusses in detail the different mathematical models developed to analyse linear and nonlinear instabilities. One way of expressing linear instability in linear differential equation form is

$$\frac{d\phi}{dt} + \frac{\phi}{\tau_{res}} = -\frac{\beta}{\tau_{res}}\phi(t - \tau_c)$$

Here,
$$\phi = \frac{p'}{p_c} \text{ and } \beta = \frac{p_c}{2\Delta p_{inj}}.$$

The perturbation in the chamber pressure is represented by $p_c = \underline{p_c} + p'$, where $\underline{p_c}$ is the steady state chamber pressure. $\tau_{res}$ is the residence time and $\tau_c$ is the combustion time. The solution for the growth rate of the pressure amplitude for the above can be expressed as

$$\phi = Ae^{(\alpha+i\omega)t}$$

where $\alpha$ is the growth of perturbation, called growth constant, A is the amplitude of the wave and $\omega$ (=$2\pi f$) is the circular frequency.

If $\alpha > 0$, the perturbation in pressure grows with time. If $\alpha < 0$, the pressure perturbation decays as the time progresses. For $\alpha = 0$, the amplitude of the oscillations remains constant.

In the linear approach of instability treatment is based on the assumption of infinitesimal perturbations. In reality, as the perturbation amplitude grows beyond a certain level (i.e., limit cycle) nonlinear effects become important and dominate the entire process. In linear instability as oscillations are growing from infinitesimal perturbations, nonlinear effects prevent the indefinite growth and resolve to a limit cycle with finite amplitude. Perturbations below this limit cycle are damped in linear instabilities. Above that limit cycle, perturbations are amplified attaining the nonlinear characteristics of instabilities. Hence, two basic problems arise dealing with combustion instabilities (a) determining the conditions for the existence and stability of limit cycles for a linearly unstable system and (b) finding the conditions under which a linearly stable system may become unstable to a sufficiently large disturbance. The two sources of nonlinear combustion instabilities are: (a) the combustion processes themselves may present important nonlinear effects and (b) independently of the behaviour of the combustion processes, the wave motion is characterized by the well-known nonlinear effects of steepening and dispersion, culminating in the appearance of shock waves in the chamber.

The problem of combustion instability is so complicated and involves so many variables and parameters that one must settle for much less. In fact, complete analytical solutions can't now and probably never will be found. Nevertheless, modelling and approximate analyses can provide useful understanding and can suggest ways of correlating and interpreting experimental observations.

## 10.4  LOW-FREQUENCY INSTABILITIES IN LRE

Low-frequency instabilities, generally referred as chug or $L^*$ instabilities, relates to a cyclic process involving propellant preparation, combustion, and exhaust of products through a choked throat. If these pressure oscillations reach high amplitude, damage the combustion chamber. The time lag concepts have been used in modelling and analyzing the observed chug instabilities in rocket engines and also to incorporate suitable criteria to suppress them in the rocket engine design and development since their origin in the 1950s. Matthew J. Casiano (2010) reviewed different time lag models in his Ph.D. thesis titled "Extensions to the Time lag Models for Practical Application to Rocket Engine Stability Design ", submitted to The Pennsylvania State University. The time lag model uses a simplified approach lumps all the processes of mixing and heat release into a single characteristic time, and presumes the acoustic time scales are much smaller than the period of chug oscillation. These time lag models consider that a time lag existed between an arbitrary fluctuation in the propellant flow and its subsequent manifestation in the combustion chamber pressure. This time delay accounts for the kinetic rates of mixing, atomization, and chemical reactions before the propellants are fully combusted. The duration of the total time lag consists of the insensitive time lag and the sensitive time lag. The insensitive time lag is unaffected by variation in physical factors including processes like atomization and mixing of the propellants. The sensitive time lag is affected by variation in physical factors and includes processes like chemical reaction and kinetics, heating and diffusion. The criterion developed by Summerfield (1951) is generally accepted in the design of rocket engine chamber. From his model he developed the following conditions for stability:

$$\tau_T < \frac{L_f \dot{m}}{p_c A_f} + \frac{2c^* L^* \Delta p_{inj}}{p_c R T_C} \qquad\qquad .....(10.10)$$

$$\Delta p_{inj} > \frac{p_c}{2} \qquad\qquad .....(10.11)$$

$\tau_P$ is the total time lag

$L_f$ is the length of the feedline

$A_f$ is the area of cross section of the feed line

$p_c$ is the chamber pressure

$\dot{m}$ is the propellant mass flow rate

$c^*$ is the characteristic velocity of gasses

$\Delta p_{inj}$ is the pressure in the injector

$R$ is the specific gas constant of combustion gasses

$T_c$ is the combustion chamber temperature

It can be observed from above equations for improving stability, increase the injector pressure drop (i.e., increasing the pressure difference between the tank and combustion chamber), increase the characteristic length (i.e., increasing the combustion chamber volume) and reduce the time lag. An interesting point that Summerfield makes is that reducing the cross-sectional area or lengthening the feedline is an attractive remedy because it takes a smaller increase in supply pressure to achieve the same degree of stability. Also, less reactive propellant combination has more lag time and is more likely to exhibit the bulk mode of instability. The injection pressure drop depends on feed system pressure. Hence, it is also known as feed system coupled combustion instability.

## 10.5 HIGH-FREQUENCY INSTABILITIES IN LRE

The high-frequency instabilities are characterized by very large acoustic pressure and velocity fluctuations and are the most dangerous in nature. As the wave length of the oscillations is smaller than the characteristic length of the chamber they propagate in the entire chamber. So, perturbations can occur at any location in the chamber. The pressure variations are functions of both the spatial coordinates and time i.e., $p' = f(x, t)$ instead of time only in bulk mode (low-frequency) combustion instabilities. The motion of waves in the chamber results in the formation of standing waves. These frequencies, highly localized throughout the combustor, lead to an increase in localized heat release rates through faster mixing of the propellants, and if the localized heat release rate becomes too large, it can ultimately cause harm to the combustor. The amplitude of the standing wave increases due to addition of mass, heat release and body force in phase. The periodic mass release creates a time dependent change in pressure. The periodic heat release creates a change in the temperature. Both these disturbances augment the existing pressure or temperature fluctuation resulting in an increase in the amplitude of the disturbance. The periodic body force, in phase with velocity disturbance, provides additional power leading to a higher value of a velocity disturbance. The rate of energy addition per unit volume due to mass, heat and body force can be represented as:

$$\sigma = \frac{\dot{m}'p'}{\underline{\rho}} + \frac{Q'T'}{\underline{T}} + \vec{F} \cdot V' \quad (W)$$

Where $\underline{\rho}$ and $\underline{T}$ are steady state density and temperature. If $\Omega$ represents the rate of energy dissipation (due to viscosity, radiation, droplets etc.) per unit volume, the net energy added over cycle of the wave is given by

$$\oint dt \oint (\sigma - \Omega)dt > 0$$

The frequencies and waveforms of the different standing wave modes can be determined from the wave equation.

The combustion instabilities associated with high-frequencies are classified as longitudinal, transverse or a combination of both. Longitudinal oscillations fluctuate along the axial direction within the combustor and are independent of chamber cross sectional shape. Transverse instabilities propagate along a direction perpendicular to the centre axis of the chamber or parallel to the injector face. These transverse instabilities can stand or spin. Longitudinal instabilities are generally less harmful than transverse mode. As such, purely transverse modes tend to be the most dangerous type of high frequency instability. They are affected by the cross-sectional shape of the combustor and the combustor wall boundary conditions. They reveal either purely tangential, purely radial or combined tangential-radial modes. Figure 10.4 shows some schematic of transverse modes of instabilities.

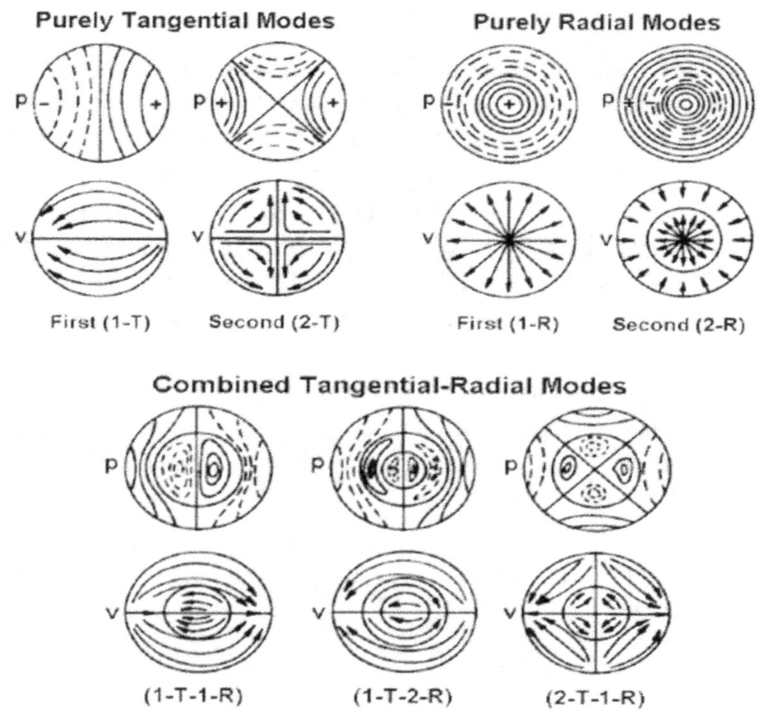


**Figure 10.4** Schematic of Some Transverse Instability Modes (Source: David T Harrje and Frederick H Reardon).

Tangential modes have pressure node lines which split the combustor in the circumferential direction. In spinning mode, the instability pressure node lines split the combustor in the circumferential direction. In standing mode, the instability nodes have a fixed orientation in the combustor. It has been observed that spinning tangential modes are able to be transformed into standing modes spontaneously and vice versa. The pressure node lines in radial modes are circular and orthogonal to the combustor axis. The combined instability modes have node lines that split the chamber in both the circumferential and radial directions. Hence, in the combined mode both spinning and standing variations of instabilities types are possible. It is observed in a liquid rocket engine, the first tangential (1-T) mode instability is the most dangerous as the pressure oscillation associated with it has the highest energy content. Figure 10.5 displays images of two variations of 1-T mode. When the spinning 1-T mode is active, combustion products spin unabated around inside of the combustor. In the standing 1-T mode a high heat pocket of combustion products which travels back and forth across the combustor, perpendicular to its node line.

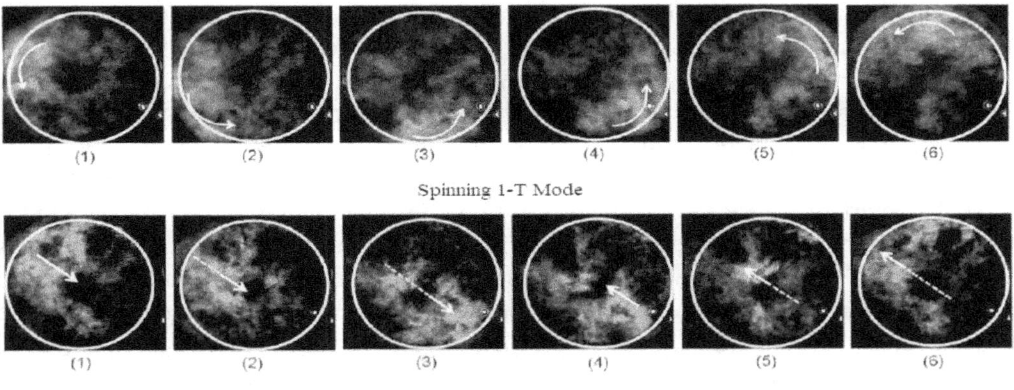


**Figure 10.5** Images of one cycle of the 1-T Modes
(Source: Bennewitz John W. and Frederick Robert A).

## 10.6 COUPLING MECHANISMS

Figure 10.6 depicts a simple outline of the various sequences of elementary processes occurring in combustor. However, several different processes may take place simultaneously in a given region of space. The discussion that follows in this section tries to explain in a modest way the context within which the elementary processes may act as controlling mechanisms for exciting and sustaining combustion instabilities. The discussion can't be complete in a practical sense as for a given engine there are inevitably characteristics peculiar to the design. For example, in the case of the F-1 engine, accumulation of liquid fuel during film cooling of the chamber was responsible for an unacceptable spiking unsteady motion.

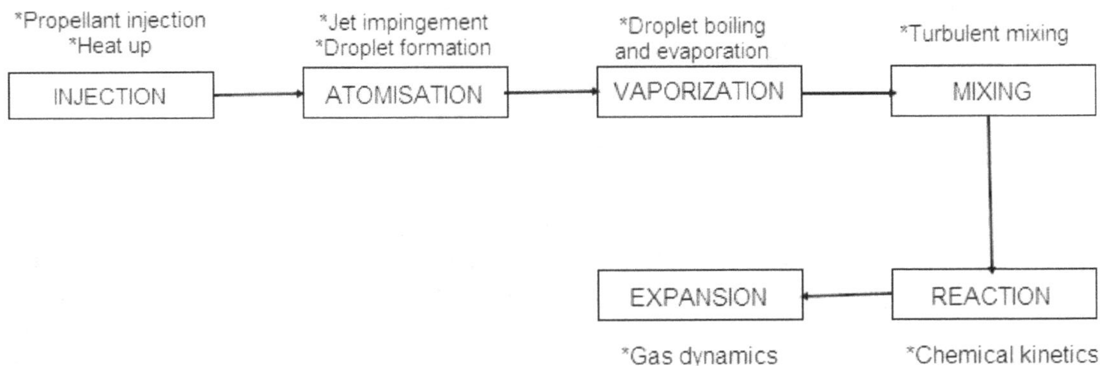


**Figure 10.6** Physicochemical Processes in Rocket Combustor.

Longitudinal modes greater than 1-L (first longitudinal frequency) and transverse modes greater than 3-T or 1-R modes are rarely observed. Longitudinal modes are generally associated with injection-coupling and they are controlled by periodic variations in injector pressure drop.

Transverse modes, associated with unsteady heat release, depend on intrinsic combustion processes (injection, atomization, vaporization, mixing and reaction) and their coupling with acoustic modes (for detailed study ref. NASA SP-194). It has been suggested that combustion instabilities are more likely to occur when the characteristic time for combustion is of the same order of magnitude of acoustic oscillation time period. For example, when acoustic frequency $f$=1000 Hz and when $\tau_{comb} \sim \tau_{acous} \sim 1ms$ then it is more likely the combustion instabilities arise. The characteristic time is mainly dictated by a process, like vaporisation, that has the longest time scale compared to other processes involved in combustion. This rate controlling process will play a dominant role in the excitation of combustion instability modes. It will play a dominant role in modifying heat release in such a way that its amplitude and phase are adequate to overcome the damping forces. Though the propellant injection doesn't have a characteristic time that allows interaction with oscillation frequency, the way it is injected to form droplet size affects the further processes of atomization, vaporization and mixing in combustion. Thus, indirectly injection can be a major factor in altering the growth of instabilities. The characteristic time of atomization is the order of 10 ns to 10 μs and hence directly not a rate controlling process. However, it is observed that the acoustic pressure and velocity oscillations can vary the initial droplet size, which in turn, modifies the vaporization and mixing processes. Vaporization of droplet occurs through two concurrent processes of droplet boiling and evaporation. Droplet boiling involves liquid heating. Evaporation is a boundary layer heating process of droplets and it takes lesser time than boiling. Generally, the characteristic time of droplet boiling is of the order of 1 ms allowing the mechanism to directly interact with the acoustic oscillations. Thus. droplet boiling is the primary rate controlling process responsible for combustion instabilities. Chemical mixing in the combustor has a characteristic time less than 1ms and hence, is not directly a rate controlling mechanism. However, it results in the spatial distribution of heat release within the combustor and this effect on driving the instability is known to be much smaller than that of vaporisation process. As the chemical reaction in a rocket combustor occurs at high pressure and temperature, its characteristic time is less than $10^{-2}$ s (1 cs) and hence, not a rate controlling mechanism. However, during fuel rich combustion chemical times become on the order of 1 ms. Thus, depending on the conditions within the chamber, chemical kinetics may be a contributing factor in driving combustion instabilities.

## 10.7 COMBUSTION INSTABILITY CONTROL DEVICES

To achieve stable and reliable combustors for a liquid rocket engine, the effective procedure is to remove energy from the oscillations or prevent certain modes of oscillation by geometric design. Extensive experiments were carried out to establish mitigation techniques to prevent the onset of instabilities. The techniques to control the high frequency instabilities are based on either to provide enough damping to take away the energy for instability driving or break the coupling between the unsteady heat release of combustion and the acoustic pressure oscillations of the chamber. A combination of these two methods is also used for effective mitigation of the instabilities. These techniques are classified as passive and active systems. The passive systems dampen oscillations in non-responsive mode while active systems are able to alter the instability response dynamically in real time. Passive systems being implemented in the liquid rocket industry due to their inclusion of non-movable parts. The most preferred techniques are.

(a) Symmetric Injector Plate Baffle (b) Symmetric Fuel Injector Distribution and (c) Resonance Absorbers.

### 10.7.1    Symmetric Injector Plate Baffle

The normal method employed to avoid the frequency instabilities, predominantly transverse modes, in the USA rocket industry is the symmetric injector plate baffle system as shown in Figure 10.7. This system contains metal barriers connected to injector plate in both tangential and radial directions. The axial length of the baffle is limited to avoid unwanted heat transfer or performance loss of the engine. Though the suppression mechanism of high frequency oscillations is not fully understood, the attachment of the baffle system modifies the acoustic resonance properties of the combustor while damping occurs with the vortex shedding and separation of the flow. Thus, the baffle is unique in that it affects both excitation and damping. The effect of baffles on the acoustic resonant frequencies in the rocket combustor shown in Figure 10.8. It can be observed that with the addition of baffles the frequencies of 1-T mode amplitudes of all the resonant frequencies are decreased drastically. Thus, it increases the damping of acoustic energy by providing more boundaries for acoustic losses. Due to addition of more physical boundaries, it is more difficult for the transverse acoustic velocity oscillations to propagate across the injector plate, limiting the unsteady oscillations between baffle compartments. Also, physical boundaries of the baffle create additional acoustic velocity nodes that reduce the velocity amplitudes to a great extent. This causes essentially unsteady heat release less likely to be onset by transverse acoustic velocity oscillations. Generally, spoke and hub design of the baffle system is used where Spokes damp tangential modes and a hub damp the radial modes. An odd number of spokes should be used, and the number of spokes should be at least one greater than the number of nodes in the mode being damped. The baffle length is taken within 20-30% of the chamber diameter. They have the disadvantage of being exposed to the combustion region and need to be cooled either by film or regenerative cooling methods.

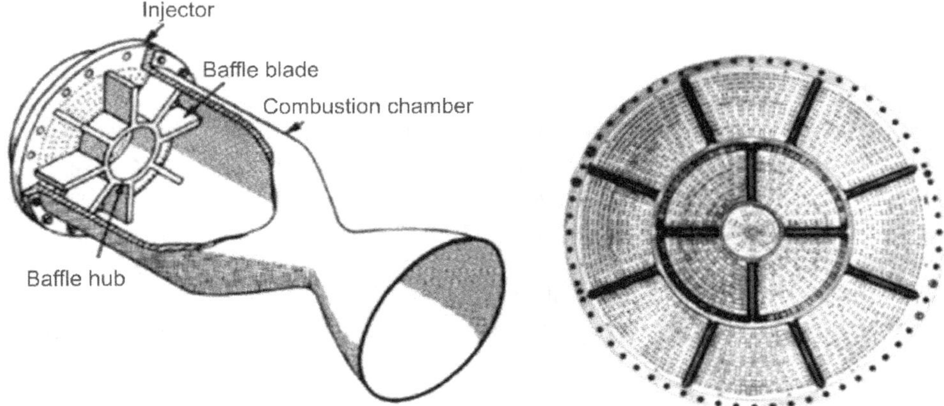


**Figure 10.7** Schematic of a Symmetric Injector Plate Baffle System (Source: David T Harrje and Frederick H Reardon).

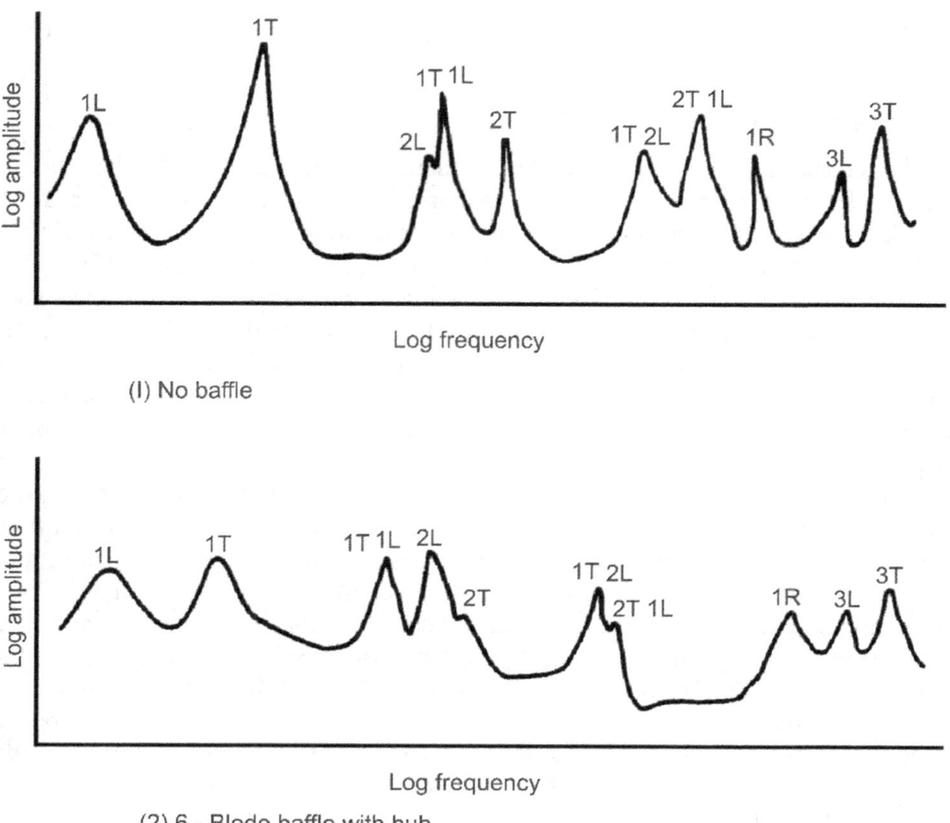


**Figure 10.8** Comparison of Acoustic Resonance Characteristics with and without an Injector Plate Baffle System in Combustor.

## Symmetric Fuel Injector Distribution

Russian developed a high frequency transverse instabilities mitigation system having a symmetric fuel injector distribution which involves strategically positioned fuel injectors with modified sprays to simulate baffle like condition through spray characteristics. Figure 10.9. show a section of the injector with modified spray distribution. The modified jet spray perturbs transverse acoustic velocity oscillations and makes it difficult for unsteady oscillations to travel between combustor segments. Also, the spray depletes some energy of these oscillations as they travel through these regions. These two effects compounded curtail the growth of instabilities in the chamber.

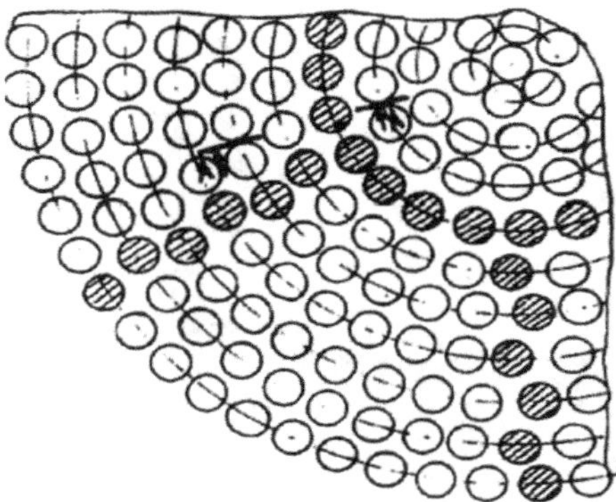

**Figure 10.9** A schematic of a typical symmetric fuel injector distribution
(Source: Dranovsky, M. L., Yang, V., Culick, F. E., and Talley, D. G).

### 10.7.2    Resonance Absorbers

Acoustic absorber, a passive device to supress the combustion instabilities, removes oscillatory energy from the system (damping) and by interacting with the overall oscillator combustion process. They are being used either alone or as a supplement to baffles to maintain stable combustion in several production engines. They are comprised of an array of acoustic resonators distributed along the wall of the combustion chamber. Three common types of resonators are Helmholtz, quarter wave and intermediate. The Helmholtz and quarter wave resonators are more frequently used ones and are simply special cases of the generalized resonator. A Helmholtz resonator consists of a resonator cavity connecting the combustion chamber through a small passage as shown in Figure 10.10. The cavity dimensions are large compared with the passage width but small compared with a wave length. In such a case, the motion of the gas in the resonator can be considered to be analogous to that of a mass-spring-dashpot system. A quarter wave resonator is a closed slot or tube with a uniform cross-sectional area. The cavity acts as a Helmholtz Resonator. The frequencies at which these absorbers resonate depend on the cavity volume and are independent of the shape. The pressure antinodes of oscillations exit at wall boundary conditions of chamber and injector. In all these locations exit zero velocity i.e., velocity nodes of oscillations. Thus, resonance absorbers are generally installed at the injector plate, near the combustor. The acoustic absorber may consist of either a limited number of resonators along the injector periphery Face, as shown in Figure 10. 10, or the entire chamber wall from the injector to the beginning of the nozzle convergence point. Usually, the open area of the cavity is chosen to be between 5 to 20% of the injector face area.

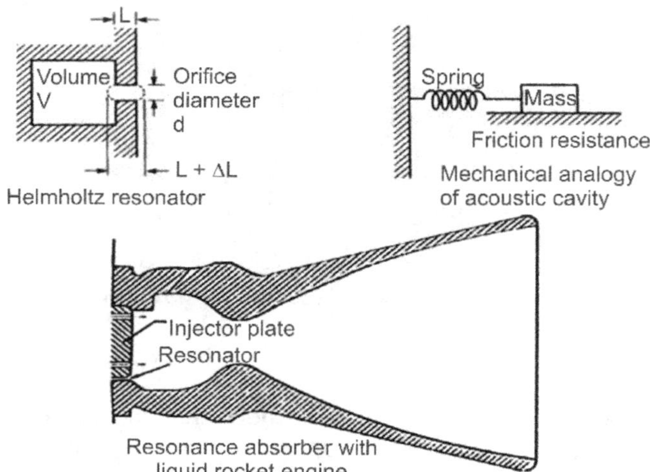


**Figure 10.10** Schematic of an Acoustic Resonance Absorber (Source: NASA SP-8113).

## 10.8 POGO INSTABILITY

POGO instability is not combustion instabilities. These instabilities occur at very low frequencies, usually less than 100 Hz, and result from coupling between the liquid rocket feed system and the vehicle structural longitudinal vibration. The vibrations, nicknamed "pogo" after the jumping stick. They have occurred usually in the first longitudinal structural mode during operation of the first liquid-propellant stage of a launch vehicle. A typical nature POGO instability is illustrated in Figure 10.11. The disadvantage of these structural vibrations is that they can produce an intolerable environment for astronauts, equipment, overload vehicle structure and sometimes may lead to loss of propulsion performance.

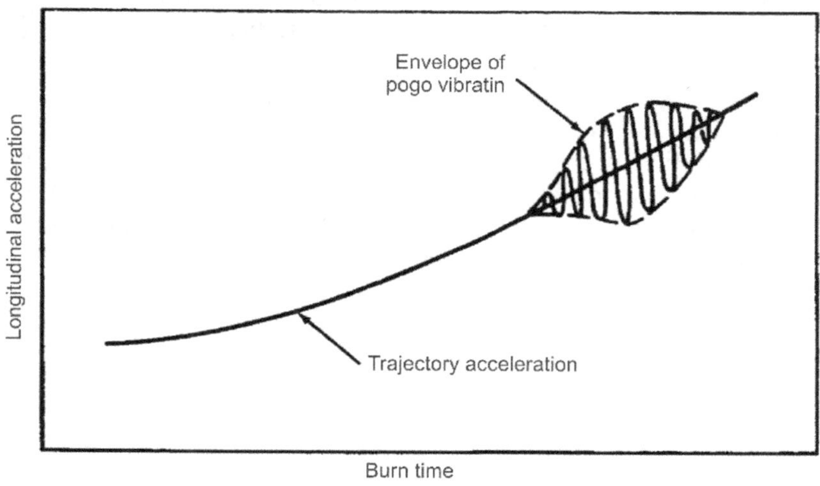


**Figure 10.11** Typical POGO Vibration in LRE (Source: NASA SP-8055).

Basically, two forms of propulsion system behaviour produce pogo instability. The normal form of pogo, called engine-coupled pogo, developed to a significant degree on certain configurations of the Thor, Titan, and Saturn space vehicles. The longitudinal vibrations of the vehicle can induce pressure oscillations at the pump inlet. The amplitude of the pressure oscillations at the pump outlet is increased according to the pressure ratio of the pump. The

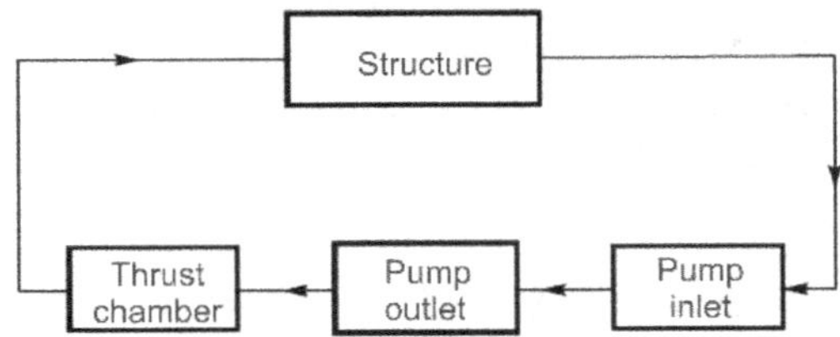


**Figure 10.12** Schematic of Engine Coupled POGO Instability.

flow oscillations lead to thrust oscillations in the engine which then act as regenerative forcing functions on the vehicle structure. The engine coupled pogo process is illustrated in Figure 10.12. POGO can occur in pressure-fed systems also. The variation of propulsion system characteristic associated with engine coupled POGO are: oscillations of propellant column in feedline, pump induced cavitation and frequency response of the engine The second form is ullage coupled POGO instability, a much less common form, results from the pneumatic behaviour of an active pressurization system for the propellant tank ullage. Figure 10.13 illustrates the ullage coupled POGO. This is also referred to as pneumatic coupled POGO or bolting. Due to longitudinal vibrations, the tank ullage volume boundaries change and result in ullage pressure oscillations. The pressure regulator responds by producing an oscillatory flow of the pressurizing gas into the supply line. This redevelops the ullage pressure oscillation that acts as an axial forcing function on the vehicle structure, and the feedback loop is closed.

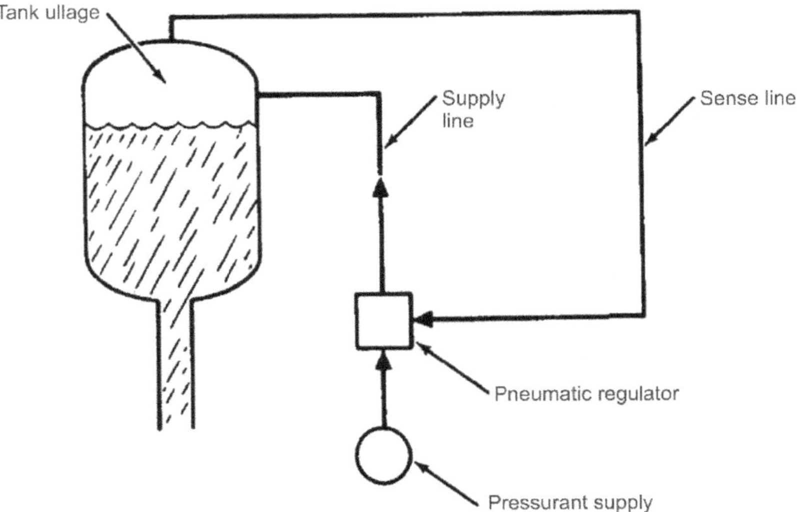


**Figure 10.13** Schematic of Ullage Coupled POGO.

The POGO related structural modes may change due to continued depletion of the propellant. The propellant oscillation in the feedlines vary significantly with variation of the pump inlet cavitation conditions. The ullage pressure response to structural vibration decreases with increasing ullage volume. The devices for suppressing engine coupled POGO are based on the separation of feedline resonant frequencies from those of significant structural modes. This is achieved by installation of a low-inertance (pressure difference between the ends of incompressible liquid column divided by the time derivative of the mass flow rate of the liquid) accumulator near the pump as shown in Figure 10.14. All devices shown are developed for the first stages of liquid rocket vehicles and flight tested successfully.

## 10.9  COMBUSTION INSTABILITY IN SOLID ROCKET MOTOR (SRM)

Two basic types of combustion instabilities occur in solid rocket motor (SRM): (a) acoustic resonances or pressure oscillations similar to liquid rocket engine and (b) vortex shedding phenomena that generally occur in large segmented motors. When instability occurs the burning rate, chamber pressure and thrust increase total burning duration to decrease. In case of liquid rocket engine chamber geometries are fixed and the feed system and injector are physically not part of the oscillating gases in the combustion chamber but may interact strongly with pressure fluctuations. In SRM, the cavity is complex since it includes igniter case, movable as well as submerged nozzles, fins, slot and star / or other shaped perforations.

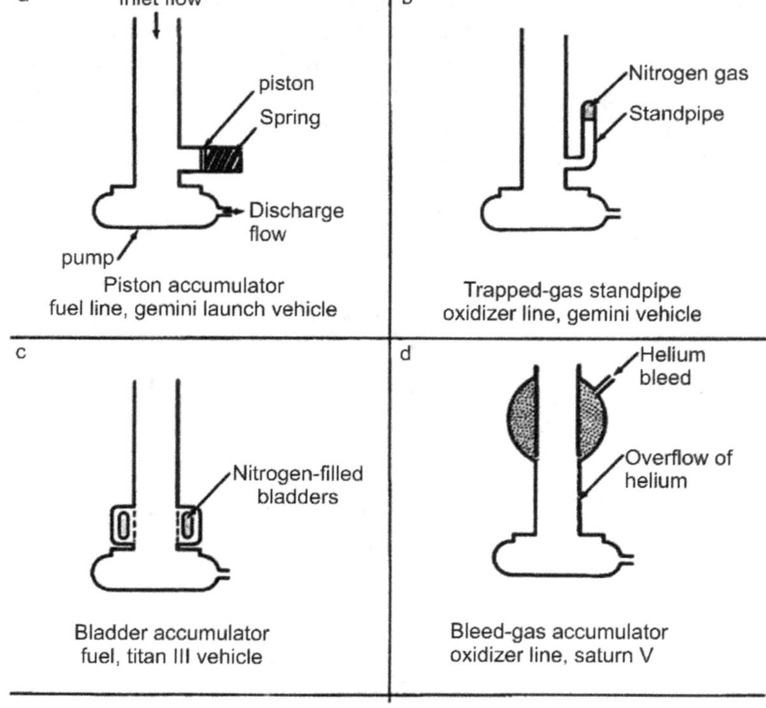


**Figure 10.14** POGO Suppression Devices.

Moreover, the resonating cavity geometry changes continually as the burning propellant surfaces recede. As the cavity volume increases the transverse oscillation frequencies are reduced. The damping factors are also changed due to the presence of solid particles present in the grain. When a brief high-pressure disturbance occurs at a burning surface, the instantaneous heat transfer and thus the burning rate increases. This results in an increase of mass flow rate at that surface. The response to pressure perturbations in SRM relates to changes in gaseous mass production and in energy release at the burning surface. The motor experiences from minor linear acoustic oscillations to violent non-linear oscillations accompanied by large ballistic pressure shifts and sometimes, leading motor destruction. Figure 10.15 shows the typical two periods of combustion instability in the pressure time history. The dashed lines show the upper and lower boundaries of the high-frequency pressure oscillations, and the dot-dash curve is the behaviour without instability. The bulk mode or $L^*$ mode occurs at relatively low frequencies (below 100 Hz) depending upon time delay. The pressure remains essential uniform throughout the cavity. Consider a SRM grain operating at steady cavity pressure of $p_c$. Let a pressure disturbance drop the pressure suddenly from A to B as illustrated in Figure 10.16. This leads to slow chemical reaction and increase in the chemical delay time. At reduced pressure (point B) the heating rate of the propellant surface is also lower. However, due to thermal inertia the heated propellant surface corresponding to pressure original steady value $p_c$ remains for some time. During this time the mass flow rate generation remains at a higher value of the steady state level. Due to the pressure reduction at B, the mass rate generation reduces but the residence time increases. This provides more time to progress chemical reactions causing more heat generation. The combined effect of thermal inertia and increase in residence time results in increased mass generation and causes the chamber pressure to increase from $B$ to $C$. The increased pressure at $C$ results a smaller residence time and a reduced thermal depth (the thermal depth of a propellant is the ratio of thermal diffusivity coefficient to the burn rate of propellant). The mass generation drops and the pressure falls to $D$. The process of oscillations, therefore, continues as shown Figure 10.16. This type of non-acoustic instability which often occurs near the beginning of motor burn immediately after ignition when the chamber volume is small and the chamber pressure is low. The occurrence of such oscillations depends on $L^*$ (= volume of the cavity/throat area), lower the L*, the more likely a motor will experience this type of instability. Figure 10.17 is a compilation of many $L^*$ values for a variety of propellants and indicates both stable and unstable combustion regions. This gives general information when the SRM might be unstable.

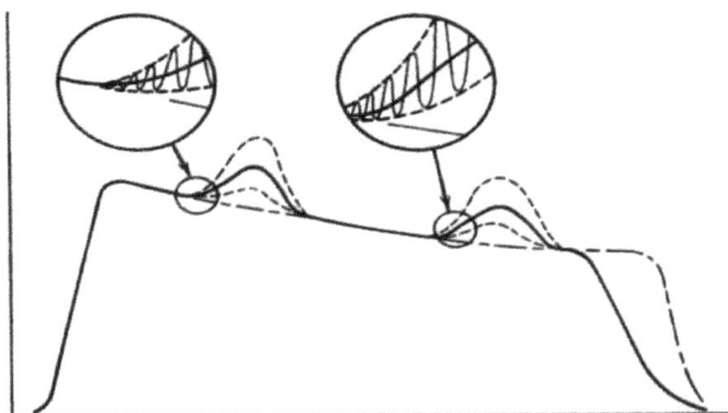

**Figure 10.15** Pressure-Time curve with Instability Region in SRM.

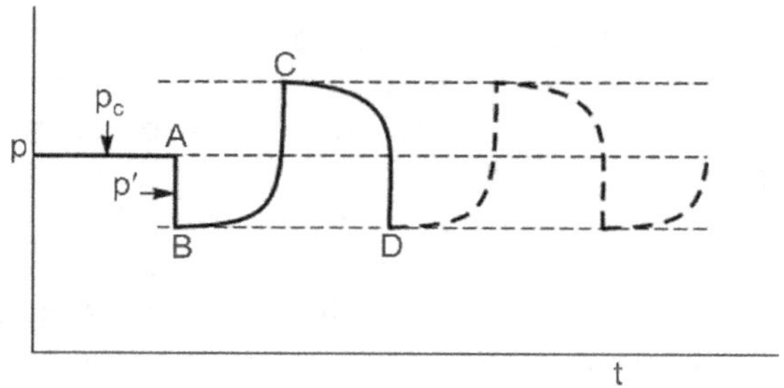

**Figure 10.16** Bulk Mode (L*) Oscillations in SRM.

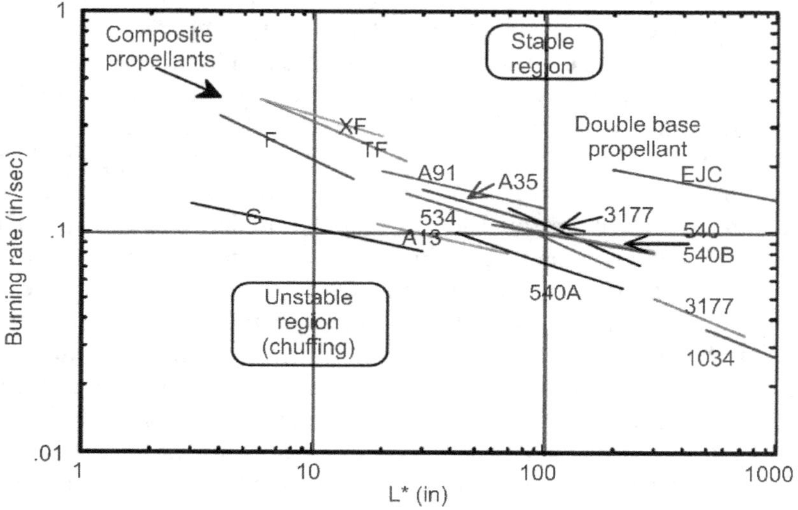


**Figure 10.1 7** $L^*$ (Bulk Mode) Instability Boundary for various Solid Propellant Grains (Source: Fred S. Blomshield).

$L^*$ instability is different from chuff, the periodic pressure surges which occur when the ignition is not effective, such as during a hang-fire, which is manifested by periodic ignition followed by extinction. These are very low frequencies, much lower than $L^*$ as illustrated in Figure 10.18.

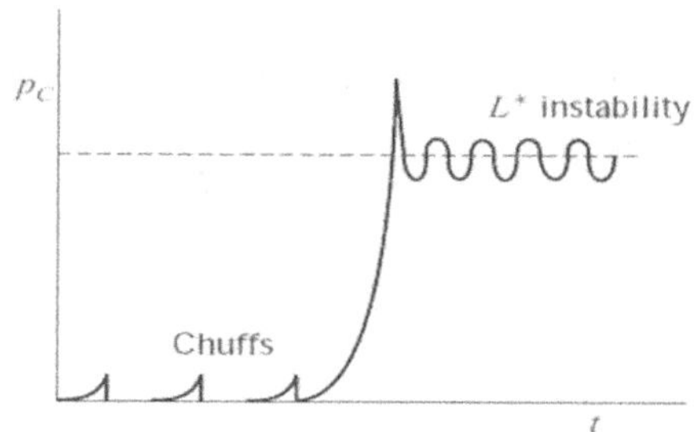


**Figure 10.18** Chuffs and L* Instability.

The combustion of solid propellant essentially occurs at the surface of the grain. The pressure-coupled response of the combustion processes, that could result in instability, takes place at the propellant surface with acoustic pressure waves. The pressure coupled response, $R_{pc}$ a non-dimensional number, is defined as the ratio between the perturbed burning rate over the mean burning rate to the perturbed pressure over the mean chamber pressure:

$$R_{pc} = \frac{r'/\underline{r}}{p'/\underline{p}}$$

It is a function of burning rate, pressure, frequency and propellant physical and mechanical properties. It can be used to compare propellants intended for the same application. The propellant with the lowest response will be less likely to drive instabilities, other factors remaining constant. The T-burner device, described later in this section, is generally used to measure the response. The velocity-coupled response, taking place at the propellant surface by the acoustic velocity, has generally negligible effect in many propellant grain configurations. Another driving mechanism is distributed combustion. It is the interaction of the acoustic field with burning metal particles taking place away from the propellant surface. It occurs in those grains whose burning characteristics allow the metal to burn in a distributed manner throughout a motor chamber.

The main damping mechanisms in solid motors are nozzle damping, particle damping, flow damping and structural damping. The nozzle damping is the usually the largest damping mechanism as waves come in contact with the nozzle throat, it radiates away part of the acoustic energy to the environment. When the propellant grain contains metallized propellants or those with acoustic stability additives, the visco-elastic nature of these species in the combustion products also absorbs some energy. The damping depends on the mass fraction of particles in the flow field and, most importantly, particle size. Larger particles damp lower frequencies, while smaller particles damp out higher frequencies. Flow losses are those losses associated with mean flow interactions with the acoustic field. It includes losses due to the flow of the combustion products turning from normal to the grain surface to axial direction and acoustic boundary layer. Structural damping mechanism is damping due to deformation of the motor structure due to acoustic pressure oscillations. This is generally very small but may be considerable in very large rocket motors where acoustic loading can be high.

Using linear stability theory, the growth of the amplitude of the pressure oscillations in the standing wave due to combustion can be expressed by:

$$p' = p_0' e^{(\alpha_M)t}$$

Here, $p_0'$ is the initial amplitude value and $\alpha_M$ is the growth constant. If $\alpha_M$ is negative, the motor is stable. If $\alpha_M$ is positive, the motor is unstable. The stability of a solid rocket motor is determined by the net sum of acoustic gains and losses in the system. Using the linear assumption this can be represented as

$$\alpha_M = \alpha_{PC} + \alpha_{VC} + \alpha_{DC} + \alpha_{NOZ} + \alpha_{PART} + \alpha_F + \alpha_S$$

The first three terms in RHS are driving forces due to pressure, velocity and distributed combustion. The last four are damping terms due to nozzle, particle, flow and structural damping.

### 10.9.1 T Burner and Pressure-Coupled Response Determination

For many years the T-Burner has been an effective standard for determining the unstable response of solid propellants. As shown in Figure 10.19, it consists of a cylindrical straight tube of variable lengths to get the different frequencies of oscillation in the fundamental longitudinal mode. The diameter of the tube varies between 35 and 150 mm. Two thin propellant disks of equal thickness are placed on each end and they are ignited simultaneously. The burning surface of the propellant disks correspond to pressure antinodes. The acoustic velocity, cross flow velocity and mean flow are all zero at the ends as to isolate the pressure-coupled response. The central vent of the tube is connected to a surge tank. The vent at the central location corresponds to a pressure node. The tube pressure can be changed by varying the surge tank pressure. The tube acts as a cavity and its fundamental longitudinal frequency $f = a/2L$, where a is the velocity of sound of the gas in the tube and is the length of the tube. The test frequency depends on the burner tube length and the combustion gas temperature. Data are obtained by pulsing the burner during the propellant burn and after burn out. Pressure transducers are used to measure the pressure amplitude rate of change of the oscillations. Phototransistors are located at each end to measure the light output at burnout. If required, pulsar at both ends is provided to introduce disturbance in the acoustic wave. The combustion response is calculated from the following equation.

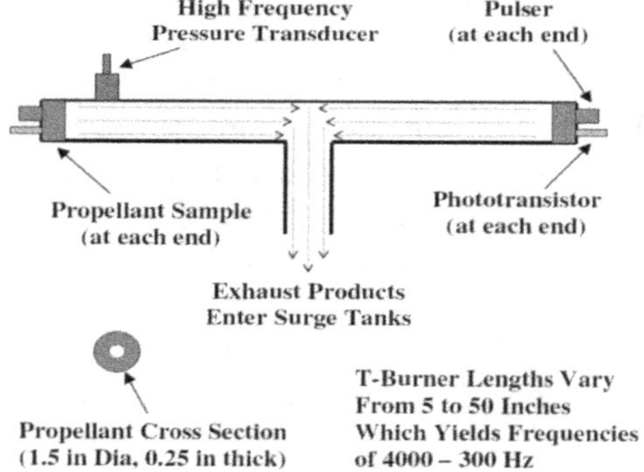


**Figure 10.19** Typical T- Burner Configuration.

$$R_{pc} = \frac{\alpha_c \underline{p}}{4fa\rho_p r_b \left(\frac{S_b}{S_c}\right)} \left(\frac{a_m}{a}\right) \qquad \qquad \dots\dots(10.12)$$

where

$\alpha_c$ = combustion growth rate = $\dfrac{\alpha_1 - \alpha_d}{\left(\frac{S_b}{S_c}\right)}$

where

$\alpha_1$ = measured growth rate during combustion

$\alpha_d$ = measured decay rate after combustion

$\dfrac{S_b}{S_c}$ = propellant burning surface area/burner cross sectional area

$\underline{p}$ = mean pressure

f = frequency

$\rho_p$ = propellant density $r_b$ = mean burning rate

$a$ = theoretical speed of sound of the gases

$a_m$ = measured speed of sound ($a_m$ = 2fL, L = burner length)

# CHAPTER 11

# Spacecraft Orbital Maneuvers

## 11.1 SPACECRAFT PROPULSION

Spacecraft propulsion is required to transfer it from launch vehicle parking orbit to designed mission orbit, orbit acquisition and orbit keeping against perturbations, orbit changes including interplanetary transfers, descent and landing on the surface of a planet or moon, and attitude control in launch, orbiting and deorbit phases of space flight. Also, propulsion is needed when a spacecraft like SpaceX Dragon or European ATV maneuver to rendezvous with another spacecraft (Ex: ISS, International Space Station). A spacecraft is always in the gravitational field of some central body (such as Earth or the Sun), it has to follow orbital-motion laws in its movement.

For most on-orbit maneuvers, the propulsive device in the spacecraft burns relatively for a very short time compared to orbital period. Hence, these burns can be treated as impulsive maneuvers with an instantaneous change in orbital velocity and no change in position. Of course, the long duration burn of a launch trajectory does not qualify as an impulsive maneuver. In the absence of any other force than thrust F,

$$m\frac{dv}{dt} = F$$

$$\Delta t = \frac{\Delta v}{F/m}$$

It can be observed from the above equation that a finite velocity change $\Delta v$ will require a relatively short burn time if the thrust acceleration $F/m$ is large. As the thrust acceleration $F/m$ tends towards infinite, the burn time shrinks to an infinitesimal value. This limiting case is impulsive thrust or an impulsive maneuver, i.e., high thrust to mass ratio instantaneously produces the desired velocity change $\Delta v$ without any change in orbital position. Once $\Delta v$ is known, the required propellant mass can be determined by using the ideal rocket equation and the propulsion system characteristics such as specific impulse.

Spacecraft propulsion can be classified according to the source of energy utilized for the ejection of propellant. Chemical propulsion uses heat energy produced by a chemical reaction to generate gases at high temperature and pressure in a combustion chamber. These hot gases are accelerated through a nozzle and ejected from the system at a high exit velocity to produce thrust force. Electric propulsion uses electric or electrostatic, electromagnetic energy to eject matter at high velocity to produce thrust. The major characteristics of spacecraft propulsion system are:

(a)   low thrust levels (1 mN to 4500 N) with low acceleration levels

(b)   high velocity increment capability (many km/s)

(c)   continuous operation mode for orbit control

(d)   pulsed operation mode for attitude control

(e)   predictable, accurate and repeatable performance impulse bit

(f)   shortest firing duration of order 0.02 sec

(g)   several hours of cumulative duration of burning

(h)   long span of life in space (15 years or more)

(i)   minimum and predictable thrust exhaust impingement effects

The low level thrust propulsive devices that are used for on-orbit maneuvers and attitude control are also referred as thrusters. The higher capacity propulsive engine is used for space trajectory travel. The spacecraft propulsion will be discussed in detail in chapter 13.

## 11.2   ATTITUDE MANEUVERS

The spacecraft, after placing it at the design mission orbit, also has to be adjusted to stay in place. For example, a communication satellite needs to point its dishes always at a ground station to facilitate transmission of signals and to maintain housekeeping activity like pointing a solar panel towards the Sun to generate electrical power, positioning a propulsive device in the right direction before it is fired to achieve the desired orbital change. The attitude of the spacecraft disturbs due to

- External Torques
  - Solar radiation pressure
  - Gravity gradient
  - Magnetic fields
  - Aerodynamics
- Vehicle-Based Torques
  - Mass movement
  - Elasticity
  - Out-gassing

For useful operation of the spacecraft, attitude control system (ACS) is required with a fixed coordinate system with respect to the spacecraft itself to control the orientation of three axes versus either inertial or Earth based coordinate systems. The ACS is concerned with the rotation of the spacecraft about its center; not concerned with the motion of the spacecraft's center along its orbit. The attitude describes the position of the spacecraft in a rotational sense, hence change in attitude implies a rotation. Therefore, a torque has to be applied to the spacecraft to bring a change in its attitude (i.e., to rotate about its center). In contrast, it is force (thrust) that changes the spacecraft's orbit along which it moves.

As illustrated in Figure 11.1(a), to apply a pure torque, it is necessary to fire two thrusters of equal thrust and equal start and stop times, placed equidistant from the center of mass. As shown in Figure 11.1 (b) spacecraft attitude must be controlled about three mutually perpendicular axes, each with two degrees of freedom (clockwise and counter clockwise rotation), giving a total of six degrees of freedom. For three axes there should be a minimum of 12 thrusters in such a system, but some spacecraft with geometrical or other limitations on the placement of these nozzles or with provisions for redundancy may actually have more than 12, as high as 24. With these thruster clusters, located at a number positions around the spacecraft, attitude and orbit control can be effectively achieved. By firing thrusters in pairs, the spacecraft can be rotated in any direction as shown in Figure 11.1 (c). Also, in some designs, the same system can, by operating different sets of thrusters, provide translation forces. As far as accuracy of the attitude, it depends on the component of the spacecraft and on the operation required: 4-10°for solar arrays and 0.1-0.5°for high gain antennas. In special cases like Hubble Space Telescope, requires a very high accuracy to maintain its orientation within 0.007 of an arcsecond.

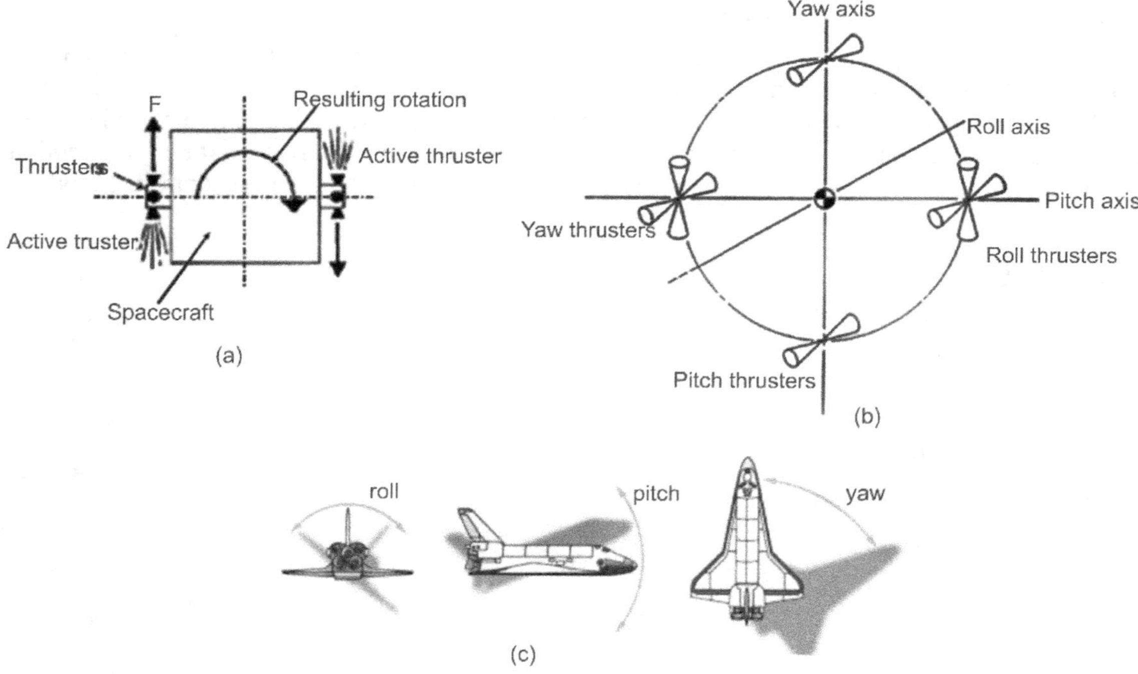


**Figure 11.1** Spacecraft Attitude in terms of Roll, Pitch and Yaw Angles.

Figure 11.2 shows the block diagram of a typical Attitude Control System (ACS). The sensing devices (Sensors), like Sun tracker, stat tracker and Earth horizon, determine the attitude and position of the vehicle with respect to some reference direction at any given time.

A control logic (Attitude Control Electronics, ACE) system compares the actual space and rotary position with the desired or programmed position and issues command signals to change the vehicle's position within a desired time period. The actuators like thrusters, momentum wheel and gyros, after receiving output signals from ACE, change the angular position as desired.

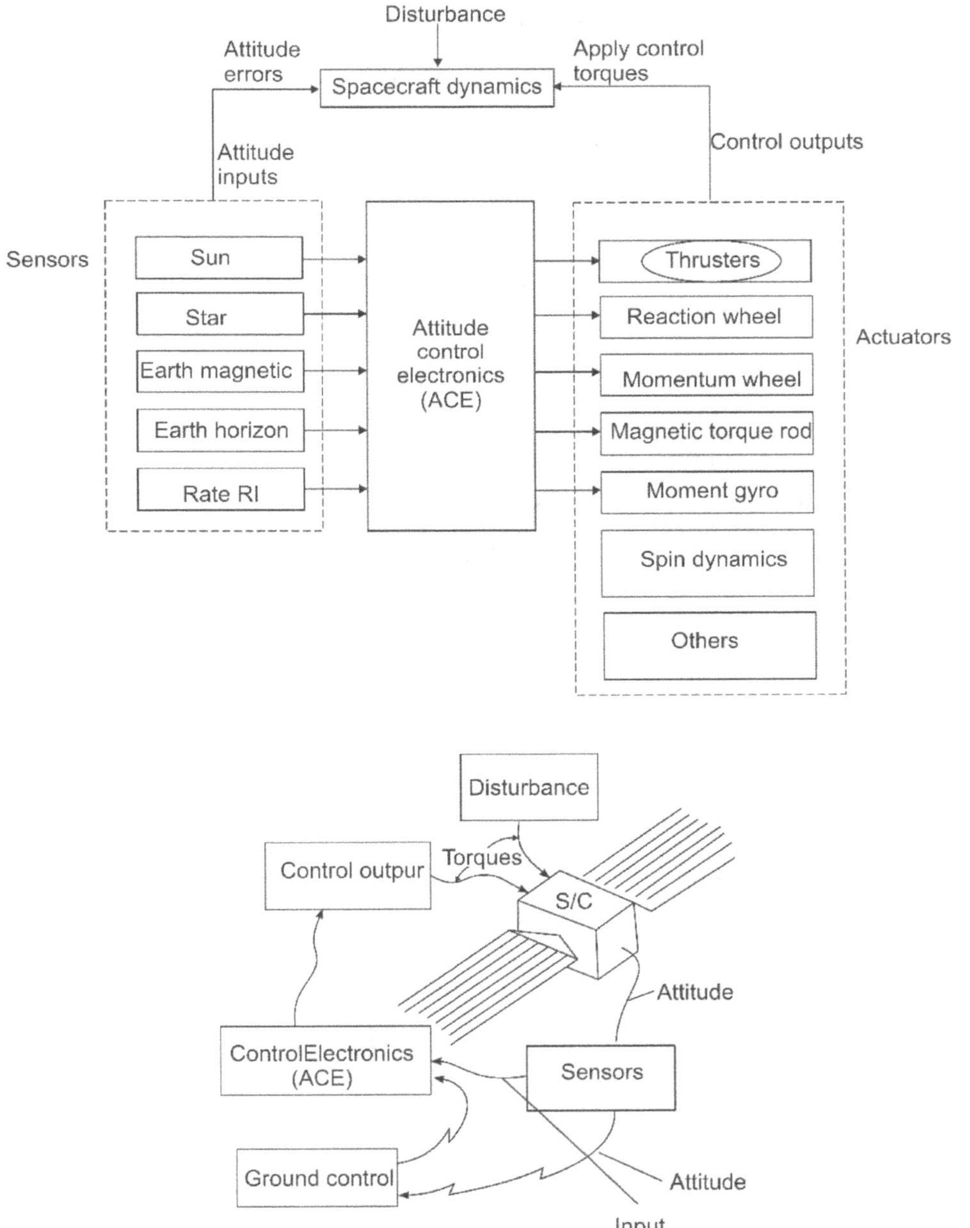


**Figure 11.2** Typical Attitude Control System Block Diagram.

## 11.3  TORQUE TO A SPACECRAFT

Some of the thrusters on a spacecraft are used to attitude control. Because these must restart frequently, only the fluid motors (bipropellant, monopropellant and cold gas) are used. To have rotational maneuver of spacecraft a torque has to be applied about an axis. To apply torque, spacecraft thrusters are fired in pairs. If a pair of thrusters ($n=2$) are fired, the applied torque is given by

$$T = nFL \qquad\qquad .....(11.1)$$

where

$T$ = torque applied to the spacecraft, Nm

$F$ = thrust of a single thruster, N

$n$ = number of motors firing usually two or must be a multiple of two for pure rotation

$L$ = radius from the vehicle center of mass to the thrust vector, m

In rotary motion torque can be expressed as

$$T = I_v \alpha \qquad\qquad .....(11.2)$$

$\alpha$ = angular acceleration of the spacecraft during a firing, rad/s$^2$

$I_v$ = mass moment of inertia of the spacecraft, kg-m$^2$

Therefore, the angular acceleration of spacecraft during burn is

$$\alpha = \frac{T}{I_v} = \frac{nFL}{I_v} \qquad\qquad .....(11.3)$$

If $t_b$ is the duration of the thruster burn, the angular velocity $\omega$ (rad/s) of the spacecraft is given by

$$\omega = \alpha t_b = \frac{nFL}{I_v} t_b \ \ rad/s \qquad\qquad .....(11.4)$$

When the thrusters are shutdown, the angular rotation $\theta$(rad) of the spacecraft is

$$\theta = \frac{\alpha t_b^2}{2} = \frac{nFL}{2I_v} t_b^2 \ \ rad \qquad\qquad .....(11.5)$$

At shutdown, acceleration goes to zero, and the spacecraft is left rotating at a velocity *of* $\omega$.

The change of spacecraft angular momentum during the firing $h$ (kg-m$^2$/s) is

$$h = I_v \omega \left( kg \ m^2 \ \frac{rad}{s} \right) = T t_b = nFL t_b \ kg \ m^2 / s \qquad ..... (11.6)$$

$$\left( kg \ m^2 \ \frac{rad}{s} = \frac{kg \ m}{s^2} m \ s \ rad = Nm \ s \right)$$

The specific impulse is thrust for unit mass of propellant $m_p$

$$I_{sp} = \frac{Total \ impulse}{Mass \ of \ propellant} = \frac{nFt_b}{m_p}$$

$$m_p = \frac{nFt_b}{I_{sp}} = \frac{h}{I_{sp}L} \qquad\qquad .....(11.7)$$

The above equation gives the mass of the propellant required to fire a pair of thrusters. It can be observed that the propellant can be saved by increasing the moment arm L. The maximum moment arm is constrained by the inside diameter of the launch vehicle payload fairing.

### 11.3.1 One Axis Manoeuvre

To control the rotation (clockwise and counter clockwise) of the spacecraft along an axis, four thrusters are to be arranged at the end of the moment arm L as shown in Figure 11.3. If a pair of thrusters is fired to have a rotation (say, counter clockwise), the total movement contains three parts as shown in Figure 11.3:

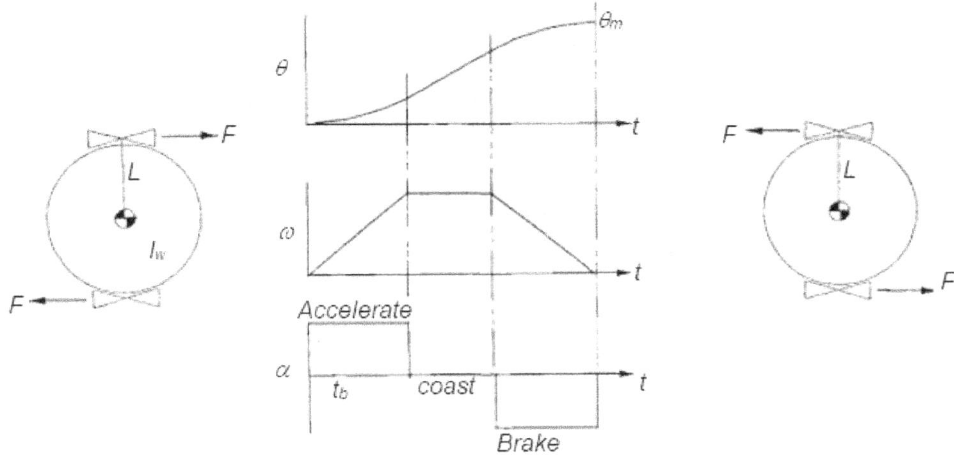


**Figure 11.3** Single Axis Manoeuvre of Spacecraft (Source: Charles D. Brown).

1. angular acceleration (produced by a thruster pair firing (1 &4)

2. coasting

3. braking (braking is caused by a firing of the opposite pair of thrusters (2&3)

The total angle of rotation is

$$\theta_m = \theta(accelerating) + \theta\ (coasting) + \theta\ (braking)$$

If $t_c$ is the duration of coasting, the rotation of spacecraft during this period using Eq. (11.4) is

$$\theta\ (coasting) = \omega t_c = \frac{nFL}{I_v} t_b t_c\ \ rad$$

The angular rotation during acceleration or braking using Eq. (11.5) is given by

$$\theta(acceleration\ or\ braking) = \frac{\alpha t_b^2}{2} = \frac{nFL}{2I_v} t_b^2\ rad$$

The total rotation during the maneuver is

$$\theta_m = 2\frac{nFL}{2I_v}t_b^2 + \frac{nFL}{I_v}t_b t_c = \frac{nFL}{I_v}\left(t_b^2 + t_b t_c\right)\ rad \qquad \dots..(11.8)$$

The maneuver time is

$$t_m = 2t_b + t_c \qquad \dots..(11.9)$$

If the coast time is zero

$$t_m = 2\sqrt{\frac{I_v\theta_m}{nFL}} \qquad \dots.(11.10)$$

The propellant required for a one-axis maneuver is twice the single burn consumption and is given by

$$m_p = 2\frac{nFt_b}{I_{sp}} = \frac{2H}{I_{sp}L} \qquad \dots..(11.11)$$

All maneuvers are three-axis maneuvers. Equation (11.11) must be applied to each axis to determine total propellant required.

**Example 11.1**

The spacecraft with moment of inertia about z-axis 600 kg-m$^2$ and moment arm 1.5 m is making a maneuver of 90 degrees turn about the z axis. To perform the maneuver the spacecraft is using thrusters generating 4 N thrust each. The specific impulse of the thruster is 200 s. Find maneuver time and the propellant consumed.

The required rotation of the spacecraft about Z-axis is

$$\theta_m = \frac{\pi}{2} = 1.5708\ rad$$

Time of maneuver, Eq. (11.10)

$$t_m = 2\sqrt{\frac{I_v\theta_m}{nFL}} = 2\sqrt{\frac{600 \times 1.5708}{2 \times 4.0 \times 1.5}} = 17.73\ s$$

Time of burn

$$t_b = \frac{t_m}{2} = 8.865\ s$$

Propellant consumed

$$m_p = 2\frac{nFt_b}{I_{sp}} = \frac{2 \times 2 \times 4 \times 8.865}{200 \times 9.81} = 0.0362\ kg$$

## 11.4  ORBIT SHAPING

Orbital maneuvers transfer a spacecraft from one orbit to another and it requires the firing of onboard thrusters to produce the required velocity change ($\Delta v$). In a coplanar orbit transfer firing thruster change orbital elements such as the semimajor axis and eccentricity. Figure 11.4 shows two impulse coplanar circular orbit transfers from inner orbit 1 to outer orbit 2. At point A of orbit 1, the thruster fires to give the spacecraft the first impulse changing its orbit to an elliptical transfer orbit shown in dotted path. The spacecraft follows the elliptical path because it is always in the gravitational field of the central body (Earth), it must follow orbital motion laws. In case, a second impulse is imparted to the spacecraft at point B of elliptical path, the velocity vector changes and enters the circular orbit 2.

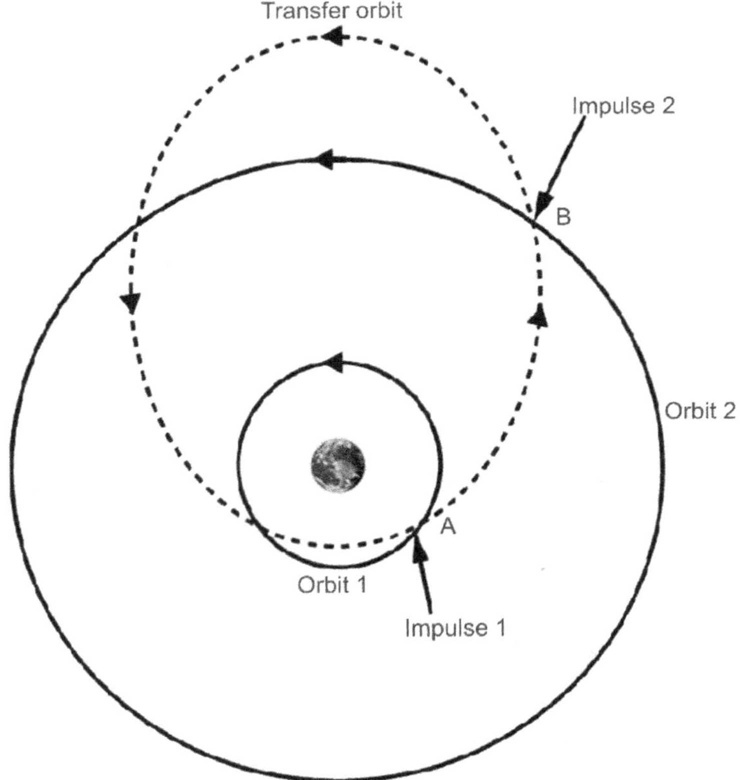


**Figure 11.4** Coplanar Orbit Transfer.

Thus, any conic orbit can be converted to any other conic orbit by adjusting velocity by changing the velocity. The velocity increments ($\Delta v$) estimation is essential to calculate the amount of propellant to be carried by the spacecraft for such orbital changes and housekeeping in its life span. In the following sections, the Energy-efficient Hohmann transfer, bielliptic Hohmann transfer, non-Hohmann transfer with and without rotation of the apse line, plane change and chase maneuvers are described briefly.

### 11.4.1   Hohmann Transfer

Hohmann transfer is the most energy-efficient for transferring between two coplanar circular orbits sharing a common focus as shown in Figure 11.5. The spacecraft is initial at orbit 1 (i.e., parking orbit) at a radius $r_1$. It is transferred to final orbit 3 at a radius $r_2$. It requires two- impulse maneuvers. The Hohmann transfer is an elliptical orbit tangent to both the inner and outer circular orbits on its apse line AB. With the first impulse, a velocity increment $\Delta v_A$ in the direction of flight boosts the spacecraft to higher energy elliptical trajectory AB. Another forward velocity increment $\Delta v_B$ at B places the vehicle on the still higher energy, outer circular orbit. Thus, the total energy increment required to transfer from orbit 1 to orbit 3 is given by

$$\Delta v_{total} = \Delta v_A + \Delta v_B$$

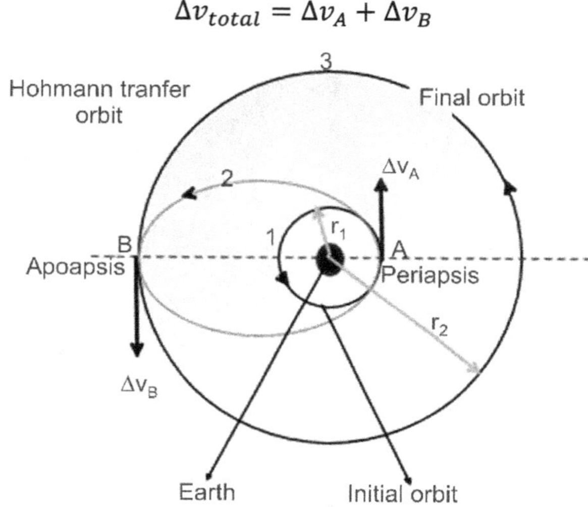


**Figure 11.5** Hohmann Transfer for Coplanar Circular Orbits.

The first velocity increment at A is

$$\Delta v_A = v_p - v_{cA}$$

$v_p$ is the periapsis velocity of Hohmann transfer

$v_{cA}$ is the circular orbit 1 velocity.

Similarly, the second velocity increment at B is

$$\Delta v_B = v_{cB} - v_a$$

$v_a$ is the apoapsis velocity of Hohmann transfer

$v_{cB}$ is the circular orbit velocity

The semimajor axis of Hohmann transfer is

$$a_t = \frac{1}{2}(r_1 + r_2) = \frac{1}{2}(r_p + r_a)$$

The specific energy of the Hohmann transfer orbit is, Eq. (2.16)

$$\varepsilon_t = -\frac{\mu}{2a_t}$$

Increasing the energy requires reducing its magnitude, to make ε less negative. Therefore, the larger the semimajor axis, the more energy the orbit has. The flight time on the Hohmann-transfer ellipse is simply one-half the elliptical period:

$$T_f = \frac{\pi}{\sqrt{\mu}}\, a_t^{\frac{3}{2}} \qquad\qquad .....(11.12)$$

Without second impulse burn $\Delta v_B$, the spacecraft remains on the Hohmann transfer ellipse, i.e., its orbit is changed from circular to elliptical. If the transfer is from outer orbit 3 to inner orbit 1, the same $\Delta v_{total}$ is required. Here, the spacecraft is moving from higher energy orbit to lower energy orbit, hence, the thrust of the maneuvering rocket is directly opposite to the flight direction to act as a brake on the motion. Though the thrust is in the opposite direction, the propellant expenditure is the same and hence only the magnitudes of $\Delta v$'s are to be considered. Obviously, only one-half of the ellipse is flown during the maneuver, which can occur in either direction, from the inner to the outer circle, or vice versa.

Figure 11.6 shows two impulse transfers between two elliptical orbits having the same apse line. The transfer can be done along path 3 (i.e., from periapsis of inner orbit 1 to apoapsis of orbit 2 along path CF) or along path 3' (i.e., from apoapsis of inner orbit 1 to periapsis of outer orbit 2 along path DE). The transfer ellipses are tangent to both the initial and target ellipses 1 and 2. To find out the lowest energy transfer, i.e., the Hohmann transfer, consider the specific energy:

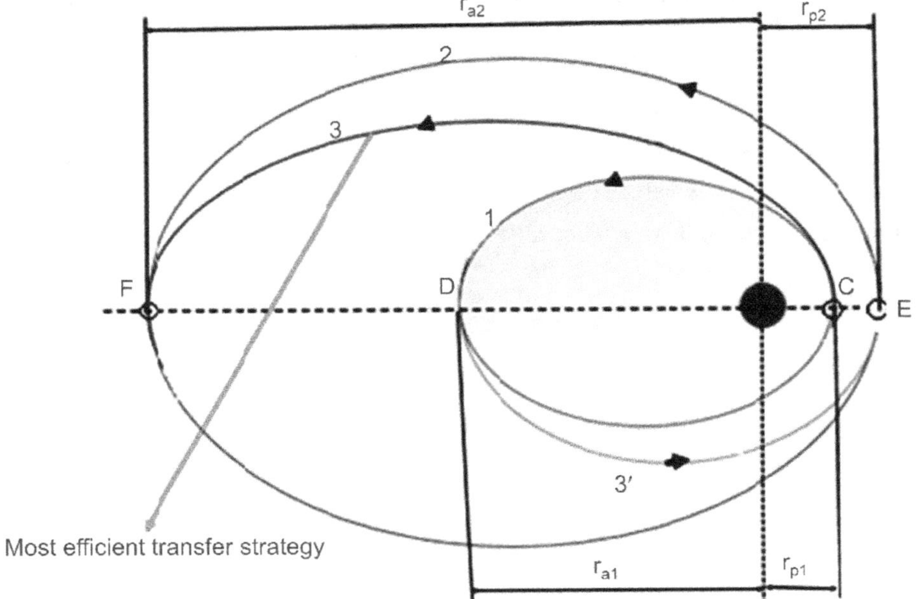


**Figure 11.6** Hohmann Transfers between two Coaxial Elliptical Orbits.

$$\epsilon_t = specific\ kinetic\ energy + specific\ potential\ energy$$

$$\epsilon_t = \frac{v^2}{2} + \frac{\mu}{r}$$

Taking differential

$$d\epsilon_t = vdv + \frac{\mu}{r^2}dr$$

For an impulsive velocity change, there is no change in position (i.e., $dr = 0$). The above equation can be written with incremental values as

$$\Delta\epsilon_t = v\Delta v$$

This shows clearly that the largest energy increase can be obtained when velocity $v$ is maximum, i.e., at periapsis. Hence, the most efficient transfer orbit begins at the periapsis on inner orbit 1, where its *kinetic energy* is greatest, regardless of the shape of the outer target orbit. Therefore, the Hohmann transfer path is 3 (CF). Similarly, from an outer ellipse to an inner ellipse the most energy-efficient transfer ellipse terminates at periapsis of the inner target orbit.

## Example 11.2

A 2000 kg spacecraft having an initial geocentric elliptical orbit with perigee radius of 7000 km and eccentricity of 0.3 is transferred to elliptical orbit having perigee 32,000 km and eccentricity 0.5. The two orbits have common apse lines and their perigee are on the same side of the earth. Find the total delta V and time of flight for transfer from the perigee of the inner orbit to the apogee of the outer orbit. Compare the values if the transfer of orbit is from the apogee of the inner orbit to the perigee of the outer orbit. Take the specific impulse as 3500 m/s.

As shown in Figure 11.7 four elliptical trajectories are involved.

It is easy to find the velocities at perigees and apogees of these orbits if angular momentum, $h$ of these orbits is known. The following equations will be used in solving the problem.

Specific angular momentum $h = r^2\dot{\theta} = wr^2 = vr$       (Ref. Eq. 2.30)

$$\text{Perigee Radius } r_p = \frac{h^2}{\mu}\frac{1}{1+e} \quad\quad\quad\text{(Ref. Eq. 2.39)}$$

$$\text{Apogee radius } r_a = \frac{h^2}{\mu}\frac{1}{1-e} \quad\quad\quad\text{(Ref. Eq.2.40)}$$

$$\text{Semimajor axis of ellipse } a = \frac{h^2}{\mu}\frac{1}{1-e^2} \quad\quad\quad\text{(Ref. Eq.2.41)}$$

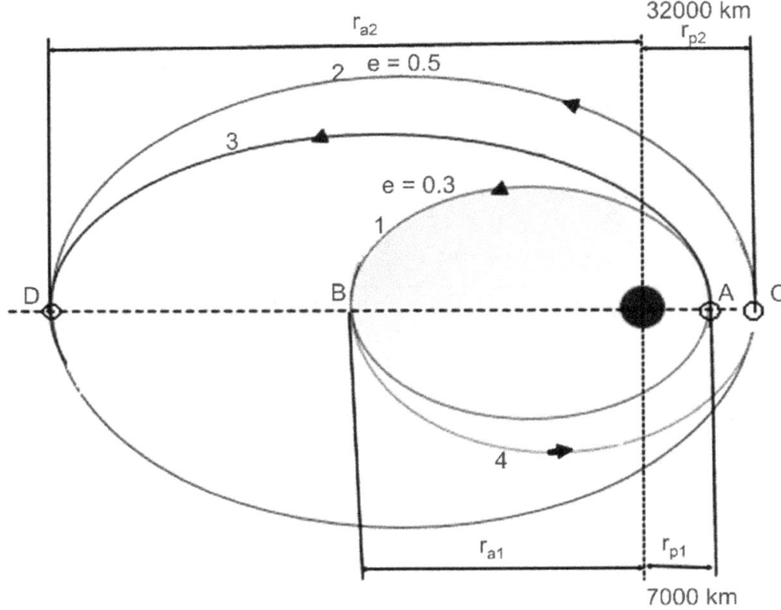


**Figure 11.7** Hohmann Transfer between Two Elliptical Orbits-Example 11.2.

## Elliptical Orbit 1:

Given $e_1 = 0.3$, $r_{p1} = 7000$ km

Angular momentum

$$h_1 = \sqrt{\mu r_{p_1}(1 + e_1)} = \sqrt{398600 \times 7000 \times (1 + 0.3)} = 60{,}226\frac{km^2}{s}$$

Apogee radius

$$r_{a1} = \frac{h_1^2}{\mu}\frac{1}{1 - e_1} = \frac{60226^2}{398600}\frac{1}{1 - 0.3} = 13{,}000\ km$$

Elliptical Orbit 2:

$e_2 = 0.5$, $r_{p2} = 32{,}000$ km

Angular momentum

$$h_2 = \sqrt{\mu r_{p_2}(1 + e_2)} = \sqrt{398600 \times 32000 \times (1 + 0.5)} = 138{,}321\frac{km^2}{s}$$

Apogee radius

$$r_{a2} = \frac{h_2^2}{\mu}\frac{1}{1 - e_2} = \frac{138321^2}{398600}\frac{1}{1 - 0.5} = 96{,}000\ km$$

Elliptical Orbit 3:

$$r_{p3} = 7000 \text{ km}$$

$$r_{a3} = r_{a2} = 96{,}000 \text{ km}$$

Eccentricity

$$e_3 = \frac{r_{a3} - r_{p3}}{r_{a3} + r_{p3}} = \frac{96000 - 7000}{96000 + 7000} = 0.864$$

Angular momentum

$$h_3 = \sqrt{\mu r_{p_3}(1 + e_3)} = \sqrt{398600 \times 7000 \times (1 + 0.864)} = 72{,}117 \frac{km^2}{s}$$

Semi major axis

$$a_3 = \frac{r_{a3} + r_{p3}}{2} = \frac{96000 + 7000}{2} = 51{,}500 \ km$$

Elliptical Orbit 4:

$$r_{a4} = 32{,}000 \ km$$

$$r_{p4} = r_{a1} = 13{,}000 \ km$$

Eccentricity

$$e_4 = \frac{32000 - 13000}{32000 + 13000} = 0.422$$

Angular momentum

$$h_4 = \sqrt{\mu r_{p_4}(1 + e_4)} = \sqrt{398600 \times 13000 \times (1 + 0.422)} = 85{,}840 \ \frac{km^2}{s}$$

Semimajor axis

$$a_4 = \frac{r_{a4} + r_{p4}}{2} = \frac{32000 + 13000}{2} = 22{,}500 \ km$$

Tabulating these values of the orbits for easy reference as below:

| Orbit | $h$ km²/s | $r_a$ km | $r_p$ km |
|-------|-----------|----------|----------|
| 1 | 60,226 | 13,000 | 7,000 |
| 2 | 138,321 | 96,000 | 32,000 |
| 3 | 72,117 | 96,000 | 7,000 |
| 4 | 85,840 | 32,000 | 13,000 |

## Part 1

Speed of Orbits at A on 1 and 3 trajectories:

Speed at point A on orbit 1

$$v_{A1} = \frac{h_1}{r_{p1}} = \frac{60226}{7000} = 8.604 \frac{km}{s}$$

Speed at point A on orbit 3

$$v_{A3} = \frac{h_3}{r_{p3}} = \frac{72117}{7000} = 10.302 \, \frac{km}{s}$$

Therefore, the $\Delta v$ required at point A to change the orbit is

$$\Delta v_{1to3atA} = 10.302 - 8.604 = 1.698 \, \frac{km}{s}$$

Speed of Orbits at D on 2 and 3 trajectories:

Speed at point D on orbit 2

$$v_{D2} = \frac{h_2}{r_{a2}} = \frac{138321}{96000} = 1.441 \, km/s$$

Speed at point D on orbit 3

$$v_{d3} = \frac{h_3}{r_{a3}} = \frac{72117}{96000} = 0.751 \, km/s$$

Therefore, the $\varDelta v$ required at point D to change the orbit is

$$\Delta v_{3to2atD} = 1.441 - 0.751 = 0.690 \, \frac{km}{s}$$

Therefore, the total delta $v$ required to change the orbit 1 to 2 with impulse firings at A and D is

$$\Delta v_{tot1to2} = 1.698 + 0.690 = 2.388 \, km/s$$

Time of flight (TOF):

The time to reach point D from A is the half of elliptical orbit time period of 3. Eq. (11.12)

$$TOF_{3AtoD} = \frac{\pi}{\sqrt{\mu}} \, a_3^{\frac{3}{2}} = \frac{1}{2} \frac{2\pi}{\sqrt{398000}} \, 51500^{\frac{3}{2}} = 58,126 \, s = 16.2 \, hours$$

Mass Propellant Required to change the orbit from 1 to 2:

Rewriting the rocket equation Eq. (3.14) as

$$\frac{M_i}{M_f} = e^{\frac{\Delta v}{V_e}} = \frac{M_i}{M_i - M_P} e^{\frac{\Delta v}{V_e}}$$

The mass of propellant expended for the impulse burn $\Delta v$ is

$$M_P = M_i \left(1 - e^{-\frac{\Delta v}{V_e}}\right) = 2000 \left(1 - exp(-\frac{2388}{3500})\right) = 988 \, kg$$

**Part 2**

Speed of Orbits at B on 1 and 4 trajectories:

Speed at point B on orbit 1

$$v_{B1} = \frac{h_1}{r_{a1}} = \frac{60226}{13000} = 4.633 \frac{km}{s}$$

Speed at point B on orbit 4

$$v_{B4} = \frac{h_4}{r_{p4}} = \frac{85840}{13000} = 6.603 \frac{km}{s}$$

Therefore, the $\Delta v$ required at point B to change the orbit is

$$\Delta v_{1to4atB} = 6.603 - 4.633 = 1.970 \frac{km}{s}$$

Speed of Orbits at C on 2 and 4 trajectories:

Speed at point C on orbit 2

$$v_{C2} = \frac{h_2}{r_{p2}} = \frac{138321}{32000} = 4.323 \frac{km}{s}$$

Speed at point C on orbit 4

$$v_{C4} = \frac{h_4}{r_{a4}} = \frac{85840}{32000} = 2.683 \frac{km}{s}$$

Therefore, the $\Delta v$ required at point c to change the orbit is

$$\Delta v_{4to2atC} = 4.323 - 2.683 = 1.640 \; km/s$$

Therefore, the total delta $v$ required to change the orbit 1 to 2 with thrustor firing at B and C is

$$\Delta v_{tot1to2via4} = 1.970 + 1.640 = 3.610 \frac{km}{s}$$

Time of flight (TOF):

The time to reach point C from B is the half of elliptical orbit time period of 4

$$TOF_{3BtoC} = \frac{1}{2} \frac{2\pi}{\sqrt{\mu}} a^{\frac{3}{2}} = \frac{1}{2} \frac{2\pi}{\sqrt{398000}} 22500^{\frac{3}{2}} = 16786 \; s = 4.66 \; hours$$

Mass Propellant Required to change the orbit from 1 to 2 via 4

The mass of propellant expended is

$$M_P = M_i \left( 1 - e^{-\frac{\Delta V}{V_e}} \right) = 2000 \left( 1 - exp(-\frac{3610}{3500}) \right) = 1286 \; kg$$

The total delta $v$ required to transfer orbit along path 3 is 2.388 km/s, whereas along path 4 is 3.160 km/s, i.e., 32.33% increase in delta $v$ expenditure. This requires additional propellant of 298 kg. Thus, the most efficient transfer orbit begins at the periapsis on inner orbit 1.

## 11.4.2    Bielliptic Hohmann Transfer

Hohmann transfer is the most efficient *two* impulse maneuver. However, for some cases, adding an additional impulse reduces the total $\Delta v$ requirement for an orbital transfer. Figure 11.8 shows bielliptic transfer from orbit 1 to orbit 2. It starts by departing the initial orbit 1 at A onto an

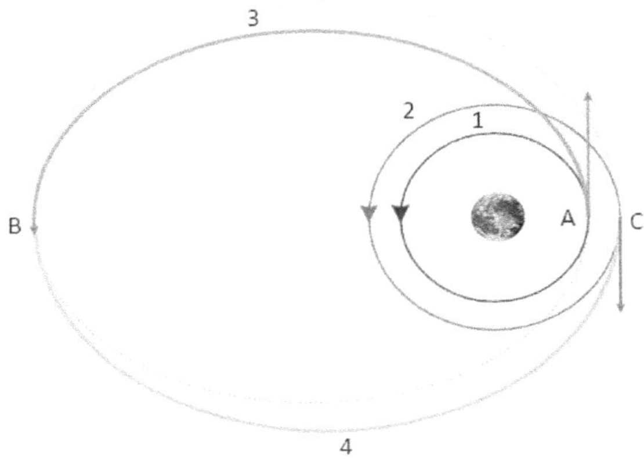


**Figure 11.8** Hohmann Bielliptic Transfer.

elliptical transfer orbit 3 whose apoapsis point B is at a higher altitude than the target orbit. At apoapsis point B of the first transfer orbit 3, the impulse burn boosts the velocity again onto a second transfer orbit 4. This orbit 4 has the same periapsis altitude as the target orbit 2. A third impulse burn at C puts the spacecraft into target orbit 2. The reason this type of transfer can be more efficient than the two-impulse Hohmann transfer is because the velocity change required to change the periapsis altitude of the second transfer orbit at point B is very small. **In fact, as the point B approaches infinity (where the orbital speed is zero) $\Delta v_B$ approaches zero.** In terms of the orbital radii $r_A$ (initial orbit 1), $r_B$ (apoapsis of the transfer orbit 3), and $r_C$ (target orbit), for:

$$\frac{r_C}{r_A} < 11.94$$

the regular Hohmann transfer is more efficient, i.e., **if the radius of the target orbit ($r_C$) is less than 11.94 times that of the initial orbit ($r_A$), then the standard Hohmann maneuver is the more energy efficient.** otherwise, if the target orbit to initial orbit ratio is:

$$\frac{r_C}{r_A} > 15.58$$

then the bielliptic transfer is more efficient. **Between those two ratios, large values of the apoapsis radius $r_B$ favor bielliptic transfer, while smaller values favor Hohmann transfer.** However, the bielliptic transfer has a much longer flight time than the standard Hohmann transfer. This is because the bielliptic transfer must traverse 360° of true anomaly on two very large ellipses.

## 11.5  PHASING MANEUVERS

When two spacecrafts' rendezvous in space, like SpaceX Dragon docks to the ISS, they must have the same position and velocity vectors. There are typically two stages to rendezvous: (a) the first stage is to bring the two spacecraft, into proximity by performing a phasing maneuver, usually by one of the spacecrafts (b) the final approach requires small adjustments to the approaching craft's velocity so that the two can match. Figure 11.9 shows that two spacecrafts, chaser and other target at phase angle $\phi$. If the target is behind the chaser in the same orbit, the chaser must speed up to enter a larger slower phasing orbit to allow the target to catch up.

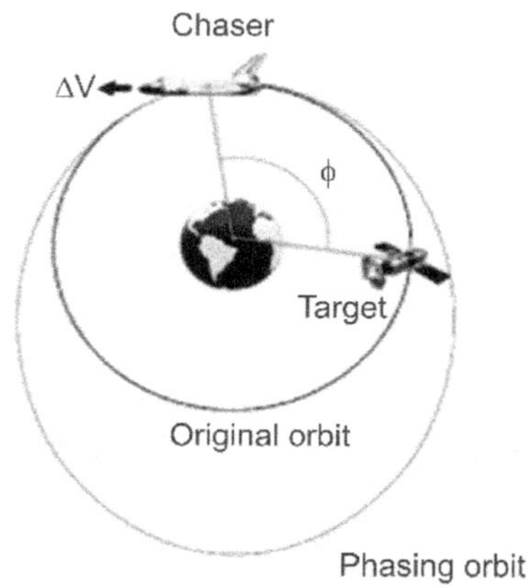


**Figure 11.9**  Phase angle between Chaser and Target Spacecrafts.

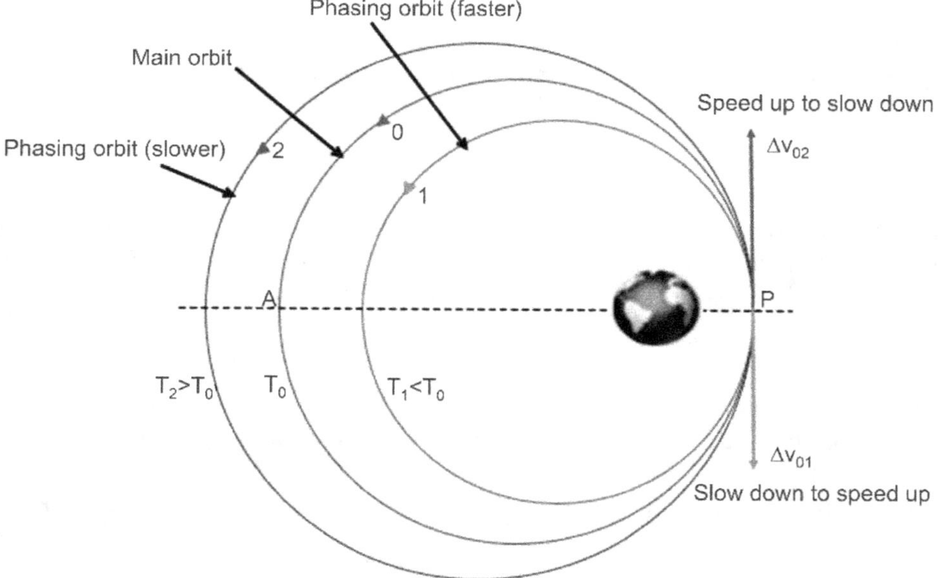


**Figure 11.10**  Phasing orbits.

Figure 11.10 shows the main orbit, 0 and two phasing orbits 1&2. The main orbit period is $T_0$. At point P the impulse burn in the direction of flight gives a velocity increment of $\Delta v_{0-2}$ to slow down the speed. Since $a_2 > a_0$ the period of the orbit is $T_2 > T_0$. An impulse burn at P opposite direction to the flight direction gives a velocity change of $\Delta v_{0-1}$ to increase the speed. i.e., the time period $T_1 < T_0$.

Therefore, referring to situation in Figure 11.9, while the chaser craft travels 360° of true anomaly in time $T_2$, the target spacecraft travels more than 360° of true anomaly. This allows the chaser spacecraft to catch up and reach the impulse point at the same time that the target returns there. The example below provides a clear understanding of phasing maneuvers.

**Example 11.3**

Chaser spacecraft A is following the target spacecraft B. Both are in the same elliptical orbit having perigee 8000 km and apogee 14000 km and phase angle difference of 90 deg. (fig 11.11) The chaser vehicle at A executes a phasing maneuver to catch the target spacecraft back at A after just one revolution of the chaser. Calculate the required total velocity change.

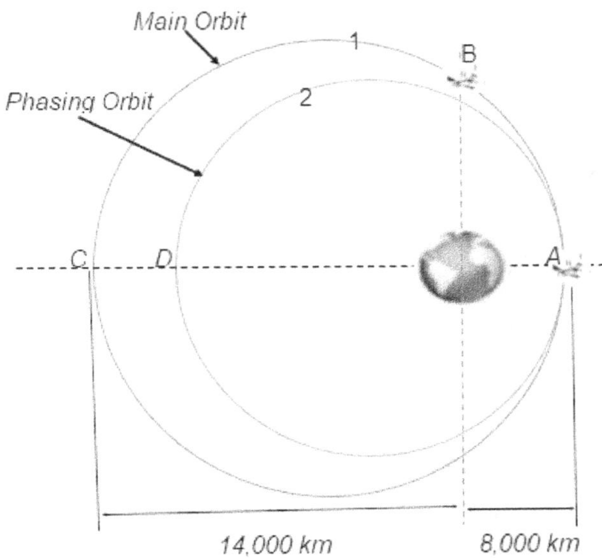


**Figure 11.11** Phasing Orbit for Example 11.2.

Main Orbit (1):

Perigee $r_{p1}$ = 8000 km

Apogee $r_{a1}$ =14,000 km

Eccentricity

$$e_1 = \frac{14000 - 8000}{14000 + 8000} = 0.273$$

Angular momentum

$$h_1 = \sqrt{\mu r_p(1+e)} = \sqrt{398600 \times 8000 \times 1.273} = 63,713 \frac{km^2}{s}$$

Semi major axis

$$a_1 = \frac{14000 + 8000}{2} = 11,000 \ km$$

Time period

$$T_1 = \frac{2\pi}{\sqrt{\mu}} a^{\frac{3}{2}}_3 = \frac{2\pi}{\sqrt{398600}} (11000)^{1.5} = 11,482 \ s$$

Velocity of chaser in orbit 1 at point A

$$v_{1atA} = \frac{h_1}{r_{p1}} = \frac{63713}{8000} = 7.964 \ km/s$$

Phasing orbit 2:

The phasing orbit must have a period $T_2$ equal to the time it takes the target vehicle at B to coast around to point A on orbit 1.

$$T_2 = T_1 - t_{AB}$$

Time from periapsis (Ref. Eqs. 2.53. 2.55 &2.56)

$$t = \frac{(E - esinE)}{n}$$

$$n = 2\pi/T$$

$$cos E = \frac{e + cos\theta}{1 + ecos\theta}$$

where E is an eccentric anomaly.

Here $\theta = 90°$; hence $\quad cos \ E = e_1 = 0.273,$

E = 74.16 deg = 1.294 rad

$$t_{AB} = (1.294 - 0.273 \times sin \ sin \ 74.16) \times 11483 \ /2\pi \ = 1885 \ s$$

$$T_2 = 11482 - 1885 = 9597 \ s$$

Semi major axis

$$a_2 = \left(\frac{\sqrt{\mu}}{2\pi} T_2\right)^{\frac{2}{3}} = \left(\frac{\sqrt{398600}}{2\pi} 9597\right)^{\frac{2}{3}} = 9752 \ km$$

Eccentricity

$$e = 1 - \frac{r_p}{a} = 1 - \frac{8000}{9752} = 0.18$$

Angular momentum

$$h_2 = \sqrt{398600 \times 8000 \times 1.18} = 61342 \frac{km^2}{s}$$

Velocity of chaser at point A of orbit 2

$$v_{2atA} = \frac{h_2}{r_{p2}} = \frac{61342}{8000} = 7.668 \ km/s$$

At the beginning of the phasing maneuver, the velocity change required to drop into phasing orbit 2 is

$$\Delta v_{start} = v_{2atA} - v_{1atA} = 7.668 - 7.964 = -0.296 \ \frac{km}{s}$$

The thruster fires in a direction opposite to flight direction to reduce the velocity. The absolute value

$$\Delta v_{start} = 0.296 \ km/s$$

At the end of the phasing maneuver, the velocity change required to return to orbit 1 is

$$\Delta v_{end} = v_{1atA} - v_{2atA} = 7.964 - 7.668 = 0.269$$

The total delta-v required for the chaser to catch up with the target is

$$\Delta v = 0.269 + 0.269 = 0.538 \ \frac{km}{s}$$

Accordingly, sufficient propellant has to be provided in the spacecraft to achieve the delta $v$.


## 11.6  GENERAL COPLANAR TRANSFER

Figure 11.4 shows the general coplanar orbit transfer between two circular orbits. For a general coplanar orbit transfer, at the impulse points the velocity vectors on the circular and elliptical transfer orbits will not necessarily be collinear. Figure 11.12 shows the velocity vectors of the circular. $v_{c1}$ and transfer orbit $v_1$ at impulse maneuver point A. The magnitude of the velocity impulse $\Delta v_1$ can be calculated using cosine law.

$$\Delta v_1 = \sqrt{v_1^2 + v_{c1}^2 - 2v_1 v_{c1} \cos\gamma}$$

where $\gamma_1$ is the flight-path angle of the elliptical-orbit velocity $v_1$.

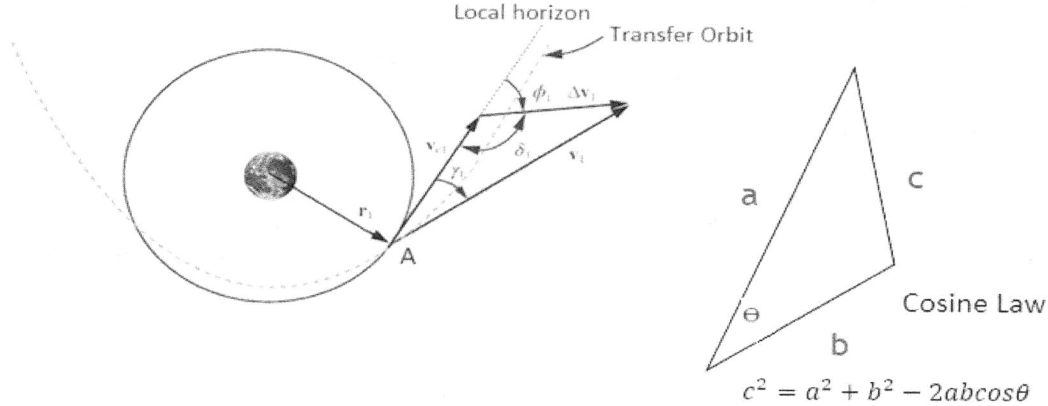


**Figure 11.12** Velocity vectors at point A.

To determine the direction of impulsive maneuver, consider with respective local horizon $\varnothing_1$ is the elevation angle of the vector $\Delta v_1$. Angle $\phi_1$ is the supplementary angle of angle $\delta 1$ as shown in Figure 11.12.

$$\varnothing_1 = \pi - \delta_1$$

From sine law

$$\frac{sin\delta_1}{v_1} = \frac{sin\gamma_1}{\Delta v_1}$$

The angles $\delta_1$ and $\varnothing_1$ can be calculated from above equations. Firing the onboard rocket at elevation angle $\phi 1$ performs the first impulsive burn. Similarly, the details of the second impulse at point B of Figure 11.4 can be calculated.

## 11.7  INCLINATION CHANGE MANEUVER

An impulse velocity increment, $\Delta v$ normal to the plane of the orbit will change the orientation of the orbital plane. In general, a $\Delta v$ component normal to the orbital plane will change the longitude of the ascending node $\Omega$ and inclination $i$ depending on where in the orbit it does the $\Delta v$ burn. With a simple plane change, only the direction of the orbital velocity changes. The velocity's magnitude (orbital speed) stays the same. Such a change is accomplished by applying an out-of-plane impulse at the intersection of the initial and final orbit plane. Figure 11.13 shows the inclination changes between to circular orbits 1 & 2. The plane-change impulse applied at the ascending node. Hence, there is no change in the orbital speed. The $\Delta v$ is in the opposite direction as the angular momentum vector $h_1$ for Orbit 1. The velocity vectors form an isosceles triangle because orbits 1 and 2 have the same circular speeds. Thus, the impulse $\Delta v$ has rotated velocity vector $v_{c1}$ to the orbit 2 plane without changing its magnitude. In addition, the impulse $\Delta v$ has rotated the angular momentum $h_1$ through angle $\Delta i$ to $h_2$ without changing its magnitude. In the vector triangle ABC, $v_{c1} = v_{c2} = v$, $\Delta i = \delta$ and AD is perpendicular BC dividing it two equal parts. Then,

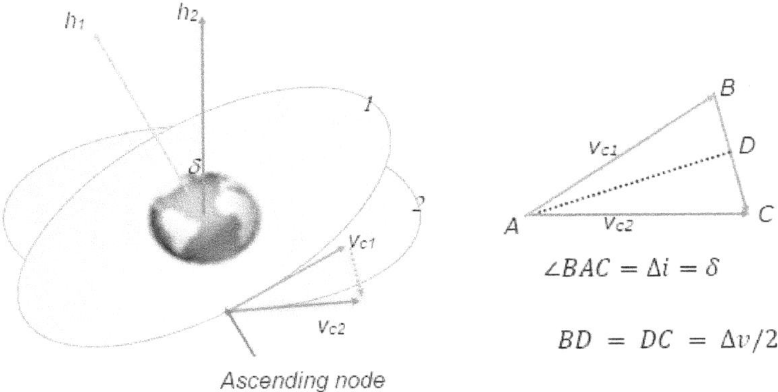


**Figure 11.13** Inclination change at the ascending node.

$$\Delta v = 2v\sin\frac{\delta}{2} \qquad\qquad .....(11.13)$$

For an elliptical orbit, the two equal velocity vectors of the isosceles triangle in Figure 11.13 must be the horizontal velocity component, i.e., $v\cos\gamma$ where $\gamma$ is the flight path angle. Hence, the velocity increment required is

$$\Delta v = 2\cos\gamma\sin\frac{\delta}{2} \qquad\qquad .....(11.14)$$

As $\Delta v$ is directly proportional to speed of the orbit, plane changes are expensive. A 10-deg plane change in low Earth orbit requires a velocity change of 1.4 km/s. It shows that to keep the energy expenditure to minimum, the number of plane changes must be restricted, important to change planes through the smallest possible angle and at the lowest possible velocity. The lowest possible velocity occurs at the greatest radius, i.e., at the apoapsis.

Figure 11.14 shows the variation of $\Delta v/v$ with orbit rotation angle $\delta$ for pure rotation. It can be observed that a plane change of about requires the same $\Delta v$ as is needed to increase the speed to the escape velocity from a circular orbit, $\Delta v/v \approx 0.414$. A plane change of 60° requires $\Delta v$ equal to the spacecraft velocity. In LEO, the velocity is approximately 7.5 km/s. A plane change of 60°, with an $I_{sp}$ of 3000 m/s, would require over 90% of the spacecraft mass to be propellant. Thus, plane changes are very expensive maneuvers, that are to be avoided whenever possible once a spacecraft is in orbit. Instead, plane changes should be accomplished during launch.

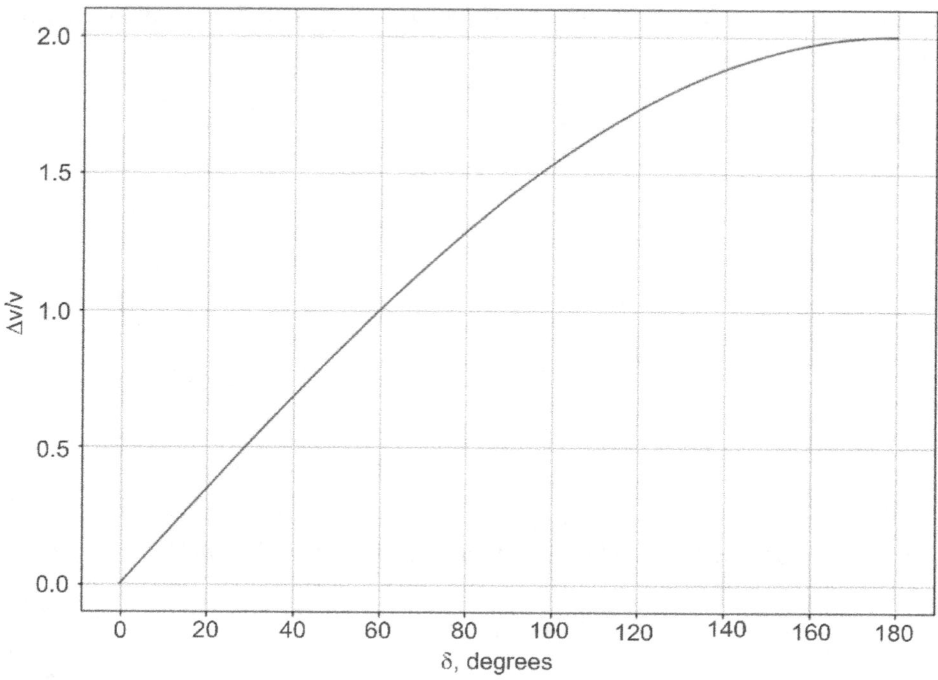

**Figure 11.14** $\Delta v/v$ as function of angle $\delta$ for pure rotation manoeuvres.

## 11.8 COMBINED MANEUVER

Major energy savings can be obtained if in-plane and out-of-plane maneuvers are combined. The savings result from the fact that any side of a vector triangle is always smaller than the sum of the other two sides. Consider plane and velocity changes that have a common focus for the initial orbit 1 and target orbit 2. For a transfer to occur with a single impulse, the orbital planes must intersect and the intersection of two planes forms a line as shown in Figure 11.15. Since the orbits have the same focus, the line of intersection of the planes must pass through the focus. The two orbits have a common apoapsis and impulse maneuver occurs at it. The speed of the orbit is minimum at apoapsis and the radial velocity components are zero in both orbits.

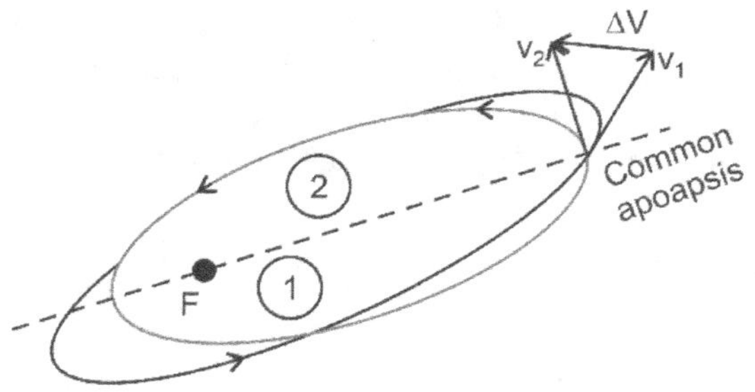


**Figure 11.15** A Single Impulsive Plane Change Manoeuvre at Apoapsis.

The velocity increment can be calculated using Eq. (11.13) or using the cosine law on velocity triangle:

$$\Delta v_1 = \sqrt{v_1^2 + v_2^2 - 2v_1 v_2 \cos \delta}$$

Using the trigonometric identity

$$\cos\delta = 1 - 2\sin^2 \frac{\delta}{2}$$

$$\Delta v_1 = \sqrt{(v_1 - v_2)^2 + 4v_1 v_2 \sin^2 \frac{\delta}{2}} \tag{11.15}$$

This equation gives the impulse velocity change required for a combined maneuver, rotating the plane $\delta$ degrees and change the velocity of the orbit from $v_1$ to $v_2$. If there is no change in the speed, so that $v = v_1 = v_2$, then Eq. (11.13) yields Eq. (11.14)

$$\Delta v = 2v\sin \frac{\delta}{2}$$

This is the delta-v for a pure rotation of the velocity vector through the angle $\delta$.

Figure (11.16) gives the velocity vector addition for three cases:

1.  Combined maneuver of rotation and change of velocity, $\Delta v_1$.

2.  Plane change followed by speed change, i.e., rotate the velocity vector and then change its magnitude. In this case delta-v is

$$\Delta v_2 = 2v_1 \sin \frac{\delta}{2} + |v_1 - v_2|$$

3.  Speed change followed by plane change, i.e., change the speed first, and then rotate the velocity vector. The delta-v required is

$$\Delta v_3 = |v_1 - v_2| + 2v_2 \sin \frac{\delta}{2}$$

It is obvious from Figure (11.16) that both $\Delta v_2$ and $\Delta v_3$ are greater than $\Delta v_1$ as the sum of the lengths of any two sides of a triangle must be greater than the length of the third side. Hence, a plane change accompanied by speed change is the most efficient of the above three maneuvers. Thus, combined maneuvers, which reduce velocity change requirements, have a very beneficial effect on the overall design as it ultimately results in reducing the amount of the propellant that the spacecraft has to carry.

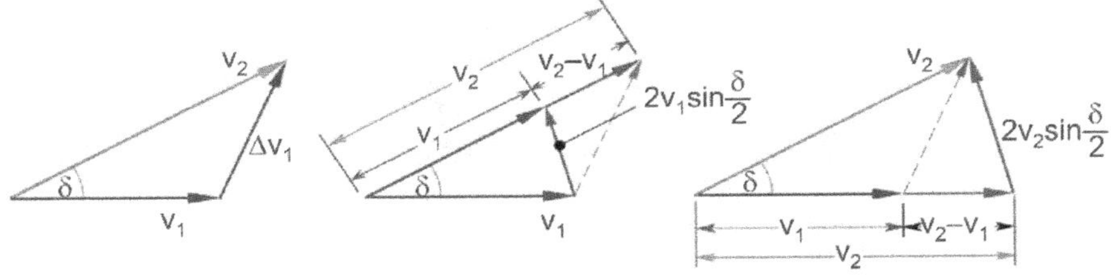


**Figure 11.16** Three cases of the orbital plane rotation about the common apse line.

## 11.9 LAUNCH AZIMUTH

Orbital plane changes are necessary in many cases when the launch location does not match the desired orbit. The latitude of the launch location affects the orbital inclination. The orbital plane intersects with the center of the earth. Therefore, if a spacecraft is launched from any latitude above or below the equator, the orbit will have some natural inclination ($i$) relative to the equatorial plane. To achieve a designed orbit like geosynchronous equatorial orbit ($i = 0°$) requires an orbital plane change depending on the launch latitude. Launching a rocket due east takes full advantage of the earth's rotational velocity. At the equator it is:

$$v_{equator} = \frac{circumference}{sidereal\ period} = \frac{2\pi\ 6378}{23.9345} = 0.4651\ km/s$$

It diminishes toward the poles according to the formula

$$v_{latitude} = v_{equator}\ cos\emptyset$$

where $\emptyset$ is the latitude. The inclination of the orbit depends on the launch azimuth, A is the flight direction at insertion, measured clockwise from north on the local meridian. Therefore, launches due East have an azimuth of 90°, due South of 180°, and due West of 270°. If the launch direction is not directly eastward, then the orbit will have an inclination greater than the launch latitude. The orbits that have an eastward velocity component, i.e., launches to the north easterly ($0° < A < 90°$) and south easterly ($90° < A < 180°$) azimuths, are called prograde orbits. Launches to the west, with $180° < A < 360°$, produce retrograde orbits with inclinations between 90° and 180° as shown in Figure 11.17. The rotating Earth correction of the azimuth is usually small enough to be neglected. The relationship between launch azimuth, latitude, and orbital inclination is given by:

$$cos\ i = cos\ \phi\ sin\ A$$

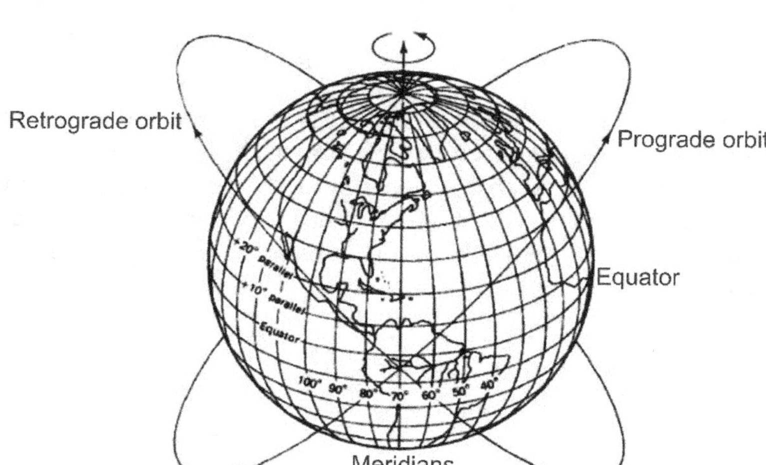


**Figure 11.17** Prograde and Retrograde Orbits.

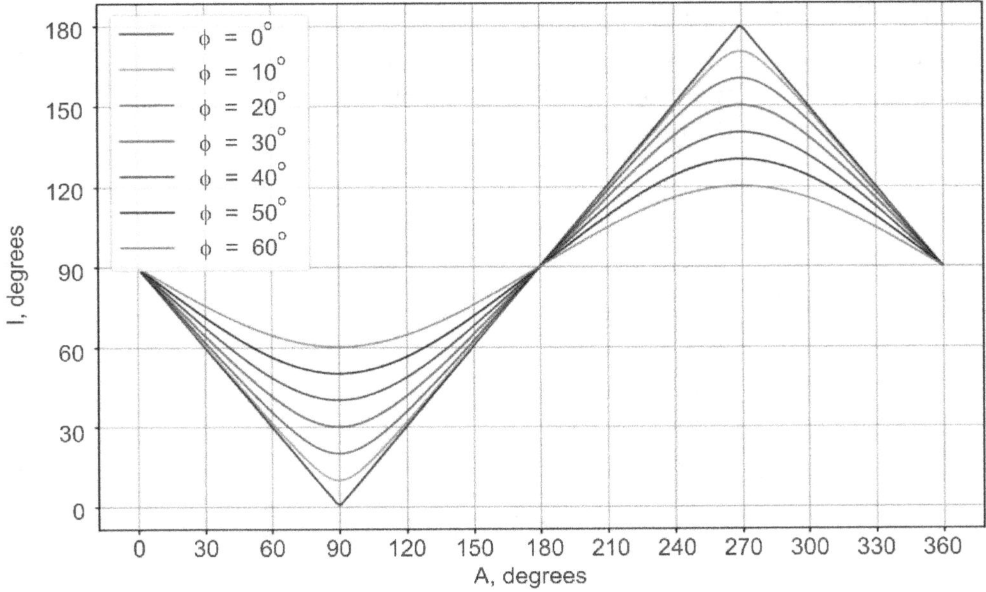


**Figure 11.18** Orbit inclination *i* versus launch azimuth *A* for several latitudes $\phi$.

Figure 11.18 shows a plot of the orbital inclination obtained for a range of latitudes and launch azimuths as per the above equation. It can be observed from the figure that the only way to achieve an inclination of 0° or 180°, an equatorial orbit, is to launch from the equator ($\phi = 0°$) with a launch azimuth of A = 90° (due East) to make a prograde orbit or A = 270° (due West) for a retrograde orbit. As a safety consideration, launch azimuths are limited by the geography surrounding the launch site as shown in Figure 11.19. **Launches from Cape Canaveral and the Kennedy Space Centre in eastern test range are limited to azimuths near due east. The azimuth limitations at Vandenberg Air Force Base in western test range only allow for southward launches.**

**Figure 11.19** Azimuth Limitations and Achievable Inclinations at US Eastern and Western Test Ranges.

### Example 11.4: Geosynchronous Orbit

Most of the world's communications satellites and meteorological satellites are positioned on this orbit. Three spacecraft positioned on this orbit can observe the Earth's entire surface except for the extreme polar region as shown in Figure 11.20.

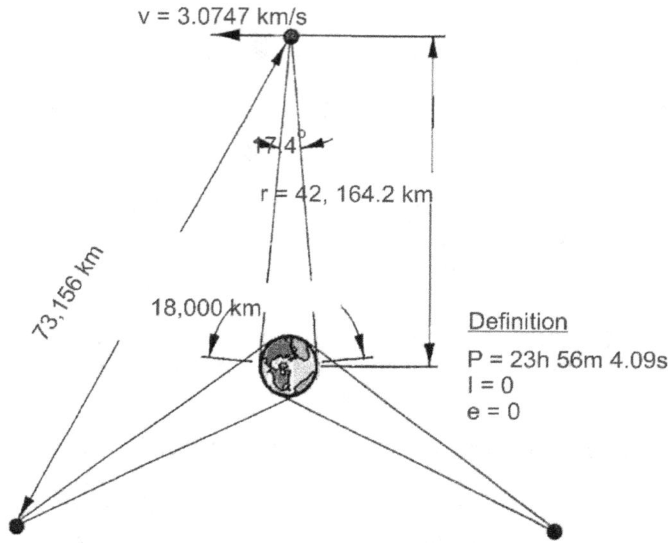


**Figure 11.20** Three Satellites in Geosynchronous Orbit.

The period of one sidereal day is 23 h 56 min 4.09 s, or 86,164.09 s

The gravitational parameter of Earth $\mu = 398{,}600\ km^3/s^2$

Radius of Earth $R_E = 6378\ km$

The period of circular orbit is

$$T = 2\pi\sqrt{r^3/\mu}$$

Rearranging

$$r = \sqrt[3]{\frac{\mu T^2}{4\pi^2}}$$

The radius of synchronous orbit of Earth is

$$r = [(398{,}600 \times 86{,}164.09)^2 /4\pi^2]^{1/3} = 42{,}164\ km$$

The height of the orbit from Earth's surface is

$$z = r - R_E = 42164 - 6378 = 35{,}786\ km$$

If a satellite is launched with the inclination $i = 28.5°$, to put it into synchronous orbit minimum four distinct velocity changes $\Delta v$ are required:

1. launch to low Earth orbit,

2. Hohmann transfer from the parking orbit to $r = 42,164$ km,

3. plane change from $i = 28.5°$ to the equatorial plane, and

4. circularization of the transfer ellipse to form the geosynchronous orbit.

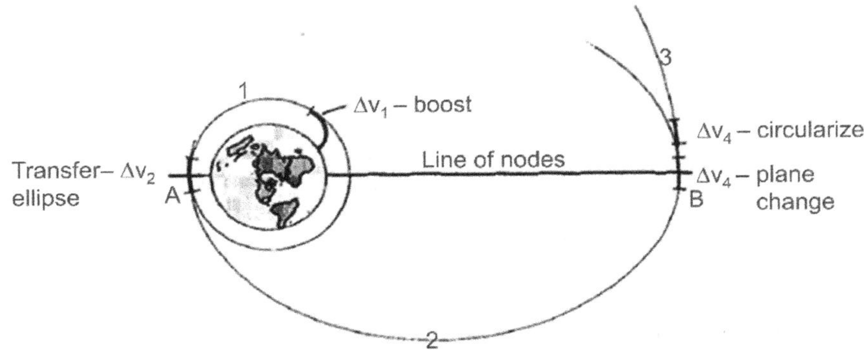


**Figure 11.21** Velocity increments required to achieve geosynchronous orbit.

Let us consider two cases, (a) combined plane change and circularization at apoapsis of transfer orbit (b) plane change at parking orbit.

**Case (a)**

Launch to low Earth orbit:

If the satellite is launched with $i = 28.5°$ eastward, it takes advantages of Earth rotational velocity.

The parking orbit radius $r_{Leo} = 280\ km$

The velocity of the spacecraft in the parking orbit is

$$v_c = \sqrt{\frac{\mu}{r_{Leo}}} = \left(\frac{398600}{6378 + 280}\right)^{0.5} = 7.737\ km/s$$

The velocity increment to put the spacecraft in parking orbit $\Delta v_1 = 7.737$ km/s

Earth rotational velocity at latitude $28.5°$ is

$$v_{latitude} = v_{equator}\ cos\emptyset = 0.4651 \times cos28.5 = 0.408\ km/s$$

The Earth rotation contributes 0.408 km/s.

The launch vehicle contribution velocity increment $\Delta v_1 = 7.737 - 0.408 = 7.329$ km/s

In addition, the launch vehicle contributes the potential energy of the spacecraft and the substantial energy to overcome gravity and drag.

Parking orbit to transfer ellipse:

The perigee radius of the Hohmann transfer ellipse must be tangent to the parking orbit.

The apogee radius of transfer orbit $r_a = 42164$ km = radius of geosynchronous orbit

Semimajor axis $a = (42164 + 6378 + 280)/2 = 24411$ km

An upper stage usually performs this burn at the first equatorial crossing.

The velocity at the perigee of the transfer ellipse is, Eq. (2.48)

$$v_p = \sqrt{\left(\frac{2\mu}{r_p} - \frac{\mu}{a}\right)}$$

$$v_p = \sqrt{\left(2 * \frac{398600}{6658} - \frac{398600}{24411}\right)} = 10.169 \; km/s$$

The required velocity increase to put the spacecraft in elliptical orbit is

$$\Delta v_2 = 10.169 - 7.737 = 2.432 \text{ km/s}$$

Plane change and circularization:

It is important to change planes through the smallest possible angle and at the lowest possible velocity. The lowest possible velocity occurs at the greatest radius, i.e., at the apoapsis. The velocity at apoapsis is, Eq. (2.47)

$$v_a r_a = v_p r_p$$

$$v_a = 10.169 \times \frac{6658}{42164} = 1.606 \; km/s$$

Significant energy savings can be made if in-plane and out-of-plane maneuvers can be combined i.e., combined maneuvers. In practice, the plane change is combined with the circularization burn because

1.  less $\Delta v$ is required with a combined burn and

2.  a solid motor can be used for the burn if restart is not required.

The solid motor that performs this burn is often called a kick stage.

The velocity on a geosynchronous orbit is

$$v_s = \sqrt{\frac{\mu}{r}} = \sqrt{\frac{398600}{42164}} = 3.0747 \frac{km}{s}$$

The $\Delta v$ for combined maneuvers is given by Eq. (9.16)

$$\Delta v_{3\&4} = \sqrt{(v_s - v_a)^2 + 4 v_s v_a \sin^2\left(\frac{i}{2}\right)}$$

$$= \sqrt{(3.0747 - 1.606)^2 + 4 \times 3.0747 \times 1.606 \times \sin^2\left(\frac{28.5}{2}\right)} = 1.831\,\frac{km}{s}$$

The total velocity change required to place a spacecraft in geosynchronous orbit is

$\Delta v_1$ ($v_c$) boost to parking orbit 1 = 7.737 km/s

$\Delta v_2$ for transfer to elliptical orbit = 2.432 km/s

$\Delta v_{3\&4}$ for plane change and circularization of orbit = 1.831 km/s

Total velocity change $\Delta v = \Delta v_1 + \Delta v_2 + \Delta v_{3\&4} = 12.000$ km/s

**Case (b)**

Suppose the plane change is made at LEO (parking orbit) instead of at GEO. In that case, the velocities at perigee of transfer orbit involved are parking circular orbit velocity and perigee velocity of the elliptical orbit at A

$$\Delta v_{1\&2} = \sqrt{\left(v_p - v_c\right)^2 + 4v_p v_c \sin^2\left(\frac{i}{2}\right)}$$

$$\Delta v_{1\&2} = \sqrt{(10.169 - 7.737)^2 + 4 \times 10.169 \times 7.737 \times \sin^2\left(\frac{28.5}{2}\right)} = 4.998\,\frac{km}{s}$$

The spacecraft travels to apoapsis B in the equatorial plane, so that when it arrives, the delta-v requirement at B to raise synchronous orbit is

$$\Delta v_3 = v_s - v_a = 3.0747 - 1.606 = 1.4687 = 1.469\ km/s$$

So total velocity increment is

$$\Delta v' = 7.737 + 4.998 + 1.469 = 14.204\ km/s$$

Considering the maneuvers made with the propellant available with the spacecraft at A and B, the net increase of velocity change that has to be provided by spacecraft is

$$= (4.998 + 1.469) - (2.432 + 1.831) = 2.204$$

This is a 51.7% increase over the total delta-v with plane change at GEO.

Clearly, it is best to do plane change maneuvers at the largest possible distance (apoapsis) from the primary attractor, where the velocities are smallest.

**Nomenclature**

$A$      azimuth angle

$F$      thrust of a single thruster (N)

$h$      angular momentum (kg-m$^2$/s)

$i$      orbital inclination

$I_{sp}$     specific impulse (s)

$I_v$     mass moment of inertia of the spacecraft (kg-m$^2$)

$L$     radius from the vehicle center of mass to the thrust vector (m)

$m$     mass (kg)

$m_p$     mass of propellant (kg)

$n$     number of thrusters

$r$     radius of the orbit (m)

$T$     torque applied to the spacecraft (Nm)

$T$     period of the orbit (s)

$t_b$     duration of the thruster burn (s)

$t_c$     duration of coasting (s)

$t_m$     maneuver time(s)

$v$     velocity (m/s)

$v_a$     apoapsis velocity (m/s)

$v_c$     circular orbit velocity (m/s)

$\alpha$     angular acceleration of the spacecraft (rad/s$^{2)}$)

$\gamma$     flight-path angle

$\varepsilon$     specific energy (J/kg)

$\theta_m$     total angle of rotation

$\omega$     the angular velocity (rad/s)

$\theta$     angular rotation (rad)

$\varphi$     latitude

## PROBLEMS

11.1   A spacecraft with thruster pair, each located at 2.0 m radius from the center of mass, has a moment of inertia 6300 kg-m$^2$ about the z-axis. The thrust of each engine is 2 N. Calculate the minimum time for the spacecraft to maneuver through a 60° rotation about the z axis? How much propellant is consumed at a specific impulse of 180 s?

11.2   A 900 kg communications satellite has attained an orbit in the equatorial plane with the apogee altitude is 41,000 km and the eccentricity is 0.07. Calculate the minimum amount of monopropellant to place the satellite in a geosynchronous orbit of 35,786 km. The specific impulse of the engine is 200 s.

11.3   Two spacecraft A and B are in the same elliptical earth orbit with perigee radius 7000 km and apogee radius 12,000 km. Spacecraft A is at perigee and spacecraft B is 30° ahead.

Calculate the total delta-v required for spacecraft A to intercept and rendezvous with spacecraft B when spacecraft B has travelled 60°.

11.4   A spacecraft is in a 400-km circular parking orbit. It is desired to increase the altitude to 800 km and change the inclination by 15°. Find the total delta-v required if the plane change is made after insertion into the higher orbit and the plane change and insertion into are accomplished simultaneously.

11.5   The space shuttle, weighing 99,800 kg, is required to make a 4° plane change in its circular orbit at an altitude of 300 km. Find the propellant required to carry out the maneuver with a propulsion system delivering the thrust at specific impulse 310 s.

# Interplanetary Trajectories

## 12.1 INTRODUCTION

The chapter 11 has dealt with maneuvers having common focus for the initial and final orbits. This chapter is focused on computing interplanetary trajectories parameters like orbital position and velocity, velocity increment ($\Delta v$) and time of flight (TOF). For interplanetary trajectories, the spacecraft will be in multiple orbits that have several foci. It also requires the spacecraft to leave the gravitational influence of the initial planet. Hence, the trajectory is influenced by the other objects in the solar system. So, the two-body theory discussed in the previous chapter cannot be applied directly. One approach is to solve the equation of motion considering all the gravitational influences. This is not usually followed as the approach is very complicated and computationally intensive. The most common method for interplanetary trajectory is known as patched conic approximation. In this approach several conic sections are patched together to compute the entire trajectory of interest. The main aim of the analysis is to calculate the total velocity increment ($\Delta v$) required to carry off the space mission. The delta-$v$ determines the propellant mass and finally estimates the payload mass that can be delivered by the space vehicle to the planetary target.

The radial distances of heliocentric (Sun centric) orbits are enormous and convenient to work with astronomical units (AU). The semimajor axis of the Earth's orbit is 149.5978707 x $10^6$ km, a distance that is called an astronomical unit (AU) as shown in Figure 12.1. Accordingly, Mars is at a distance of 1.524 AU from the center of the Sun. As can be seen from Table 12.1 most of the planetary orbits have small eccentricities and their orbits in the solar system lie very close to the earth's ecliptic orbital plane, the exceptions being Mercury and Pluto. The innermost planet, Mercury and the dwarf planet Pluto differ most in inclination. The orbital planes of the other planets lie within 3.5° degrees of the ecliptic. The velocities shown in Table 12.1 are the mean velocities of the planets around the sun; since the orbits are elliptical, the velocities are time variant.

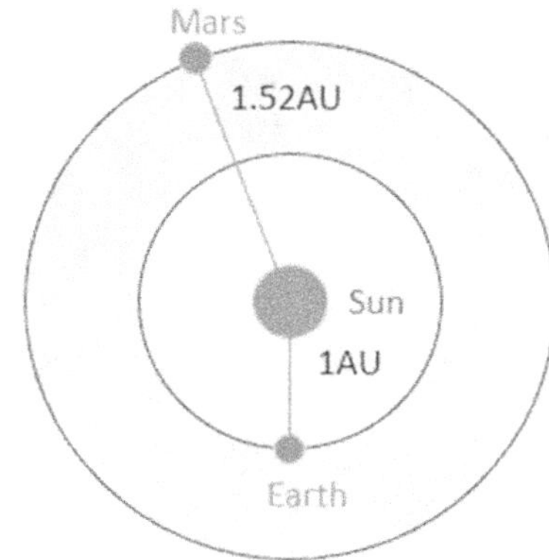


**Figure 12.1** Astronomical Unit.

**Table 12.1** Mean Orbital Parameters of the Planets.

| Planet | $a$, AU | $r_p$, $\times 10^6$ km | $e$ | $i$, deg | Velocity, km/s |
|---|---|---|---|---|---|
| Mercury | 0.387 | 45.99 | 0.2056 | 7.005 | 47.89 |
| Venus | 0.723 | 107.437 | 0.0068 | 3.395 | 35.05 |
| Earth | 1.000 | 147.10 | 0.0167 | 0.001 | 29.77 |
| Mars | 1.524 | 206.72 | 0.0933 | 1.850 | 24.13 |
| Jupiter | 5.203 | 740.84 | 0.0482 | 1.305 | 13.05 |
| Saturn | 9.516 | 1345.02 | 0.0552 | 2.487 | 9.64 |
| Uranus | 19.166 | 2729.29 | 0.0481 | 0.772 | 6.80 |
| Neptune | 30.011 | 4447.85 | 0.0093 | 1.772 | 5.43 |
| Pluto | 39.557 | 4436.42 | 0.2503 | 17.150 | 4.73 |

Figure 12.2 shows the heliocentric interplanetary cruise between the Earth and an outer planet. The major part of the trajectory is in heliocentric space with the Sun as the primary gravitational body. Once the spacecraft escapes Earth, it becomes an independent member of the solar system and follows an elliptical trajectory. Therefore, its motion can be first approximated by laws of two-body problem with the Sun as the central body. The spacecraft to begin the interplanetary cruise, however, must escape Earth's gravitational force and transit from near-Earth space to interplanetary space in a hyperbolic trajectory. In a similar manner spacecraft enter Mars' gravitational field in a hyperbolic trajectory. Thus, the spacecraft is under the gravitational influence of the departure and arrival planets for a relatively small period.

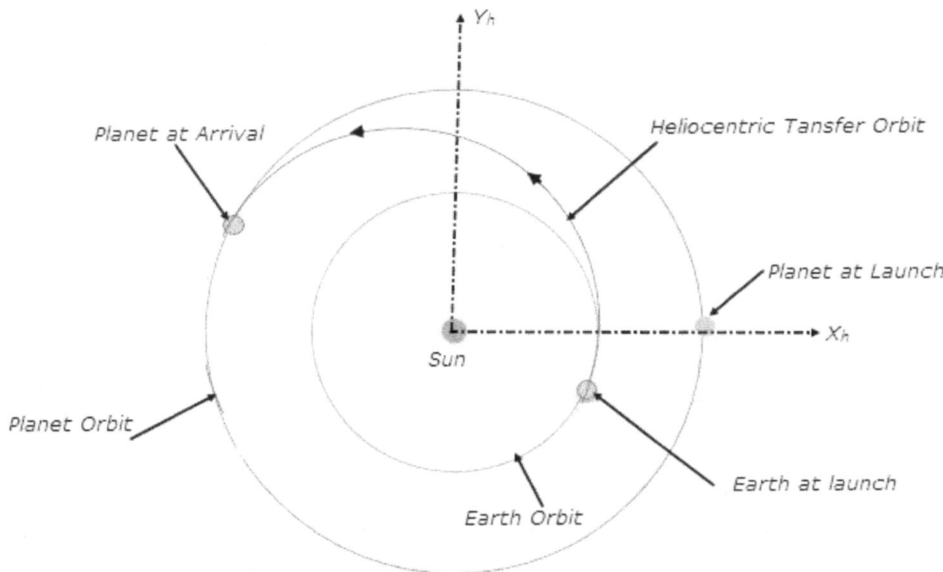


**Figure 12.2** Heliocentric Trajectory from Earth to an outer Planet.

## 12.2  SPHERE OF INFLUENCE (SOI)

In the two-body theory, the main assumption is that only one heavenly body gravity force acts on the spacecraft (a second body). In the Earth's orbital spacecraft, the only force acting on it is Earth gravitational force. To deal with gravity from only one body at a time, consider how gravity operates in space. The Sun's gravitational force, the dominant body in the solar system, holds all the planets in its place according to Newton's law of gravitation. However, near a given planet, the influence of its own gravity exceeds that of the sun: ex., at its surface the earth's gravitational force is over 1600 times greater than the sun's. Theoretically, as per the inverse square law of gravitation, a body's gravitational attraction reaches out to infinity, but practically, it's effective only within a certain volume of space. Figure 12.3 shows that the gravitational force of a planet drops off rapidly with distance $r$ from the center of attraction. The gravitational force $F_r$ at a radius twice of the planet radius is almost around 0.25 times force on the surface.  At 10 body radii, the force is 1% of its value at the surface. Eventually, the sun's gravitational field overwhelms that of the planet.

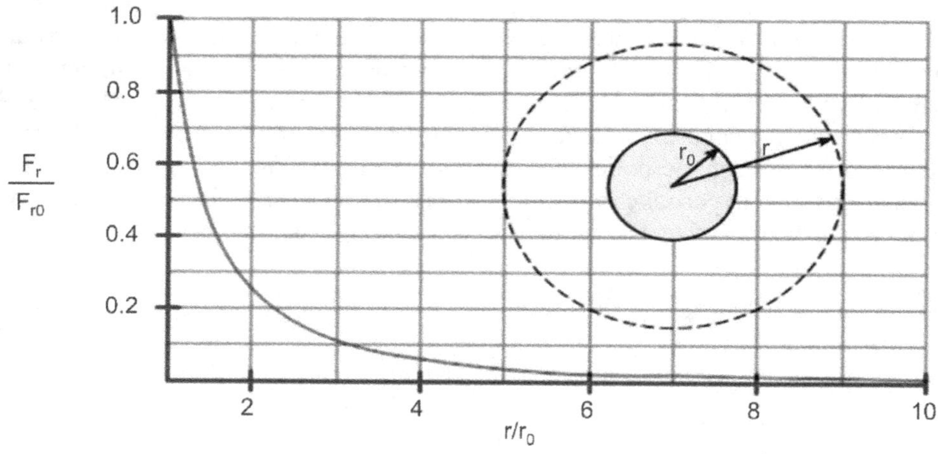


**Figure 12.3**  Variation of gravitational force with distance from a planet's surface. (Source: Howard D. Curtis).

The **sphere of influence (SOI)** of a planet represents the boundary of where the planet's gravitational influence on the spacecraft is stronger than the Sun's. It is a fictitious boundary for two-body motion about a planet. It is not an exact quantity. It is simply a reasonable estimate of the distance beyond which the sun's gravitational attraction dominates that of a planet. This concept allows if a spacecraft within the planet's SOI, then its trajectory can be analyzed as a two-body problem in the inertial planet-centered frame. As a spacecraft crosses its SOI, its trajectory can be analyzed as a two-body problem in the heliocentric-ecliptic coordinate frame. The size of the SOI depends on the planet's mass (a more massive planet has a longer "gravitational reach") and how close the planet is to the Sun. The radius of the sphere of influence suggested by Laplace is

$$R_s = R \left(\frac{M_{Planet}}{M_{Sun}}\right)^{\frac{2}{5}}$$

.....(12.1)

where

$R_s$ = radius of the sphere of influence of a planet

$R$ = mean orbital radius of the planet

$M_{planet}$ = mass of the planet

$M_{sun}$ = mass of the sun

Table 12.2 shows the sphere of influence values found by using the Laplace method.

**Table 12.2** Planetary sphere of influence radii.

| Planet | $R_S$, x$10^6$ km |
|---|---|
| Mercury | 0.111 |
| Venus | 0.616 |
| Earth | 0.924 |
| Mars | 0.577 |
| Jupiter | 48.157 |
| Saturn | 54.796 |
| Uranus | 51.954 |
| Neptune | 80.196 |
| Pluto | 3.400 |
| Moon | 0.0662 |

## 12.3  PATCHED CONIC METHOD

In two-body or Keplerian space, orbits are conic sections with the focus at the attracting body. Figure 12.4 shows a schematic diagram of an interplanetary mission from Earth to planet. The entire interplanetary mission can be watched into a sequence of three distinct phases where each phase is a separate two-body problem. The spacecraft leaves the parking orbit of the planet Earth

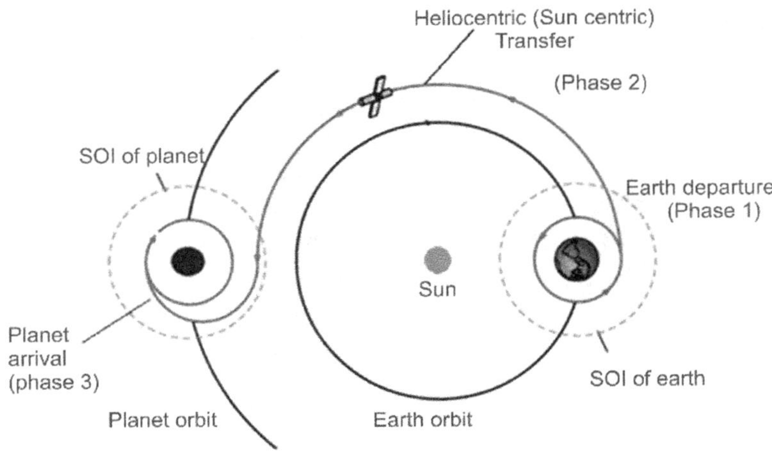


**Figure 12.4** Patched Conic Interplanetary Trajectory.

to Start the interplanetary cruise to reach the target planet. Within each planetary sphere of influence, the spacecraft travels an unperturbed Keplerian path about the planet. When the spacecraft is outside the sphere of influence of a planet it follows an unperturbed Keplerian orbit around the sun.

The spacecraft is transferred to orbit around the Sun. This means that it needs some excess velocity to break out of the planet's orbit (i.e., planet's gravitational force). As shown in Figure 12.5 if the orbit velocity, $v$ is less than the escape velocity $v_{esc}$ it will remain in the elliptical orbit around the planet. In the case $v$ is equal the $v_{esc}$ it will be on a parabolic trajectory. It will arrive at the sphere of influence with a relative speed of zero. Hence, the spacecraft remains in the same orbit as the planet and does not embark upon a heliocentric elliptical path. To escape the gravitational pull of a planet, the spacecraft must travel a hyperbolic trajectory relative to the planet, arriving at its sphere of influence with a relative speed $v_\infty$ (hyperbolic excess speed) greater than zero.

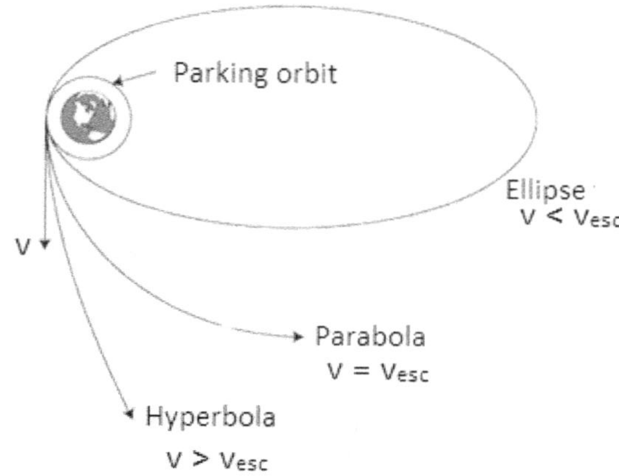


**Figure 12.5** Departure Trajectories from Earth.

On the scale of the solar system, the sphere of influence looks as mere a spot, whereas from the point of view of the planet it is very large indeed and may be considered to lie at infinity. In short, the three phases are:

1. *Earth departure phase*: The spacecraft leaves the Earth's gravity field along a hyperbolic escape trajectory. From the point of the planet the SOI surface at infinity. Thus, $v_\infty$ is denote the speed at which a body on a hyperbolic path arrives at infinity, i.e., $r = \infty$. The spacecraft motion in the departure phase is analyzed using a planetocentric inertial frame.

2. *Heliocentric phase*: Once the spacecraft crosses the SOI of Earth, it follows an elliptical transfer orbit where the sun is the primary gravitational body. The motion is analyzed using two body motion in the heliocentric-ecliptic coordinate frame.

3. *Planetary-arrival phase*: The spacecraft crosses the target planet's SOI where the planet's gravity prevails. The arrival trajectory inside the planet's SOI is a hyperbola analyzed in an inertial planetocentric frame.

The three phases of interplanetary mission require tracking of velocities relative to a reference frame that changes from phase to phase. It is important to understand and evaluate the velocities properly with respect to the planets and Sun reference frames. Figure 12.6 shows the conversion of the planetocentric velocity of the spacecraft to heliocentric velocity. The superscript represents the heliocentric and subscripts S for spacecraft and P for planet velocities.

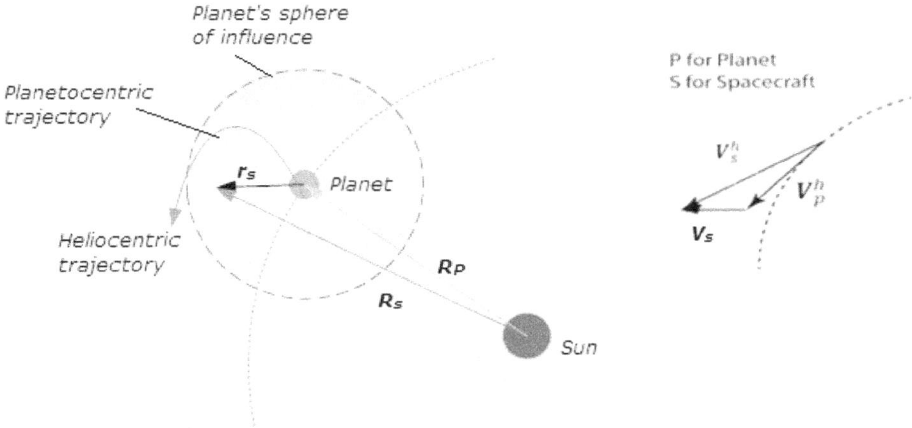


**Figure 12.6**  Heliocentric velocity.

The difference between the planet frame and the heliocentric frame is just the velocity of the planet with respect to the Sun. To convert from the planet frame to the heliocentric frame, simply add the velocity of the planet to the spacecraft velocity vectorially. In the case the spacecraft leaves a sphere of influence (departure from a planet)

$$v_S^h = v_s + v_p^h$$
.....(12.2)

*Heliocentric velocity of spacecraft*
*= Planetocentric velocity of spacecraft*
*+ Heliocentric velocity of the planet*

In the case the spacecraft enters a sphere of influence (arrival at a target planet)

$$v_s = v_S^h - v_p^h$$
.....(12.3)

*Planetocentric velocity Spacecraft = Difference between the heliocentric movement of the spacecraft and the planet*

## 12.4  HELIOCENTRIC TRANSFER

In an interplanetary cruise mission, the major portion of the space travel is a heliocentric phase that may take several months to several years. In contrast, the planet centered departure and arrival phases take only a few days. Hence, the heliocentric phase is typically determined first in order to establish the mission launch and arrival dates and the duration. The three primary results of the calculations of this trajectory are:

(a) The spacecraft velocity at the edge of the departure planet's SOI to place it on the appropriate transfer trajectory.

(b) The spacecraft velocity at arrival.

(c) **The timing of the transfer.** It determines when the spacecraft should depart the planet Earth so that it arrives at the target planet's orbit in the same location as the target planet.

*To calculate the orbital parameters for a transfer* **the following assumptions are made**:

1. All the planetary orbits are coplanar, with the exceptions of Mercury and Pluto.

2. The planetary orbits are circular with the exceptions of Mercury and Pluto.

3. The radius of the circular orbit is equal to the mean semimajor axis of the orbit as given in Table 12.1.

The most energy-efficient way for a spacecraft to transfer from one planet's orbit to another is to use a Hohmann transfer ellipse. In the Hohmann transfer, the velocity vectors of the spacecraft at departure and arrival are parallel to the initial and final orbits, and that the transfer is made of 180° of an ellipsis. Consider heliocentric transfer from Earth to an outer planet as shown in Figure 12.7. The departure point D is at periapsis (perihelion) of the transfer ellipse and the arrival point A is at apoapsis (aphelion). The circular orbital speed of planet 1 relative to the sun is given by Eq. (2.7)

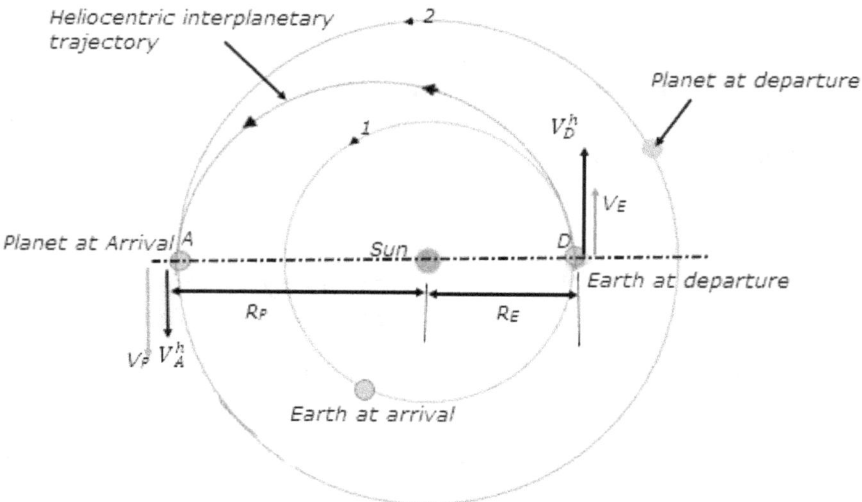


**Figure 12.7** Hohmann Transfer from Earth to Outer plane.

$$v_E = \sqrt{\frac{\mu_{sun}}{R_E}} \qquad\qquad .....(12.4)$$

The heliocentric speed $v_D^h$ of the spacecraft on the transfer ellipse at the departure point D is given by Eq. (2.48)

$$v_D^h = \sqrt{\frac{2\mu_{sun}}{R_E} - \frac{\mu_{sun}}{a}}$$
.....(12.5)

$$v_D^h = \sqrt{2\mu_{sun}} \sqrt{\frac{1}{R_E} - \frac{1}{2a}}$$

$$v_D^h = \sqrt{2\mu_{sun}} \sqrt{\frac{1}{R_E} - \frac{1}{(R_E + R_P)}}$$

$$v_D^h = \sqrt{2\mu_{sun}} \sqrt{\left(\frac{R_P}{R_E(R_E+R_P)}\right)}$$
.....(12.6)

This spacecraft velocity $v_D^h$ is greater than the speed of the planet Earth $v_E$. Therefore, the required spacecraft delta-$v$ at D is

$$\Delta v_D = v_D^h - v_E = \sqrt{2\mu_{sun}} \sqrt{\left(\frac{R_P}{R_E(R_E + R_P)}\right)} - \sqrt{\frac{\mu_{sun}}{R_E}}$$

$$\Delta v_D = \sqrt{\frac{\mu_{sun}}{R_E}} \left(\sqrt{\frac{2R_P}{R_E+R_P}} - 1\right)$$
.....(12.7)

Similarly, the $\Delta v_A$ at the point of arrival is

$$\Delta v_A = v_P - v_A^h = \sqrt{\frac{\mu_{sun}}{R_P}} \left(1 - \sqrt{\frac{2R_E}{R_E+R_P}}\right)$$
.....(12.8)

This velocity increment, like that at departure point D, is positive since outer planet 2 is traveling faster than the spacecraft at point A.

For a spacecraft's heliocentric trajectory to an inner planet of Earth, as illustrated in Figure 12.8, the departure point D and the arrival point A are now at aphelion and perihelion, respectively.

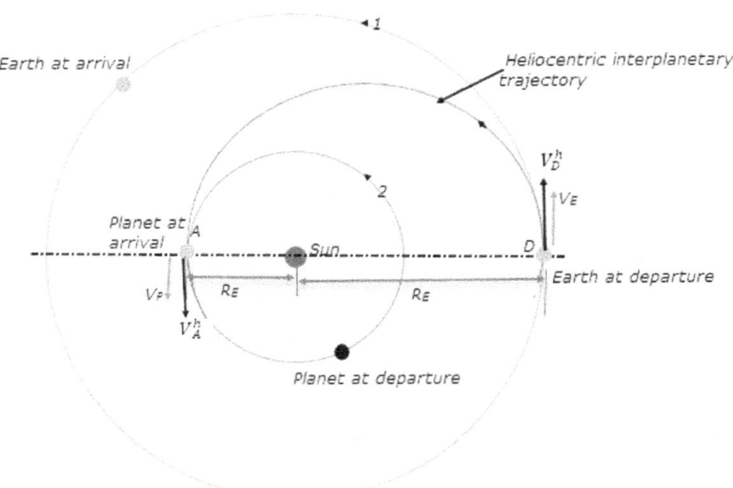


**Figure 12.8** Hohmann Transfer from Earth to inner planet.

At departure point D, the spacecraft is entering to the lower energy elliptical orbit, its speed must be reduced. Again, at point A to arrive in the lower energy circular orbit of planet, its speed must be reduced. Hence, the delta-v's calculated with above equations will be both negative. However, the absolute values are used in estimating the propellant requirement of the spacecraft.

## 12.5  PLANETARY DEPARTURE

As explained in the previous section, the spacecraft to transfer into heliocentric orbit needs excess velocity relative to the parabolic trajectory to break out of the planet's gravitational force. The only type of trajectory with excess velocity is hyperbolic trajectory (Eq. 2.72). Clearly, the heliocentric delta-v, $\Delta v_D$ calculated in Eq. (12.7) is the excess velocity, $v_\infty$ greater than the speed of the planet Earth $v_E$. The spacecraft does not receive maneuver impulse when it crosses the SOI. It must have the correct $v_\infty$ on its geocentric hyperbolic trajectory so that it can coast onto the heliocentric transfer trajectory as shown in Figure 12.9. To minimize the delta-v requirement from parking orbit hyperbolic escape trajectory, the impulse is to be applied at the periapsis, P of the hyperbola. The radius of this parking orbit equals the periapsis radius $r_p$ of the departure hyperbola.

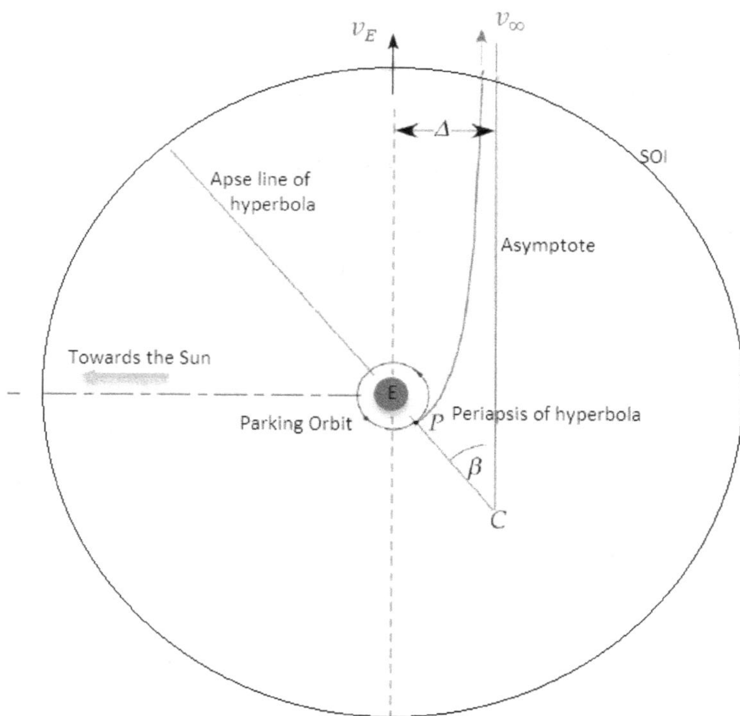


**Figure 12.9**  Departure Trajectory from Earth to Outer Planet.

The hyperbolic excess speed of the departure hyperbola from Eq. (12.7) is

$$v_\infty = \Delta v_D = \sqrt{\frac{\mu_{sun}}{R_E}}\left(\sqrt{\frac{2R_P}{R_E + R_P}} - 1\right)$$

The velocity at periapsis of the hyperbola from Eq. (2.72) is

$$v_p = \sqrt{v_\infty^2 + v_{esc}^2} = \sqrt{v_\infty^2 + \frac{2\mu_E}{r_p}}$$

The speed of parking circular orbit is

$$v_c = \sqrt{\frac{\mu_E}{r_p}}$$

The $\Delta v$ required to put the vehicle onto the hyperbolic departure trajectory,

$$\Delta v = v_p - v_c = \sqrt{v_\infty^2 + \frac{2\mu_E}{r_p}} - \sqrt{\frac{\mu_E}{r_p}} = v_c\left(\sqrt{\left(\frac{v_\infty}{v_c}\right)^2 + 2} - 1\right)$$

The angle $\beta$ gives the orientation of the apse line of the hyperbola to the planet's heliocentric velocity vector.

$$\beta = \frac{1}{e}$$

From hyperbola relations from section 2.10.4

$$r_p = a(e - 1)$$

Or

$$e = \frac{r_p}{a} + 1$$

Also, the specific energy of hyperbolic orbit is

$$\epsilon = \frac{\mu_E}{2a}$$

The conservation of energy for hyperbolic trajectory is

$$\frac{v_\infty^2}{2} - \frac{\mu_E}{R_s} = \frac{\mu_E}{2a}$$

The radius of $R_s$ is very high and is considered as infinity,

$$a = \mu_E/v_\infty^2$$

$$e = \frac{r_p v_\infty^2}{\mu_E} + 1 \qquad\qquad .....(12.9)$$

## 12.6  PLANETARY ARRIVAL

A spacecraft arrives at the sphere of influence of the target planet with a hyperbolic excess velocity $v_\infty$ relative to the planet. As illustrated in the Figure 12.10, the spacecraft approach is from the Earth to an outer planet (e.g., earth to Mars), the spacecraft's heliocentric approach velocity $v_A^h$ is smaller in magnitude than that of the planet $v_P$. The spacecraft crosses the sphere of influence in front of the planet. From Eq. (12.8) the magnitude of the hyperbolic excess velocity is,

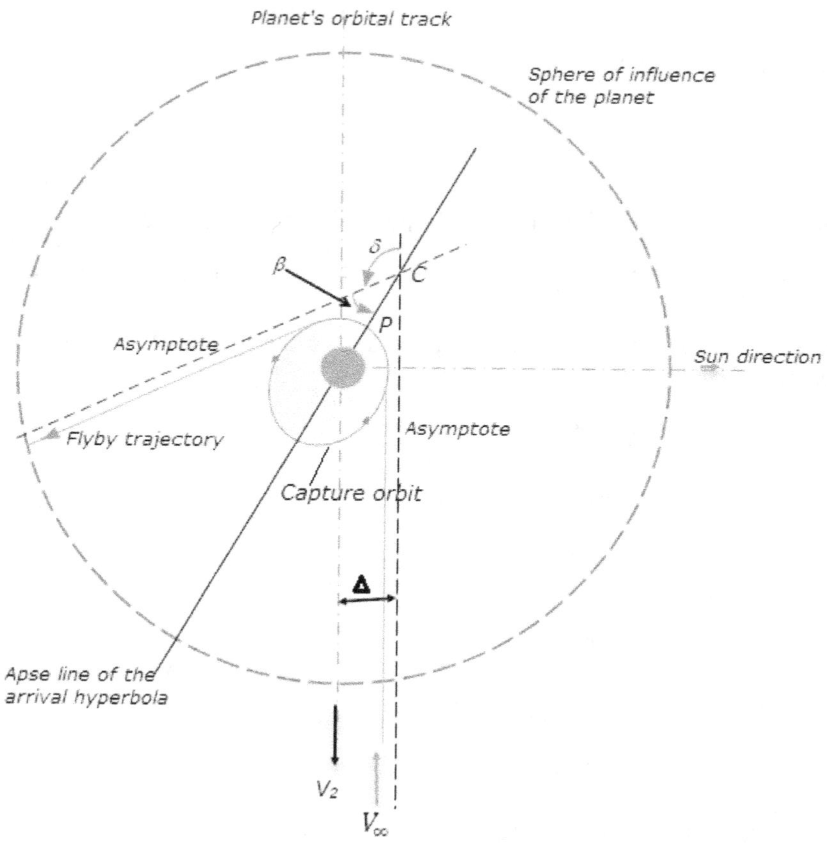


**Figure 12.10**  Arrival Trajectory from Earth to outer Planet.

$$v_\infty = \Delta v_A = v_P - v_A^h = \sqrt{\frac{\mu_{sun}}{R_P}}\left(1 - \sqrt{\frac{2R_E}{R_E + R_P}}\right)$$

If the mission is from an outer planet to an inner one (e.g., Earth to Venus), then $v_A^h$ is greater than $v_P$. In that case the spacecraft crosses the sphere of influence behind the planet.

$$v_\infty = \Delta v_A = v_A^h - v_P$$

Depending upon the mission nature, three trajectory option are available inside the SOI of the planet:

1. Impact the planet
2. Go into orbit around the planet
3. Flyby the planet, using the planet's angular momentum to change the spacecraft's heliocentric trajectory

If the mission is to impact the target planet, the hyperbola's periapsis radius $r_p$ equals essentially the radius of the planet. If an atmosphere is present, then an altitude at the approximate edge of the atmosphere would be more appropriate. If the intent is to go into orbit around the planet, then $\Delta$ must be chosen so that the $\Delta v$ burn at periapsis will occur at the correct altitude above the planet. If there is no impact with the planet and no capture orbit around the planet, then the spacecraft will take a flyby trajectory.

The eccentricity of the approach hyperbola can be calculated by Eq. (12.9)

$$e = \frac{r_p v_\infty^2}{\mu_P} + 1$$

The aiming radius $\Delta$ is given by Eq. (2.70)

$$\Delta = a\sqrt{e^2 - 1}$$

The aiming radius in terms of the periapsis radius and the hyperbolic excess speed is

$$\Delta = r_p \sqrt{1 + \frac{2\mu_P}{r_p v_\infty^2}}$$

Consider the purpose of the mission is to enter an elliptical orbit of eccentricity $e$ around the planet. This will require a delta-v maneuver at periapsis P which is also periapsis of the ellipse.

The speed in the hyperbolic trajectory at periapsis is given by

$$v_{p-hyp} = \sqrt{v_\infty^2 + \frac{2\mu_P}{r_p}}$$

The velocity at periapsis of the capture elliptical orbit from Eq. (2.48) is

$$v_{p-elli} = \sqrt{\frac{2\mu_P}{r_p} - \frac{\mu_P}{a}}$$

Substituting for a and simplifying

$$a = r_p/(1 - e)$$

$$v_{p-elli} = \sqrt{\frac{\mu_P(1 + e)}{r_p}}$$

Therefore, the required $\Delta v$ to be applied at periapsis is

$$\Delta v = v_{p-hyp} - v_{p-elli} = \sqrt{v_\infty^2 + \frac{2\mu_P}{r_p}} - \sqrt{\frac{\mu_P(1+e)}{r_p}}$$

Clearly, $\Delta v$ depends on the choice of periapsis radius $r_p$ and capture orbit eccentricity e. Also, $\Delta v$ is maximum for a circular capture orbit ($e=0$) and decreases with increasing eccentricity until $\Delta v = 0$, which, of course, means no capture (flyby). The optimum value of the $\Delta v_{opt}$ will be obtained by differentiating the above equation. Here, the equations for optimum values are presented.

The minimum value of $\Delta v$ is

$$\Delta v_{opt} = v_\infty \sqrt{\frac{1-e}{2}}$$

The optimal periapsis radius is

$$r_{popt} = \frac{2\mu_P}{v_\infty^2} \frac{1-e}{1+e}$$

The apoapsis radius is

$$r_{aopt} = \frac{2\mu_P}{v_\infty^2}$$

The expression for the aiming radius required for minimum $\Delta v$ is

$$\Delta = \sqrt{\frac{2}{1-e}} r_p$$

**Example 12.1**

A spacecraft is launched on a mission to Mars starting from a 400-km circular parking orbit. The spacecraft will be in elliptical orbit with periapsis of 6000 km and eccentricity 0.4. Analyze the interplanetary travel mission to the planet Mars. Assume specific impulse as 3000 m/s.

(a) *Heliocentric Trajectory*: The orbital parameters of planets Earth and Mars

Planets Data:

Gravitational parameters

$$\mu_{sun} 1.327 \times 10^{11} \; km^3/s^2$$

$$\mu_{earth} = \mu_E = 398{,}600 \; km^3/s^2$$

$$\mu_{Mars} = \mu_P = 42830 \frac{km^3}{s^2}$$

Orbital Radii

$$R_{earth} = R_E = 149.6 \times 10^6 \; km$$

$$R_{Mars} = R_P = 227.9 \times 10^6 \ km$$

Planet Radii

$$Earth \ radius \ r_E = 6378 \ km$$

$$Mars \ Radius \ r_M = 3396 \ km$$

Heliocentric transfer begins when the spacecraft crosses the Earth's SOI. Orbital speed of Earth relative to the sun is

$$v_E = \sqrt{\frac{\mu_{sun}}{R_E}} = \sqrt{\frac{1.327 \times 10^{11}}{149.6 \times 10^6}} = 29.78 \frac{km}{s}$$

The heliocentric speed $v_D^h$ of the spacecraft on the transfer ellipse at the departure point D is given by Eq. (12.6)

$$v_D^h = \sqrt{2\mu_{sun}} \sqrt{\left(\frac{R_P}{R_E(R_E + R_P)}\right)}$$

$$= \sqrt{2 \times 1.327 \times 10^{11}} \sqrt{\left(\frac{227.9}{149.6 \times 10^6(149.6 + 227.9)}\right)} = 32.72 \ km/s$$

The spacecraft in order to begin the heliocentric Hohmann transfer must achieve a velocity increment

$$v_\infty = \Delta v_D = v_D^h - v_E = 32.72 - 29.78 = 2.94 \ km/s$$

(b) *Departure Trajectory:* The velocity of parking circular orbit at an altitude 400 km

$$v_c = \sqrt{\frac{\mu_E}{r_p}} = \sqrt{\frac{398600}{6378 + 400}} = 7.669 \ \frac{km}{s}$$

The velocity at periapsis P of the hyperbola is

$$v_p = \sqrt{v_\infty^2 + v_{esc}^2} = \sqrt{v_\infty^2 + \frac{2\mu_E}{r_p}}$$

$$v_p = \sqrt{2.94^2 + \frac{2 \times 398600}{6778}} = 11.\ 245 \frac{km}{s}$$

The delta-v required to put the vehicle onto the hyperbolic departure trajectory or the velocity impulse at LEO for "trans-Mars injection" (TMI) is

$$\Delta v_{TMI} = v_p - v_c = 11.245 - 7.669 = 3.576 \ \frac{km}{s}$$

TMI velocity impulse applied in LEO is the only rocket burn required to send the spacecraft on its coasting trajectory to Mars.

Amount of propellant required as a percentage of the spacecraft mass:

Specific impulse $I_{sp} = v_e = 3000 \; m/s$ . From rocket equation

$$\frac{\Delta m}{m} = \left(1 - exp\left(-\frac{\Delta v_{TMI}}{v_e}\right)\right) = 1 - exp\left(-\frac{3576}{3000}\right) = 0.696$$

i.e., ~70% of the spacecraft mass must be propellant.

The specific energy of hyperbolic orbit is

$$\epsilon = \frac{\mu_E}{2a}$$

The conservation of energy for hyperbolic trajectory is

$$\frac{v_\infty^2}{2} - \frac{\mu_E}{r} = \frac{\mu_E}{2a}$$

The SOI radius of Earth is very high and is considered at infinity, $i.e., r = R_{searth} = \infty$.

Therefore, the semimajor axis of departure hyperbola

$$a = \frac{\mu_E}{v_\infty^2} = \frac{398600}{2.94^2} = 41,115 \; km$$

$$e = \frac{r_p}{a} + 1 = \frac{6778}{41115} + 1 = 1.165$$

Orientation of the apse line of the hyperbola to the planet's heliocentric velocity vector.

$$\beta = \frac{1}{e} = \frac{1}{1.165} = 30.866 \; deg$$

It is also of interest to note that the rocket burns to establish the escape hyperbola occurs when the spacecraft is in the Earth's shadow for prograde orbit (west to east).

(c) *Arrival Trajectory*: The spacecraft will follow a hyperbolic trajectory after it crosses Mars' SOI. Mars-relative velocity at Mars' SOI; that is, the second "patch point"

$$v_\infty = V_A^h - V_2$$

Circular orbital velocity of Mars relative to the sun

$$v_P = \sqrt{\frac{\mu_{sun}}{R_P}} = \sqrt{\frac{1.327 \times 10^{11}}{227.9 \times 10^6}} = 24.13 \frac{km}{s}$$

The heliocentric speed $v_A^h$ of the space vehicle on the transfer *ellipse* at the arrival point A (the aphelion velocity of the spacecraft on the Hohmann transfer relative to the sun) is

$$v_A^h = \sqrt{\frac{2\mu_{sun}}{R_P} - \frac{\mu_{sun}}{a}}$$

$$v_A^h = \sqrt{2\mu_{sun}} \sqrt{\frac{1}{R_P} - \frac{1}{2a}}$$

$$v_A^h = \sqrt{2\mu_{sun}} \sqrt{\frac{1}{R_P} - \frac{1}{(R_E + R_P)}} = \sqrt{2\mu_{sun}} \sqrt{\frac{R_E}{R_P(R_E + R_P)}}$$

$$v_A^h = \sqrt{2 \times 1.327 \times 10^{11}} \sqrt{\frac{149.6 \times 10^6}{227.0 \times 10^6 \times (149.6 + 227.9) \times 10^6}} = 21.48 \; km/s$$

Therefore, $v_\infty$ at arrival is

$$v_\infty = v_A^h - v_P = 21.48 - 24.13 = -2.65 \frac{km}{s}$$

As the spacecraft's sun-relative velocity at aphelion ($v_A^h$) is less than Mars' sun-relative velocity ($v_P$), the spacecraft's hyperbolic trajectory approaches Mars along its leading edge, i.e., the minus sign indicates that the Mars-relative velocity vector $v_\infty$ is in the opposite direction as Mars' heliocentric velocity vector $v_P$.

The spacecraft is in the elliptical orbit around Mars with $r_p = 6000$ km and $e = 0.4$

The speed in the hyperbolic trajectory at periapsis is given by

$$v_{p-hyp} = \sqrt{v_\infty^2 + \frac{2\mu_P}{r_p}} = \sqrt{2.65^2 + \frac{2 \times 42830}{6000}} = 4.615 \; km/s$$

The elliptical velocity at periapsis P is

$$v_{p-elli} = \sqrt{\frac{\mu_P(1 + e)}{r_p}} = \sqrt{\frac{42830 \times (1 + 0.4)}{6000}} = 3.161 \frac{km}{s}$$

Therefore, the required delta-v to be applied at periapsis $P$ to place the spacecraft in elliptical orbit around Mars is

$$\Delta v = v_{p-elli} - v_{p-hyp} = 3.161 - 4.615 = -1.454 \; km/s$$

The semimajor axis of the elliptical orbit is

$$a = \frac{r_p}{1 - e} = \frac{6000}{1 - 0.4} = 10{,}000 \; km$$

The time period of the orbit is

$$\tau = \frac{2\pi}{\sqrt{\mu}} a^{\frac{3}{2}} = \frac{2\pi}{42830} 10000^{1.5} = 8.43 \; hrs$$

## 12.7 LUNAR ORBIT

The method of patched conics is not applicable for 3 body problems. For example, consider an Earth–moon trajectory (such as the Apollo lunar missions), where the spacecraft's motion is simultaneously influenced by the Earth and lunar gravitational forces. In the translunar trajectory, clearly, there will be a point where both the Earth and moon gravitational forces are equal and acting on the spacecraft. In such a scenario, it should be treated as a three-body problem and requires complicated numerical solutions. The Earth and moon masses are somewhat comparable. Thus, their center of mass lies almost three-quarters of the Earth radius from its center as shown in Figure 12.11. So, the motion of the moon cannot be accurately described as rotating around the center of the earth. However, with a few assumptions an analytical solution can be arrived at for spacecraft motion in translunar trajectory. The method is known as circular restricted three-body problem (CR3BP). The two major assumptions are required for the CR3BP are:

(1) the two gravitational bodies move in circular orbits about their center of mass; and

(2) the mass of the third body (the satellite) is negligible and does not influence the motion of the two primary bodies.

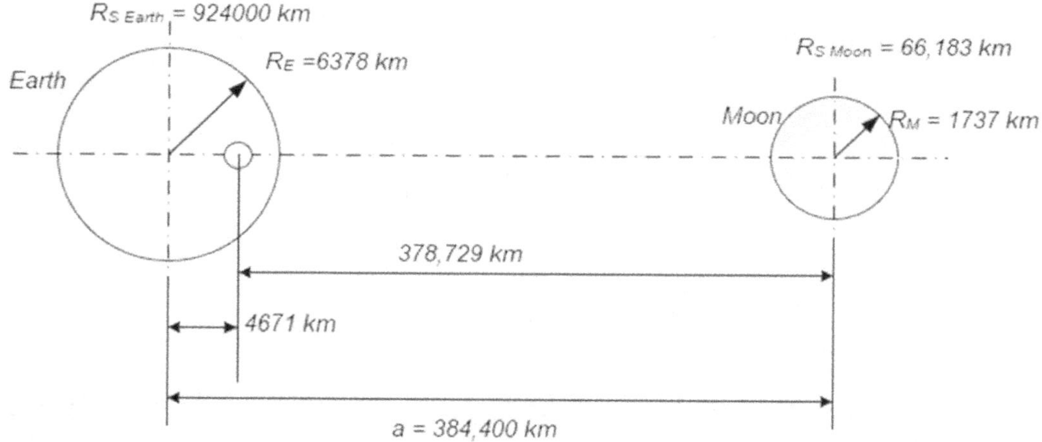


**Figure 12.11** Earth -Moon Orbit Characteristics.

The moon's orbit about the Earth has an eccentricity, $e = 0.0549$. Thus, the first assumption leads to some loss of accuracy in the calculation of trajectory. However, the CR3BP method yields preliminary results for translunar trajectory.

In the analysis of the heliocentric trajectory of interplanetary motion, it is assumed that the sphere of influence of the planet is negligibly small compared with the transfer ellipse and its influence is ignored. For a lunar trajectory, this assumption is not valid and it is necessary to acknowledge it and make a trajectory patch at its boundary. The SOI of the moon is

$$R_{smoon} = R \left(\frac{M_{moon}}{M_{earth}}\right)^{\frac{2}{5}}$$

where

$R_{smoon}$ = radius of the lunar sphere of influence

$R$ = distance between centres of mass for Earth and moon, 384,400 km

$M_{moon}/M_{earth}$ = ratio of mass for moon and Earth, 1/81.3

$$R_{smoon} = 384400 \left(\frac{1}{81.3}\right)^{\frac{2}{5}} = 66{,}183 \; km$$

The sphere of influence of the Earth, $R_{searth} = 924{,}000 \; km$. So, the orbit of the moon is determined primarily by the Earth whose sphere of influence extends well beyond the moon's orbital radius. The perigee and apogee radii of $r_p$ is 363,400 km and $r_a$ is 405,500km respectively. Therefore, the semimajor axis of the moon's orbit is

$$a = \frac{r_a + r_p}{2} = 384{,}400 \; km$$

A circular orbit of radius equal to the semi-major axis of the moon's orbit has the same period as the moon's elliptical orbit. Assuming that the moon's path around the earth is a circle of radius $r_m = a = 384{,}400$ km, the circular orbital speed $v_m$ of the moon is

$$v_m = \sqrt{\frac{\mu_e}{r_m}} = \sqrt{\frac{398600}{384400}} = 1.0183 \; km/s$$

Let a spacecraft be placed in a circular earth orbit of 320 km altitude and that its orbit is coplanar with that of the moon. The speed of the spacecraft in this circular parking orbit is

$$v_c = \sqrt{\frac{\mu_e}{r_c}} = \sqrt{\frac{398600}{6378 + 320}} = 7.7143 \, \frac{km}{s}$$

Consider a Hohmann transfer orbit from parking orbit to moon circular orbit. At its perigee it is tangent to the circular low earth orbit and at apogee it is tangent to the moon's orbit. Then,

$$r_p = 66987 \; km$$

$$r_a = 384{,}400 \; km$$

Therefore, the semimajor axis of this transfer ellipse is

$$a = \frac{r_a + r_p}{2} = \frac{384400 + 6698}{2} = 195{,}549 \; km$$

The eccentricity e of the orbit is

$$e = \frac{r_a - r_p}{r_a + r_p} = \frac{384400 - 6698}{384400 + 6698} = 0.96575$$

Its period T is given by

$$T = \frac{2\pi}{\sqrt{\mu_e}} a^{\frac{3}{2}} = \frac{2\pi}{\sqrt{398600}} \times (195549)^{\frac{3}{2}} = 8.6059 \times 10^5 \; sec = 239.05 \; hr.$$

It follows that the time-of-flight $T_F$ for this Hohmann transfer is

$$T/2 = 119.52 \; hr = 4.98 \; days.$$

The angular momentum of the transfer ellipse is given by

$$h = \sqrt{2\mu_e} \sqrt{\frac{r_a r_p}{r_a + r_p}} = 72{,}444 \; km^2/s$$

The speeds $v_p$ and $v_a$ at perigee and apogee

$$v_p = \frac{h}{r_p} = \frac{72444}{6698} = 10.815 \frac{km}{s}$$

$$v_a = \frac{h}{r_a} = \frac{72444}{384400} = 0.18846 \frac{km}{s}$$

The $\Delta v_p$ required to transfer from the initial circular parking orbit to the Hohmann transfer trajectory at its perigee is

$$\Delta v_p = v_p - v_c = 10.815 - 7.7143 = 3.1007 \frac{km}{s}$$

The delta-v required to finally transfer from the Hohmann ellipse to the moon's orbit is

$$\Delta v_a = v_m - v_a = 1.0183 - 0.18846 = 0.82984 \frac{km}{s}$$

The total $\Delta v$ is the sum of the two

$$\Delta v = 3.1007 + 0.82984 = 3.9305 \frac{km}{s}$$

By increasing the injection speed at perigee speed, $v_p$ higher than 10.815 km/s the flight time can be reduced. However, they require higher delta-v. On the other hand, perigee speeds with less than 10.815 km/s can not reach the lunar orbit. In the above simple calculations, the moon's gravity is neglected. The lunar gravity will increasingly affect the spacecraft's trajectory, bending it more and more toward the moon. However, it will be helpful for preliminary design. For accurate work, a numerical analysis is required which is out of the scope of this book on propulsion requirements.

## 12.8  GRAVITY ASSISTS

The use of the relative movement planet around the Sun and gravity of the planet to change the velocity in direction and speed of the spacecraft is known as gravity assist maneuver or swing-by or gravitational slingshot.  Interplanetary spacecraft often use a sequence of gravity assist maneuvers to impart the delta-v requirements to reach regions of the solar system that would be inaccessible using only existing propulsion technology.

In the planetocentric hyperbolic trajectory, as the spacecraft approaches the planet, it gains the kinetic energy (i.e., speed) and loses gravitational potential energy. As it moves out of the planet's gravity along the trajectory, it loses whatever kinetic energy it gained during the approach, ending up with the same final speed it started with. However, the direction of the spacecraft during the encounter and leaves the planet in a different direction as shown in Figure 10.12, i.e., the magnitude of the velocity, $v_\infty$ at the entry and exit of the trajectory is the same but the direction changes. To convert from the planetocentric frame to the heliocentric frame, add the velocities of the planet to the spacecraft vectorially as shown in Figure 12.12.

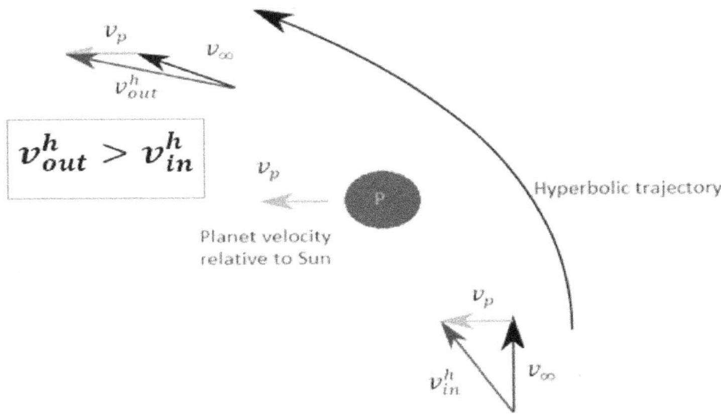


**Figure 12.12** Gravity Assist Velocity Increment.

It can be observed that the outgoing speed is not the same as the incoming speed, and the spacecraft can either speed up or slow down. If the spacecraft moves in the planet's movement direction, it acquires some planet's orbital energy and its speed increases. To decrease speed, the spacecraft flies against the movement of the planet to transfer some of its own orbital energy to the planet. The planet's losses or gains are insignificant but the spacecraft's velocity and direction may change by a large amount. For example, during the Voyager encounter with Jupiter in 1979, the planet slowed down by roughly $10^{-24}$ km/s, that is very negligible compared to the Voyager's large velocity gain of 10 km/s. In general, after the encounter, the spacecraft is pointing more along the planet's direction than it was before the encounter, its speed will increase.

Consider a hypothetical case of the maneuver where the spacecraft of mass, $m$ is approaching a planet of mass $M$ head-on at speed $V$ and the planet itself is moving at a speed $U$ as shown in Figure 12.13. The spacecraft is making an extremely eccentric hyperbolic orbit, making a virtual 180-degree turn. The conservation of kinetic energy and momentum before and after the interaction gives:

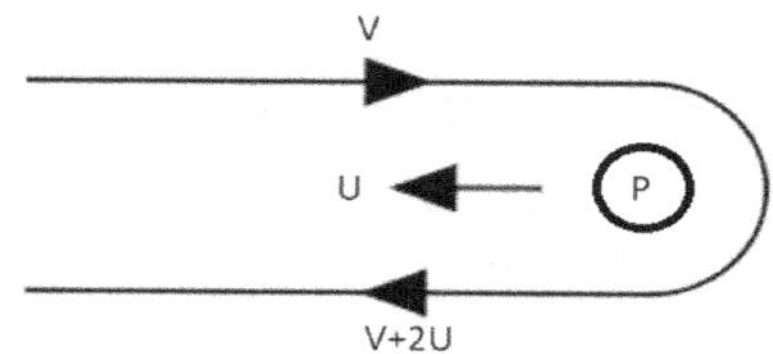



**Figure 12.13** Hypothetical Case of Planet Encounter of Spacecraft.

$$MU_1^2 + mV_1^2 = MU_2^2 + mV_2^2$$

$$MU_1 - mV_1 = MU_2 - mV_2$$

where subscripts 1 and 2 denote before and after, respectively. Solving the above equations for $V_2$:

$$V_2 = \frac{\left(1 - \frac{m}{M}\right)V_1 + 2U_1}{1 + \frac{m}{M}}$$

Since $M >> m$, the ratio $m/M$ is negligible. Then, the spacecraft return with a velocity $V_2$ given by

$$V_2 = V_1 + 2U_1$$

That means after fly by so close, in a maneuver that approaches the spacecraft head on to planet, the speed increases to two times of the planet speed plus original speed of the spacecraft.

The turning angle $\delta$ is solely a function of the eccentricity of the hyperbolic trajectory:

$$\delta = 2\frac{1}{e}$$

Decreasing aiming radius $\Delta$ (offset distance) as shown in Figure 12.14 will decrease angular momentum and eccentricity $e$; decreasing eccentricity (toward unity) enhances the effect of the gravity assist. Increasing the $\Delta$ increases the periapsis radius, angular momentum, and eccentricity $e$; thus, decreases the turning angle $\delta$ and flattens out the hyperbolic trajectory.

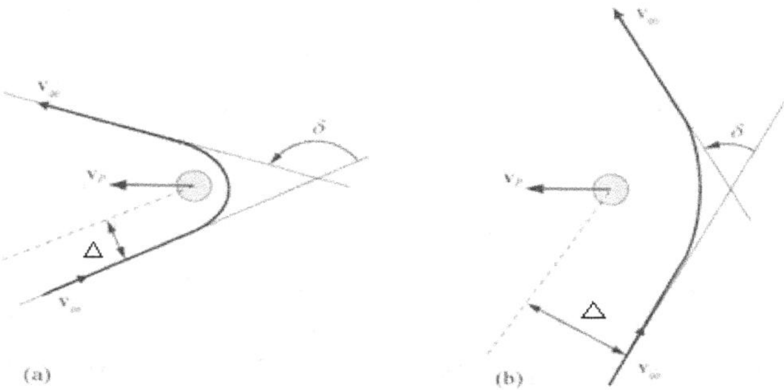



**Figure 12.14** Effect of Turning angle on eccentricity of Hyperbola.

Hence, the closer the planetary flyby, the greater the change in heliocentric energy. With a smaller value of $\Delta$, the danger of colliding with the planet increases. Flybys of Earth and Venus are typically limited to about 300 km altitude to avoid their atmospheres. To minimize exposure to Jupiter's hazardous radiation environment, Jupiter flyby altitude is limited to roughly 500,000 km.

## 12.9  SYNODIC PERIOD

The interplanetary transfer orbits, like Hohmann transfer, take long periods from a few months to years to reach the target orbits. During this period the target planet may complete a significant fraction of its orbit. The main purpose of interplanetary trajectories is to correctly time the transfer so that the spacecraft not only intercept a planet's orbit but also to rendezvous with the planet at arrival: i.e., target planet arrives at the apse line of the transfer ellipse at the same time as the spacecraft does. The time duration that this favorable position occurs is said to offer a launch opportunity. To determine the launch time of an interplanetary mission, the total time of transfer orbit and the angular distance between departure and target planets (phase angle) relative to the Sun are required.

Consider two planets, departure planet 1 and target planet 2, in circular orbits around the Sun as shown in Figure 12.15. AD is the common horizontal apse line from which to measure the true anomaly $\theta$ of planets 1 and 2. The mean motion, i.e., the angular velocity $n$ of the planet is given by

$$n_1 = \frac{2\pi}{T_1} \; ; \quad n_2 = \frac{2\pi}{T_2}$$

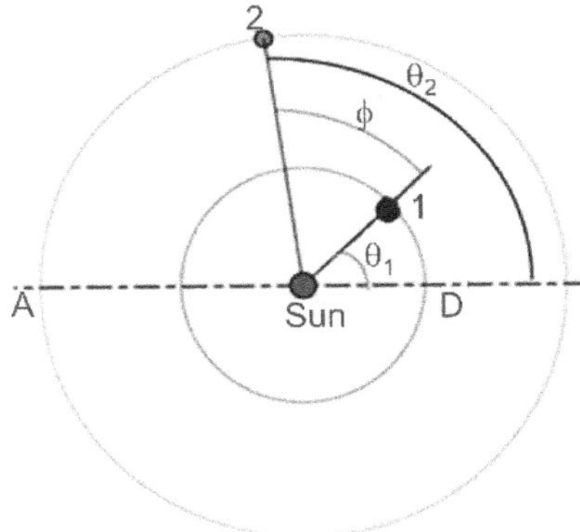


**Figure 10.15**  Planets phase angles.

The true anomaly of a planet at a given time t is

$$\theta = \theta_0 + nt$$

where $\theta_0$ is the value of the true anomaly at $t=0$. Applying this equation to the two planets, the true anomalies of the planets at time t are:

$$\theta_1 = \theta_{1,0} + n_1 t$$

$$\theta_2 = \theta_{2.0} + n_2 t$$

The phase angle $\phi$, the angular position of the planet 2 relative to 1, can be expressed as

$$\emptyset = \theta_2 - \theta_1 = (\theta_{2.0} - \theta_{1.0}) + (n_2 - n_1)t = \emptyset_0 + (n_2 - n_1)t \qquad (12.10)$$

Here, $\emptyset_0$ is the phase angle at time zero. Note that if the planet 2 is an outer planet of planet 1, as shown Figure 12:15, $n_1 > n_2$ the value of relative angular velocity $n_2 - n_1$ is negative. On the other hand, if the planet is an inner planet of the value of $n_2 - n_1$ is positive. The phase angle varies linearly with time as per the Eq. (12.10). The question is, if the phase angle is $\emptyset_0$ at $t=0$, how long will it take to become $\emptyset_0$ again? Obviously, it happens when the position vector of planet 2 rotates through $2\pi$ radians relative to planet 1. The time required for the phase angle to return to its initial value is called the synodic period of the two planets, and is denoted $T_{syn}$. For the configuration shown in Figure 12.15, $T_{syn}$ is the time required for $\emptyset$ to change from $\emptyset_0$ to $\emptyset_0 - 2\pi$. From Eq. (12.10)

$$\emptyset_0 - 2\pi = \emptyset_0 + (n_2 - n_1)T_{syn}$$

$$T_{syn} = \frac{2\pi}{n_1 - n_2} \quad (n_1 > n_2)$$

In case $n_2 > n_1$, the synodic time period is the time required for $\emptyset$ to change from $\emptyset_0$ to $\emptyset_0 + 2\pi$. Therefore,

$$T_{syn} = \frac{2\pi}{n_2 - n_1} \quad (n_2 > n_1)$$

Combining both the cases

$$T_{syn} = \frac{2\pi}{|n_1 - n_2|} \qquad \ldots\ldots(12.11)$$

Substituting the value of $n$, the synodic time period of orbit 2 relative to 1 is

$$T_{syn} = \frac{T_1 T_2}{|T_1 - T_2|} \qquad \ldots\ldots(12.12)$$

Table 12.3 shows the synodic period of the planets relative to Earth based on circular orbits.

It can be observed that the synodic period of the outer planets is essentially one year because their orbital motion is very slow in comparison to that of the Earth.

**Table 12.3** Synodic periods of planets relative to Earth.

| Planet | $T_{syn}$ (days) |
|---|---|
| Mercury | 116 |
| Venus | 584 |
| Mars | 780 |
| Jupiter | 399 |
| Saturn | 378 |
| Uranus | 370 |
| Neptune | 367 |
| Pluto | 367 |
| Moon | 30 |

Figure 12.16 illustrates the interplanetary mission from departure planet 1 and rendezvous with outer target planet 2 following a heliocentric Hohmann transfer. The spacecraft returns to planet 1 through another Hohmann transfer.

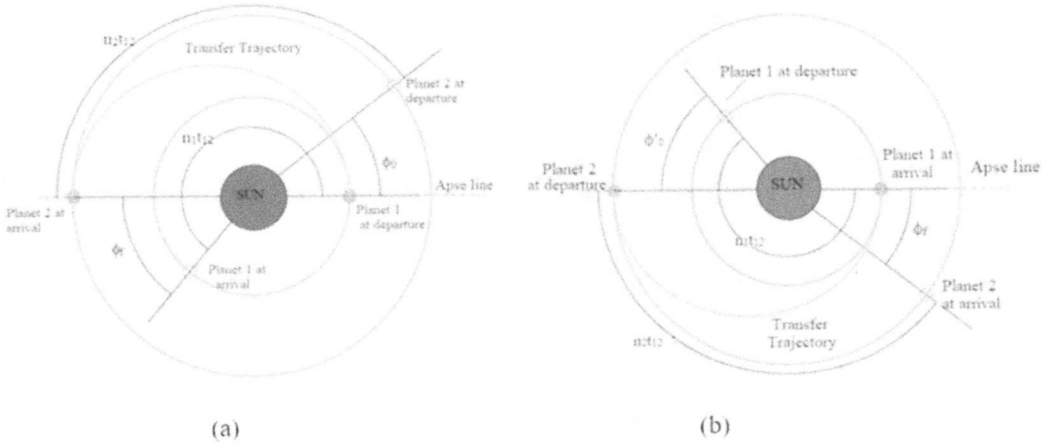


**Figure 12.16** Interplanetary Mission between 2 planets
(a) forward trip Planet 1 to 2, (b) return trip from planet 2 to 1.

Time required to cover heliocentric Hohmann transfer is

$$t_{12} = \frac{\pi}{\sqrt{\mu_{sun}}} \left(\frac{R_1 + R_2}{2}\right)^{\frac{3}{2}}$$

The initial phase angle $= \emptyset_0$

The spacecraft travels an angular distance during the period $t_{12} = \pi$ radians.

The planet 2 during the time $t_{12}$ covers an angular distance of BD equal to $n_2 t_{12}$ on orbit 2 and reaches point D on the common apse line. The planet 1 covers an angular distance of DC and occupies the position C. Hence, the initial phase angle $\emptyset_0$ can be expressed as

$$\emptyset_0 = \pi - n_2 t_{12}$$

When the spacecraft rendezvous with plant, the final phase angle between plants 1 and 2 is $\emptyset_f$. This can be expressed as

$$\emptyset_f = \emptyset_0 + (n_2 - n_1)t_{12} = \pi - n_2 t_{12} + (n_2 - n_1)t_{AD}$$

$$\emptyset_f = \pi - n_1 t_{12}$$

The planet 2 ends up being behind planet 1 by an amount equal to the magnitude of $\emptyset_f$. Now, assume that the spacecraft makes a return trip to planet 1 as shown in Figure 10.16 b. As the transfer time is the same for the return trip, the phase for the return trip must be

$$\emptyset_0' = -\emptyset_f$$

The phase angle at the beginning of the return trip must be the negative of the phase angle at arrival from planet 1. The time required for the phase angle to reach its proper value is called the wait time, $t_{wait}$. Setting $t = 0$ at the instant of arrival at planet 2,

$$\emptyset = \emptyset_f + (n_2 - n_1)t$$

After the waiting time $t_{wait}$, $\emptyset$ becomes $-\emptyset_f$

$$-\emptyset_f = \emptyset_f + (n_2 - n_1)t_{wait}$$

$$t_{wait} = \frac{-2\emptyset_f}{(n_2-n_1)} \qquad\qquad .....(12.13)$$

Depending on the value of $\emptyset_f$, above equation may give a negative result. Therefore, we must add or subtract integer multiples of $2\pi$ until $t_{wait}$ becomes positive.

$$t_{wait} = \frac{-2\emptyset_f - 2\pi N}{(n_2 - n_1)} \qquad (n_1 > n_2)$$

$$t_{wait} = \frac{-2\emptyset_f - 2\pi N}{(n_2-n_1)} \qquad (n_1 > n_2) \qquad .....(12.14)$$

$$t_{wait} = \frac{-2\emptyset_f + 2\pi N}{(n_2 - n_1)} \qquad (n_1 < n_2)$$

where N = 0, 1, 2,... is chosen so that $t_{wait}$ is positive.

**Example 12.2**

Find the travel time from Earth to Mars and return for a spacecraft.

Radii of planets

$$R_{earth} = 149.6 \times 10^6 \ km$$

$$R_{mars} = 227.9 \times 10^6 \ km$$

Gravitational parameter of the Sun

$$\mu_{sun} = 132.71 \times 10^9 \frac{km^3}{s^2}$$

The time of flight from Earth to Mars on heliocentric Hohmann transfer orbit is

$$t_{12} = \frac{\pi}{\sqrt{\mu_{sun}}} \left(\frac{R_1 + R_2}{2}\right)^{\frac{3}{2}} = \frac{\pi}{\sqrt{132.71 \times 10^9}} \left(\frac{149.6 \times 10^6 + 227.9 \times 10^6}{2}\right)^{1.5}$$

$$= 2.231 \times 10^7 s$$

$$t_{12} = 258.22 \ days$$

The angular velocity of Earth and Mar

$$n_{earth} = \frac{2\pi}{365.26} = 0.017202 \frac{rad}{day}$$

$$n_{mars} = \frac{2\pi}{1.881 \times 365.26} = 0.009145 \frac{rad}{days}$$

For a spacecraft to depart on a mission to Mars by means of a Hohmann transfer, the phase angle between earth and Mars is

$$\varphi_0 = \pi - n_{mars}t_{12} = \pi - 0.009145 \times 258.22 = 0.7814 \, rad = 44.77°$$

The phase angle between earth and Mars when the spacecraft reaches Mars is given

$$\varphi_f = \pi - n_{earth}t_{12} = \pi - 0.017202 \times 258.22 = -1.3003 = -74.5°$$

The wait time is given by

$$t_{wait} = \frac{-2\varphi_f - 2\pi N}{n_{mars} - n_{earth}} \, for \, n_1 > n_2$$

$$= \frac{-2 \times (-1.3003) - 2\pi N}{0.009145 - 0.017202} = -332.78 + 779.84N$$

Taking N= 1 to get positive value

$$t_{wait} = 447.06 \, days = 1.23 \, years$$

The return time is the same as forward journey

$$t_{12} = 258.22 \, days$$

The minimum total time for a manned Mars mission, using Hohmann transfers is

$$t = 258.22 + 447.06 + 258.22 = 963.5 \, days = 2.64 \, yrs$$

## Nomenclature

| | |
|---|---|
| $a$ | semi major axis of elliptical orbit (m or AU) |
| AU | astronomical unit (1AU=149.5978707 $\times$ $10^6$ km) |
| $e$ | eccentricity of orbit |
| $i$ | orbit inclination (deg) |
| $M_{planet}$ | mass of the planet (kg) |
| $M_{sun}$ | mass of the sun s(kg) |
| $R$ | mean orbital radius of the planet (km) |
| $r_a$ | apogee radius (m or km) |
| $R_E$ | radius of the Earth (m) |
| $r_p$ | periapsis radius (m or km) |
| $R_P$ | radius of the planet (m) |
| $R_s$ | radius of the sphere of influence of a planet (km) |
| $T_{syn}$ | synodic period (s) |
| $v$ | velocity (m/s) |

$v_\infty$     hyperbolic excess speed (m/s)

$v_m$     circular orbital speed of moon (m/s)

$v_s$     Planetocentric velocity of spacecraft (m/s)

$v_p^h$     heliocentric velocity of the planet (m/s)

$v_s^h$     Heliocentric velocity of spacecraft (m/s)

$\delta$     turning angle of orbit (deg)

$\theta$     true anomaly angle (deg)

$\phi$     phase angle (deg)

$\Delta$     aiming radius of hyperbolic orbit (m or km)

$\mu_{sun}$     gravitational parameter of the Sun ($km^3/s^2$)

## PROBLEMS

12.1   Find the synodic period of Mars relative to that of the earth.

12.2   Calculate the synodic period of Venus relative to the earth.

12.3   Calculate the total delta-v required for a Hohmann transfer from earth's orbit to Saturn's orbit.

12.4   A spacecraft is launched on a mission to Mars starting from a 250-km circular parking orbit. Calculate the delta-v required, the location of perigee of the departure hyperbola, and the amount of propellant required as a percentage of the spacecraft mass before the delta-v burn, assuming a specific impulse of 250 s.

12.5   Find the total delta-v required for a Hohmann transfer from earth's orbit to Saturn's orbit.

12.6   A spacecraft in a circular 160-km earth orbit, coplanar with that of the moon, is launched at perigee onto an elliptical trajectory whose apogee lies on the moon's SOI. Calculate the $\Delta v$ required to boost the spacecraft onto this trajectory.

# CHAPTER 13  Chemical Thrusters for Spacecraft Maneuvers

## 13.1  INTRODUCTION

This chapter covers the chemical propulsion systems used to change the velocity of spacecraft and artificial satellites. The spacecraft engine includes clusters of small thrusters that are installed at different locations with a common pressurized feed system. They are also known as auxiliary propulsion systems (APS) or reaction control systems (RCS) or attitude control systems (ACS). These thrusters produce low thrust (0.001 to 4000 N), use storable liquid propellants, have accurate repeatable pulsing, a long life in space and/or long-term storage with loaded propellants in the flight tanks. Thrusters using chemical Propulsion comprise propellant storage system, feed system, thrust chamber and control unit. Feed system consisting pressurized gas tanks, valves, piping that connects the propellant storage system with the thruster. Electric control unit is to operate electrically the valves and thruster. They use a pressurized feed system with high pressure gas supply with tank pressure of 0.7 to 18 MPa. The chamber thrust pressure is around 2.0 MPa. They are suitable for several thousands of starts and stop operations in their long period of life in space. The propulsion technologies that are available are:

1. Cold gas system

2. Monopropellant system

3. Bipropellant system

4. Solid motor system

The solid motor system is useful for single firing. The other systems can be used for multiple burn usage in a space mission.

## 13.2  COLD GAS SYSTEM

Cold gas propulsion systems are one of the best choices to perform orbit maneuvers that require small velocity changes. They are relatively simple systems that provide limited spacecraft propulsion and are one of the most mature technologies for small spacecraft. They are simple, less expensive, provide a small impulse bit and suitable for multiple restarts and pulsing.

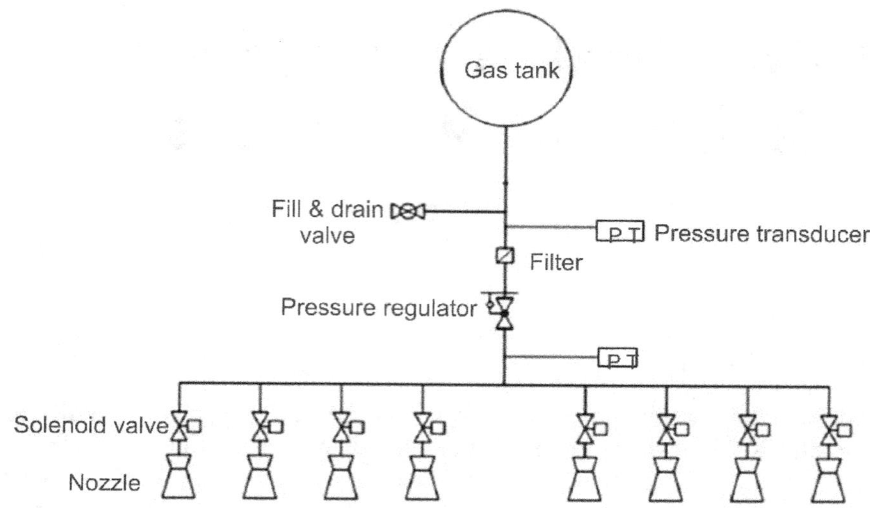


**Figure 13.1** Typical Cold Gas Thruster Schematic.

The typical system operating with cold gas consists of a propellant tank, fill & drain valve, filter, pressure regulator, line pressure transducers and thrusters with solenoid valves as shown in Figure 13.1 The pressure regulator provides propellant at constant pressure as the tank pressure drops. The cold gas is stored at high pressures in a tank. Nitrogen, argon, krypton, Freon 14, ammonia and propane has been employed in operational spacecraft, but nitrogen has been the most common cold-gas propellant. Nitrogen is popular as leakage is less likely. Table 13.1 gives the specific thrust developed with different pressurized gasses. The developed thrust level is around 5 mN to 10 N and impulse bits around 5-10 Ns. The actual or measured impulse is 90% theoretical. The measured performance is lower than theoretical value is primarily due to conical nozzles commonly used, as opposed to the contoured nozzles in larger thrusters and launching vehicles.

**Table 13.1** Cold Gas Performance.

| Propellant | Molecular Weight (Kg/Kmole) | Density (g/cm³) | Specific Thrust (s) | |
|---|---|---|---|---|
| | | | Theoretical | Measured |
| Hydrogen | 2.0 | 0.02 | 296 | 272 |
| Helium | 4.0 | 0.04 | 179 | 165 |
| Nitrogen | 28.0 | 0.28 | 80 | 73 |
| Ammonia | 17.0 | Liquid | 105 | 96 |
| Carbon dioxide | 44.0 | Liquid | 67 | 61 |

Cold gas thrusters are available off-the-shelf from different manufacturers like Rafael, Marotta, DASA and Moog with specifications to design orbit maneuvers. Figure 13.2 shows the typical Moog cold gas thrusters with specifications.

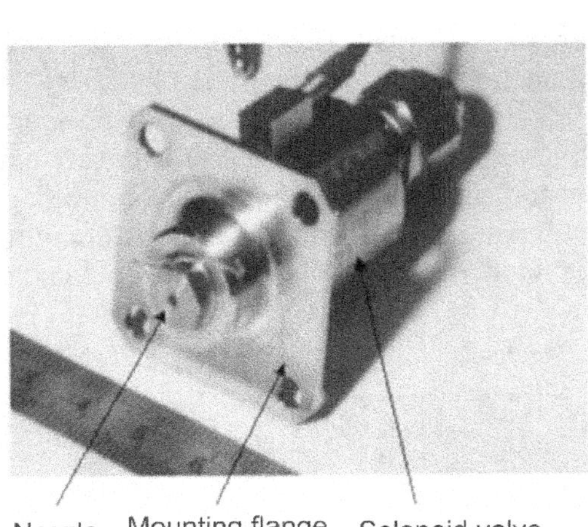

| Typical Performance Specifications | | |
|---|---|---|
| Pressures | Operating | 200 to 2500 psig (13.7 to 172.4 bar) |
| | Proof | 3750 psig (258.6 bar) |
| | Burst | 6,250 psig (431 bar) |
| Atmospheric Thrust | | Thruster 1&3: 11.75lbf (52N) at 2,000psi thruster 2:23.5lbf (105N) at 2,000psi |
| Operating voltage range | | 24 to 34 vdc |
| Response time | Open | <10msec |
| | Close | <10msec |
| Power consumption | | 6 to 12 watts |
| Leakage | Internal | ≤ 10 sec/min GN per seat |
| | External | < 30 sec/min GN per seat |
| Cycle life | | > 6,000 |
| Weight | | 0.95 lb (0.43 kg) |
| Thermal capability (operating) | | 0°F to + 120°F (−17°C to +48°C) |
| Wetted Materials | | Aluminum, Stainless steel and vespel® |

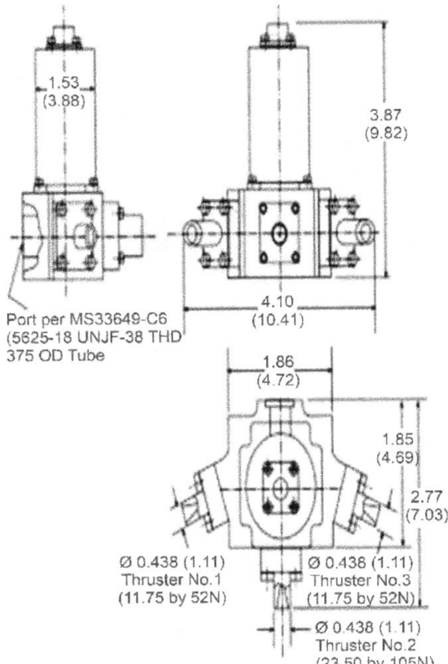


**Figure 13.2** Typical Cold Gas Thruster Off-the-shelf Product.

## 13.3 MONOPROPELLANT SYSTEM

A monopropellant is an unstable chemical that decomposes exothermically to produce a hot gas. It is injected into a catalyst bed, where it decomposes and generates hot gases at around 1500 K. These gases then are expelled through a converging-diverging nozzle generating thrust. High-temperature alloys are used for the nozzle, and supplemental cooling is not required. Figure 13.3 illustrates the main components of the monopropellant thrust. The propellant flow valve, an integral part of the thruster, controls the propellant flow in to the catalytic chamber. They are designed for operation in both steady state and pulse mode operation over a wide pressure range and are thus ideal for propulsion systems operating in blow-down mode. Thrusters are equipped with an internally redundant catalyst bed heater and thermal insulation to guarantee optimum reproducible start-up conditions. Monopropellant thrusters have flown in the sizes ranging from 0.5 to 2600 N and blow down ratios up to 6. Pulse widths as low as 7 ms have been demonstrated as have thrust levels in the 4500 N class.

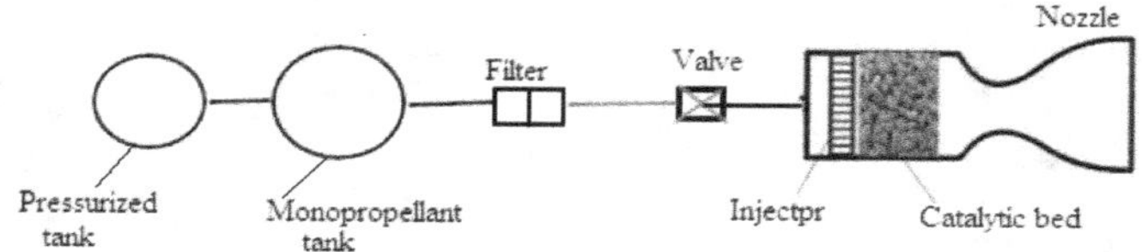


**Figure 13.3** Main Components Monopropellant System.

Only three monopropellants, hydrazine, hydrogen peroxide, propyl nitrate have ever been used on flight vehicles. Shock sensitivity eliminated propyl nitrate after limited use for jet engine starters. Hydrogen peroxide with significant use up to the 1970s, current use in the Soyuz spacecraft, has the persistent problem of slow decomposition during storage. This increases the storage tank pressure and causes water dilution of the propellant, leading to performance degradation. Recent interest is on LMP-103S, a Swedish developed monopropellant based on ammonium dinatramine (ADN) having similar properties to the hydroxylammonium nitrate (HAN). It is not yet fully flight qualified. Hydrazine, $N_2H_4$ is the monopropellant in use presently due its stability, restart capability, high performance, clean exhaust, and low flame temperature. The Shell 405 iridium pellet catalyst bed results in almost unlimited spontaneous restart capability in hydrazine thrusters. Generally, hydrazine propulsion systems are supplied by manufacturers like ArianeGroup Gmbh and Moog Inc., as self-contained 'plug-in' propulsion modules. The module comprises the complete propulsion system together with all the associated sensors, heaters, thermistors, pressure transducers and harness elements.

### 13.3.1 Steady State and Pulsed Performance

The propellant flow into the chamber is controlled by a propellant valve. The propellant is injected into a catalyst bed, where it decomposes into hydrogen, nitrogen, and ammonia. The decomposition of hydrazine into hydrogen and ammonia is exothermic. The further decomposition of ammonia into hydrogen and nitrogen is endothermic and leads to reduction of flame temperature and specific impulse as shown in Fig 13.4. Ammonia decomposition is held around 50% in current design of thrusters. A steady-state, theoretical, vacuum specific impulse of about 240 s can be expected at an area ratio of 50:1. Real system specific impulse will be about 93% theoretical. The adiabatic flame temperature is about 1700 K. The pulsing mode of operation, to control attitude of the spacecraft, covers a wide range of duty cycles and pulse widths. The performance depends on the propellant valve, feed tubing and chamber, i.e., the valve response characteristics, hydraulic delay (the flow time from valve to the injector), ignition delay and pressure rise time. A 20 ms pulse typical characteristic is shown in Figure 13.5 with the signal and valve response time. For all practical purposes the impulse bit can be estimated as Ft (N), where $t$ is the pulse width, time from valve on to valve off and $F$ is the steady-state thrust reached.

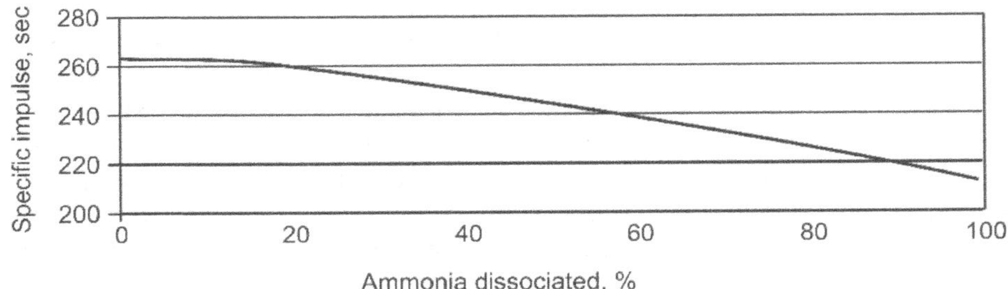

Figure 13.4 Hydrazine monopropellant performance vs ammonia dissociation (Nozzle area ratio=50).

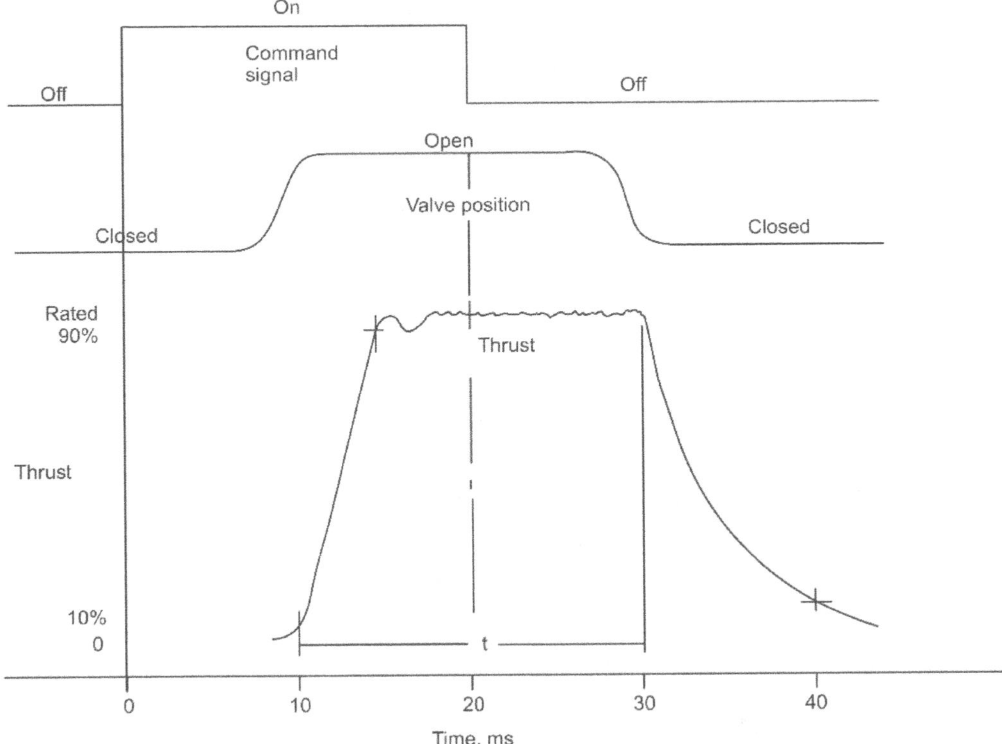


**Figure 13.5** Typical pulse mode operation characteristics of thruster (20 ms pulse) (Source: Charles D. Brown).

The main causes for degradation of the thruster are due to mechanical failure of the catalyst and impurities accumulated on the surface of it. The impurities can be avoided by using super pure hydrazine. The damage of catalyst pellets is mainly due to cold starts. During the cold start the propellant accumulates in the catalyst bed due to ignition delay and thermally stress the pellets. When ignition occurs, an overpressure spike stresses the pellets. To avoid the cold start damage, the catalyst bed and propellant tank are maintained at elevated temperature of around 600 K using electrical heaters.

### 13.3.2 Blowdown System

In a blowdown system, as shown in Figure 13.6, the tank is initially pressurized to required level around 20 to 30 bars. The pressure is allowed to decay as propellant is used up to around 6 bar. The performance of the thruster also depends on the amount of propellant expelled as shown in Figure 13.6. When most of the fuel has been consumed, the fuel has sufficient volume to slosh and possibly disturb the flight trajectory. Also, in the zero-gravity environment, the propellant does not remain in the bottom of the tank and it will take a shape in the tank itself. To avoid these problems, the pressurized gas and the propellant are separated with a metallic/Teflon diaphragm or bladder made of Teflon or compressible bellows made of corrosion resistant steel. The pressurizing gas is an inert gas have a lower molecular weight.

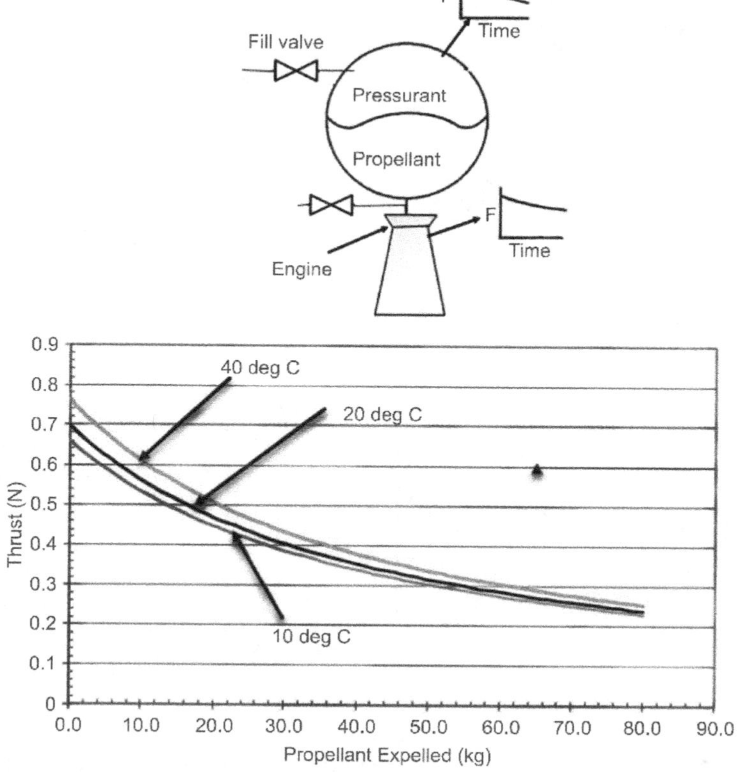


**Figure 13.6** Monopropellant Blowdown Pressurization System.

The inert gasses, either nitrogen or helium, are generally used. Helium provides the lightest system, but it is difficult to prevent helium leakage. Nitrogen is generally used in spacecraft. The blowdown system is simple, reliable and less expensive and is used with monopropellant exclusively in modern spacecraft designs.

For short duration operation of a thruster, the flow of propellant is considered as an isothermal process. In this case the temperature of the gas stored is equal to the temperature propellant. In such case the mass of the pressurant, $m_g$ $(kg)$ can be computed by using the law of perfect gas, as follows

$$m_g = \frac{p_g V_u}{R_g T_g} \qquad \qquad .....(13.1)$$

where

$p_g$ is the gas pressure (N/m$^2$)

$V_u$ is the ullage volume (the gas volume over the propellant) (m$^3$)

$R_g$ is the specific gas constant (N m/kg K)

$T_g$ is the temperature of gas (K)

At any given time, the gas in the ullage will be an equilibrium mixture of pressurant and propellant vapor. The propellant vapor will saturate the tank gas and reach a partial pressure equal to the propellant vapor pressure. Then, the sum of the partial pressure of the pressurant gas and the vapor pressure of the propellant is the total pressure of the tank. The vapor affects the tank pressure due to permeability of the bladder material. The vapor pressure of the fuels is lower than the oxidizers. Another important parameter in the design is the blowdown ratio (B), the ratio of initial pressure ($p_{gi}$) to final pressure ($p_{gf}$) of the pressurant gas.

$$B = \frac{p_{gi}}{p_{gf}} = \frac{V_{uf}}{V_{ui}} \qquad \qquad .....(13.2)$$

where $V_{ui}$ and $V_{uf}$ are initial and final ullage volumes. The difference between these ullage volumes give the volume of the useful propellant ($V_{p,u}$).

$$V_{p,u} = V_{uf} - V_{ui} = V_{ui}(B - 1) \qquad \qquad .....(13.3)$$

Generally, a blowdown ratio of 3 to 4 is common to use. The specific impulse decreases with the pressure decay in a monopropellant thruster and is due to the dissociation of ammonia in the case of hydrazine. The loss can be estimated by

$$I_{sp\ at\ pressure\ 2} = I_{sp\ at\ pressure\ 1}(1 - 0.005\ B) \qquad \qquad .....(13.4)$$

In the case the propellant is drawn from the tank rapidly the process cannot be considered as isothermal. The gas temperature will drop below the propellant temperature following the isentropic expansion. After the burn, the temperature of the gas approaches the propellant temperature asymptotically with time. The isentropic expansions can be used to find the final conditions of the gas.

$$p_{gi} V_{gi}^{\gamma} = p_{gf} V_{gf}^{\gamma} \qquad \qquad .....(13.5)$$

$$\frac{T_f}{T_i} = \left(\frac{V_{gi}}{V_{gf}}\right)^{\gamma-1}$$

In the blowdown system, the performance parameters of thrusters like chamber pressure, propellant flow rate and thrust are function of tank pressure that decay with time. The flow rate of the propellant ($\dot{m}_p$) is a function of hydraulic resistance and can be expressed for incompressible flow as

$$\dot{m}_p = \sqrt{\frac{\Delta p}{R}} \qquad\qquad .....(13.6)$$

where $\Delta p = p_t - p_c$, is the pressure drop from tank pressure to chamber pressure against the flow resistance, R in the feed system. It is the sum of flow resistance of feed lines, fittings, thrust chamber valve and catalyst bed. The chamber pressure, $p_c$ produced by the flow rate $\dot{m}_p$ can be expressed from Eq. (4.36) & (4.45) as

$$p_c = \frac{I_{sp}\dot{m}_p}{A_t C_F} \qquad\qquad .....(13.7)$$

The propellant flow rate at a given tank pressure is calculated by solving the Eq. (13.6) and Eq. (13.7) simultaneously. The pressure drop across a catalyst bed is designed to be around 2 to 3 bar. The thrust chamber valve pressure drop is usually around 3 to 4 bars.

In the regulated pressurization system, as shown in Figure 13.7, the pressure in the propellant tank is maintained at a pre-set value. The pressurant is stored in a separated tank at a high pressure around 200 to 400 bars. The regulated pressurant gas is ducted to the propellant tank and the pressure is maintained at around 10 to 20 bar. The advantage of regulation is a constant thrust as a function of time. The regulator valve requires an inlet pressure around 5 bar higher than the outlet pressure to operate properly. A relief valve is necessary to protect the tank in case the regulator fails to open. The minimum ullage volume to have a stable response for the regulator when outflow starts must be about 3% of the propellant tank volume.

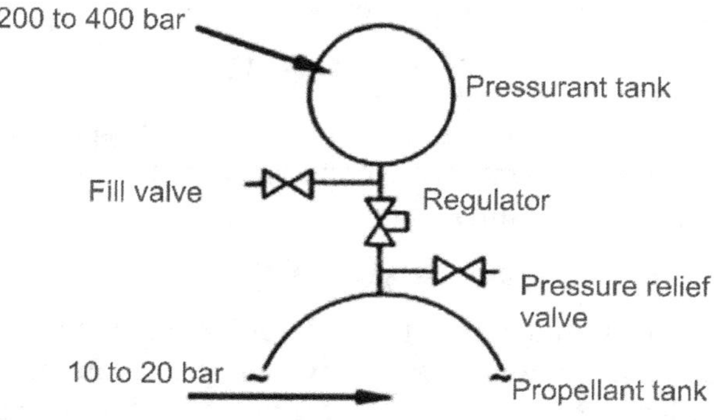


**Figure 13.7** Regulated pressurization system.

The blowdown system is not suitable for bipropellant systems due to variation of flow rates of oxidizer and fuel to maintain the correct mixture ratio. For these systems, regulation is essential for both oxidizer and fuel tanks to keep the flow rate of each propellant constant and at the correct mixture ratio. Each propellant tank ullage must be isolated to prevent vapor mixing.

### 13.3.3    Propellant Inventory

The loaded propellant weight in the system can be divided into three categories:

1. Useful propellant: It is the calculated propellant that is required to carry all maneuvers and all attitude control functions under worst-case conditions.

2. Trapped propellant: It is the amount of propellant trapped or remaining in the feed lines, tanks and valves, hold-up in expulsion devices, and retained vapor left in the system with the pressurizing gas. Trapped propellant can be taken as about 3% of the usable propellant

3. About 0.5% of usable propellant is taken as measurement uncertainty in propellant loading in unknown worst conditions.

Table 13.1 shows the propellant estimation for a typical space mission

**Table 13.2** Typical propellant estimation.

|  | $\Delta V$ (m/s) | Propellant Usage (kg) |
|---|---|---|
| Primary mission TCM | 110 | 22.3 |
| Altitude control |  | 29.3 |
| Primary mission margin | 132 | 25.2 |
| *Original margin allocation* | *91* | *17.5* |
| *Additional margin obtained from unused* | *41* | *7.7* |
| *spacecraft dry mass allocation* |  |  |
| **Total navigation** $\Delta V$ | **242** |  |
| **Total propellant load** |  | **76.8** |

In a mission, having propellant reserves are useful as

1. Extend the life of the spacecraft.

2. Used for unplanned maneuvers in unexpected situation or emergency conditions

3. The mission objectives can be extended after completing primary mission like in Voyager spacecraft.

### 13.3.4    Main Components of the Monopropellant System

Figure 13.8 illustrates the typical monopropellant space propulsion system consisting of multiple thrusters under one propellant tank. Several other components associated with conditioning, transferring and controlling of the propellant are included. The details of the propellant tank are described in section 13.3.2. Either spherical or cylindrical shape tank is used. It consists of a built-in propellant management device which is used to ensure a steady supply of gas-free propellant under all orientations of the satellite. The single tank use does not allow any disturbance torque about the center of inertia of the satellite platform. The propellant tank is externally covered by multilayer insulation with heaters to keep the temperatures of pressurant and propellant nearly constant during the mission. The delivery of the propellant from tank to the thrusters is through titanium or titanium alloy tubes.

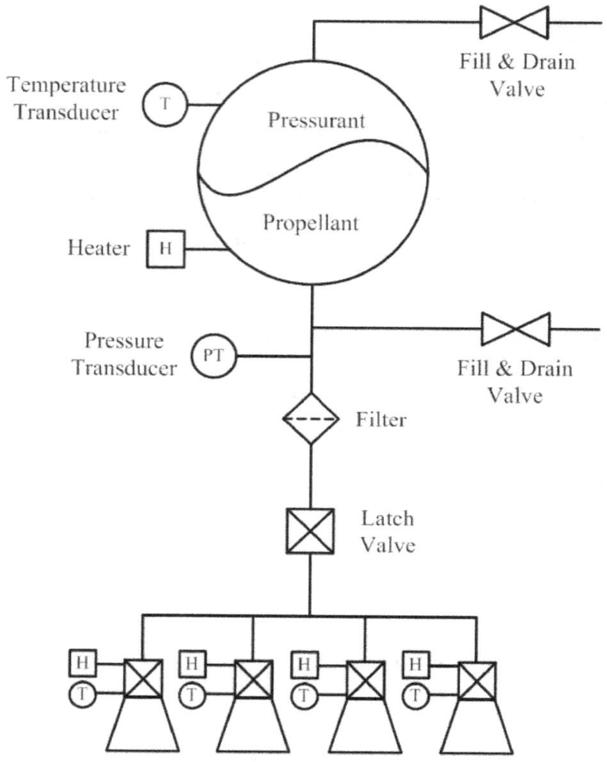


**Figure 13.8** Schematic sketch of Monopropellant Propulsion System.

The thruster is composed of two major subassemblies: the solenoid flow control valve and the engine. Thrust levels can be achieved by means of adjusting the opening ratio of the flow control valve. Heaters are used to ensure vapor phase propellant was expelled from the thrusters. They are configured in clusters and each cluster generally consists of four numbers. Generally, four-thruster configuration is used. Six or more thruster configurations are also possible. A filter is used at the downstream of the tank to catch any dirt, particles, or debris in the propellant which could block the flow in valves and injectors. It ensures that the thrusters are fed with propellant with a high purity level.

Two service valves are used for filling and draining the propellant and the pressurant tank before launch. Latching valve is essentially a switch valve for flow control; a valve which stays in whatever position it was last commended (open or close). These valves are only opened during planned $\Delta v$ maneuvers, and then closed after the maneuver(s) is performed. Solenoid Valve controls the propellant flow rate to fire the thrusters. Each thruster has an independent solenoid valve, and it is suitable for more than hundred thousand cycles.

Pressure and temperature transducers are used for their monitoring on propellant lines. Using two transducers in different locations, the pressure difference can be determined. Figure 13.9 shows the ArianeGroup plug-in monopropellant thrusters of capacity 1N and 400N thrusters with their specifications.

## 13.4  ESTIMATION OF MASS OF THE PROPULSION SYSTEM

In the spacecraft propulsion system, the weight of the propellant and pressurization tanks is the major dry weight component. To increase the system design flexibility the present trend is to use dual identical tanks. The advantage of the dual tank is reduction in tank size, bladder or diaphragm and more choice in the tank location. With a single tank, balance dictates a location at the center of the spacecraft. The mission requirements such as steady state impulses, pulsing impulses and maneuvering thrust level required for automatic control systems determines the propellant requirement and in turn the volume of the tank.

Consider the following example to estimate the volume of the propellant tank. For a spacecraft mission, let the steady state burn impulse for station keeping and other control units is 300,000 Ns  and the pulsing impulse required for operation like Spin/de-spin/nutation is 250,00 Ns. Let the specific impulse for short burn be 2100 m/s and pulsing be 1000 m/s.

Propellant required for short burn = 300000/2100 = 143 kg

Propellant required for impulse burn = 25000/1000 = 25 kg

Taking a reserve propellant around 40% = 68 kg

Total propellant requirement = 143 + 25 + 68 = 236 kg

Trapped propellant around 3% = 7 kg

Loading uncertainty 0.5% = 1 kg

Loaded propellant = 236 + 7 + 1 = 244 kg

Density of the propellant hydrazine at 50° C = 982 kg/m$^3$

Volume of the loaded propellant = 244/982 = 0.249 m$^3$ = 249 lit

Volume of useful propellant = 236/982 = 0.240 m$^3$ = 240 lit

Initial ullage volume is around 33% (1/3) of useful propellant volume = 80 lit

Tank volume required = 249 + 80 = 329 lit

Providing around 1% volume for diaphragm or bladder the required tank volume is 332 lit. There are several manufacturers supplying flight tested off the shelf tanks of different volumes and shapes, spherical or cylindrical for both monopropellant and bipropellant systems. As shown in Figure 13.10, the manufacturer provides the weight and required parameters for design. The amount of pressurization gas can be calculated from the gas equation knowing the adjusted ullage

| 1N Monopropellant Thruster Key Technical Characteristics | |
|---|---|
| **Characteristics** | |
| Thrust Nominal | 1 N |
| Thrust Range | 0.320 ... 1.1 N |
| Specific Impulse, Nominal | 220 s |
| Pulse, Range | 200 ... 223 s |
| Mass Flow, Nominal | 0.44 g/s |
| Mass Flow, Range | 0.142 ... 0.447 g/s |
| Inlet Pressure Range | 5.5 ... 22 bar |
| Minimum Impulse Bit | 0.01 ... 0.043 Ns |
| Nozzle Expansion Ratio | 80 |
| Mass, Thruster with valves | 290 g |
| Propellant | Hydrazine ($N_2H_4$), High-Purity Grade |
| **Qualification** | |
| Total Impulse | 135,000 Ns |
| Cycle Life | 59,000 cycles |
| Propellant Throughput | 67 kg |
| Single Burn Life | 12 h |
| Accumulated Burn Life | 50 h |
| No of Cold Starts <10°C | 10 |

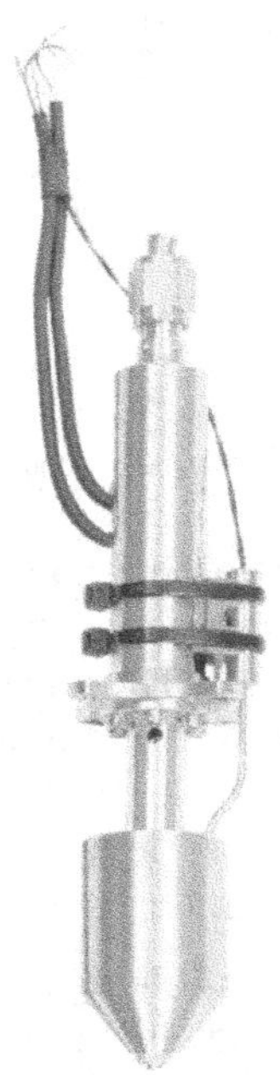

| 400N Mono-Propellant Thruster Key Technical Characteristics | |
|---|---|
| **Characteristics** | |
| Thrust Range | 120-420N |
| Supply Pressure Range | 5,5-26bar |
| Nominal Mass Flow Range | 58-190 g/s |
| Nominal Specific Impulse Range | 2080 - 2155 Ns/kg |
| Minimum Impulse Bit Range | < 9Ns |
| Shortest On-Time | 16ms |
| Nozzle area ratio | 30 |
| Propellant | Monopropellant grade Hydrazine ($N_2H_4$) |
| Mass | P1-design 2,7 kg<br>P2-design 3,8 kg |
| **Qualification** | |
| Total Impulse | < 188kNs |
| Total number of pulses | >3900 pulses |
| Total hydrazine throughput | ~300kg |
| Total operating time | >850s |
| Longest steady state burn | 450s |
| Number of cold starts < 25°C | 19 |

**Figure 13.9** ArianeGroup off-the-shelf 1N and 400N Monopropellant Thrusters.

volume from a selected tank dimension. The standard thrust capacity thrusters are available readily. The initial schematic can be made with selected thrusters, tanks and other control valves to estimate the overall weight of the system. Figure 13.11 shows the schematic of the two-tank monopropellant system. Fault tolerance requirements dictate redundant branches of 12 engines each to accommodate an engine failure. If the dual tanks are cross-trapped, a failure of either propellant system can be accommodated. The initial mass estimate of the propulsion system can be done as shown in Table 13.3. Taking the initial mass determined for the spacecraft and subtracting the propulsion system mass gives the spacecraft propulsion system payload mass that can then be allocated to the other spacecraft subsystems for the mission.

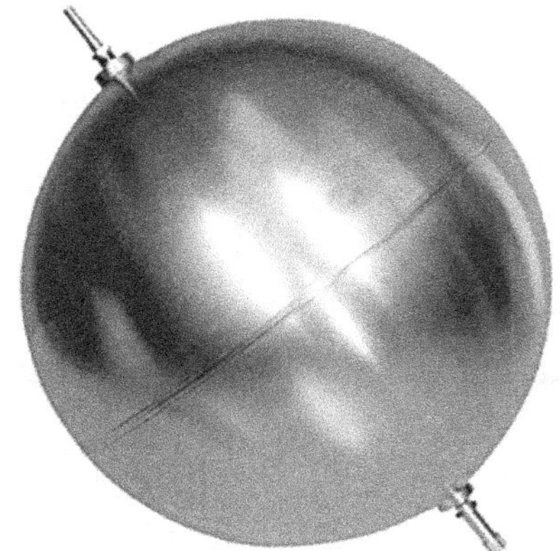

| 104 to 177 Litre Hydrazine Tank | |
| --- | --- |
| Volume (net) | 104 to 177 litres |
| Propellant volume | 78 litres (Blow down 4:1) |
| Propellant | Hydrazine (N2H4) |
| MEOP | 24.6 bar |
| Proof pressure | 36.9 bar (1.5 × MEOP) |
| Burst pressure | 49.2 bar (2 × MEOP) |
| Mass | 6.4 kg |
| Pressurant gas | Helium or nitrogen |
| Geometry | Spherical |
| Interface fixation | Pole mounting rigidly fixed at propellant port boss |
| Materials | |
| -pressure vessel | Ti6AlV st (3.7167.7) |
| -fluid ports | Ti6AlV st (3.7167.1) |
| -PMD | All titanium |
| Dimensions | Diameter: 586 mm |
| Heritage | Globalstar, Theos, Pleiades, RocSat |

**Figure 13.10** Typical Hydrazine propellant Tank.

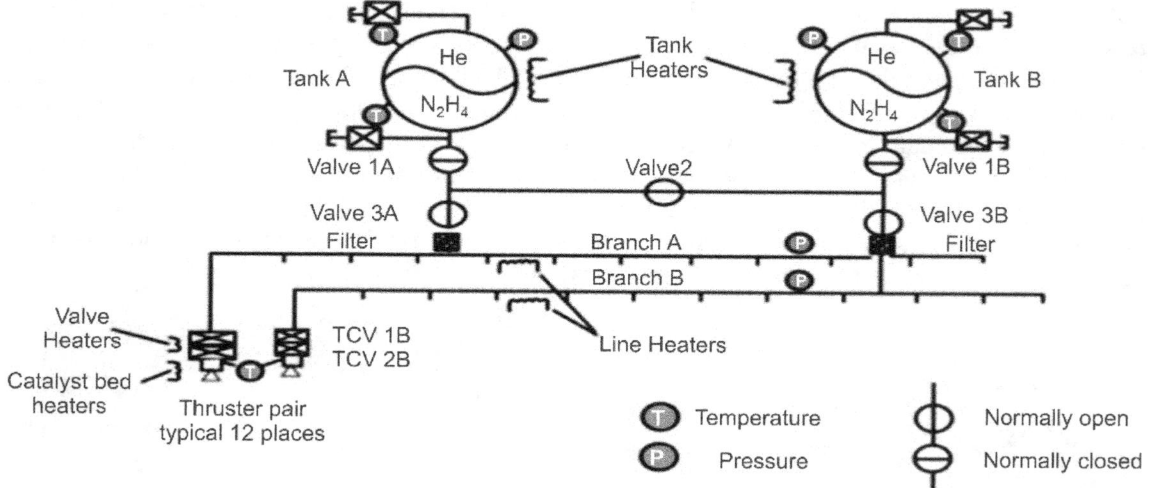


**Figure 13.11** Schematic a Functional Propulsion System.

**Table 13.3** Propulsion System Mass Estimation.

| Component | No. of units | Unit Mass(kg) | Total Mass (kg) |
|---|---|---|---|
| Propellant | | | |
|     Usable | | | |
|     Trapped | | | |
| Pressurant | | | |
| Feed System | | | |
|     Tanks | | | |
|     Valves | | | |
|     Filters | | | |
|     Feed lines | | | |
|     Filters | | | |
|     Instruments | | | |
|     Heaters | | | |
|     Others | | | |
| Thrusters with Valves | | | |
|     Type 1 | | | |
|     Type 2 | | | |
| Wet Mass | | | |
| Burnout Mass | | | |
| Dry Mass | | | |

**Example 13.1**

The catalyst bed of a hydrazine monopropellant rocket is designed to give 30 % of dissociation of ammonia at temperature of 25° C. Find the heat liberated per kg of hydrazine, temperature of products of decomposition and characteristic velocity. Take standard heats of formation of hydrazine and ammonia as + 55 kJ/mole and − 46 kJ/mole respectively. The specific heats of ammonia, hydrogen and nitrogen are 0.04, 0.03 and 0.035 kJ/mole K respectively.

Take $x$ is the fractional dissociation of ammonia. The decomposition equation of hydrazine is

$$3N_2H_4 \rightarrow 4(1-x)NH_3 + (1+2x)N_2 + 6H_2$$

With $x = 0.3$, the above equation becomes

$$3N_2H_4 \rightarrow 2.8NH_3 + 1.6N_2 + 6H_2$$

From Eq (5.3). the heat of formation

$$H_r^0 = \sum n_i \left(\Delta H_f^0\right)_{products} - \sum n_i\left(\Delta H_f^0\right)_{reactants}$$

So, the heat released during the reaction $= -H_r^0 = -[2.8 \times (-46) + 1.6 \times (0) - 3 \times (55)]$

$$= 293.8 \ kJ$$

Therefore, the heat released by one kg of hydrazine

$$= \frac{293.8}{3 \times (2 \times 14 + 4 \times 1) \times 10^{-3}} = 3060 \ kJ$$

Let $T$ K be the temperature of products of decomposition. Then,

$$Heat \ released = 2.8 \ c_{pNH_3}(T - 298) + 1.6 \ c_{pN_2}(T - 298) + 1.8 c_{pH_2}(T - 298)$$

$$293.8 = (2.8 \times 0.04 + 1.6 \times 0.035 + 1.8 \times 0.03)(T - 298)$$

$$T = 1620 \ K$$

The molecular mass of the products, i.e., combustion gases is:

$$\mathcal{M} = \frac{Molar \ masses \ of \ individual \ components}{Number \ of \ moles}$$

$$\mathcal{M} = \frac{2.8 \times 17 + 1.6 \times (14 \times 2) + 1.8 \times 2}{2.8 + 1.6 + 1.8} = 15.5 \ kg/mole$$

The specific heat of the products at constant pressure $c_{p,product}$ is:

$$c_{p,product} = \frac{2.8 \times 0.04 + 1.6 \times 0.035 + 1.8 \times 0.03}{2.8 + 1.6 + 1.8} = 0.0358 \ kJ/(mole \ K)$$

$$c_{v,product} = c_{p,product} - R_u = 0.0358 - 0.008314 = 0.02748 \frac{kJ}{mole \ K}$$

The ratio of specific heats of products $\gamma = \frac{0.0358}{0.02748} = 1.302$

The characteristic velocity is given by Eq. (4.37)

$$c^* = \frac{1}{\Gamma}\sqrt{\frac{R_u T_c}{\mathcal{M}}}$$

$$\Gamma = \sqrt{\gamma}\left(\frac{2}{\gamma+1}\right)^{\frac{\gamma+1}{2(\gamma-1)}} = 0.668$$

$$c^* = \frac{1}{0.668}\sqrt{\frac{8.314 \times 1620}{15.5 \times 10^{-3}}} = 1396 \; m/s$$

## 13.5  BIPROPELLANT SYSTEM

Bipropellant systems provide a very versatile and high performance with specific impulse around 300 s   and a wide range of thrust. They can be used in steady state and pulsating modes. They are complex and expensive compared to monopropellant systems. They are generally used where high impulse and restart of the engine are required. As shown in Figure 13.12, the system basically consists of a pressurizing-gas system, propellant tanks, propellant feed lines and thrusters. The propellants are injected into combustion chamber where they react spontaneously (hypergolic propellant) to form high-temperature, high pressure, low molecular weight combustion products, which are then expelled through the nozzle. The common propellant for spacecraft is nitrogen tetroxide ($N_2O_4$) as oxidizer and monomethyl hydrazine (MMH) as fuel.

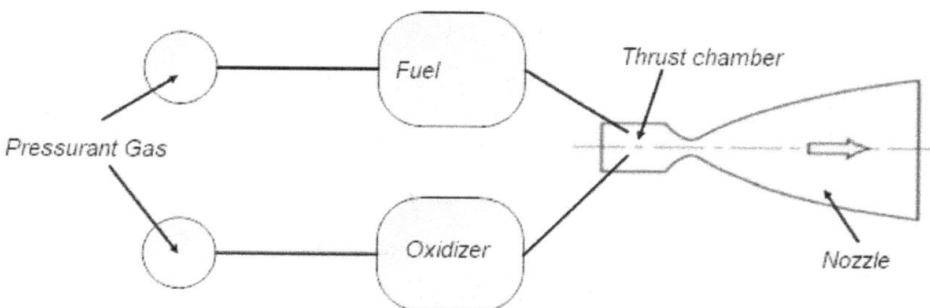


**Figure 13.12** Bipropellant Space Propulsion System.

Bipropellant thrusters accept only a limited range of propellant inlet pressure variation. A regulated system as shown in Figure 13.13 controls the pressure in the propellant tanks at a pre-set pressure. Pressure regulation is essential to keep the flow rate of each propellant constant and at the correct mixture ratio. The pressurant is stored at a high pressure 210 to 345 bar. The engine is supplied with propellant at a tightly controlled lower pressure and flow rate. Either helium or nitrogen systems are used although helium is the more frequent choice. Figure 13.14 shows a schematic of a bipropellant system The system contains check valves upstream of the propellant tanks to prevent possible back-flow, mixing, and combustion of the propellant vapors in the common pressuring gas line. In the event of a pressure regulator failure, relief valves prevent system rupture. Filters prevent clogging of the injector or damage of the valve seat by entrained foreign material. For long duration flights, proper care should be taken to avoid infiltration of propellant vapor between the check valves and check valve seals.

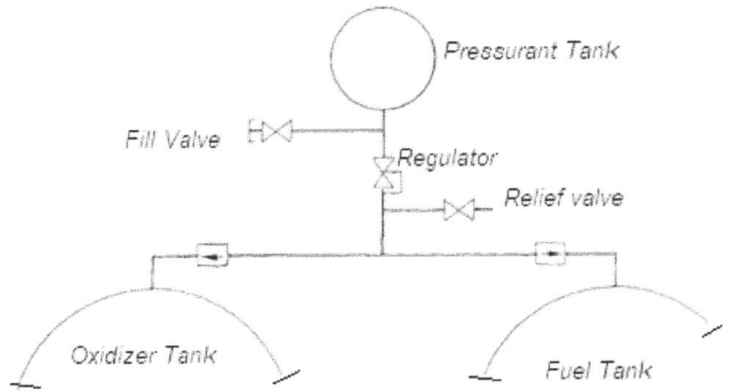


**Figure 13.13** Regulated Pressurant System for Bipropellant Thruster.

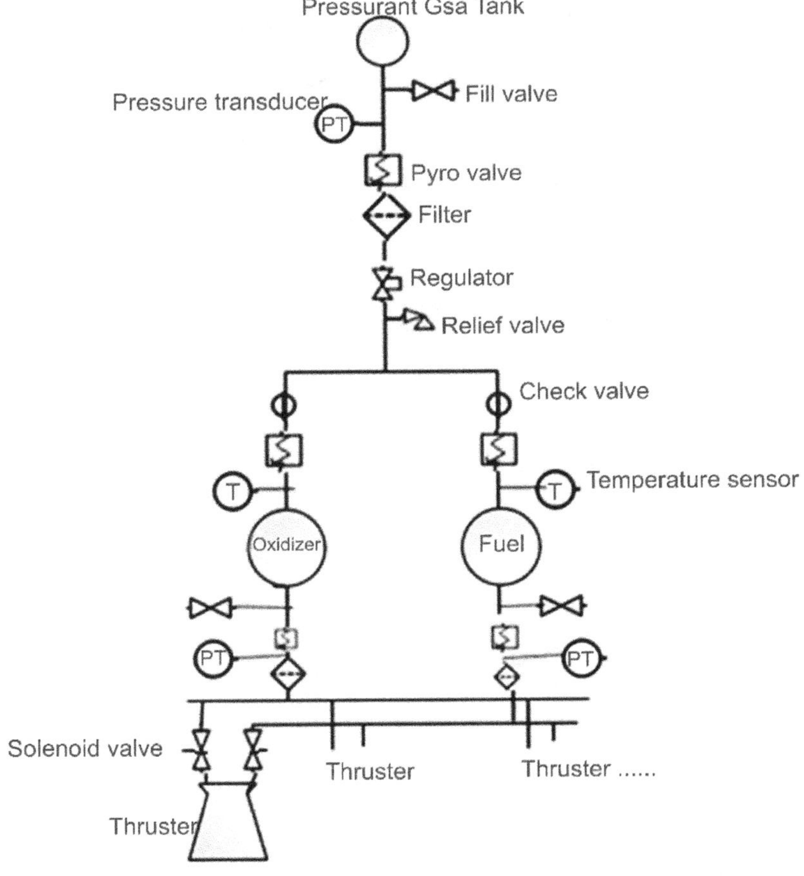


**Figure 13.14** Schematic of Bipropellant System.

*The typical pressure levels at various locations of the system are:*

| | |
|---|---|
| Maximum pressurant tank pressure | 31 MPa (310 bar) |
| Minimum pressurant tank pressure | 4.5 MPa (45 bar) |
| Minimum regulator inlet pressure | 4.1 MPa (41 bar) |
| Nominal propellant tank pressure | 3.45 MPa (34.5 bar) |
| Minimum engine valve inlet pressure | 3.1 MPa (31 bar) |
| Nominal chamber pressure | 2.4 MPa (24 bar) |

The trust level of bipropellant is 4 to 500N for satellites, up to 45 000 N for general spacecraft application. The impulse bit is $\geq 10^{-2}$ Ns (with thruster minimum on-time of 20 ms). The specific impulse is $\leq 2850$ Ns/kg for thrust levels $F \leq 25$ N (for steady state operation), $\leq 3110$ Ns/kg for $F \leq 400$ N. For pulse mode operation the specific impulse is $\geq 1000$ Ns/kg.

The unusable propellant is greater with bipropellant systems than monopropellant systems. Table shows in 13.4 an example propellant inventory for 1000 kg of usable propellant at a mixture ratio of 1.6. Figure 13.15 shows typical bipropellant systems marketed by ArianeGroup.

**Table 13.4** Bipropellant System Inventory.

| Propellant use | Fuel | Oxidizer | Total |
|---|---|---|---|
| Usable propellant | 384.6 | 615.4 | 1000 |
| Trapped propellant, 3% | 11.5 | 18.5 | 30 |
| Outage, 1% | 3.8 | 6.2 | 10 |
| Loading error, 0.5% | 1.9 | 3.1 | 5 |
| *Loaded propellant* | 401.8 | 643.2 | 1045 |

| 200N Bi-Propellant Thruster Key Technical Characteristics | |
|---|---|
| Thrust Nominal | 216N ± 10N |
| Thrust range | 180N ± 15N to 270N ± 15N |
| Specific impulse at nominal point | > 2650 Ns/kg (> 270s) |
| Flow rate nominal | 78 g/s |
| Flow rate range | 60 to 100 g/s |
| Mixture Ratio nominal | 1.65 ± 0.035 |
| Mixture ratio range | 1.2 – 1.9 |
| Chamber pressure nominal | 8 bar |
| Inlet pressure range | 17 ± 7 bar |
| Minimum on time | 28 ms |
| Minimum off time | 28 ms |
| Minimum impulse bit | < 8 Ns at 28 ms |
| Pulse frequency | 1 to 5 Hz |
| Throat Diameter (inner) | 12 mm |
| Nozzle end diameter (inner) | 95 mm |
| Nozzle expansion ratio (by area) | 50 |
| Injector type | Impingement with film cooling |
| Mass, Thruster with valves and instrumentation | 1,9 kg |
| Chamber / Nozzle Material | SiCrFe coated niobium alloy |
| Fuel | MMH (qualified) / UDMH (demonstrated) |
| Oxidizer | MON-3 (qualified) / N204 (demonstrated) |
| Valve | Monostable dual coil solenoids, 32w |
| Cumulated on time | 46500 s |
| Cumulated number of pulses | 320.000 |
| Number of full thermal cycles | 250 |
| Max. t_on (single burn) | 7600 s |

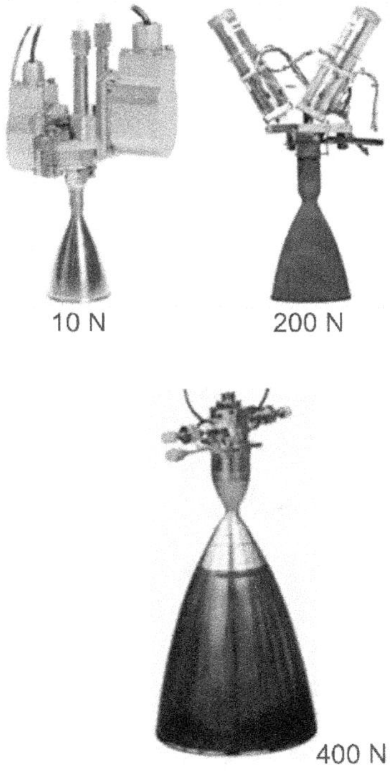

**Figure 13.15** Typical Bipropellant Thrusters (ArianeGroup).

## 13.6 DUAL MODE SYSTEM

Many missions require a high-impulse velocity change burn as well as pulse mode attitude control operation. For example, a planetary orbiter requires a high impulse single burn for orbit insertion and pulse mode for attitude control. Geosynchronous spacecraft also require a high-impulse burn to establish the required orbit as well as pulsing performance for ACS. The dual-mode system uses the hydrazine as a monopropellant in attitude control thrusters and as the fuel in bipropellant main engines. Figure 13.16 shows schematic of dual mode system.

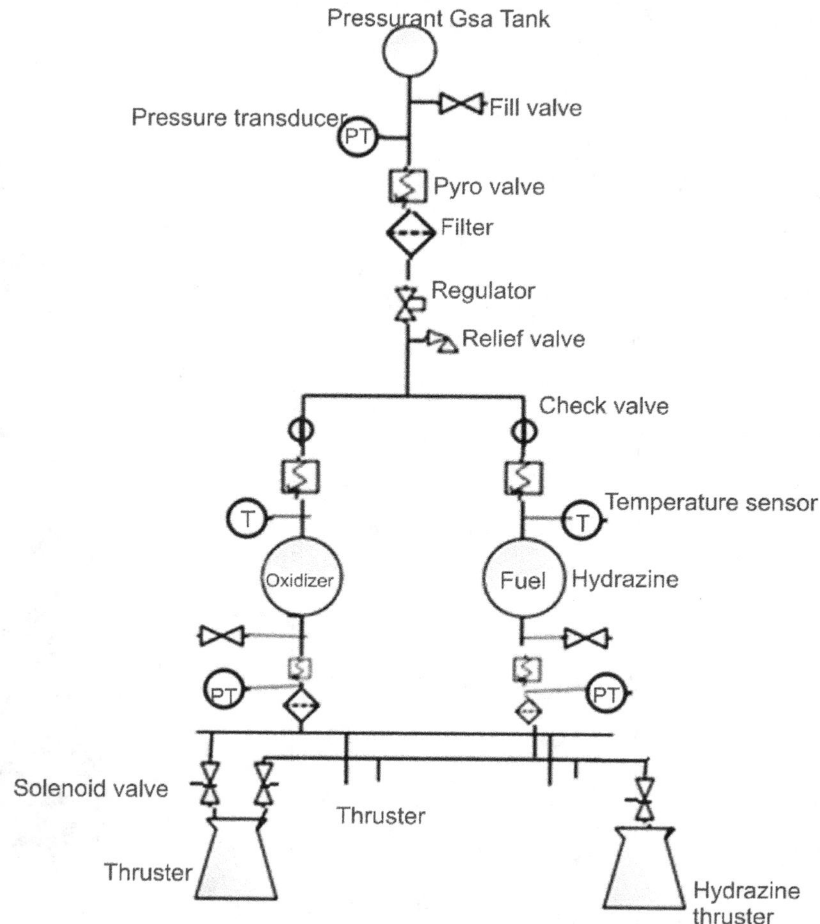



**Figure 13.16** Dual Mode Propulsion System.

## 13.7 TYPICAL MISSIONS WITH CHEMICAL PROPULSION

### (A) Galileo Propulsion System

The Galileo propulsion system is a highly redundant bipropellant MMH/nitrogen tetroxide system shown in Figure 13.17. It was the first spacecraft to examine Jupiter and its moons for an extended period launched from the Space Shuttle Atlantis in October 1989 and arrived at Jupiter in December 1995. The spacecraft flew by Venus once and Earth twice as it built up energy to reach Jupiter. The spacecraft plunged into Jupiter's crushing atmosphere on Sept. 21, 2003 and it was destroyed intentionally.

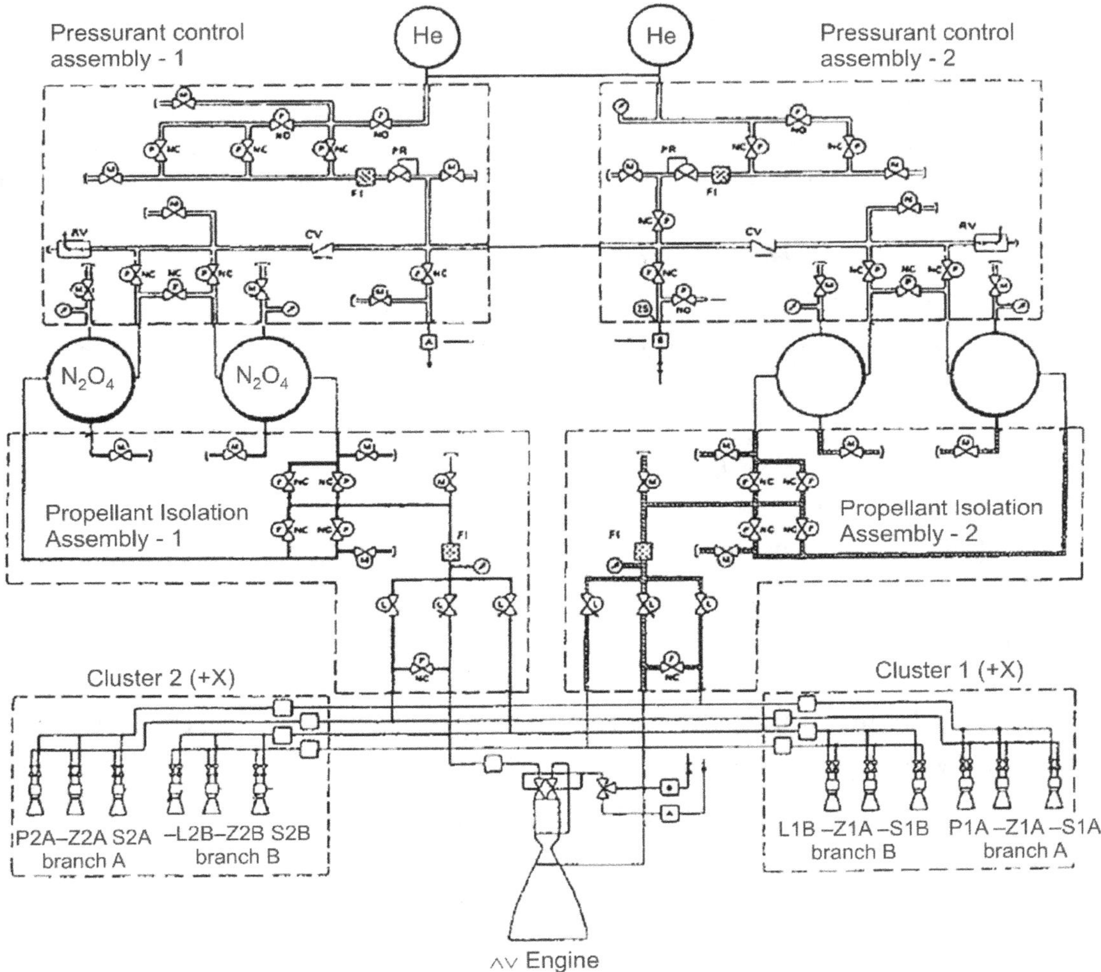


**Figure 13.17** Galileo propulsion system (Source: Rainer Killinger and Klaus Bohnhoff).

The propulsion module design provides redundancy against single point failures, with only a few active components exempted: check valves, system filters and 400 N main engine. All other feed system components including 10 N thrusters are redundant or can be bypassed in case of a failure. The propulsion module details are:

| | | |
|---|---|---|
| Number of thrusters | 1 | 400 N (Maneuvering propulsion) |
| | 12 | 10 N (2 clusters of 6 thrusters each, attitude control) |
| Pressurant Tanks | 2 (He) | |
| Propellant Tank | 4 | (2 oxidizer $N_2O_4$ and 2 fuel MMH) |
| Launch Mass | 1,1413 kg | |

| | |
|---|---|
| Dry Mass | 179.6 kg |
| Propellant capacity | 959 kg |
| Usable Propellant | 925 kg |

The 400-N engine can be fired for up to 70 min; the 10-N thrusters can be used in pulse mode or steady state. Special low leakage check valves were designed to prevent propellant vapor mixing between the two check valves.

*The propulsive module provided all propulsive maneuvers with 10 N thrusters*:

- Attitude control (Sun Pointing, Earth Pointing, Science Turns)

- Spin control and spin-up/down maneuvers (3 to 10 rpm) for probe release and 400 N engine maneuvers

- $\Delta V$ maneuvers -- Trajectory Correction Maneuvers (TCM) during interplanetary cruise and Orbit Trim Maneuvers (OTM)

The 400 N main engine provided major $\Delta v$ maneuvers

o Orbiter Deflection Maneuver (ODM)

o Jupiter Orbit Insertion (JOI)

o Peri-Jove Raise (PJR) to avoid repeated passages through high intensity radiation close to Jupiter.

Up to May 1$^{st}$, 1997 the Galileo RPM performed about 350 propulsive maneuvers. The summary of the highlights is listed in Table 13.5.

**Table 13.5** Galileo Propulsion Module Performance Summary.

| | |
|---|---|
| Total Propellant Consumed. | 863 kg (= 93% of usable load) |
| Number of pulses 10 N thruster | Approx. 65,000 |
| Trajectory change Manoeuvres (TCM) interplanetary. | 26 |
| Average accuracy of all TCMs | 0.78% |
| Total of attitude control and sun Acquisition manoeuvres | 232 |
| Orbit trim manoeuvre (OTM) Including PJR | 22 |

**(B)** **Mars Global Surveyor Spacecraft;** The dual-mode system is used on the Mars Global Surveyor spacecraft launched in 1996. The propulsion schematic shown in Figure 13.18 provides Mars orbit insertion, attitude-control pulsing, and several orbit trim maneuvers. In dual mode, hydrazine was used as fuel for bipropellant during $\Delta v$ burns and for monopropellant pulsing operations. Table 13.6 gives detail of the propulsion module.

**Table 13.6** Mars Global Surveyor Propulsion Module.

| | | |
|---|---|---|
| Launched | Nov 1996 | |
| Mission Ended | Jan 2007 | |
| Purpose | Mapping entire Mars | |
| | | |
| Bipropellant thruster | 1 No | 490 N |
| Monopropellant | 12 Nos | 4.45 N |
| Propellant Tanks | 3 | 2 fuel $N_2H_4$, one oxidizer $N_2O_4$ |
| Pressurant gas | He | 1.36 kg at 255 bars |
| Initial mass of fuel | 216.5 kg | |
| Initial mass of oxidizer | 144 kg | |
| Launch Mass | 1030 kg | |
| Power | 980 W | |

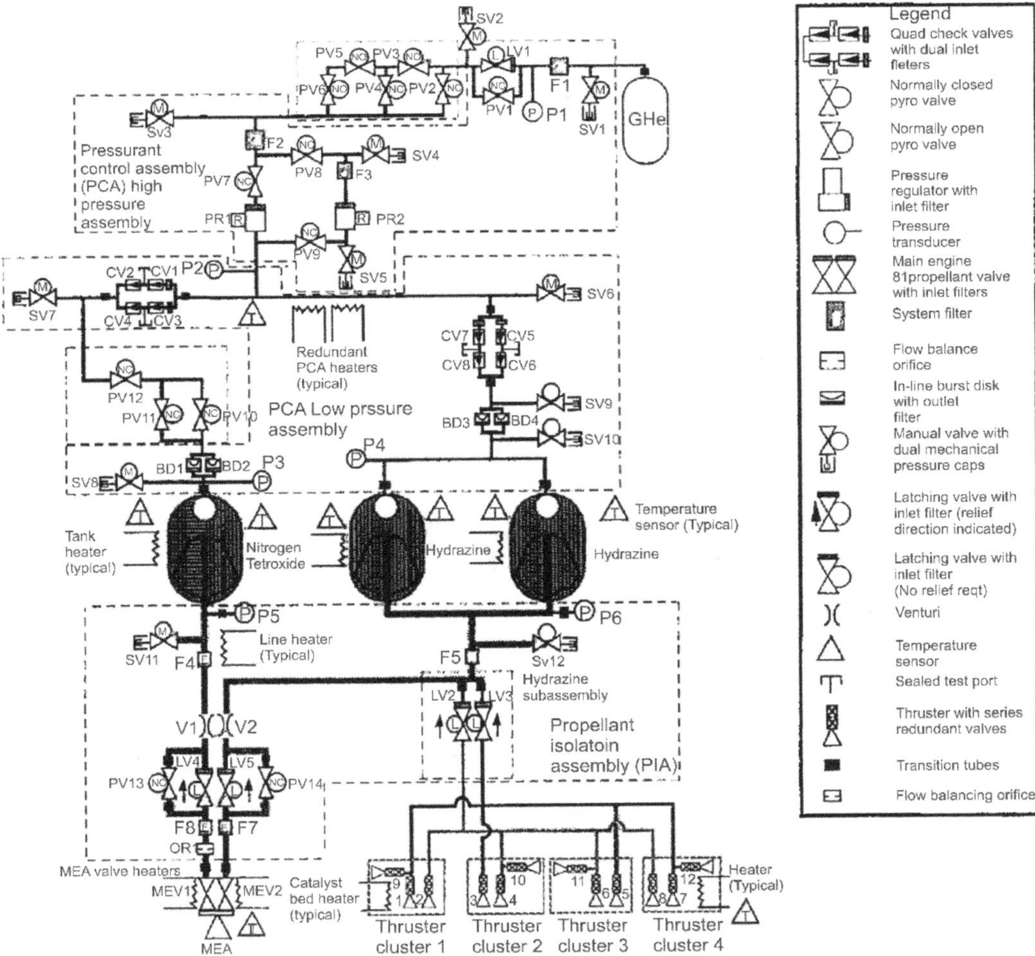


**Figure 13.18** Mars Global Surveyor dual-mode propulsion system (Source: Dominick S).

**Nomenclature**

$\dot{m}_p$     mass flow rate of propellant (kg/s)

$R_g$     specific gas constant (N m/kg K)

$T_g$     temperature of gas (K)

$V_u$     ullage volume (the gas volume over the propellant) (m³)

$V_{ui}$     initial ullage volume (m³)

$p_g$     gas pressure (N/m²)

$\Delta p$     pressure (bar, Pa)

B     blowdown ratio

$m_g$     mass of the pressurized gas (kg)

$p_{gf}$     final pressure of the pressurant gas

$p_{gi}$     initial pressure of the pressurant gas

$T_f$     final temperature of pressurant gas (K)

$T_i$     initial temperature of pressurant gas (K)

$V_{uf}$     final ullage volume (m³)

## PROBLEMS

13.1    A monopropellant system for a spacecraft has the following requirements:

    1) Translational maneuver $\Delta v = 200$ m/s at an average specific impulse of 250 s.

    2) Spacecraft wet weight = 500 kg

    3) Attitude-control total impulse = 71000 Ns in pulse mode at an average specific impulse of 150 s.

    4) Propellant reserve = 30% of usable propellant.

   Prepare a propellant inventory.

13.2    A spacecraft has a monopropellant propulsion system that delivers a specific impulse of 220 s. How much propellant would be consumed to trim the orbit if a $\Delta v$ of 200 m/s were required and the spacecraft weighed 950 kg at the end of the burn?

13.3    A thruster pair, each developing a thrust of 1 N, located at 1500 mm radius from a center of the mass of a spacecraft. Its moment inertia about Z axis is 2500 kg m². Find time required for the spacecraft to maneuver through 60 degrees rotation about the z axis. If the specific impulse is 200 s what is the amount of propellant consumed for the maneuver.

13.4    The catalyst bed of a hydrazine monopropellant rocket is designed to give 20 % of dissociation of ammonia at temperature of 25° C. Find the heat liberated per kg of hydrazine, temperature of products of decomposition and characteristic velocity. Take standard heats of formation of hydrazine and ammonia as + 50.3 kJ/mole and − 45.9 kJ/mole respectively. The specific heats of ammonia, hydrogen and nitrogen are 0.04, 0.03 and 0.035 kJ/mole K respectively.

# Electric Propulsion Systems

## 14.1 INTRODUCTION

Electric propulsion (EP) achieves thrust with high exhaust velocities compared to chemical propulsion. This results in a reduction of propellant mass and launch mass of the spacecraft. The electrical thrusters work on the same basic principle as chemical thrusters, i.e., accelerating mass and ejecting it from the vehicle. The propellant is ejected up to twenty times faster than from a classical chemical thruster and therefore the overall system is many times more mass efficient. The ejected mass from electric thrusters, however, is primarily in the form of energetic charged particles. An electrically charged atom or molecule is an ion. In the process of ionization an atom or molecule is charged electrically by adding (negative ion) or removing (positive ion) electrons. A gas is considered ionized when some or all the atoms or molecules contained in it are converted into ions. Plasma is an electrically neutral gas in which all positive and negative charges from neutral atoms, negatively charged electrons, and positively charged ions add up to zero. It is a good conductor of electricity and forms the building block for all types of electric propulsion. The applied electric and/or magnetic fields push the charged ions in the plasma to generate thrust.

The addition of energy to the working propellant in EP is from electrical sources through physically different devices based on principles of electrothermal, electrostatic, electromagnetic, or mixed. The propellant can be a noble gas, a chemical monopropellant, or even a solid. In general, electric propulsion encompasses any propulsion technology in which electricity is used to increase the propellant exhaust velocity. Though the concept was tested in 1960s, its development as an alternative to chemical thruster was hindered due to:

- o Lower priority than chemical propulsion as it is useful in space application and not useful for space launch and military applications.
- o Required mature space access to be of use.
- o Technical limitations of on-orbit power generation: low efficiency solar cells and heavy, inefficient power conversion.
- o Political concerns around space-based nuclear power.

With the maturity of space technology and ever-increasing demands of space applications, now major attention is given to Electric Propulsion. The thrust levels in electric propulsion thrusters are small compared to chemical propulsion systems. Presently, they provide low accelerations and are not sufficient for overcoming the high-gravity fields of planetary launches. They operate exclusively in space, which also matches the near-vacuum exhaust pressures required for electrostatic and electromagnetic systems. The major applications are orbital plane change, attitude control, station keeping and landing.

## 14.2 TYPES OF ELECTRIC THRUSTERS

Electric thrusters are generally described in terms of methods used to accelerate charged particles and produce the thrust.

1. *Electrothermal.* Propellant is heated electrically and expanded thermodynamically; i.e., gas is accelerated to supersonic speeds through a converging/diverging nozzle, as in chemical rocket propulsion systems.

   Ex: Arcjets, Resistojets

2. *Electrostatic.* Acceleration is achieved by the interaction of electrostatic fields with charged propellant particles such as atomic ions, charged droplets, or colloids.

   Ex: Gridded Ion Engines, Field Emission Electric Propulsion, Colloidal Thrusters

3. *Electromagnetic.* Acceleration is achieved by the interaction of electric and magnetic fields within a plasma. Some devices add a nozzle to enhance performance

   Ex: Hall Effect Thrusters, HEMPTs, Pulse Plasma Thrusters

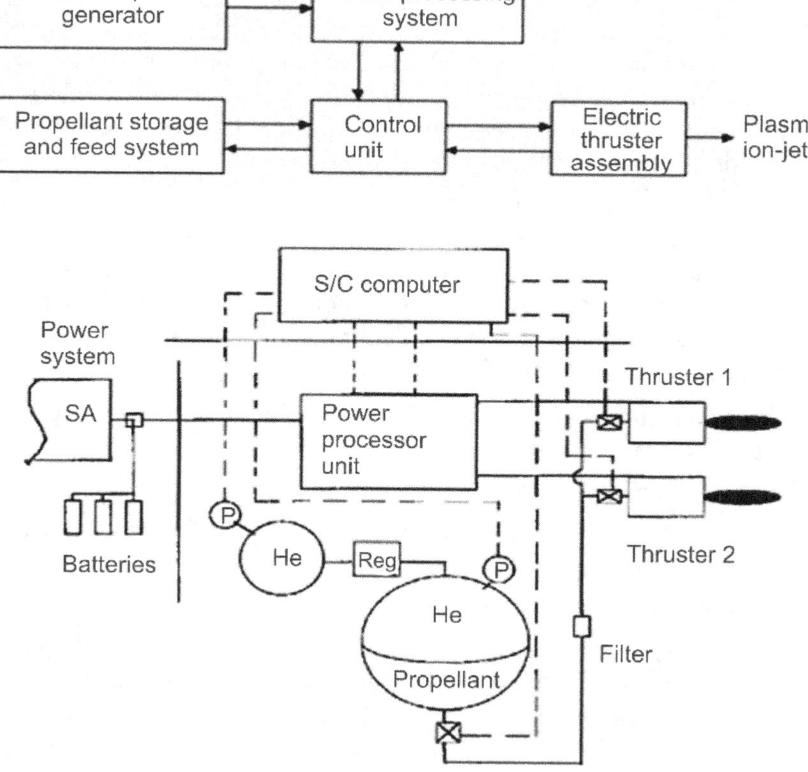


**Figure 14.1** Schematic of a typical Electric Propulsion System (Source: M. Martinez-Sanchez andJ. E. Pollard).

Figure 14.1 shows the basic subsystems of the electric propulsion unit. A propellant subsystem is used for storing, metering and delivering the propellant and/or propellant fill provisions. Valves, piping which connects the propellant storage system with the thruster are part of the feed system. Electric power supply and power processing unit (PPU) is to transform the spacecraft's electrical power to voltages, frequencies, pulse rates, and currents suitable for particular electrical propulsion systems. The PPU processes the raw power into the specific form required by the thrusters and is usually one of the most complex and challenging EP components. Electric Propulsion is not limited in energy, but is only limited by the available electrical power on-board the spacecraft. The control unit controls the valves and thruster operations. Table 14.1 gives the required power range and achievable specific impulse of different types of electrical thrusters.

**Table 14.1** Types of EP Thrusters and Power Range.

| Thruster | Power Range | Specific Impulse (s) |
| --- | --- | --- |
| Electrothermal | 100s of W | 300 to 400 |
| Resistojet | | |
| Arcjet | | |
|    Hydrazine | kWs | 500 to 600 |
|    Hydrogen | 10s of kW | 900 to 1200 |
|    Ammonia | kW to 10s of kWs | 600 to 800 |
| Electrostatic | | |
|    Ion Engine | W to 100 kW | 2000 to 10,000 |
|    Stationary Plasma Thruster (SPT) | 100s W to 10s of kW | 1000 to 2500 |
| Electromagnetic | | |
|    Magnetoplasmadynamic (MPD) | | |
|      Pulsed | kW (average) | 1000 to 4000 |
|      Steady State | 10s of kWs to MWs | 3000 to 7000 |
|    Pulsed Plasma Thruster (PPT) | 10s to 100s W | 1000 to 1500 |

## 14.3  ELECTRIC THRUSTER EFFICIENCY

The ability to transform electric power $P$ into thrust $F$ is expressed by the efficiency $\eta_t$. As shown in Figure 14.2, $\dot{m}$ is the rate of propellant flow through the thruster and $P_i$ is the input electric power. The output power $P_e$ imparts kinetic energy to the propellant mass. The input power diminishes to from $P_i$ to $P_e$ is due to

1. losses in the power conversion, such as from solar or nuclear into electrical energy;

2. conversions into the forms of electric power (voltage, frequency, etc.) required by the thrusters;

3. losses of the conversion of electric energy delivered to the thruster into propulsive jet energy (kinetic energy) in the output jet.

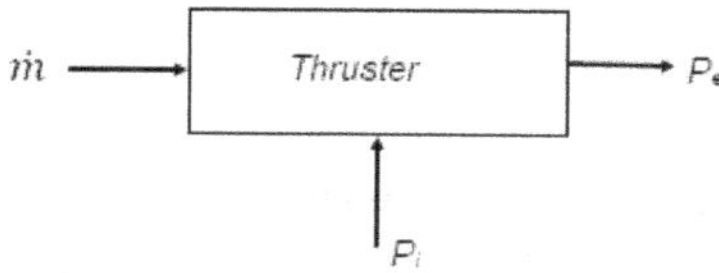


**Figure 14.2** Thruster Efficiency.

The output power is

$$P_e = \frac{1}{2}\, \dot{m} v_e^2$$

where $v_e$ is the exhaust velocity of jet. The thrust F is

$$F = \dot{m} v_e$$

Therefore,

$$Kinetic\ power\ per\ unit\ thrust = \frac{Kinetic\ power\ P_e}{Thrust\ F} = \frac{1}{2}\, v_e = \frac{1}{2}\, g I_{sp}$$

Thruster efficiency $\eta_t$ is the ratio of the thrust producing kinetic power of the exhaust beam (axial component) to the total electrical power applied to the thruster $P_i (= \Sigma(IV)$, where $I$ is current and $V$ voltage).

$$\frac{F}{P_i} = \frac{2\eta_t}{g I_{sp}}$$

Thus, $\eta_t$ measures how effectively electric power and propellant are used in the production of thrust. Because of their low thrust, flight regimes for space vehicles propelled by electric thrusters are quite different from those using chemical rockets. Accelerations tend to be relatively low ($10^{-4}$ to $10^{-6}$ g) and thrusting times are typically long.

## 14.4  ELECTROTHERMAL THRUSTERS

The propellants in electro-thermal thrusters are heated to high temperatures using electrical power, and the hot gases are expanded in a convergent-divergent nozzle to provide the thrust.

There are two different configurations of electro-thermal thrusters:

1. The *resistojet*, in which high electrical resistance dissipates power and in turn heat the propellant, largely by convection.

2. The *arcjet*, in which current flows through the bulk of the propellant gas ionizing it with an electrical discharge.

### 14.4.1 Resistojet Thruster

As illustrated in Figure 14.3 resistojet operates by passing the gaseous propellant around one or more ohmically heated surfaces, such as coils of heated wire and heated hollow tubes. The heated gases then expanded using a conventional nozzle to generate thrust. Though theoretically any gas can be used, the most successful application has been based on the superheating of catalytically decomposed hydrazine, used in monopropellant chemical thrusters. The catalytic decomposition of hydrazine that produces approximately 1 volume of $NH_3$ and 2 volumes of $H_2$, has the advantage of being compact and the catalytic decomposition preheats the mixed gases to about 900 K prior to their being heated electrically to higher temperatures. The maximum temperature is generally limited to around 2000 K due to the material limitation of the heating coil. These resistojet thrusters can operate over the wide pressure range in the blowdown system. They don't require special power conditioning units as the input voltage is low and hence possess the lowest overall system empty mass. Their plumes are not harmful as they are uncharged. The performance characteristics are:

- Specific Impulse: 200-300 s
- Thrust: 2-350 mN
- Efficiency: 65-90 %
- Power: 10-5000 W

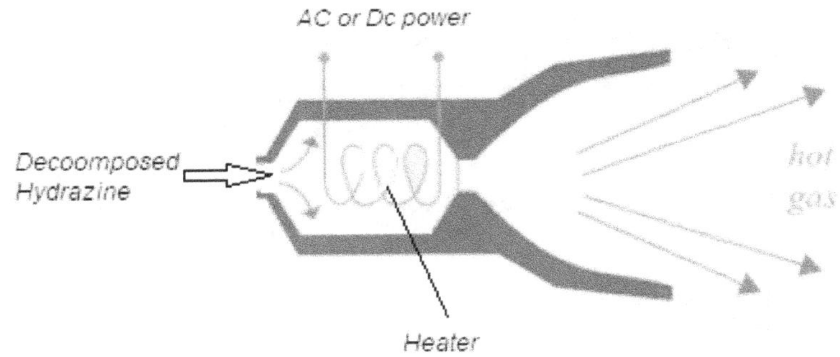


**Figure 14.3** Resistojet Operating Principle.

Even with its comparatively lower value of specific impulse, the resistojet's superior efficiency contributes to far higher values of thrust/power ratio than any of its competitors. Figure 14.4 illustrates a commercial resistojet MR 501 from Aerojet Rocketdyne. Resistojets have flown on Intelsat V, Satcom 1-R, GOMS, Meteor 3-1, Gstar-3 and Iridium S/C.

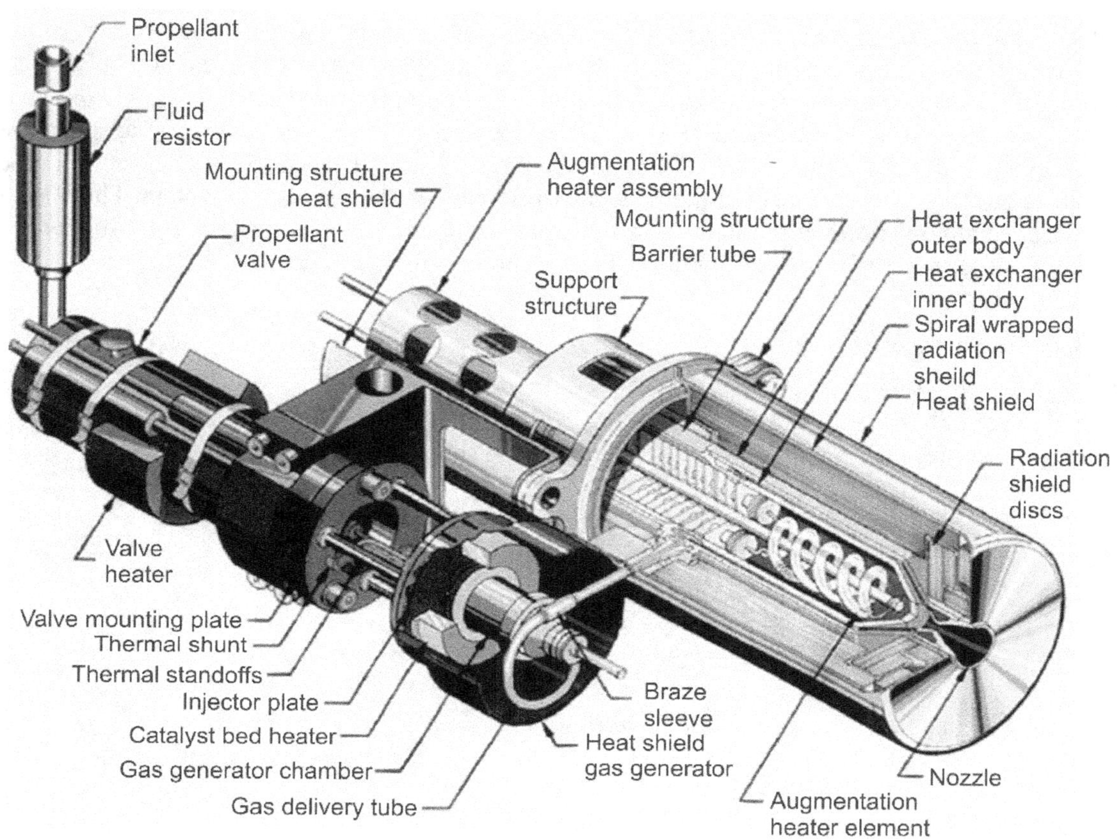


Selected Performance Values of a Typical MR 501 Resistojet

| | |
|---|---|
| Propellant for resistojet | Hydrazine liquid, decomposed by catalysis |
| Inlet pressure (MPa) | 0.689–2.41 |
| Catalyst outlet temperature (K) | 1144 |
| Resistojet outlet temperature (K) | 1922 |
| Thrust (N) | 0.18–0.33 |
| Flow rate (kg/sec) | $5.9 \times 10^{-5} - 1.3 \times 10^{-4}$ |
| Specific impulse in vacuum (sec) | 280–304 |
| Power for heater (W) | 350–510 |
| Power for valve (max.) (W) | 9 |
| Thruster mass (kg) | 0.816 |
| Total impulse (N-sec) | 311,000 |
| Number of pulses | 500,000 |
| Status | Operational |

**Figure 14.4** Electrothermal Resistojet MR 501.

## 14.4.2   Arcjet Thruster

In arcjet thrusters, the wall temperature limitation of the resistojet is overcome by depositing power in the form of an electric arc between a concentric upstream rod cathode and a downstream anode that also serves as the supersonic nozzle. As shown in Figure 14.5, the arc stretches from the tip of a central cathode and to the anode nozzle that accelerates the propellant flow as it is

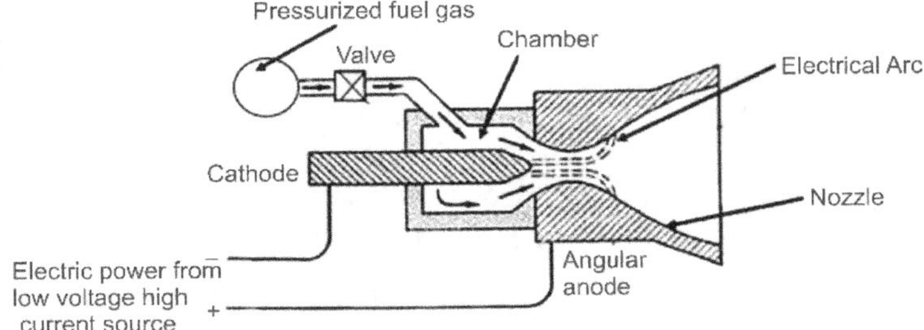


**Figure 14.5**  Arcjet Thruster Working Principle.

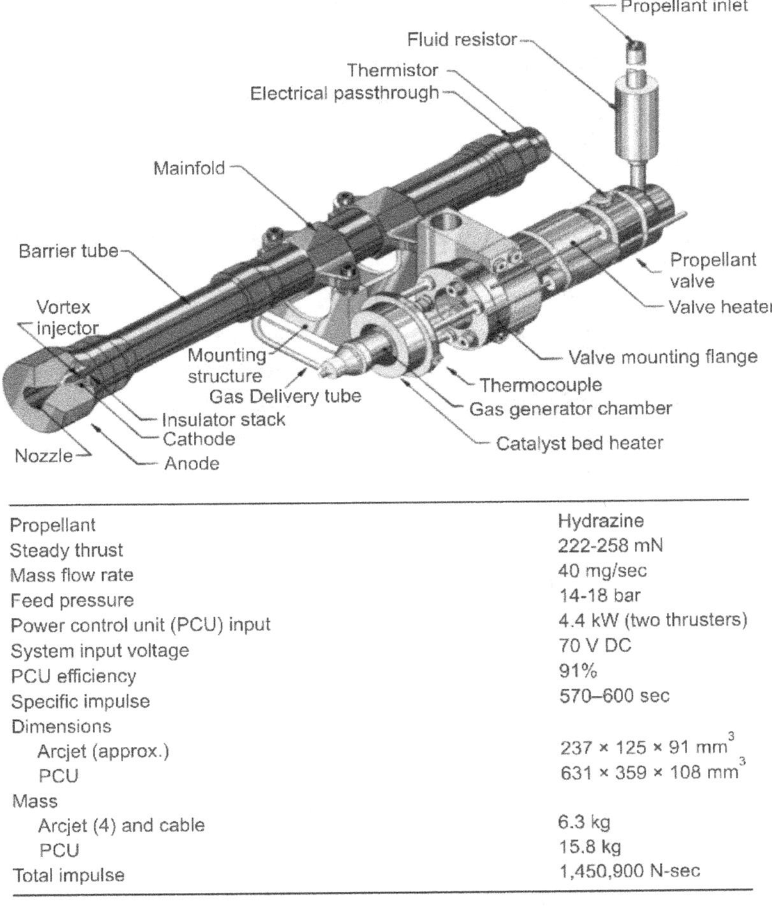


| Propellant | Hydrazine |
|---|---|
| Steady thrust | 222-258 mN |
| Mass flow rate | 40 mg/sec |
| Feed pressure | 14-18 bar |
| Power control unit (PCU) input | 4.4 kW (two thrusters) |
| System input voltage | 70 V DC |
| PCU efficiency | 91% |
| Specific impulse | 570–600 sec |
| Dimensions | |
|    Arcjet (approx.) | $237 \times 125 \times 91 \text{ mm}^3$ |
|    PCU | $631 \times 359 \times 108 \text{ mm}^3$ |
| Mass | |
|    Arcjet (4) and cable | 6.3 kg |
|    PCU | 15.8 kg |
| Total impulse | 1,450,900 N-sec |

**Figure 14.6**  2 kW MR-510 Hydrazine Arcjet System.

being heated. The components must be electrically insulated from each other and must be designed to withstand high temperature gas environments. The discharge is either generated by applying a low DC voltage (100 V) and high current (hundreds of Amperes) or by a high frequency high voltage field or microwave or AC discharge. Temperatures in the order of 30,000-50,000 K are achieved in the center line which fully ionizes the propellant. The heat is transmitted through conductivity and radiation down to the buffer layer near the nozzle walls causing a strong temperature gradient to keep the nozzle wall temperature not more than 2000 K to avoid the melting of the nozzle material. Swirling of the propellant and the column shape of the anode increases the time of exposure to the arc and therefore improves performance. Non-uniform heated flow, thermal losses towards electrodes due to high local heat and power loss due to recombination gasses make the overall efficiency much lower than resistojets at 30-50%. Electrode erosion limits lifetime to a typical 1,500 hours of operation.

Hydrazine is the preferred propellant for arcjet thrusters and achieves specific impulse of range 500-800 s, a significant improvement from resistojet thrusters. The power processing unit (PPU) for an arcjet is significantly more complex than that for a resistojet. The discharge voltage is higher than most bus voltages, e.g., 80 120 V, requiring at a minimum dc-dc conversion. It is much heavier than the thruster itself. Figure 14.6 illustrates the commercial 2 kW MR-510 Hydrazine Arcjet System. The plume is similar to those of conventional thrusters with a few percent ionization. Arcjet thrusters will continue to compete with plasma thrusters in geostationary applications, where they present fewer spacecraft integration problems. The Telstar 4 and GE-1 satellite series have used these thrusters.

## 14.5  ELECTRIC FIELD AND ELECTROSTATIC FIELD

In thermal thruster hot propellant gases are accelerated using supersonic nozzles. With the use of electric forces to accelerate the propellant nozzles are not essential. These forces accelerate a suitably ionized propellant to speeds ultimately limited by the speed of light. Coulomb's law, in a form similar to Newton's law for gravitation, gives the relation for the force between two charge particles as shown in Figure 14.7. The law states that the force F between a pair of point charges is directly proportional to the product of the charges $q_1$, $q_2$ and inversely proportional to the square of the distance $r$ separating them.

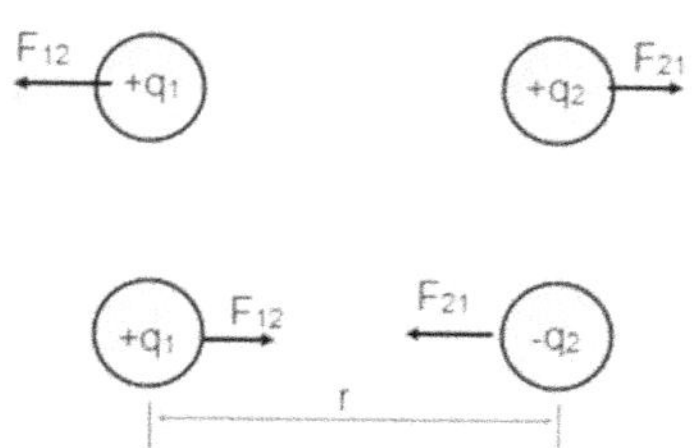


**Figure 14.7**  Force between Two Charges.

$$\vec{F} = K \frac{q_1 q_2}{r^3} \vec{r}$$

.....(14.1)

where $q_1$ and $q_2$ are charges measured in Coulomb (C) and force $F$ in Newton N, and distance $r$ in meter. The value of the constant K

$$K = 9 \times 10^9 \ Nm^2/C^2$$

The constant, K is generally expressed in terms of $\epsilon_0$ electrical permittivity of free space. Permittivity denotes the ability to store a charge in the medium.

$$K = \frac{1}{4\pi\epsilon_0}$$

The unit of permittivity $\epsilon_0$ is

$$\frac{1}{K} = \frac{C^2}{Nm^2} = \frac{C}{\left(\frac{Nm}{C}\right)m} = \frac{C}{Vm} = \frac{F}{m}$$

The potential difference expressed in Volts ($V$) corresponds to the work done in transporting unit charge, i.e., ($Nm/C$) and Farad ($F$) which denotes the charge in Coulomb per volt ($C/V$). The permittivity of free space is determined from the constant $K$ is

$$\epsilon_0 = 8.852 \times 10^{-12} \ \frac{F}{m}$$

The permittivity of any medium, $\epsilon$ is expressed as a function of the permittivity of free space and is given by

$$\epsilon = \epsilon_r \epsilon_0$$

$\epsilon_r$ is called relative permittivity. The region around a charge $q_1$ has a field, called electrical field, given by

$$\vec{E} = K\frac{q_1}{\vec{r^3}} \vec{r} \ \ V/m$$

If a charge $q_2$ is placed in this field, the force acting on it is given by:

$$\vec{F} = \vec{E}q_2 \ \ N \hspace{4cm} .....(14.2)$$

The unit of the electrical field is

$$\frac{N}{C} = \frac{Nm}{Cm} = \frac{V}{m}$$

The electric field is produced by stationary charges, and the magnetic field $\vec{B}$ by moving charges (current). A charge $q$ moving with velocity $\vec{v}$ experiences a force perpendicular to the direction of $\vec{v}$ and field $\vec{B}$. The force is given by Lorentz law:

$$\vec{F} = q\vec{v} \times \vec{B} \hspace{4cm} .....(14.3)$$

If the charge is moving in the direction of the magnetic field, the force is zero. The unit for magnetic field is

$$\frac{Ns}{Cm} = \frac{N}{Am} = T \ (Tesla)$$

The flow of charge for sec is Ampere A. The unit of magnetic field is also mentioned as Weber/m$^2$. In the presence of both the electric field $E$ and magnetic field B, the net force (Lorentz force) experienced by a charge of $q$ Coulomb is a combination of electrostatic and electromagnetic forces given by adding Eq. (14.2) & (14.3)

$$\vec{F} = q\vec{E} + q\vec{v} \times \vec{B} \qquad \dots\dots(14.4)$$

The vector product is illustrated simply in Figure 14.8.

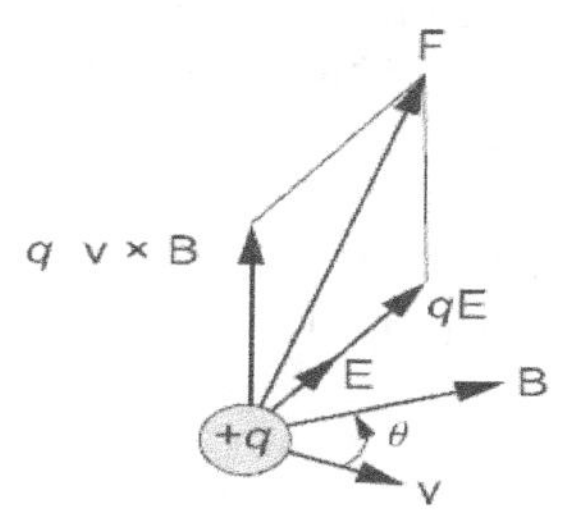


**Figure 14.8** Lorentz Force on a Charged Particle.

## 14.6  ELECTROSTATIC THRUSTERS

### 14.6.1    Ion Thruster

In a gridded electrostatic ion accelerator, as shown in Figure 14.9, ions are produced in a magnetically confined ionization chamber and then accelerated to high velocities using an electrostatic field. The velocity $v_e$ gained by an ion of mass $m$ due to acceleration by the electrostatic field, then the kinetic energy achieved is

$$KE = \frac{1}{2}mv_e^2 \; J$$

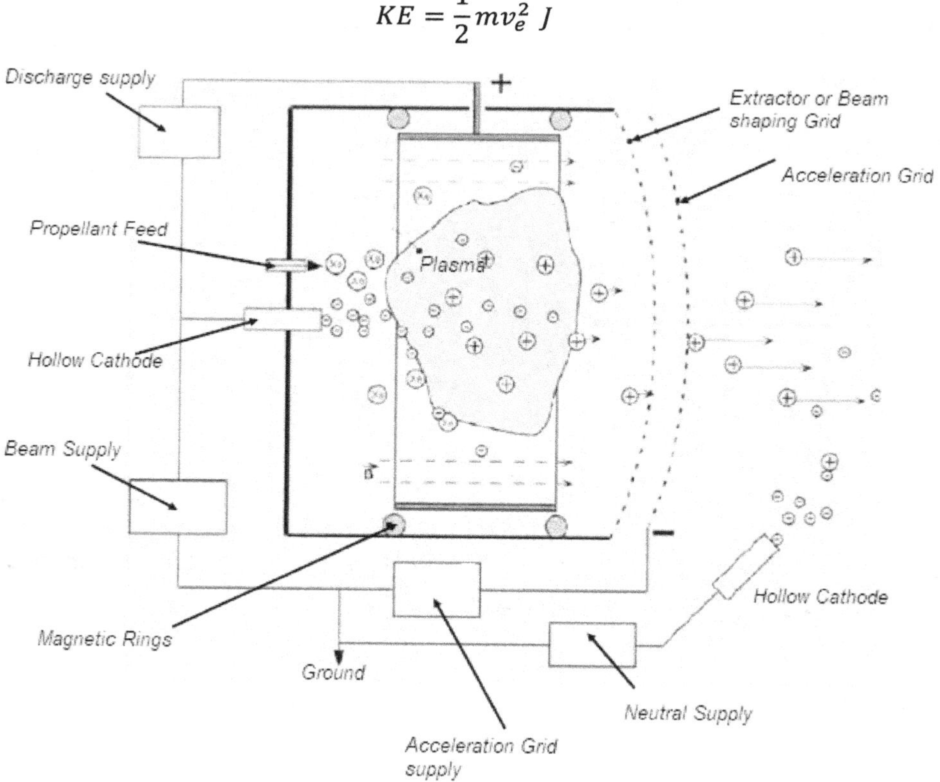


**Figure 14.9** Schematic of Ion Thruster.

This is equal to the work done $(qV)$ by the moving charge $q$ Coulomb from lower to higher potential, $V$ volts, in the electric field $E$. In the ideal condition

$$\frac{1}{2} m v_e^2 = qV$$

$$v_e = \sqrt{\frac{2qV}{m}} \quad \frac{m}{s}$$

The thrust, F generated by ejecting efflux of charged particles, $\dot{m}$ at the velocity $v_e$

$$F = \dot{m} v_e$$

The mass flow rate of the charged particles, $\dot{m}$ is related to current, $i$

$$\dot{m} = \frac{mi}{q}$$

$i$ is the current in Amperes associated with the motion of the charge $q$ Coulomb as it is being accelerated. Then thrust is

$$F = \frac{mi}{q} v_e = \frac{mi}{q} \sqrt{\frac{2qV}{m}} = i \sqrt{\frac{2mV}{q}} \tag{14.5}$$

From the above equation it is obvious that to have a higher value of thrust the mass of charged propellant must be high. As the mass of an electron is 1/1873 times of proton, the thrust F developed by negatively charged ions is relatively negligible even at nonrelativistic high velocities, $v_e$. Hence, positive ions, which have larger mass than the negative ions (electrons), are used and the electrostatic thrusters are referred to as ion thrusters. Table 14.3 gives typical propellants that can be used to generate ions in electrostatic thrusters. Mercury has the one of the highest masses among the mentioned propellant in the Table 14.3, the contamination problems of the optical surface of the spacecrafts by its vapors and toxicity have been a problem and its use has been discontinued. Cesium is readily ionized by heated tungsten and has been used for spacecraft propulsion. However, it is very reactive. The current trend is to use xenon, which is a noble gas having atomic mass near to cesium.

**Table 14.3** Propellants for electrostatic and electromagnetic thrusters

| Propellant | Molecular weight | Storability | Compatibility | Comments |
|---|---|---|---|---|
| Cesium | 133 | Good | Poor | Condensing on S/C |
| Mercury | 201 | Very good | Good | Toxicity limits application |
| Xenon | 131 | Very good | Excellent | Common use today |
| Krypton | 84 | Good | Excellent | Can be used in blend with Xe |
| Iodine | 254 | Excellent | Poor | Condensing on S/C |

*Four basic and separable processes occur in an ion thruster as shown in Figure 14.9:*

1.  Energetic Electron Production,

2. Ion Production,

3. Ion Extraction and Acceleration,

4. Ion Beam Neutralization.

Electrons are emitted into the main discharge chamber using orifice hollow cathodes and accelerated towards an anode maintained at typically high potential of 1100 V. Positive ions are produced by bombarding with the propellant gas, usually Xenon, by these energetic electrons. To increase ionization efficiency, a magnetic field is applied to facilitate a gyro-movement of the electrons and thus a higher chance of ionizing a neutral atom along its way to the anode. Hollow cathodes are evolved as the preferred electron source due to low power consumption and long lifetime. Figure 14.10 illustrates the working of the hollow cathode. They consist of a low work function material, such as Barium oxide, which is heated to cause thermionic emission of electrons. An anode, called a keeper, is located just downstream of the orifice. A small part of the

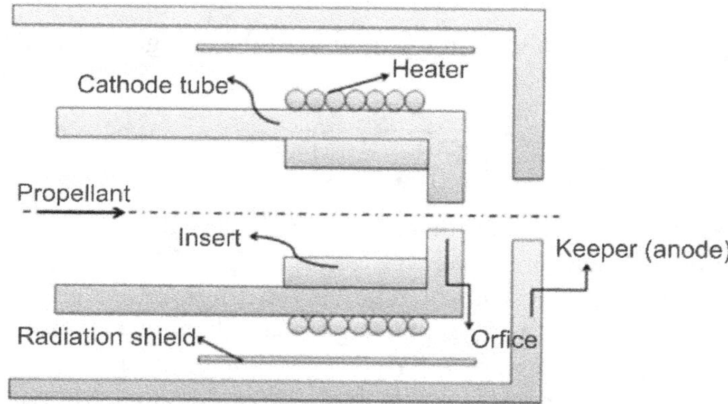


**Figure 14.10** Hollow Cathode of Ion Thruster.

propellant gas needs to be flowing through the cathode as well. The electrons hit the propellant gas and ionize it, creating an additional electron. The electrons are drawn towards the positive potential of the keeper. Ions inside the cathode bombard the and cathode interior surfaces, thereby heating them and sustaining the thermionic emission process. The radio frequency Ion thruster uses a high-frequency electromagnetic field to ionize xenon gas atoms to form a plasma containing free 'light' electrons and 'heavy' positive ions.

The other side of the main chamber is covered by a double-grid structure (perforated plates), extractor grid and accelerating grid as shown in Figure 14.11 with spacing of the order of 1 to 1.5 mm across which the ion acceleration voltage is applied. The first grid extracts or takes in the charged particles (ion). The second grid, maintained at a high potential difference with respect to the first, and of opposite polarity to the charge, accelerates the charge to a high velocity, $v_e$. The impact of the charged particles on the grids leads to their erosion. The erosion, known as sputtering, is to be avoided if the ion thruster has to operate over a long period of time. Molybdenum is used as the material of the grids due to its low sputter.

The negative ions accumulate in the chamber after positive ions leave the thruster. The negatively charged thruster attracts back the ions leaving the thruster and prevents the generation of thrust. The ions leaving the thruster, therefore, require to be neutralized. This is done by supplying electrons through hollow cathode placed downstream of the accelerator as shown in Figure 14.9

In the operation of the ion thruster, the ion source produces continuously ions at sufficient rate and by applying a negative screen field extracts more ions. This increases the positive space charge in the gap of extractor and accelerator. As the positive ions are accelerated, a beam current flows opposite direction to positive ion flow. This will reduce the negative screen field, with which ions are getting attracted and the process of extraction of ions stops when this field is driven to near zero. Thus, there exists a limit to the beam current density, beyond which the acceleration of the charged particles by the acceleration grid is not possible. This limiting value of the beam current density is known as the space charge-limited current density ($j$) and is given by the Child-Langmuir law:

$$j = \frac{i}{A} = \frac{4\epsilon_0}{9}\sqrt{\frac{2q}{m}}\frac{V^{\frac{3}{2}}}{L^2} \qquad \qquad .....(14.6)$$

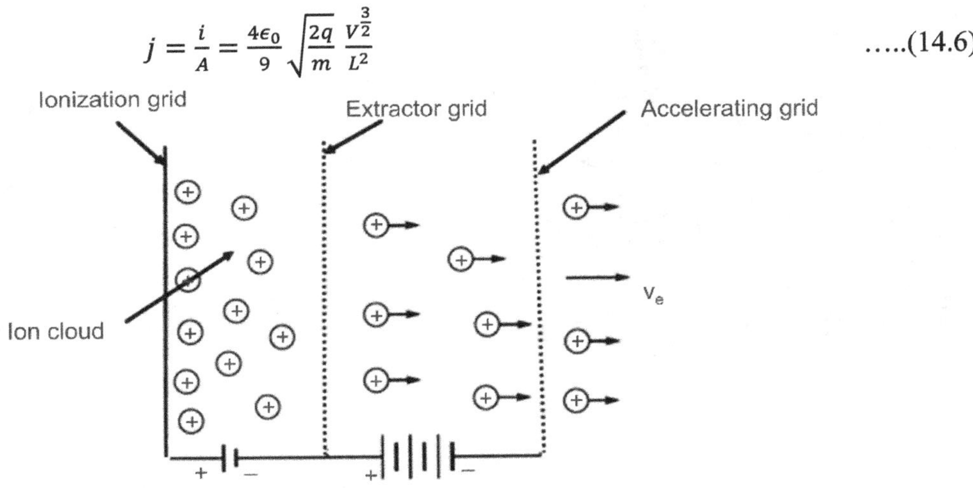


**Figure 14.11** Acceleration of ions.

$\epsilon_0$ is the permittivity of free space in F/m,

$L$ is the distance between the extractor and acceleration grid in m

$V$ is the voltage across the grids in Volts

$q$ is the charge in Coulomb

$m$ is the mass of the charged particle in kg.

Substituting the value of $i$ from Eq. (14.6) into Eq. (14.5) and simplifying, the maximum thrust is

$$F_{max} = \frac{4\epsilon_0}{9}A\sqrt{\frac{2q}{m}}\frac{V^{\frac{3}{2}}}{L^2}\sqrt{\frac{2mV}{q}} = \frac{8\epsilon_0}{9}\frac{V^2}{L^2}A$$

If the diameter of the current beam is $d$ and the aspect ratio of the beam $R = d/L$, the maximum thrust is

$$F_{max} = \frac{2\pi}{9} \epsilon_0 V^2 R^2$$

The thrust depends upon the potential difference between extractor and accelerating grids. The spacing between the grids (L) should be kept small as possible (1mm approx.). The accelerating potentials are between 1.5 to 2.5 kV. Figure 14.12 shows a typical RF (radio frequency) ion thruster developed by ArianeGroup.

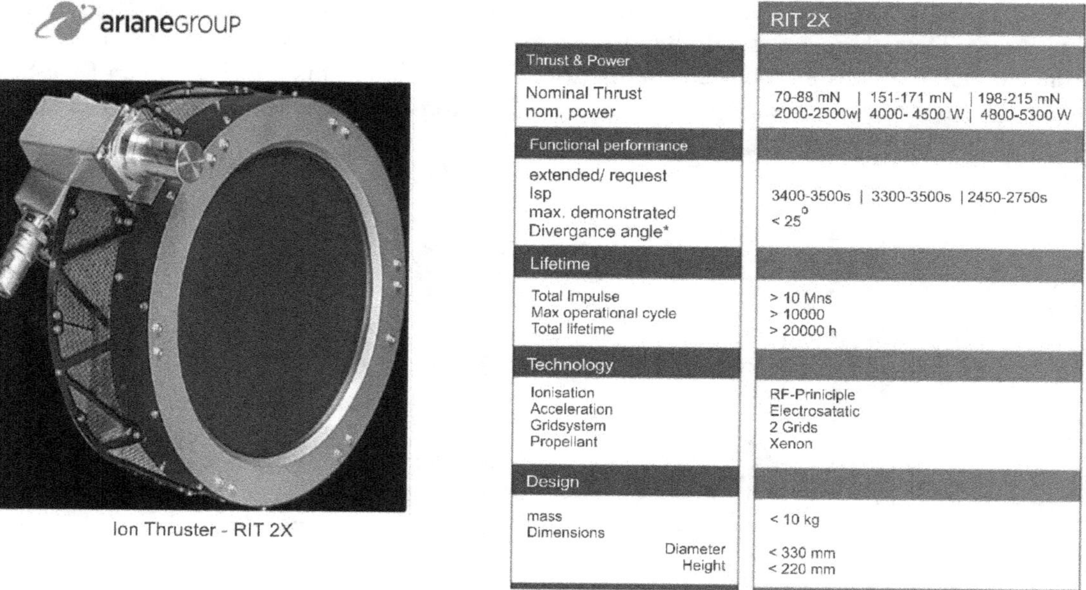


**Figure 14.12** Ion Thruster- RIT 2X – RF Thrusters-Ariane Group.

## 14.6.2 Hall Effect Thrusters

A Hall effect thruster is a girdles electrostatic propulsion system, i.e., it operates on a similar principle as the gridded ion thruster, but does not require physical grids to accelerate the plasma. Hall Effect states that when the current flows in a conductor in the presence of a magnetic field perpendicular to the flow of current, the combination of the current and the magnetic field generates a voltage perpendicular to both the current and the field. As shown in Figure 14.13 they are relatively simple devices consisting of cylindrical channel with an interior anode, magnetic circuit that generates a primarily radial magnetic field across the channel and a cathode external to the channel. Electrons generated from an external hollow cathode enter a ring-shaped anode with a potential difference of about 300 Volts. Radial magnetic field is produced along the annulus by positioned magnets. Due to Hall Effect the electrons travelling through the annulus towards anode are deflected and follow the spiral path. Propellant, xenon, is introduced through the annular space from anode. The spirally moving electrons having large residence time interact with xenon producing ions. These ions continue to be pushed by the electrostatic field away from the anode along the length of the annulus and produce the thrust. The Hall Effect thrusters are also known as Stationary Plasma Thrusters (SPT). Ceramic material is used for the construction of the annular chamber. The efflux ions are neutralized by the negatively charged electrons from the cathode. The efficiency and specific impulse of flight-model Hall thrusters are typically lower than that achievable in ion thrusters, but the thrust-to-power ratio is higher and the device requires fewer power supplies to operate. An optimum specific impulse around 1,500 seconds and moderate power consumption around 1.5 kW make this thruster very attractive. Hall thruster

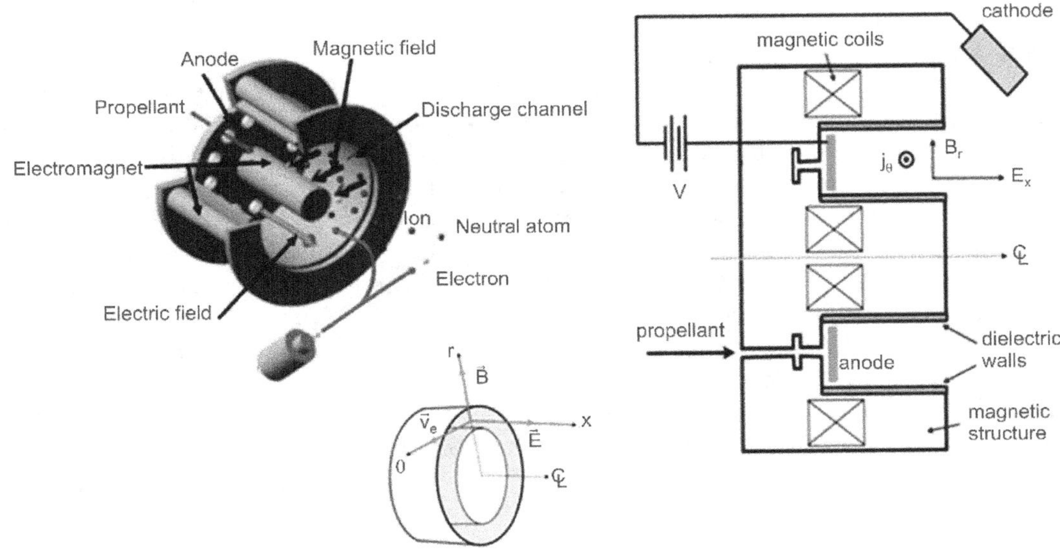


**Figure 14.13** Hall Effect Thruster Working Principle.

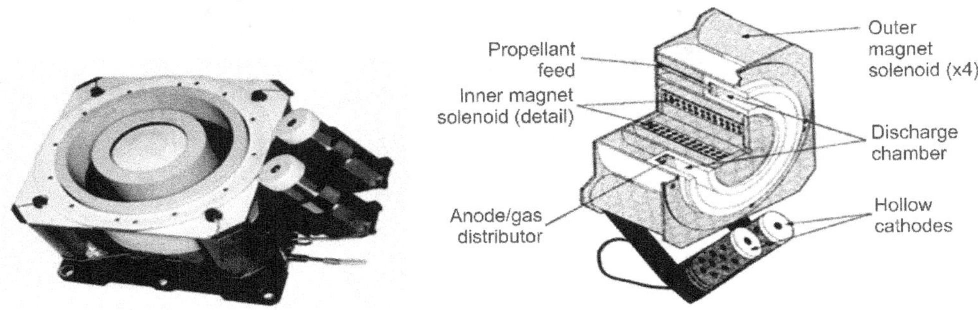


| Parameter | SPT-50 | SPT-70 | SPT-100 | SPT-140 |
|---|---|---|---|---|
| Slot diameter (cm) | 5 | 7 | 10 | 14 |
| Thruster input power (W) | 350 | 700 | 1350 | 5000 |
| Average Isp (s) | 1100 | 1500 | 1600 | 1750 |
| Thrust (mN) | 20 | 40 | 80 | 300 |
| Total efficiency (%) | 35 | 45 | 50 | >55 |
| Status | Flight | Flight | Flight | Qualified |

**Figure 14.14** SPT-100 Hall thruster.

technology was first developed to flight status in Russia and has only recently been developed and flown in Europe and USA. They are extensively used for North - South station keeping of telecommunication satellites. Figure 14.14 shows the STP-100 Hall effect thruster from OKB Fakel (Russian) and performance.

### 14.6.3 Field-Effect Electrostatic Propulsion (FEEP) Thruster

A field emission thruster, also called Field Emission Electric Propulsion (PEEP) thruster, is used for very high specific impulse, low thrust applications. It uses an electric field to extract and accelerate atomic ions directly from the surface of a metal exposed to vacuum by applying suitable voltages to a closely spaced electrode configuration. The strong surface electric field extracts individual ions from a metal lattice, particularly the easily ionizable metallic liquid. The two most common propellants are Cesium and Indium. Rubidium and Gallium are also suitable as alternate propellants for FEEP thruster. Figure 14.15 shows the basic components and schematic layout of FEEP thruster. The main components are an emitter, an accelerator and a neutralizer. The Cesium liquid propellant is stored in a reservoir in the emitter. It flows towards the emission slit via a capillary suction effect. The electric field is formed by applying a positive potential to an emitter, and a negative potential to an accelerator. When the free surface of the liquid metal is exposed to a high electric field, 8 to 15 kV, it is distorted into conical or series of conical protrusions called Taylor cones. At a threshold value of $10^9$ V/m, the surface layer of the atoms is ionized and accelerated by the same field that created them, producing a thrust. One capillary can generate a maximum thrust of 100 μN. For higher thrust requirements, a number of capillaries are clustered. The neutralizer generates an electron beam across the ionized thrust exhaust path, thereby neutralizing the exhaust.

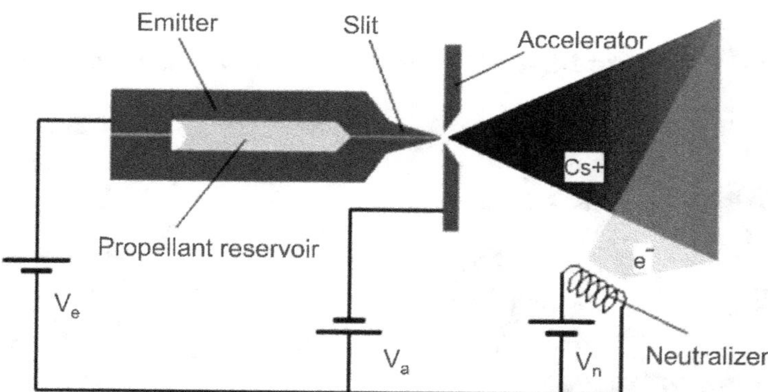


**Figure 14.15** Schematic layout of FEEP Thruster (Source: Muhammad Ali Badam).

## 14.7  ELECTROMAGNETIC THRUSTERS

Electromagnetic thrusters generate thrust through an interaction of crossed electric and magnetic fields as per Lorentz law shown in Eq. (14.3). It states that a charge $q$ moving with velocity $\vec{v}$ experiences a force perpendicular to the direction of $\vec{v}$ and magnetic field $\vec{B}$. The field could either be an externally applied magnetic field or else it could be an induced magnetic field due to a current. The current I is flowing in the radial direction and the applied magnetic field B in the tangential direction is normal to current $I$ as shown in Figure 14.16. Then, the

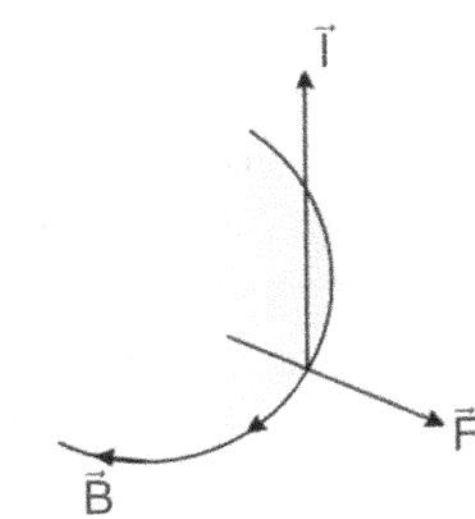


**Figure 14.16**  Lorentz Law.

Lorentz force, that accelerates the ions, is generated in the axial direction perpendicular to the direction of $I$ and $B$. To ionize the propellant gas to produce plasma commonly a high voltage to create an arc discharge or radio frequency (RF) field to add energy to free electrons which ionize the gas via collisions is used. The high electric currents needed to generate thrusts require high power levels of order of hundreds of kW up to MW or pulsed-mode operation with heavy capacitors. Once the plasma is generated, electromagnetic thrusters accelerate ionized propellant through either self-induced and/or applied magnetic fields.

### 14.7.1 Magneto Plasma Dynamic (MPD) Thrusters

In a magneto plasma dynamic (MPD) thruster a high current discharge is applied between two metal coaxial Electrodes, a central rod cathode and a surrounding cylindrical anode as shown in Figure 14.17. Due to high current an electric arc forms between the anode and cathode. The cathode heats and emits electrons which collide with the propellant gas to form plasma. The current generates a self-induced radial magnetic field that interacts with the currents, resulting in a Lorentz force acceleration. An additional external magnetic field can be added to help stabilize and accelerate the plasma discharge. In this case the thruster is known as applied-field MPD thruster. Non-oxidizing propellants such as hydrogen, hydrazine, Argon and alkali metals like Lithium are generally used. The formation of oxides degrades the cathode life. Being a condensable propellant, lithium generally is not considered as it may coat spacecraft surfaces and power arrays.

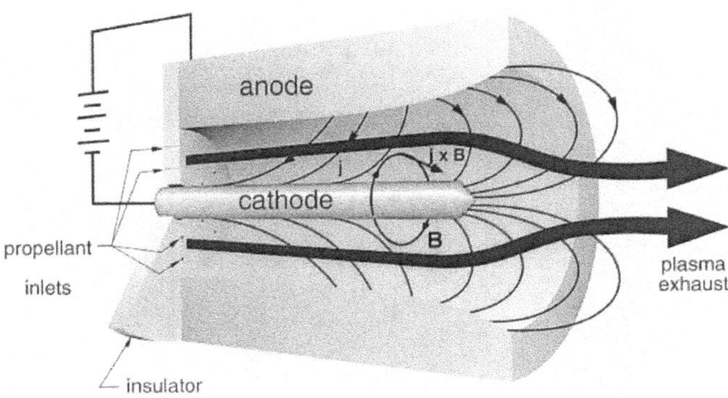


**Figure 14.17**  MPD Thruster Operating Principle (Source: Dan Lev).

MPD thrusters can generate very high thrusts up to 200 N. They have the ability to efficiently convert megawatts of electric power into thrust. The higher the power, the better the efficiency, which is typically around 35% and can increase up to 75%. MPD thrusters are currently the most powerful form of ion propulsion system as they process high power and generate high thrust while maintaining high exhaust velocities at a reasonable efficiency. It has been considered as a leading technology candidate, in combination with nuclear Power plants, for future space missions such as delivery of lunar and Mars cargo, outer planet rendezvous, and ventures in deep space robotic and piloted planetary exploration. These missions will require exhaust velocities more than 100,000 m/s. For perspective, a 100,000 m/s velocity of MPD thruster will allow a spacecraft to travel roughly 11 times the top speed of the space shuttle, 28,000 km/hr. MPD is still an experimental technology and research efforts are concentrated to develop high-specific-impulse, megawatt-class, hydrogen-fueled thruster technology.

### 14.7.2    Pulsed Plasma Thruster (PPT)

Pulsed plasma thruster, as shown in Figure 14.18, operates by taking power from the satellite bus and charges a capacitor to several thousand Volts. It is a very simple design and the most common configuration uses a solid propellant Teflon bar, moving with the tension of a spring. Teflon bar is then fed between a pair of electrodes charged by the capacitor. A spark plug igniter is then fired and as the arc discharge passes from the cathode to the anode the face of the Teflon bar ablates. It creates a small amount of plasma between the electrodes. The flowing current sets up a magnetic field and the interaction of the current and the magnetic field produces a force on the plasma which accelerates it out at high velocity. The process is repeated with the recharging of the capacitor. They operate in short pulses of around 10 ms that makes them attractive for attitude control or fine pointing applications. The advantages are zero warm-up time, zero standby power, no propellant tanks and feed lines. The very low efficiency of the order 5 to 15% is the main disadvantage of PPT.

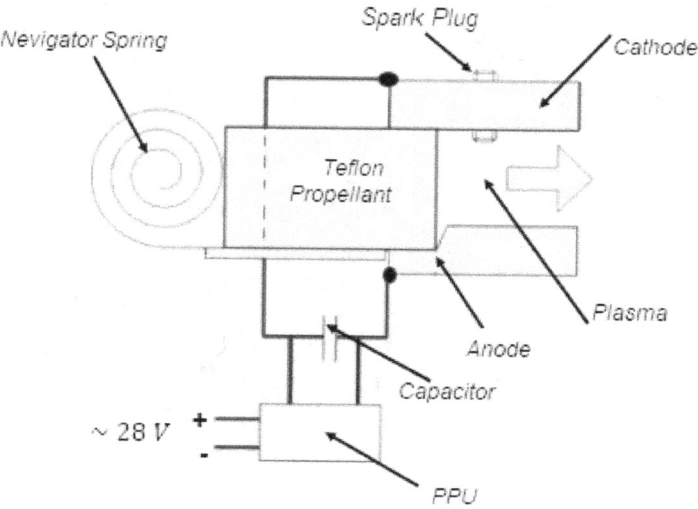


**Figure 14.18** Pulsed Plasma Thruster Schematic.

## 14.8  ELECTRIC PROPULSION APPLICATIONS

Though many electrical propulsion applications are possible, the predominant mission-oriented uses are station keeping in GEO orbit, orbit raising or lowering, orbital repositioning and deep space propulsion.

### 14.8.1 Station Keeping

Satellite orbit is perturbed due to small forces like solar pressure, Earth oblateness effects, atmospheric drag and outgassing. These disturbances may be secular, causing drift of the orbit in the same direction, or periodic causing drift first in one direction then in the opposite direction. Thrusters are fired to keep the satellite in the desired position to achieve proper mission targets. In the planned lifetime, 10 to 20 years, the GEO satellites are maintained within typically $\pm 0.15°$ east west and $\pm 0.15°$ north/south of its nominal orbit position. North–south Station keeping is performed in the vertical axis relative to the orbit plane, while east–west station keeping is performed tangentially to the orbit. To maintain the positional accuracy of the orbit, propulsive maneuvers are required, amounting in a year the velocity change $\Delta v$ of 40 to 50 m/s. In the designed life period of the satellite, the saving of fuel due to use of electric propulsion is substantial. The saving can extend the lifetime of a satellite 200 to 300% over what was possible with chemical thrusters.

### 14.8.2  Orbit Raising

Orbit raising maneuvers are required in variety situations like deployment to high LEO orbits from low delivery orbits, total or partial LEO–GEO transfers, interplanetary transfers, LEO altitude changes and deorbiting of spacecraft at the end of life. It will allow in some cases to use low-cost smaller launchers. Using EP thrusters in on-orbit maneuvers such as the transfer from where the launcher drops the satellite off to its final destination (Ex: GTO to GEO for communications satellites) can save a substantial amount of propellant. However, use of EP for on-orbit transfer takes significantly more time. While chemical apogee engines accomplish such transfer in hours to a few days, EP thrusters take a longer period like 4-8 months. For orbit raising maneuvers, thrusters with higher T/P (thrust/power) are more appropriate than with higher specific impulse. Presently, EP thrusters that can operate in two modes – one which maximizes T/P during orbit raising and then operates in a higher $I_{sp}$ mode for station keeping are readily available from different manufacturers.

### 14.8.3 Deep Space Propulsion

A great example for deep space exploration is NASA's first truly interplanetary spaceship, Dawn. The Dawn mission, which travelled to two very different extra-terrestrial bodies: giant asteroid Vesta and dwarf planet Ceres in the main asteroid belt and orbited both. It used 3 gridded ion engines each weighing 8.9 kg having thrust levels of $19 - 91$ millinewtons to accomplish this feat. The important events of the mission are:

- 2007 — Launch (September)
- 2009 — Mars Gravity Assist (February)
- 2011 — Vesta Arrival (July)

- 2012 — Vesta Departure (September)
- 2015 — Ceres Arrival (March)
- 2016 — End of prime mission (June)
- 2016 — Start of first extension (July)
- 2017 — Start of second extension (November)
- 2018 — End of mission (October)

In the process of performing these maneuvers, its ion engines delivered more than 9 km/s of velocity increase to the spacecraft which was more than the Delta II launch vehicle which was originally used to send Dawn into space.

## 14.9 VARIABLE SPECIFIC IMPULSE MAGNETOPLASMA ROCKET (VASIMR)

Variable Specific Impulse Magnetoplasma Rocket (VASIMR) engine is a steady state electrothermal thruster. It is a plasma-based propulsion system. Figure 14.19 illustrates the working of VASIMR. The Helicon source ionizes the propellant (a neutral gas like hydrogen, krypton, xenon and argon). Helicon is a device capable of generating high density plasma ($10^{13}$ $cm^{-3}$ and higher) at high-efficiency using helical radio frequency antennas. The superconductor magnets create magnetic fields to isolate the plasma from the chamber walls so that the temperature can exceed material melting limits. In the main chamber, ion cyclotron heating (ICH) coupler is used by applying resonant radiofrequency (RF) fields to heat the plasma to high temperatures of the order of 1 million degrees. The heated plasma expanded through a magnetic nozzle into vacuum thus producing a moderate thrust and a high specific impulse. The magnetic nozzle can control the thrust by varying the nozzle entrance section. The specific impulse can be varied by changing RF heating power. This allows it to vary thrust and specific impulse independently of each other and the performance of the unit can be modified to suit a specific mission. The exhaust velocity varies from 15 km/s to 100 km/s depending on the technology and propellant used.

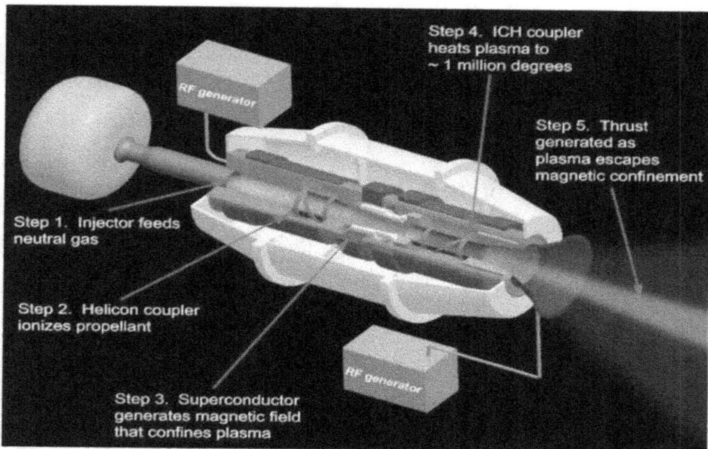


**Figure 14.19** VASIMR Working Principle (Source: Christopher S. Olsen et al).

To achieve the strategy to send crew to Mars in the 2030s, NASA plans to develop the technology needed for long-duration space flight, including advanced transportation systems and the systems needed to work and live in deep space. Among the electric propulsion technologies, VASIMR has been identified as an important propulsion system to efficiently transfer large masses and crew through interplanetary space. It occupies the high-power niche needed for sustained human exploration. VASIMR propulsion system power levels range from 10s of kW to MWs. Electrical power sources are solar or nuclear. For long-range missions, say to Mars, nuclear power is definitely a must. The conversion efficiency is high and > 70%. The solar arrays are based on novel structures that provide a large deployed solar collection area with a small packed volume. To accomplish this, they use tensioned mesh backplanes rather than rigid composite panels to reduce both mass and stowed volume. Two solar arrays were designed and built: (1) a rectangular roll-out design called the Roll-Out Solar Array (ROSA) from Deployable Space Systems and (2) a circular fan-fold design called MegaFlex from Orbital ATK (Alliant Techsystems Inc.). On Nov. 24, 2021, DART (Double Asteroid Redirection Test, is a planetary defense driven test of technologies for preventing a hazardous asteroid from impacting Earth) launched with two ROSAs with autonomous deployment capabilities, marking the first planetary mission powered by ROSAs – in this case, on a journey through deep space to encounter the asteroid system at almost seven million miles away. Gateway's (Gateway will be an orbiting outpost, crucial for NASA's Artemis program, functioning as a way station between Earth, the Moon, and deeper space destinations) Power and Propulsion Element (PPE) is a solar electric propulsion spacecraft that will provide power, high-speed communications, orientation control, and the capability to move Gateway to different lunar orbits. The PPE will be powered entirely by two ROSAs, generating 60 kW

The work on development of VASIMR was first started in MIT and is presently going on at academic institutes (MIT, Princeton, UMD, UTexas, UMich), USA National laboratories (ORNL, LANL), NASA, in some international labs and private industries.

## 14.10    ELECTRICAL PROPULSION THRUSTERS POWER REQUIREMENT

Electrical thrusters require power from the spacecraft bus bar. The overall efficiency of the EP system is > 60%. PPU (power Processing Unit) efficiency is around 90%. As the power level increases, the amount of waste heat to be rejected from the limited path. Thermal considerations play an important role in EP systems integration, especially as power levels increase. Also, care should be taken to avoid the interaction of the plumes with the spacecraft surfaces. The plumes contain charged particles moving at high velocities. The impact of high-velocity particles, whether charged or neutral, can result in sputtering of surfaces. The cold surfaces relative to plumes, such as radiators, are coated with slow moving neutral particles that can degrade the performance of such devices.

Figure 14.20 shows the performance of various electric propulsion systems in terms of specific impulse with the required power. It can be observed that resistojet and PPT (Pulsed Plasma thruster) require less power, of the order of 1 kW. PPT has higher specific impulse than the resistojet. However, their thrust levels are much lower. The ion and Hall thrusters have higher specific impulse than PPT but require higher power (1- 10 kW). The arcjet demands more power than the resistojet but has a higher specific impulse. The power levels of magnetoplasma thrusters are about a few MW and their performance is higher with higher power.

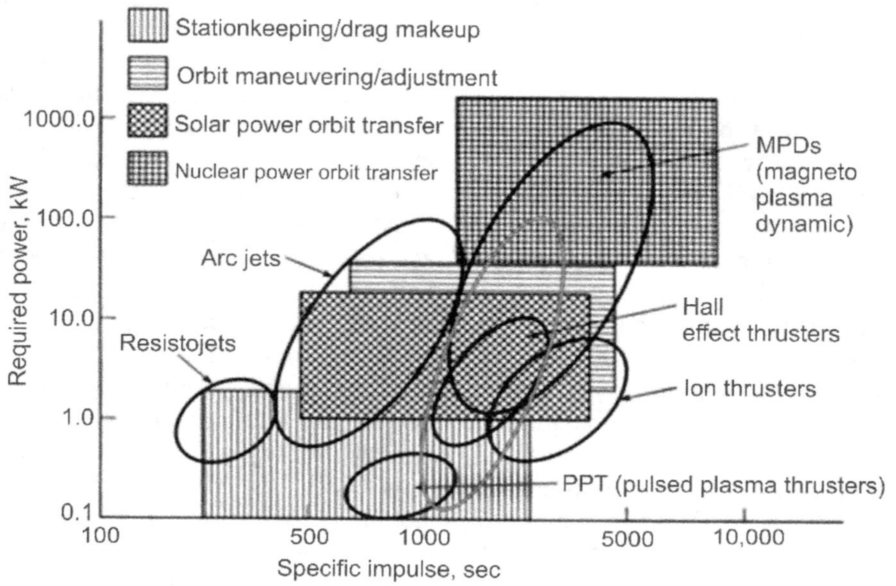


**Figure 14.20** EP Thrusters Power Requirement.

As illustrated in Figure 14.1 the main subsystems of the spacecraft are propellant storage, flow controls, power processing unit and thruster with pointing (orientation) mechanism. PPU matches voltage, frequency, power rate, and other electrical properties between the space-power generating unit and the electric thruster requirement. Figure 14.21 shows the block diagram of a typical EP thruster with power sources. Electric power generation units are classified as either direct (no moving mechanical parts) or dynamic. Direct energy conversion methods considered include photovoltaic, thermoelectric, thermionic, and electrochemical, while methods with moving parts include the Brayton, Rankine, and Stirling cycles. The output of solar-cell arrays is typically 28 to 300 V DC. The electrical power is converted to the required voltage and current in a Power Processing Unit (PPU). The limitations and recent trends in solar array are discussed in section 14.9.

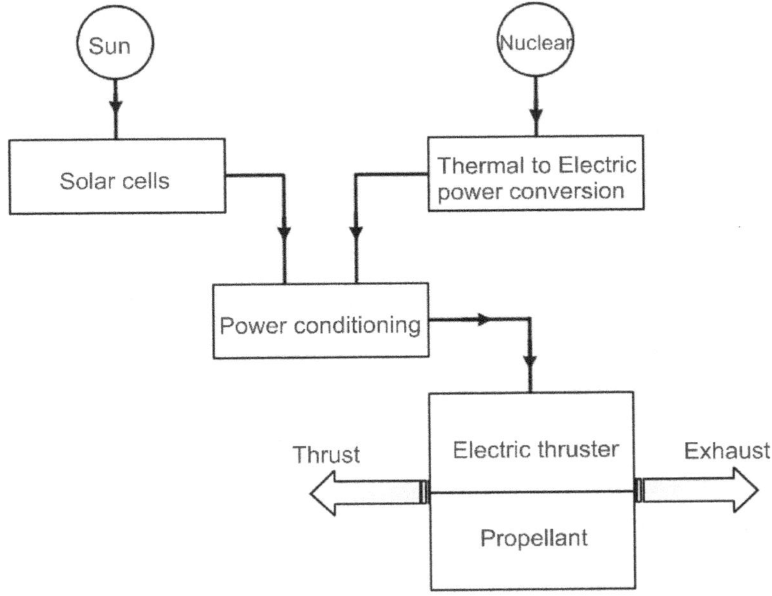


**Figure 14.21** Power sources for Spacecraft Propulsion.

The larger the power, higher would be the mass of the power source and the PPU. Hence, the initial and final mass of the electrical propulsion system will therefore include an additional mass due to the power source and PPU. The power plant and the PPU are supported by the structure of the rocket and a larger power source will demand a larger structural weight to support it. Therefore, the important quantity in the selection is the System-specific Impulse, $I_{ssp}$, which is the total impulse, $I_{tot}$ delivered by the system, divided by the total mass, $m_{PS}$, of the propulsion system, that is, all of the propulsion system and not only the propellant as for $I_{sp}$.

$$I_{ssp} = \frac{I_{tot}}{m_{PS}} \frac{Ns}{kg}$$

$I_{ssp}$ = system-specific impulse [Ns/kg]

$I_{tot}$ = total impulse delivered by the propulsion system [Ns]

$m_{PS}$ = mass of propulsion system (all of the propulsion system and not only the propellant as for $I_{sp}$) [kg]

The optimum values of thruster exhaust velocity, $v_{e\ opt}$ depends on total system mass as shown in Figure 14.22. With increasing exhaust velocity, the combined mass of propellant and tank is decreasing while the mass of the power supply is increasing. The point of intersection of the two curves determines the minimum of the system mass by $v_{e\ opt}$ resulting in a maximum value of $I_{ssp}$.

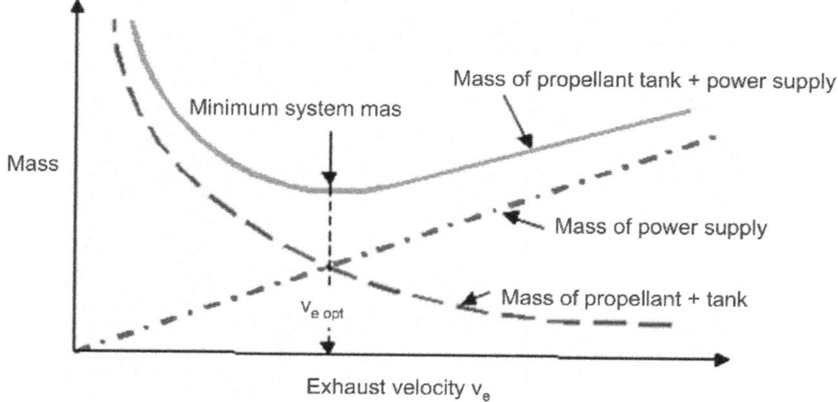


**Figure 14.22** Optimum exhaust velocity of EP System.

## 14.11 NUCLEAR PROPULSION

Nuclear propulsion has high potential in space propulsion, particularly deep space exploration and long duration missions, due to its high $I_{sp}$ performance and very high specific energy capability. Significant developments were reported in the US and Russia between 1955-72. Major and popular nuclear project in 1960's was NERVA (Nuclear Engine for Rocket Vehicle Applications). But following the "Atmospheric Test Ban Treaty" in 1972, all activities were stopped. Until 1972, many nuclear propulsion systems were ground tested.

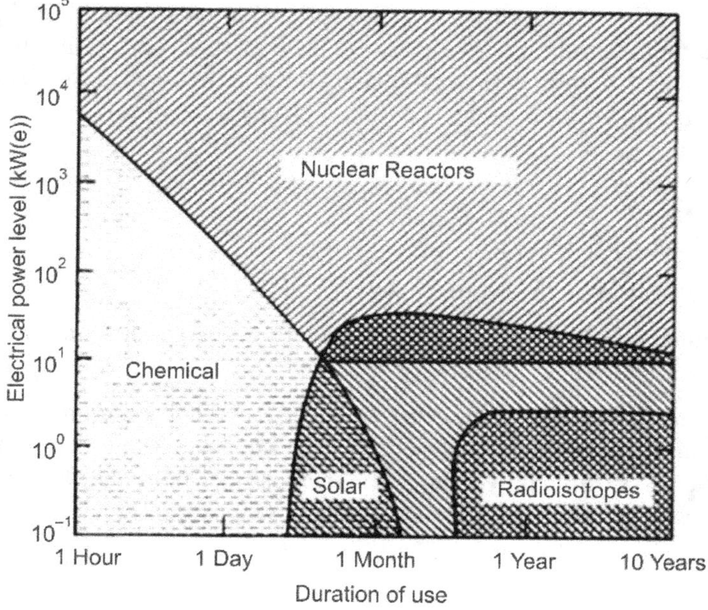


**Figure 14.23** Regimes of Space Power Applicability.
**(Source: IAEA)**

Spacecrafts require their own power source to fulfil mission requirements. For Earth orbiting satellites the power is taken from silicon or germanium solar arrays with an efficiency about 18%. As the intensity of sunlight varies inversely with the cube of the distance from the Sun, deep space probes to Jupiter or Saturn would only receive a few percent of the sunlight it would receive were it in earth orbit. Hence, large solar arrays would be required that are totally impractical to deploy. The use of nuclear power system (NPS), completely independent of solar energy, meets the requirement. Figure 14.23 shows the regimes of possible space power applicability. solar power is most efficient for power levels of some 10–50 kW for as long as it is needed. But they can't provide rapid surges of large amounts of energy. Nuclear reactors can provide almost limitless power for almost any duration. However, they are not practicable for applications below 10 kW. For long interplanetary missions, radioisotope power generators are best choice for continuous supply of low levels up to 5 kW; higher power can be achieved in combination of up to many times of this value.

The nuclear process can either be a critical reactor or radioisotope fuel source such as plutonium oxide. The heat generated is converted into electrical power through either a static process (no moving parts) using thermoelectric or thermionic converter, or dynamic process using a heat cycle (Rankine, Stirling, Brayton). The static power generators using radioisotopes are generally referred to as radioisotope thermoelectric generators (RTGs), thermoelectric generators (TEGs) and radioisotope heater units (RHUs). These units have been employed on US missions like Apollo, Viking, Pioneer, Voyager, Galileo, Ulysses and Cassini. Many Russian spacecraft have used these power generators in their missions.

The thermoelectric generator is based on Seebeck effect: when two dissimilar wires are connected at two junctions, and if one junction is kept hot while the other is cold, an electric current will flow in the circuit. In a radioisotope thermoelectric generator, the heat can be supplied from an isotope such as $^{238}$Pu. The heat production is continually decaying but the radioisotope is custom selected to fit the intended use of the electricity and for its planned mission duration. RTGs have been used in many lunar and planetary missions. It can be customized and designed for specific applications. It is a very versatile unit that can be custom designed for very specific applications. The Cassini mission to Jupiter and Saturn was equipped with 3 RTGs which produced 885 W at the beginning of the mission and 633 W at the end. Cassini also had 82 small RHUs and there were 35 more on the Huygens probe, each producing 1 W of heat to keep nearby electronics warm. These contained a total of about 0.32 kg of 238Pu. The general-purpose heat source (GPHS) developed in US comprises $^{238}$Pu fuel pellets encased in iridium shells (4 pellets each weighing 151 g) and 572 multiply redundant thermocouples made of silicon–germanium, each can produce around 0.5 W, as shown in Figure 14.24.

Thermionic converter is a static device with a very hot emitter surface around 1800 K that 'boils' electrons across a small space about 0.5 mm to a cooler collector surface typically at 1000 K. Its development has been hindered by the space charge effects between the plates, radiant heat transfer between the hot emitter and the cool collector, energy losses to the environment and the unavailability of materials with low work function. It is only recently that researchers have started to look back into this technology as recent advances in manufacturing technology techniques have made it possible to solve these problems, making TECs a viable option.

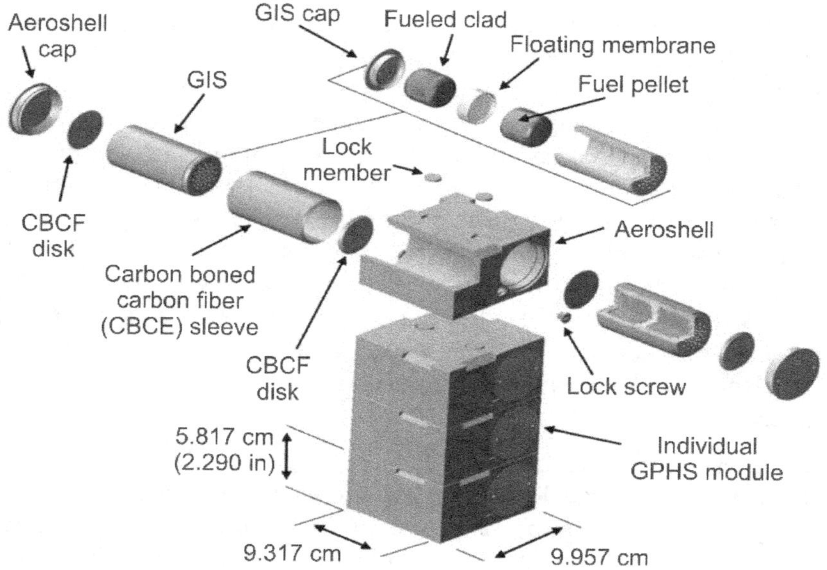



**Figure 14.24** GPHS module (Source: Wikimedia Commons).

While the radioisotope powered systems are ideal for long term low power functions, nuclear reactors have the capability of producing almost unlimited power. The nuclear thermal propulsion (NTP) units will be capable of transferring significantly heavier payloads into earth orbit than is currently possible using conventional chemical propellants. Figure 14.25 illustrates the typical working of NTP systems. The main components are propellant tank, turbopump (including the gas generator), nuclear reactor Chamber and nozzle. The reactor power is used to heat a single propellant, hydrogen that is forced through the rocket nozzle to provide the required thrust. It achieves, when compared to chemical propulsion, twice the specific impulse with a solid core having the same mass ratio of 3.2. With different cores, the specific impulse can be as much as seven times greater with a mass ratio of only 1.2.

Though the nuclear rocket program was considered a technical success, it was terminated and the US program has since used chemical propellants. However, the program is renewed for proposed manned missions to Mars and other deep space activities. Figure 14.26 gives one of the proposed road maps for the future manned Mars program. A 1990 assessment of the program concluded that a Mars mission could be performed with NERVA (Nuclear Engine for Rocket Vehicle Application) (Ref. Figure 14.27) technology developed in US during 1963-73. The NERVA programs accomplished maximum thermal power 4500 MW, 1.11 MN thrust, specific impulse 850 s and 90 min of burn. The fuel is enriched Uranium, moderator graphite and propellant hydrogen. For the on-going research work the goals are developing a rapid transit nuclear thermal propulsion technology utilizing low-enriched uranium that could potentially provide 20 percent shorter travel time to Mars while substantially improving mission flexibility. The NTP development partners include some NASA centers, academia, industries like Aerojet Rocketdyne & BWXT technologies and National labs like Los Alamos National Lab & Idaho National Lab.

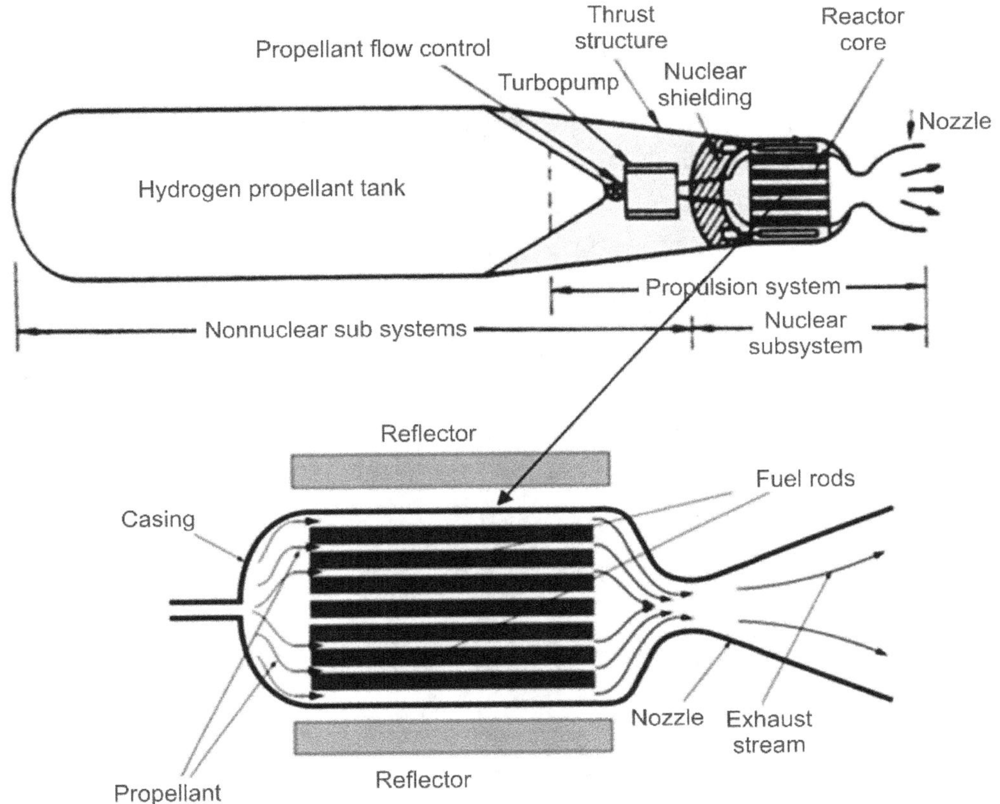


**Figure 14.25** Typical NTP layout with main components.

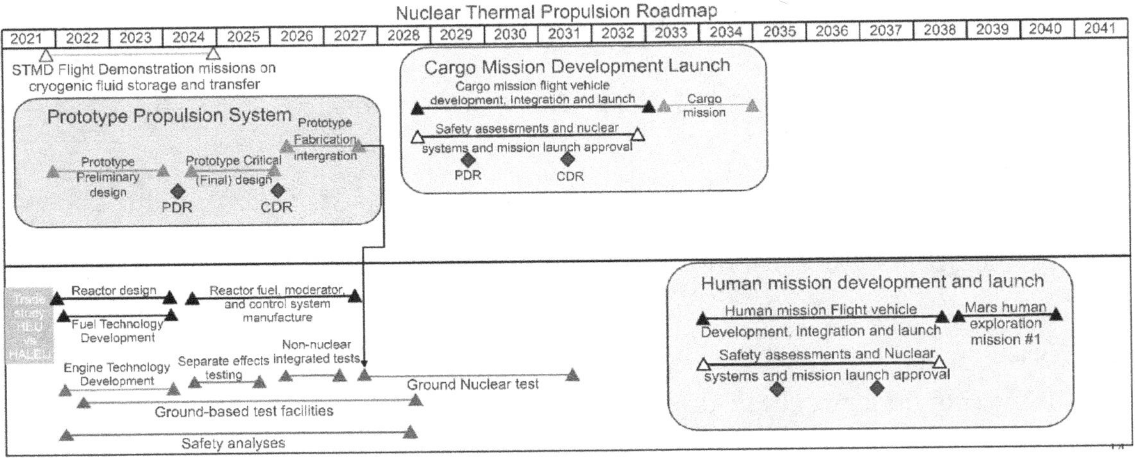


**Figure 14.26** NTP Road Map (Source: **Space Nuclear Propulsion for Human Mars Exploration, The National Academies of Sciences, Engineering & Medicine, February 2021**).

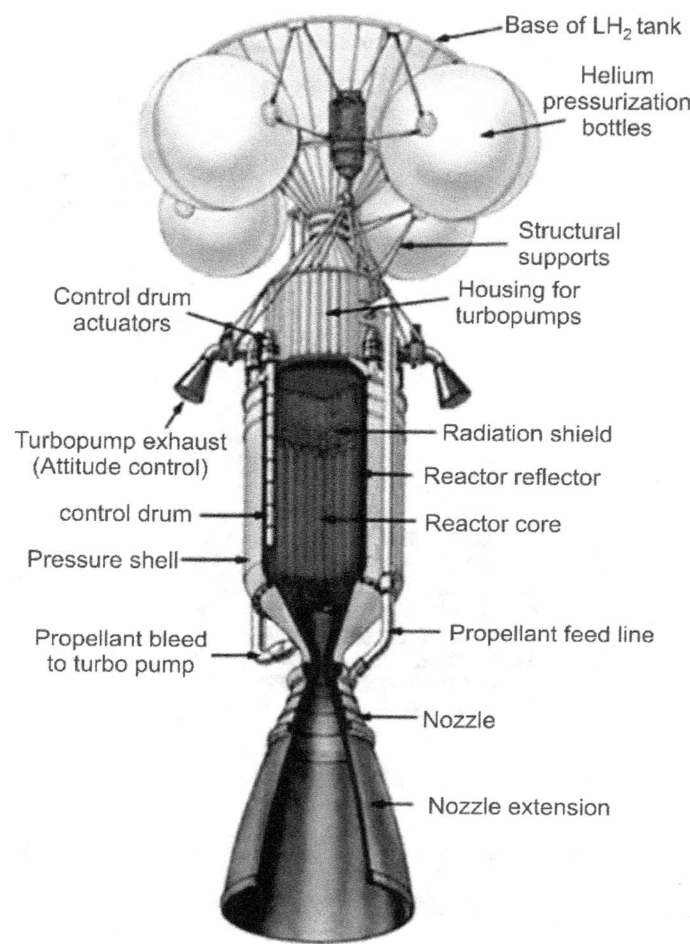


**Figure 14.27** NERVA Experimental Engine.

## Nomenclature

$\vec{B}$      magnetic field (A/Nm, Tesla)

$\vec{E}$      electrical field (V/m)

$\dot{m}$      mass flow rate (kg/s)

$v_e$      exhaust jet velocity (m/s)

$d$      diameter of the current beam (m)

$F$      thrust (N)

$I, i$      current (A)

$I_{sp}$      specific impulse (s)

| | |
|---|---|
| $I_{ssp}$ | system-specific impulse (Ns/kg) |
| $I_{tot}$ | total impulse delivered by the propulsion system [Ns] |
| $j$ | space charge-limited current density (A/m$^2$) |
| $L$ | distance between the extractor and acceleration grid (m) |
| $m$ | mass of the charged particle (kg) |
| $m_{PS}$ | mass of propulsion system (all of the propulsion system) (kg) |
| $P_e$ | output electrical power of a thruster (W) |
| $P_i$ | input electrical power of a thruster (W) |
| $q$ | electrical charge (C) |
| $R$ | aspect ratio of the beam ($R = d/L$) |
| $V$ | voltage (V) |
| $\epsilon_0$ | electrical permittivity of space (Farad/meter) |
| $\epsilon_r$ | relative permittivity of space (Farad/meter) |
| $\eta_t$ | thruster efficiency |

## PROBLEMS

14.1   A satellite with a payload of 120 kg is using electric thruster to raise the LEO in 45 days. The thruster specific impulse is 2000 s and thrust developed is 0.25 N. The specific mass of the power plant is 90 W/kg. Take thruster efficiency as 0.55. Calculate the required propellant mass, electric power and its mass. Find the velocity increment under vacuum and zero-g condition. What is acceleration of the vehicle.

14.2   A spacecraft with payload of 100 kg is to be placed in low earth orbit using an ion thruster. The specific impulse is 2500, thruster efficiency 0.65and thrust of 0.40 N. If is operates 15 days and ratio between electrical power and mass of power plant is 150 W/kg, then determine the total propellant mass, velocity increase, and acceleration of vehicle considering the ideal vacuum condition.

14.3   A resistojet with throat diameter of 2.4 mm is used with helium as propellant gas for maneuverings of a satellite. The propellant at the chamber pressure of 0.40 MPa and temperature 1300 K is expanded in a CD nozzle with expansion ratio of 90. Find mass flow rate, exit velocity, characteristic velocity, and specific impulse for this rocket.

14.4   The xenon gas, in an ion thruster, with charge density of $8 \times 10^5$ C/kg is passing through the accelerator grids with a gap of 0.8 mm and a voltage difference of 1100 V. Find the exit velocity, specific impulse, efficiency, and thrust-to-power ratio. If the maximum voltage difference between the grids happens to be 2000 V, determine the charge current density limit of this thruster. Determine the exit velocity and specific impulse.

# References


1. Adam Okninski, Wioleta Kopacz, Damian Kaniewski, Kamil Sobczak, Hybrid rocket propulsion technology for space transportation revisited - propellant solutions and challenges, FirePhysChem 1 (2021) 260–271.

2. Alessandro de Iaco Veris, Fundamental Concepts of Liquid-Propellant Rocket Engines, Springer Aerospace Technology.

3. Alliot P, Delange J F, E. Edeline and P. Sabin, The VINCI® upper stage engine: toward the demonstration of maturity, 49th AIAA/ASME/SAE/ASEE Joint Propulsion Conference and Exhibit July 2013, San Jose, California.

4. Altman D. and A. Holzman, "Overview and History of Hybrid Rocket Propulsion," in Fundamentals of Hybrid Rocket Combustion and Propulsion, vol. 218, M. J. Chiaverini and K. K. Kuo, Eds., USA: AIAA, 2006.

5. Baloni B D, Kumar S P and Channiwala S A 2017 Computational analysis of bell nozzles. 4th International Conference of Fluid Flow, Heat and Mass Transfer, Toronto, Canada, August 21 – 23, 2017. 110.

6. Ballard Richard, Launch Vehicle Propulsion & Systems, AIAA/ASME/SAE/ASEE Joint Propulsion Conference, San Jose, CA July 2013.

7. Beckstead M.W., Meredith K.V. and Blomshield F.S., Examples of Unsteady Combustion in Non-Metallized Propellants, 36th AIAA/AMSE/SAE/ASEE Joint Propulsion Conference & Exhibit July 16-19, 2000 Huntsville, Alabama.

8. Bennewitz John W. and Frederick Robert A., Overview of Combustion Instabilities in Liquid Rocket Engines - Coupling Mechanisms& Control Techniques, AIAA Conference Paper · July 2013 DOI: 10.2514/6.2013-4106.

9. Brian R. Donius and Joshua L. Rovey Analysis and Prediction of Dual-Mode Chemical and Electric Ionic Liquid Propulsion Performance, 48th AIAA Aerospace Sciences Meeting Including the New Horizons Forum and Aerospace Exposition, 4 - 7 January 2010, Orlando, Florida.

10. Byron K. Wood, Propulsion for the 21$^{st}$ Century – RS-68, AIAA 38th Joint Liquid Propulsion Conference 8–10 July 2002 Indianapolis, Indiana.

11. Calabro M., Overview on Hybrid Propulsion, Progress in Propulsion Physics 2 (2011) 353-374.

12. Cannon, J.L., Turbomachinery for Liquid Rocket Engines, "Liquid Propulsion Systems - Evolution and Advancements, American Institute of Aeronautics and Astronautics Professional Development Short Course, 2003. (BC98-04 Space Shuttle Main Engine Orientation, Rocketdyne).

13. Charles D. Brown, Elements of Spacecraft Design, AIAA Education Series.

14. Charles D. Brown, Spacecraft Propulsion, AIAA Education Series.

15. Christopher Glaser, Jouke Hijlkema and Jérôme Anthoine, Evaluation of Regression Rate Enhancing Concepts and Techniques for Hybrid Rocket Engines, Aerotecnica Missili & Spazio (2022) 101:267–292.

16. Christopher S. Olsen, Maxwell G. Ballenger, Mark D. Carter, Franklin R. Chang Díaz, Matthew Giambusso, Timothy W. Glover, Andrew V. Ilin, Jared P. Squire, Benjamin W. Longmier, Edgar A. Bering, III, and Paul A. Cloutier, Investigation of Plasma Detachment from a Magnetic Nozzle in the Plume of the VX-200 Magnetoplasma Thruster, IEEE Transactions on Plasma Science, VOL. 43, NO. 1, Jan 2015.

17. Cornelisse J.W, Schöyer H.F.R, and K.F. Wakker, Rocket Propulsion and Spaceflight Dynamics, Pitman Publishing.

18. Craig A. Kluever, Space Flight Dynamics, John Wiley & Sons Ltd.

19. Culick F.E.C., Combustion Instabilities in Solid Propellant Rocket Motors, Lectures, Internal Aerodynamic is in Solid Rocket Propulsion, Von Karman Institute, 27-31 May 2002.

20. Curt A Henry, An Introduction to the Design of the Cassini Spacecraft, Space Science Reviews 104: 129–153, 2002.

21. Curtis A. Woodruff, Darren M. King, Rodney L. Burton and David L. Carroll, Fiber-fed Pulsed Plasma Thruster (FPPT) for Small Satellites,36th International Electric Propulsion Conference University of Vienna, Vienna, Austria.

22. Dan Lev, Investigation of Efficiency in Applied Field Magneto Plasma Dynamic Thrusters, Ph.D Thesis, Princeton University.

23. Dan M. Goebel and Ira Katz, Fundamentals of Electric Propulsion: Ion and Hall Thrusters, JPL SPACE SCIENCE AND TECHNOLOGY SERIES.

24. Dario Pastrone, Approaches to Low Fuel Regression Rate in Hybrid Rocket Engines, International Journal of Aerospace Engineering, Volume 2012, Article ID 649753, 12 pages.

25. David T Harrje and Frederick H Reardon (Editors), Liquid Propellant Rocket Combustion Instability, NASA SP-194.

26. Davis K, Fortner E, Heard M, McCallum H and Putzke H2015 Experimental and computational investigation of a dual-bell nozzle., 53rd AIAA Aerospace Sciences Meeting 2015.

27. Dexter C E et al, Scaling Techniques for Design, Development, and Test, Progress in Astronautics and Aeronautics Series, Vol. 200, AIAA.

28. Dieter K Huzel and David H Huang, Modern Engineering for Design of Liquid Propellant Rocket Engines, AIAA 1992.

29. Dominick, S., Design, Development, and Flight Performance of the Mars Global Surveyor Propulsion System, 35th AIAA Joint Propulsion Conference, A99-31052, Los Angeles, California, 1999.

30. Dranovsky, M. L., Yang, V., Culick, F. E., and Talley, D. G., Combustion Instabilities in Liquid Rocket Engines: Testing & Development Practices in Russia, Progress in Astronautics & Aeronautics, Reston, VA, 2007.

31. Fred S. Blomshield, Lessons Learned in Solid Rocket Combustion Instability, AIAA, Missile Sciences Conference, Monterey, CA November 2006.

32. Gang Zheng, Wansheng Nie, Songjiang Feng, Gaoyang Wu, Numerical Simulation of the Atomization Process of a Like-doublet Impinging Rocket Injector, 2014 Asia-Pacific International Symposium on Aerospace Technology, APISAT201.

33. Gary A. Flandro, Sean R. Fischbach, and Joseph Majdalani, Nonlinear rocket motor stability prediction: Limit amplitude, triggering, and mean pressure shift, PHYSICS OF FLUIDS 19, 094101 (2007).

34. George P Sutton, Oscar Biblarz, "Rocket Propulsion Elements", John Wiley & Sons, Inc., Hoboken, New Jersey (9th Edition).

35. George P Sutton, "History of Liquid Propellant Rocket Engines", AIAA 2005.

36. Giovanni Pace, Dario Valentini, Angelo Pasini, et al., Effects of Geometry on Flow Instabilities of Different Three Bladed Axial, 15th International Symposium on Transport Phenomena and Dynamics of Rotating Machinery, ISROMAC-15 February 24 - 28, 2014, Honolulu, HI, USA.

37. Gordon A. Dressler and J. Martin Bauer, TRW Pintle Engine Heritage and Performance Characteristics, AIAA 2000-3871 Page 1 of 22.

38. Haidn, O.J. (2008) Advanced Rocket Engines. In Advances on Propulsion Technology for High-Speed Aircraft (pp. 6-1 – 6-40). Educational Notes RTO-EN-AVT-150, Paper 6. Neuilly-sur-Seine, France.

39. Hill, P.G. and Peterson, C.R., Mechanics and Thermodynamics of Propulsion, Addison Wesley Publishing Company, Reading, MA.

40. Howard D. Curtis, Orbital Mechanics for Engineering Students, Elsevier Ltd.

41. Humble, R. W., Henry, G. N. and Larson, W. J. (1995) Space Propulsion Analysis and Design, McGraw Hill Publishers.

42. Huzel Dieter K and David H. H liang, Design of Liquid Propellant Rocket Engines, NASA SP-125.

43. IAEA, The Role of Nuclear Power and Nuclear Propulsion in the Peaceful Exploration of Space, 2005.

44. Larson, W. J and Wertz, J. R. (eds.), Space Mission Analysis and Design. Kluwer Academic.

45. Les Johnson, Michael Meyer, Bryan Palaszewski et al., Development priorities for in-space propulsion technologies, Acta Astronautica 82 (2013) 148–152.

46. Lohner K, J. Dyer, E. Doran and Z. Dunn: Fuel Regression Rate Characterization Using a Laboratory Scale Nitrous Oxide Hybrid Propulsion System, AIAA 2006-4671, 2006.

47. Kubota, N., Propellants and Explosives, Thermochemical Aspects of Combustion, Wiley-VCH, Weinheim, Germany, 2002.

48. Martin Tajmar, Advanced Space Propulsion Systems, Springer-Verlag Wien GmbH.

49. Martinez-Sanchez M., Pollard J. E., Spacecraft Electric Propulsion—An Overview, Journal of Propulsion and Power, Vol. 14, No. 5, September –October 1998.

50. Matthew T. Domonkos, Michael J. Patterson and S. Jankovsky, Ion Engine and Hall Thrusters Development at the NASA Glenn Research Center, ASME International Mechanical Engineering Congress and Exposition, Nov 17-22, 2002, New Orleans, Louisiana.

51. Michael Binder, Thomas Tomsik and Joseph P. Veres, RL10A-3-3A Rocket Engine Modeling Project, NASA Technical Memorandum 107318.

52. Miguel SANGREGORIO, Kan XIE, Ningfei WANG, Ning GUO, Zun ZHANG, Ion engine grids: Function, main parameters, issues, configurations, geometries, materials and fabrication methods, Chinese Journal of Aeronautics, (2018), 31(8): 1635–1649.

53. Mishra D.P., Fundamentals of Rocket Propulsion, Taylor & Francis Group, LLC.

54. Mojtaba Ghasemi and Abdol rahim Rezaeiha, Design and Development of a Flight-model Cold Gas Propulsion System, The 28th International Symposium on Space Technology and Science (ISTS2011), 5-12 June, 2011, Okinawa, Japan.

55. Molina-Cabrera P., Herdrich G, Lau M., Fausolas S.., Pulsed Plasma Thrusters: a worldwide review and long yearned classification, 32nd International Electric Propulsion Conference, Wiesbaden, Germany September 11–15, 2011.

56. Muhammad Ali Badam, Design of a FEEP Thruster for Micro-/Nanosatellites, Department of Computer Science, Electrical and Space Engineering Luleå University of Technology 2019.

57. Mukund R. Patel, Spacecraft Power Systems, CRC Press.

58. Nan Zong and Vigor Yang, Cryogenic fluid dynamics of pressure swirl injectors at supercritical conditions, Physics of Fluids 20, 05, 2008.

59. NASA SP 8076, Solid Propellant Grain Design and Internal Ballistics, March 1972.

60. NASA SP-8025, Solid Rocket Motor Metal Cases, April 1970.

61. NASA SP-8039, Solid Rocket Motor Performance Analysis and Prediction, May 1971.

62. NASA SP-8051, Solid Rocket Motor Igniters.

63. NASA SP-8055, Prevention of Coupled Structure-Propulsion Instability (POGO).

64. NASA SP-8100, Liquid Rocket engine turbopump Gears.

65. NASA SP-8112, Pressurisation Systems for Liquid Rockets, Oct 1975.

66. NASA SP-8113, Liquid Rocket Engine Combustion Stabilisation.

67. NASA SP-8115, Solid Rocket Motor Nozzles, June 1975.

68. Oguz Korkmaz, Ali Enes Ozturk and ve Murat Celik, Design Process of the Hollow Cathode Manufactured for Use in Electric Propulsion Systems as Electron Source, V. NATIONAL AERONAUTICS AND ASTRONAUTICS CONFERENCE UHUK-2014-090 8-10 September 2014, Erciyes University, Kayseri, Turkey.

69. Oskar J. Haidn, Advanced Rocket Engines, Institute of Space Propulsion, German Aerospace Center (DLR) Germany.

70. Paccagnella, E., Barato, F., Gelain, R., Pavarin, D., CFD simulations of self-pressurized nitrous oxide hybrid rocket motors, 2018 Joint Propulsion Conference, p. 4534. AIAA, Cincinnati, Ohio (2018).

71. Paul A. Czysz • Claudio Bruno, Bernd Chudoba. Future Spacecraft Propulsion Systems and Integration, Springer Praxis Books.

72. Pilinski C and Nebbache A 2004 Flow separation in a truncated ideal contour nozzle; J. Turbul. 5 1–3.

73. Price T. W. and Evans D. D., The Status of Monopropellant Hydrazine Technology, Technical Report 32-7227, NASA.

74. Rainer Killinger and Klaus Bohnhoff. The Galileo Propulsion System: Mission Summary and Long-Term Performance, Proc. Second European Spacecraft Propulsion Conference, 27-29 May 1997 ESA SP-398, Aug.1997.

75. Raji R, Jancy Rose K, P R Murali and R Neetha, POGO Stability Analysis of a Typical Launch Vehicle for Studying the Effectiveness of POGO Corrector in Cyro Stage, Journal of Physics: Conference Series **1355** (2019) 012020.

76. Ramamurthi K., "Rocket Propulsion", Trinity Press New Delhi India.

77. Rao G V R, Exhaust nozzle contour for optimum thrust; J. Jet Propulsion, 28 377–382.

78. Schoenman L., Low-Thrust ISP Sensitivity Study, NASA CR-165621 September 15-20, 2019.

79. Shivang Khare and Ujjwal K Saha, Rocket nozzles: 75 years of research and development, Sådhanå (2021) 46:76.

80. Stephen D. Heister, W E Anderson, T Pourpoint and R J Cassady, Rocket Propulsion, Cambridge Press UK 2019.

81. Summerfield Martin, A Theory of Unstable Combustion in Liquid Propellant Rocket Systems, 1951TECHNICAL REPORT No. 26 Princeton University Princeton, New Jersey.

82. Turner, M.J.L., Rocket and Spacecraft Propulsion, Springer Verlag, Heidelberg, Germany, 2001.

83. Varghese T L. and Krishnamurthy V N., The chemistry and Technology of Solid Rocket Propellants, Allied Publishers Pvt Ltd. India.

84. Von Braun, W., and Ordway, F. I., Space Travel: A History, Harper and Row.

85. Wintenberger E. and Shepherd J. E., Thermodynamic Analysis of Combustion Processes for Propulsion Systems, 42nd AIAA Aerospace Sciences Paper 2004-1033 January 5-8, 2004, Reno, NV.

86. Winter, Jerry M., Pavli, Albert J., and Shinn, Arthur M. Jr.: Design and Evaluation of an Oxidant-Fuel-Ratio-Zoned Rocket Injector for High Performance and Ablative Engine. NASA TN D-6918, 1972.

87. Yash Pal, Sri Nithya Mahottamananda, Sasi Kiran Palateerdham, Sivakumar Subha and Antonella Ingenito, Review on the regression rate-improvement techniques and mechanical performance of hybrid rocket fuels, FirePhysChem 1 (2021) 272–28.

## M

Magneto plasma dynamic thruster (MPD) 379

Mass ratios of rocket 52

Mach number 73, 75

Magnetic field 368, 373, 376

Magneto plasma dynamic (MPD) thruster 11, 376

Metal Fuels 135

Methane ($CH_4$) 132

Missile 3, 6, 144

Mixture ratio (MR) 113

Mole fraction 110, 123

Momentum 18,19, 30, 50

Mono-methyl hydrazine (MMH) 179, 246

Monopropellant 340, 345, 359

Multistage Rockets 60, 65

## N

Newton's Laws of Motion 15

Nitrogen tetroxide 130, 131

Non-hypergolic propellants 217

Nonlinear instabilities 260

Nonimpinging injectors 215

Nozzle area ratio 79,86,199.

Nozzle shape 75, 97

Nuclear propulsion 1, 383

## O

Ohnesorge number 212, 213

Optimum thrust coefficient 86, 87

Orbit changes 8, 278

Orbit equation 28, 31, 40

Over-expansion 92

Orbit Maintenance or Station Keeping 8

Orbit Types 19

Orbital elements 13, 46

Orbitalvelocity 22, 23, 33

Orbital energy 23, 328

Over-expanded nozzles 86

Oxidisers 229, 231

## P

Parabolic orbit 20, 25, 38, 39

Patched conic method 312

Performance Parameters 49, 58, 150

Periapsis 288, 306, 320

Perigee 41, 288, 306

Phasing Maneuvres 296

Phasing orbits 294

Pintle injector 221, 218, 219

Piobert's law 165

Planetocentric velocity 314, 335

Planetary arrival 39, 313

Planetary departure 317

POGO instability 270, 271

POGO suppression devices 272

Polar orbit 20, 47, 48

Polar Satellite Launch Vehicle (PSLV) 65

Polybutadiene acrylic acid acrylonitrile (PBAN) 11, 138, 135

Polyurethane 138

Power processing unit (PPU) 362, 367, 382

Prandtl number 235, 239, 241, 254

Pressure-fed System 183, 185, 271

Progressive burning 163, 167

Propellant grain 147, 150, 161